Advanced Intermetallic-Based Alloys for Extreme Environment and Energy Applications

MATERIALS RESEARCH SOCIETY
SYMPOSIUM PROCEEDINGS VOLUME 1128

Advanced Intermetallic-Based Alloys for Extreme Environment and Energy Applications

Symposium held December 1–4, 2008, Boston, Massachusetts, U.S.A.

EDITORS:

Martin Palm
Max-Planck-Institut für Eisenforschung GmbH
Düsseldorf, Germany

Bernard P. Bewlay
General Electric Company
Schenectady, New York, U.S.A.

Yue-Hui He
Central South University
Changsha, P.R. China

Masao Takeyama
Tokyo Institute of Technology
Tokyo, Japan

Jörg M.K. Wiezorek
University of Pittsburgh
Pittsburgh, Pennsylvania, U.S.A.

Materials Research Society
Warrendale, Pennsylvania

CAMBRIDGE UNIVERSITY PRESS
Cambridge, New York, Melbourne, Madrid, Cape Town,
Singapore, São Paulo, Delhi, Mexico City

Cambridge University Press
32 Avenue of the Americas, New York NY 10013-2473, USA

Published in the United States of America by Cambridge University Press, New York

www.cambridge.org
Information on this title: www.cambridge.org/9781107408425

Materials Research Society
506 Keystone Drive, Warrendale, PA 15086
http://www.mrs.org

First published 2009
First paperback edition 2012

Single article reprints from this publication are available through
University Microfilms Inc., 300 North Zeeb Road, Ann Arbor, MI 48106

CODEN: MRSPDH

ISBN 978-1-107-40842-5 Paperback

CONTENTS

*INTERMETALLICS FOR HYDROGEN
STORAGE AND THERMOELECTRIC
APPLICATIONS*

*Invited Paper

*TITANIUM ALUMINIDES II —
STRUCTURE, PROPERTIES AND COATINGS*

*Invited Paper

ix

NICKEL/COBALT SUPERALLOYS AND NICKEL ALUMINIDES

*Invited Paper

NIOBIUM AND MOLYBDENUM
SILICIDE-BASED ALLOYS

LAVES PHASES — STRUCTURE
AND PROPERTIES

*Invited Paper

xiv

FUNDAMENTAL ASPECTS OF
INTERMETALLICS — PHASE STABILITY,
DEFECTS, THEORY

*Invited Paper

PREFACE

Symposium U, "Advanced Intermetallic-Based Alloys for Extreme Environment and Energy Applications," held December 1–4 at the 2008 MRS Fall Meeting in Boston, Massachusetts, continues the series of symposia devoted to the progress in research on intermetallic phases held bi-annually at the MRS Fall Meeting. The increasing demand for materials which enable a more efficient energy conversion, i.e., materials allowing higher operating temperatures and a lower weight of the components with preferably better corrosion resistance, has put intermetallic-based alloys back into focus. This resurgence in technology-demand driven interest in intermetallic-based alloys is reflected in the increasingly large number of contributions to this year's symposium compared to the number of contributions at the turn of the millennium. This proceedings volume presents a representative cross-section of the research papers discussed during the symposium.

In the wake of the use of TiAl-based alloys for automotive and aerospace applications, research on other intermetallic-based alloys has again significantly increased. Among others, this lead to the recent discovery of the $Co_3(Al,W)$ intermetallic compound with the $L1_2$ structure which may yield a major breakthrough in the development of corrosion and wear resistant Co-based alloys. Also, aluminides and silicides have become a major research topic as this class of materials may allow even higher operating temperatures. For the same reason Laves phases are of considerable interest, though knowledge on this the most abundant of intermetallic phases is still limited.

The current volume offers a recent overview of the current research topics and shows how far intermetallic-based alloys have matured towards application.

Martin Palm
Bernard P. Bewlay
Yue-Hui He
Masao Takeyama
Jörg M.K. Wiezorek

May 2009

ACKNOWLEDGMENTS

It is a pleasure to acknowledge financial support for this symposium from the following organizations:

CBMM, Reference Metals Co., Brazil
JEOL Ltd, Japan
JEOL USA Inc.
Leistritz Turbinenkomponenten GmbH, Germany
Materials Design Technology Co., Ltd., Japan
Mitsubishi Materials Corp., Japan
Plansee SE, Austria
Tokyo Institute of Technology, Japan

The organizers of this symposium would like to thank all of the participants, authors, reviewers and the MRS staff for their efforts in preparing this proceedings.

MATERIALS RESEARCH SOCIETY SYMPOSIUM PROCEEDINGS

MATERIALS RESEARCH SOCIETY SYMPOSIUM PROCEEDINGS

Prior Materials Research Society Symposium Proceedings available by contacting Materials Research Society

Intermetallics for Hydrogen Storage
and Thermoelectric Applications

Mater. Res. Soc. Symp. Proc. Vol. 1128 © 2009 Materials Research Society   1128-U01-04

Hydrogen Storage Properties, Metallographic Structures and Phase Transitions of Mg-Based Alloys Prepared by Super Lamination Technique

Nobuhiko Takeichi[1], Koji Tanaka[1], Hideaki Tanaka[1], Nobuhiro Kuriyama[1], Tamotsu T Ueda[2], Makoto Tsukahara[2], Hiroshi Miyamura[3], and Shiomi Kikuchi[3]
[1]National Institute of Advanced Industrial Science and Technology (AIST),
1-8-31 Midorigaoka, Ikeda, Osaka 563-8577, Japan
[2]IMRA Material R&D Co., Ltd., 5-50 Hachiken-cho, Kariya, Aichi 448-0021, JAPAN
[3]The University of Shiga Prefecture, 2500 Hassaka-cho, Hikone, Shiga 522-0057, Japan

ABSTRACT

We have prepared Mg/Pd laminate composites with (Mg/Pd)=6, 3 and 2.5 atom ratios, by a super lamination technique. The homogeneous Mg-Pd intermetallic compounds, Mg_6Pd, Mg_3Pd and Mg_5Pd_2, are formed during the initial activation process. We investigated the hydrogen storage properties of these materials. The compounds can reversibly absorb and desorb a large amount of hydrogen, up to 1.46~0.9 H/M, at 573 K. Except for the Mg_5Pd_2-hydrogen system, the pressure composition-isotherms show two plateaux. The mechanism of the phase transition during hydrogenation/dehydrogenation was analyzed by in-situ XRD measurements. These intermetallic compounds absorb and desorb hydrogen through reversible multistage disproportionation and recombination processes.

INTRODUCTION

From the viewpoint of hydrogen storage systems, Mg [1] is a promising material because it can absorb a large amount of hydrogen up to 7.6 mass%, as MgH_2. However, the hydrogen absorption/desorption kinetics is too low for practical use and requires the use of a high temperature such as 573 K. To improve the reaction kinetics and diffusion properties, reduction of the grain size and the addition of various catalysts have been investigated [2, 3].

Also, various Mg-based alloys and intermetallic compounds have been investigated to improve the rate of hydrogenation and the hydrogenation temperature [4, 5]. The Mg_2Ni intermetallic compound is well known to form a Mg_2NiH_4 hydride, which has a hydrogen storage capacity of 3.6 mass% [5]. However, it is difficult to cast Mg-containing alloys accurately with desirable composition by conventional melt-cast methods because of the high vapor pressure of Mg, etc. In addition, single phase Mg-containing compounds cannot be obtained simply by casting, due to phase separation during solidification.

Accordingly, various methods, such as mechanical alloying [6], vapor phase processing [7], and combustion synthesis [8], have been investigated. Recently, Ueda et al. have reported initial investigations on the synthesis and hydrogen storage properties of Mg-based laminate composites prepared by a super lamination technique [9]. A 67 at%Mg-33 at%Ni laminate composite was shown to transform due to a heat treatment after repetitive rolling to single phase Mg_2Ni, a phase which could not be obtained simply by casting.

Palladium reacts rapidly with hydrogen, even after it is placed in air, because its surface and diffusion properties provide a high catalytic activity. Therefore, adding Pd to Mg is expected to improve the activation properties and kinetics for hydrogen reaction. We have prepared Mg/Pd laminate composites by the super lamination technique. In this work, we investigated the hydrogen storage properties of Mg/Pd laminate composites and the phase transition during hydrogenation/dehydrogenation.

EXPERIMENT

A commercial Pd sheet (99.99 % purity, 10 μm in thickness) and, a Mg sheet (99.9% purity, rolled to 94, 47 or 39 μm in thickness in our laboratory) were used as starting materials. The sheets, cut into pieces of 20 mm x 30 mm in size, were stacked alternately. By combining the Pd sheet with a Mg sheet of different thickness, stacks with molar volume ratios Mg/Pd of 6, 3 and 2.5 were prepared. The specimens were then prepared by the super lamination technique as described previously [9].

Investigation of the as-rolled sample and specimens used for hydrogenation/dehydrogenation was carried out using a scanning electron microscope (SEM) equipped with secondary electron imaging (SEI), backscattered electron imaging (BEI) and an energy dispersive X-ray spectrometer (EDX). The crystal structure of samples was observed by powder X-ray diffraction (RINT2000, Rigaku Co., Ltd.) with Cu-Kα radiation.

The pressure-composition (P-C) isotherms were determined with a Sieverts' apparatus. Before starting the measurements, each sample was subjected to one cycle of hydrogenation and dehydrogenation, which is a so-called "activation-process". The process was actually carried out by keeping the sample at 573 K, under 5 MPa of hydrogen for 86.4 ks, followed by degassing with a rotary pump for 86.4 ks. After this activation-process, the P-C isotherms were determined. Because the hydrogenation/dehydrogenation rates of Mg-Pd compounds are extremely slow, the equilibriation periods for each point on the P-C isotherms were set to those longer than 28.8 ks.

Phase transformations of the Mg/Pd laminate composites during hydrogen absorption and desorption were analyzed by in-situ XRD measurements made using a special cell, which can control the temperature and pressure up to 5 MPa at 573 K. The in-situ XRD measurements were carried out after the temperature and hydrogen pressure of the sample cell had equilibrated.

RESULT AND DISCUSSION

We prepared three Mg/Pd laminate composites, Mg/Pd=6, 3 and 2.5, by the super lamination technique. Figure 1(a) shows a BEI of a cross-section of the as-rolled Mg/Pd laminate composite with (Mg/Pd)=6. The specimen consists of the Mg matrix (attributed to the dark areas) and Pd fragments (white) dispersed and compressed in the matrix; the image contrast is consistent with the result of the XRD profile. We also confirmed the composition of each phase by EDX. No other phase was observed in the specimen by SEM. Figure 2(a) shows the XRD profile of the same as-rolled laminate composite. It indicated the existence solely of Mg and Pd, with no other phase being observed. However, if an intermediate phase composed with a nano-ordered crystalline size could be formed though solid-solid reaction during the sample preparation, the crystalline size is too small to be detected by our XRD experiments. The diffraction peaks associated with intermediate phase were not observed. The highest peak of the Mg phase of the specimen corresponds to a (002) reflection. Usually, the highest intensity peak of Mg is that from a (101) reflection. This intensity change means that the metallographic structure of the specimen has a preferred orientation.

4

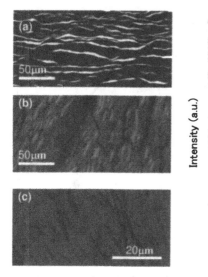

Figure 1. Backscattered electron image (BEI) of a cross-section of Mg/Pd=6 laminate composite (a) as-rolled, (b) after the activation process at 573 K and 5 MPa hydrogen pressure (c) after dehydrogenation at 573 K.

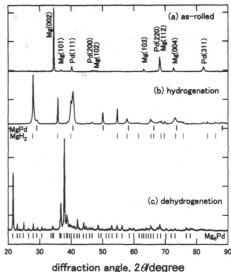

Figure 2. XRD profiles of Mg/Pd=6 laminate composite (a) as-rolled, (b) after the activation process at 573 K and 5 MPa hydrogen pressure (c) after dehydrogenation at 573 K.

The XRD profile in Fig. 2(b), taken after hydrogenation at 573 K and 5 MPa hydrogen pressure, shows the presence of two phases assigned to MgH_2 and MgPd. The corresponding BEI in Fig. 1(b) also shows that the specimen is mainly composed of two phases, which are thought to be MgPd (bright grey) and MgH_2 (dark grey) in order of brightness. After dehydrogenation at 573 K, the XRD pattern in Fig. 2(c) shows that the Mg/Pd laminate composite consists of only the Mg_6Pd intermetallic compound. This compound presumably forms by decomposition of MgH_2 to Mg, and then Mg and MgPd form Mg_6Pd through a recombination process. This intermetallic compound is a homogeneous phase and has the laminated structure shown in Fig.1(c). In the Mg/Pd laminate composites with (Mg/Pd)=3 and 2.5, the intermatallic compounds Mg_3Pd and Mg_5Pd_2 are respectively formed. It appears that the initial activation process for the Mg/Pd laminate composites results in the formation of intermetallic compounds.

Figure 3 shows the P-C isotherms for Mg_6Pd, Mg_3Pd and Mg_5Pd_2 at 573 K obtained with the Sieverts' apparatus. These compounds can reversibly absorb and desorb a large amount of hydrogen. The maximum hydrogen contents of the compounds at 573 K and 10 MPa were 1.59, 1.4 and 0.9 H/M, respectively, where H/M means the atomic ratio of hydrogen to metal. Except for Mg_5Pd_2, the P-C isotherms show two plateaus, at pressure P_L and P_H. During hydrogenation, the lower plateau pressure, P_L Mg_6Pd: 0.42 MPa between H/M=0 and 0.8 and Mg_3Pd: 0.7 MPa between H/M=0 and 0.2, increases with increasing Pd composition. The higher plateau pressure, P_H, of Mg_6Pd and Mg_3Pd is 2 MPa, is almost the same value as the single plateau pressure for

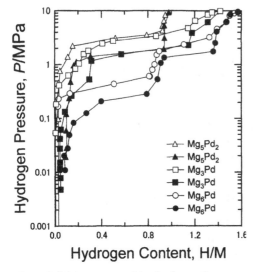

Figure 3. Pressure-composition isotherms for hydrogen absorption (open symbols) and desorption (filled symbols) for Mg_5Pd_2, Mg_3Pd and Mg_6Pd intermetallic compounds at 573 K.

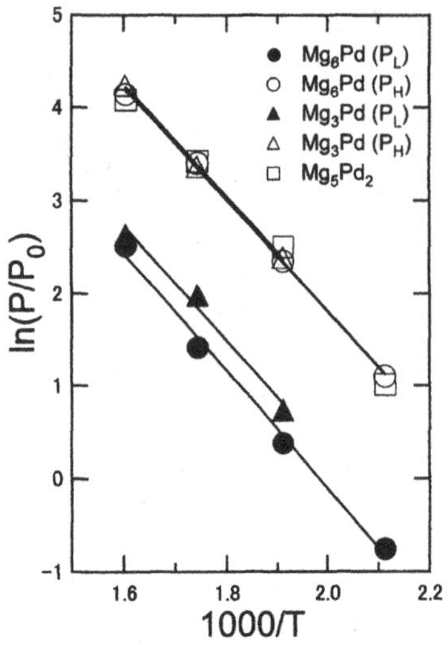

Figure 4. Logarithmic plot of equilibrium pressure vs. reciprocal temperature for the plateau region in desorption isotherms.

Table 1 Changes in enthalpy ΔH and entropy ΔS for the reaction of several Mg-Pd intermetallics with hydrogen.

	ΔH (kJ/mol-H_2)	ΔS (J/K mol-H_2)
Mg_6Pd (P_L)	-52.9	105
Mg_6Pd (P_H)	-50.1	115
Mg_3Pd (P_L)	-51.8	108
Mg_3Pd (P_H)	-50.1	115
Mg_5Pd_2	-50.2	115

Mg_5Pd_2. The dissociation pressures of Mg-Pd intermetallic hydrides are higher than that of Mg hydride. The logarithms of the equilibrium pressures for the plateau region are plotted in Figure 4 vs. the reciprocal temperature, where P means a plateau pressure in the hydrogen absoption process and P_0 means a standard pressure, P_0=0.1 MPa. Using the van't Hoff equation, the thermodynamic quantities per mole of hydrogen were calculated. They are listed in Table 1. The absolute values of ΔH for the reaction of hydrogen with Mg-Pd intermetallic compounds (~50-53 kJ/mol H_2) are smaller than those for several other intermetallic hydrides including MgH_2, for example, MgH_2: -74.4 kJ/mol H_2 [1], Mg_2Cu: -72.8 kJ/mol H_2 [4], Mg_2Ni-H: -64.4 kJ/mol H_2 [5].

Figure 5(a) shows the XRD profiles of Mg_6Pd under vacuum, 0.1, 1 and 4 MPa hydrogen pressure during hydrogen absorption. The XRD profile of the specimen under 0.1 MPa pressure was identical to that of the specimen under vacuum. We did not observe a low-pressure solid

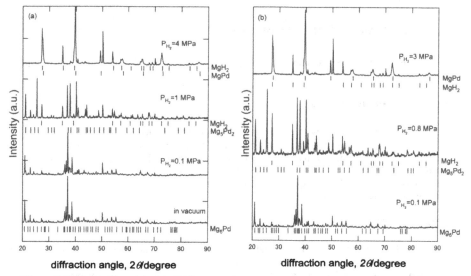

Figure 5. XRD profiles of the Mg$_6$Pd intermetallic compound: (a) during hydrogenation at 573 K and (b) during dehydrogenation at 573 K.

solution phase, such as Mg$_6$PdH$_x$, in this study. This is in contrast to the report of Huot [10] that Mg$_6$Pd decomposed to Mg$_{3.65}$Pd and MgH$_2$ at 0.39 MPa hydrogen pressure. Under a 1 MPa hydrogen pressure, the diffraction peaks associated with Mg$_5$Pd$_2$ and MgH$_2$ were observed, while those associated with Mg$_6$Pd disappeared. On gradually increasing the pressure to 4 MPa, the diffraction peaks associated with Mg$_5$Pd$_2$ disappeared and those associated with MgH$_2$ became stronger than those observed at 1 MPa. Also, diffraction peaks from MgPd were observed at this highest pressure.

Figure 5(b) shows the XRD profiles of a specimen under hydrogen pressures of 3, 0.8, 0.1 MPa during hydrogen desorption. The diffraction peaks of MgPd disappeared and those of MgH$_2$ became weak, while the diffraction peaks of Mg$_5$Pd$_2$ appeared at 0.8 MPa. From 0.8 MPa down to 0.1 MPa, the diffraction peaks from MgH$_2$ and Mg$_5$Pd$_2$ disappeared, leaving only the peaks associated with Mg$_6$Pd appearing.

Taking into account the absorption/desorption processes displayed in the P-C isotherms for Mg$_6$Pd in Fig.3, the in-situ XRD results clarify the correlation between the hydrogen storage properties and the phase transitions. The P-C isotherm of the Mg$_6$Pd intermetallic compound exhibits two plateaux at 0.2 MPa and 2 MPa at 573 K. In the lower plateau pressure region, the Mg$_6$Pd reacted to form Mg$_5$Pd$_2$ and MgH$_2$.

$$Mg_6Pd + \tfrac{7}{2}H_2 \leftrightarrow \tfrac{1}{2}Mg_5Pd_2 + \tfrac{7}{2}MgH_2 \tag{1}$$

The amount of hydrogen consumed, 1 H/M, corresponds to the lower pressure plateau of the P-C isotherms. In the higher plateau pressure region, Mg$_5$Pd$_2$ reacts to produce MgH$_2$ and MgPd.

$$\tfrac{1}{2}Mg_5Pd_2 + \tfrac{7}{2}MgH_2 + \tfrac{3}{2}H_2 \leftrightarrow MgPd + 5MgH_2 \tag{2}$$

Combining reaction equations (1) and (2), we write an overall reaction that agrees with the P-C isotherms of Mg_6Pd in Fig.3 as follows.

$$Mg_6Pd + 5H_2 \leftrightarrow MgPd + 5MgH_2 \tag{3}$$

In our previous study, we have shown that a micro/nano-structure is introduced into Mg-based compounds by the super lamination technique. This unique structure improves the diffusion rate of the metallic elements and hydrogen and increases the kinetics of hydrogen absorption/desorption [8]. Therefore, Mg-Pd inetermetallic compounds can absorb and desorb hydrogen reversibly at 573 K, in spite of the complex multistage disproportionation and recombination reactions of Mg and Pd.

CONCLUSIONS

The hydrogen storage properties and metallographic structures of Mg/Pd laminate composites were studied. Several Mg-Pd intermetallic compounds are formed during the initial activation process: Mg_6Pd, Mg_3Pd and Mg_5Pd_2. These compounds can reversibly absorb and desorb a large amount of hydrogen at 573 K through reversible multistage disproportionation and recombination processes. The absolute value of enthalpy change for the reaction of hydrogen with these Mg-Pd intermetallic compounds is smaller than that of Mg.

ACKNOWLEDGMENTS

The studies were administrated through the New Energy and Industrial Technology Development Organization (NEDO) as a part of the Development for Safety Use and Infrastructure of Hydrogen Program, with funding from the Ministry of Economy, Trade and Industry of Japan (METI).

REFERENCES

1. J. F. Stampfer, Jr., C. E. Holley, Jr., and J. F. Suttle: J. Am. Chem. Soc. **82**, 3504 (1960).
2. S. Orimo, H. Fujii, and K. Ikeda: Acta Mater. **45**, 331 (1998).
3. G. Liang, J. Huot, S. Boily, and R. Schulz: J. Alloy. Compd. **305**, 239 (2000).
4. J. J. Reilly and R. H. Wiswall: Inorg. Chem. **6**, 2220 (1967).
5. J. J. Reilly and R. H. Wiswall: Inorg. Chem. **7**, 2254 (1968).
6. L. Zaluski, A. Zaluska, J.O. Str¨om-Olsen, J. Alloys Compd. **253–254**, 70 (1995).
7. S.E. Guthrie, G.J. Thomas, in Proceedings of the 43rd International SAMPE Symposium, edited by H.S. Kliger (Advancement of Materials and Process Engineer, Covina 1998), p. 1105.
8. T. Akiyama, H. Isogai, J. Yagi, Int. J. Self-Propagating High-Temp. Synth. **4(1)**, 69 (1995).
9. T. T. Ueda, M. Tsukahara, Y. Kamiya, and S. Kikuchi : J. Alloy. Compd. **386**, 253 (2005).
10. J.Huot, A.Yonkeu, J.Dufour: J. Alloy. Compd. (2008) doi:10.1016/j.jallcom.2008.07.034.
11. N. Takeichi, K. Tanaka, H. Tanaka, T. T. Ueda, Y. Kamiya, M. Tsukahara, H. Miyamurac, S.Kikuchi: J. Alloy. Compd. **446-447**, 543 (2007).

Mater. Res. Soc. Symp. Proc. Vol. 1128 © 2009 Materials Research Society   1128-U01-07

Change in the Thermoelectric Properties With the Variation in the Defect Structure of ReSi$_{1.75}$

Shunta Harada, Katsushi Tanaka, Kyosuke Kishida, Norihiko L. Okamoto, Haruyuki Inui
Department of Materials Science & Engineering, Kyoto University,
Sakyo-ku, Kyoto 606-8501, Japan

ABSTRACT

Crystal structure variation and thermoelectric properties of binary rhenium silicide with different heat treatment conditions are investigated. In quenched rhenium silicide, dense planar defects are observed and the crystal structure is identified as crystallographic shear structure with crystallographic shear operation of ($\bar{1}$ 09)$_{C11b}$ / [100]$_{C11b}$. The crystallographic shear structure observed in quenched samples is not thermally stable. The structure is annealed out by prolonged heat treatment at relatively low temperature. Thermoelectric properties of quenched and annealed rhenium silicide are significantly different. The quenched sample is an n-type semiconductor, while the annealed sample is a p-type semiconductor. Planar defects in the quenched sample are expected to introduce a donor level in the band gap and the electrical conduction becomes n-type.

INTRODUCTION

Semiconducting transition-metal silicides have attracted great interest due to their applications as a thermoelectric device [1]. Some semiconducting transition-metal silicides are characterized by the number of valence electrons per transition-metal atom (valence electron concentration, VEC) being equal to approximately 14 (MnSi$_{1.75}$, Ru$_2$Si$_3$, ...). ReSi$_{1.75}$ belongs to this group, however, its crystal structure and electrical transport properties are controversial.

Initial investigations of crystal structures for the binary rhenium silicide found that rhenium silicide had a body-centered tetragonal C11$_b$ structure and a composition of ReSi$_2$ [2, 3]. Later, Siegrist et al. have reported a body-centered orthorhombic structure but found the composition to be close to ReSi$_2$ [4]. Recent investigations by Gottlieb et al. have reported that the stoichiometry of the silicide is ReSi$_{1.75}$ and the unit cell is monoclinic [5]. In their assessment of the crystal structure, the occupancy of half the Si sites is assumed to be 75 %, resulting in the displacement of both Re and Si atoms from their positions corresponding to the underlying C11$_b$ lattice sites. More recently, however, we have determined the crystal structure of ReSi$_{1.75}$ to be monoclinic with the space group Cm due to the ordered arrangement of vacancies on Si sites in the underlying C11$_b$ lattice [6, 7, 8].

Thermoelectric properties of the binary rhenium silicide have been reported by several researchers and are also controversial. Siegrist et al. have reported the value of Seebeck coefficients ranging from -90 to -130 μV/K at 310 K [4]. On the other hand, Neshpor et al. have reported the value, which is in fair agreement with that reported by Siegrist et al., but the sign is reversed [9, 10]. Recently, we have reported the temperature and crystal orientation dependence of electrical resistivity and Seebeck coefficients of a ReSi$_{1.75}$ single crystal [11], where ReSi$_{1.75}$ is a p-type semiconductor and values of both properties significantly depend on both temperature and crystal orientation. The values of thermal conductivity at room temperature are about 6 W/mK and gradually increase with temperature. The value of the dimensionless figure of merit

(ZT) along one direction is relatively high (0.7 at 1173K, which is comparable to that of Si-Ge alloys currently used as a thermoelectric material for power generation).

Recently, Misra *et al.* have reported other structural anomalies in binary rhenium silicide by transmission electron microscopy [12, 13]. They have reported several kinds of structures including monoclinic crystal structure. They have inferred that such structural anomalies are caused by variation in cooling rate and other heat treatment conditions. More recently, we have reported the crystal structure and thermoelectric properties of rhenium silicide with ternary additions and a relationship between crystal structure and thermoelectric properties has been implied [14]. In the present study, we focus on structural anomalies in binary rhenium silicide and investigate the change in the thermoelectric properties with the variation in the crystal structure.

EXPERIMENTAL PROCEDURE

Rods of polycrystalline $ReSi_{1.75}$ with dimensions of 10 mm in diameter and 50 mm long were prepared by argon arc-melting of high purity rhenium and silicon. Single crystals of binary $ReSi_{1.75}$ were grown from the rods, using our ASGAL FZ-SS35W optical floating-zone furnace at a growth rate of 2.5 mm per hour under argon gas flow. After crystal growth, samples were quenched to room temperature by rapidly shutting down the lamp power of the floating-zone furnace (named quenched sample). Some samples were subsequently annealed at 1173 K for 48 hours in argon (named annealed sample). Specimens with dimensions of $2 \times 2 \times 7$ mm^3 for measurement of transport properties were cut from both quenched and annealed samples after determining their crystallographic orientations by the back-Laue X-ray diffraction method. Two different orientations, $[100]_{C11b}$, $[001]_{C11b}$, were chosen to investigate the transport properties. Microstructures and crystal structures were examined by transmission electron microscopy (TEM) with JEM-2000FX and JEM-4000EX electron microscopes operated at 200kV and 300kV, respectively.

RESULTS AND DISCUSSION

Microstructure evolution during heat treatment

Figure 1 shows the microstructures of quenched and annealed $ReSi_{1.75}$. Both samples possess twinned microstructure with twin thickness of 100-500nm. However, the quenched

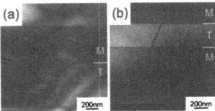

Figure 1. Bright field TEM images of the quenched sample (a) and the annealed sample (b). The quenched sample contains two twinned domains (labeled "M" and "T") and dense planar defects (a). The annealed sample also contains two twinned domains (b).

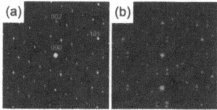

Figure 2. Selected area electron diffraction patterns taken from defected domain in quenched sample (a) and annealed sample (b). Zone-axis is $[010]_{C11b}$. Both diffraction patterns are superposition of two twinned domains.

sample shows other domains which contain dense planar defects. Such defected domains are not observed in the annealed sample. The selected area electron diffraction (SAED) patterns taken from this domain of quenched and annealed samples are shown in Figure 2. The position of the diffraction spots from quenched and annealed samples is identical, but the diffraction spots from quenched sample exhibit streaks.

In our previous study [15], we identified crystal in rhenium base silicide containing planar defects as crystallographic shear structure. Crystallographic shear structures are observed in some oxides of transition metal, such as V_nO_{2n-1}, Ti_nO_{2n-1} etc. (n=2, 3, 4, ...) [16]. Crystallographic shear structure can be formed by operating the following geometrical operation to the mother lattice [16].

Operation 1: The mother lattice is divided into blocks on the boundary of a certain crystal face called crystallographic shear plane with plane indices (hkl) (Figure 3 (a)).

Operation 2: One of the blocks is translated by a certain vector called crystallographic shear vector with [uvw].

Operation 3: If there are spaces between the translated blocks, atomic layers are inserted (Figure 3 (b)). If there are overlaps between blocks, atomic layers are eliminated (Figure 3 (c)).

In our study, two types of crystallographic shear structure have been observed in ternary substituted rhenium silicide, ($\overline{1}$ 09)$_{C11b}$ / $[100]_{C11b}$ and (107) $_{C11b}$ / $[100]_{C11b}$ [15]. The

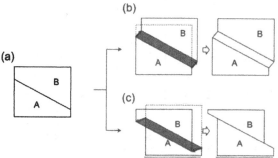

Figure 3. Schematic illustrations of crystallographic shear operation.

11

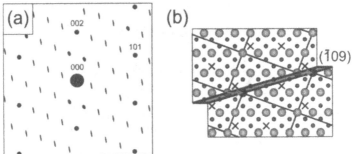

Figure 4. (a) Assumed diffraction pattern take from one domain. (b) Schematic illustration of the crystallographic shear structure model for quenched rhenium silicide.

crystallographic shear plane can be determined by the direction of the streaks in the diffraction pattern. The diffraction pattern shown in Figure 2 is a superposition of diffraction patterns from two twinned domains. Figure 4 (a) shows the simulated diffraction pattern for one domain and Figure 4 (b) shows the schematic illustration of monoclinic rhenium silicide. The direction of streaks corresponds to $(\bar{1}09)_{C11b}$, thus we predict that the crystallographic shear operation of $(\bar{1}09)_{C11b} / [100]_{C11b}$ is introduced to the quenched sample.

In the annealed sample, no such dense planar defect arrays are observed. Thus, the crystallographic shear structure is not thermally stable at low temperature. Crystallographic shear structure is only stable at high temperature up to 1173 K. In order to determine the temperature at which crystallographic shear structure exists stably, annealed samples were again heat treated at 1173K, 1473K, 1673 K and near melting temperature (~2173 K) for 30 minutes and then quenched. Crystallographic shear structure is observed only in sample quenched from temperature near melting point. Therefore the temperature at which crystallographic shear structure is considered to form at temperatures higher than 1673K.

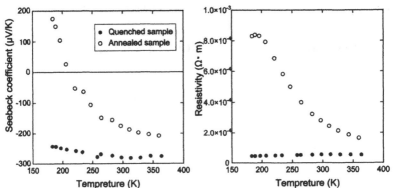

Figure 5. Temperature dependence of Seebeck coefficient and electrical resistivity along $[001]_{C11b}$ from 173 K-373 K for quenched sample and annealed sample.

12

Change in the thermoelectric properties with the variation in microstructure

The temperature dependence of Seebeck coefficient and electrical resistivity along $[001]_{C11b}$ from 173 K-373 K for quenched and annealed samples is shown in Figure 5. The sign of Seebeck coefficient for annealed sample is positive at temperatures below about 200 K, but turns to negative value at higher temperatures. On the other hand, the value of Seebeck coefficients for quenched sample remains almost constant in the temperature range investigated. This indicates that the major carrier for electrical transport in the quenched and annealed samples is different at temperatures below 200 K. The major carrier are electrons for quenched sample for the temperature range investigated, while the major carrier are holes for annealed sample at temperatures below 200 K. Planar faults in the quenched sample are expected to introduce a donor level in the band gap and the electrical conduction turns to n-type.

The quenched sample possesses significantly lower electrical resistivity than the annealed sample for the temperature range investigated, especially below 200 K. This is also well explained by changing the carrier type. Ivanenko et al. reported that the effective mass of electrons is much smaller than that of holes along $[001]_{C11b}$ [17]. Therefore, the resistivity along $[001]_{C11b}$ for n-type quenched samples is smaller than p-type annealed samples.

CONCLUSION

Two types of crystal structure are observed in rhenium silicide with different heat treatments: one is crystallographic shear structure with crystallographic shear operation of ($\bar{1}$ 09) $_{C11b}$ / $[100]_{C11b}$ and the other contains no dense planar defects. Thermoelectric properties are also different with different heat treatment. Quenched samples possessing crystallographic shear structure are n-type semiconductor, while annealed samples containing no dense planar defects are p-type semiconductor. Planar defects in quenched samples are expected to introduce donor levels in the band gap and, therefore, the electrical conduction turns to n-type.

ACKNOWLEDGMENTS

This work was partly supported by Grant-in-Aid for Scientific Research (A) from the Ministry of Education, Science and Culture and Technology (MEXT), Japan (No.19656179) and in part by the Global COE (Center of Excellence) Program on International Center for Integrated Research and Advanced Education in Material Science from the MEXT, Japan. One of the authors (S. Harada) greatly appreciates the supports from Grant-in-Aid for JSPS Fellows.

REFERENCES

1. A. Heinrich, H. Griebmann, G. Behr, L. Ivanenko, J. Schumann and H. Vinzelberg, Thin Solid Films **381**, 287 (2001).
2. V. S. Neshpor, G. V. Samsonov, Sov. Phys. Solid State **2**, 1966 (1961).
3. V. S. Neshpor, G. V. Samsonov, Phys. Met. Metallogr. **11**, 146 (1961).
4. T. Siegrist, F. Hulliger and G. Travaglini, J. Less-Common Met. **92**, 119 (1983).
5. U. Gottlieb, B. L.Andron, F. Nava, M. Affronte, O. Laborde, A. Rouault and R. Madar, J. Appl. Phys. **78**, 3902 (1995).

6. Y. Sakamaki, K. Kuwabara, Gu Jiajun, H. Inui, M. Yamaguchi, A. Yamamoto and H. Obara, Mater. Sci. Forum **426-432**, 1733 (2003).
7. H. Inui, MRS Symp. Proc., **219**, 886 (2006).
8. K. Kuwabara, H. Inui and M. Yamaguchi, Intermetallics **10**, 129 (2002).
9. V. S. Neshpor and G. V. Samsonov, Izv. Akad. Nauk S.S.S.R., Neorg. Mater. **1**, 655 (1965) [Inorg. Mater. (U.S.S.R.) **1**, 599 (1965)].
10. V. S. Neshpor and G. V. Samsonov, Fiz. Met. Metalloved. **11**, (4) 638 (1961) [Sov. Phys.- Phys. Met. Metallogr. **11**, (4) 146(1961)].
11. J. J. Gu, M. W. Oh, H. Inui and D. Zhang, Phys. Rev. B **71**,113201 (2005).
12. A. Misra, F. Chu and T. E. Mitchell, Phil. Mag. A **79** 1411 (1999).
13. T. E. Mitchell and A. Misra, Mater. Sci. Eng. A **261** 106 (1999).
14. S. Harada, K. Tanaka, K. Kishida, H. Inui, Adv. Mater. Res., **26-28**, 197 (2007).
15. S. Harada, K. Tanaka, K. Kishida, H. Inui, MRS Symp. Proc., 980, II05-40 (2007).
16. K. Kosuge, in Chemistry of nonstoichiometric compounds (Oxford University Press, Oxford, 1994), pp. 115–129.
17. L. Ivanenko, V. L. Shaposhnikov, A. B. Filonov, D. B. Migas, G. Behr, J. Schumann, H. Vinzelberg, and V. E. Borisenko, Microelectron. Eng., **64**, 225 (2002).

Mater. Res. Soc. Symp. Proc. Vol. 1128 © 2009 Materials Research Society   1128-U01-09

Phase Stability and Thermoelectric Properties of
Half-Heusler (M^a, M^b)NiSn (M^a, M^b = Hf, Zr, Ti)

Yoshisato Kimura, Hazuki Ueno, Takahiro Kenjo, Chihiro Asami and Yoshinao Mishima
Tokyo Institute of Technology, Interdisciplinary Graduate School of Science and Engineering,
Department of Materials Science and Engineering, 4259-G3-23 Nagatsuta, Midori-ku,
Yokohama 226-8502, Japan.

ABSTRACT

Aiming to improve thermoelectric properties of half-Heusler (M^a,M^b)NiSn alloys (M^a,M^b = Hf, Zr, Ti), phase equilibria in the (M^a,M^b)NiSn systems were investigated focusing on the phase separation of TiNiSn from ZrNiSn and HfNiSn while (Zr,Hf)NiSn forms all proportion miscible solid solution. Diffusion couples consisting of liquid Sn and solid (Ti,Zr)Ni were used to examine the partitioning behavior which is associated with T-rich and Ti-poor half-Heusler phase separation during the reaction at the interface. Thermal conductivity can be reduced in ($M^a_{0.5}$,$M^b_{0.5}$)NiSn and ($Ti_{0.13}$,$Zr_{0.87}$)NiSn alloys due to the solid solution effect of M-site substitution. ($Ti_{0.13}$,$Zr_{0.87}$)NiSn alloy has high potential as a ecological thermoelectric material.

INTRODUCTION

Intermetallic-based alloys have high potentials as both structural and functional materials since they provide us wide variations of ordered crystal structures and wide selections of chemical compositions. Thermoelectric power generation is an appealing approach for conserving energy and preserving the global environment, for which intermetallic-based alloys can be quite attractive candidate materials. We have focused on half-Heusler compounds [1-10] that can be used at around 1000 K to directly convert waste heat into clean electrical power [3-6]. The ordered crystal structure of half-Heusler, ABX, consists of four interpenetrating fcc sub-lattices of elements A, B, X and vacancy, where A and B are transition metals and X is typically Sn and Sb. A half of B sites in (full-)Heusler AB_2X are vacant in half-Heusler ABX. In general, half-Heusler compounds have excellent electrical properties, i.e., high absolute values of Seebeck coefficient and low electrical resistivity. Many research groups, including ours, pay much attention to excellent n-type MNiSn (M = Hf, Zr, Ti) systems. Thermoelectric properties can be generally improved by the optimization of carrier concentrations for electrical properties [2,7,11,13]. On the other hand, relatively high thermal conduction is a disadvantage of half-Heusler compounds. Many research groups reduce the lattice thermal conductivity by substituting M site elements between the group 4 elements Ti, Zr and Hf [7,8,10-12]. It is called "solid solution effect" since the differences in atomic mass and atomic size in a solid solution reduce the lattice thermal conduction through enhancing phonon scattering [13]. High values of dimensionless thermoelectric figure of merit, ZT, for instance exceeding 1.5 were reported for multi-component systems of (Hf,Zr,Ti)Ni(Sn,Sb) [11]. It has been believed that Ti, Zr and Hf are fully miscible with each other in the M site of MNiSn systems. We confirmed that Zr and Hf are all proportion miscible in (Zr_x, Hf_{1-x})NiSn (x = 1 ~ 0) using nearly single-phase alloys fabricated by the directional solidification [4]. On the other hand, our group also confirmed the phase separation between TiNiSn and HfNiSn, and between TiNiSn and ZrNiSn, using arc-melt alloys including impurity phases and directionally solidified half-Heusler single-phase alloys [3].

The objective of the present work is to establish the basis of thermoelectric material design for half-Heusler MNiSn (where M consists of any of Hf, Zr, Ti) systems based on the phase stability and phase equilibria. It is important to understand the substitution behavior of M site among Ti, Zr and Hf to improve thermoelectric performance. We particularly focus on a (Ti, Zr)NiSn system from the economical and ecological viewpoint.

EXPERIMENTAL DETAILS

Alloys with nominal compositions of (M^a, M^b)NiSn, where M^a and M^b are any two of Hf, Zr and Ti, were prepared by arc-melting under an argon atmosphere. Then the directional solidification (DS) was conducted using the optical floating zone (OFZ) melting method in flowing argon gas atmosphere under a slightly positive pressure. The solidification rate was controlled in a range from 2.0 to 10.0 mm/h. Microstructures were observed by scanning electron microscopy using a back scattered electron image (BEI). The chemical compositions of constituent phases were qualitatively measured using electron probe microanalysis (EPMA) to evaluate the phase equilibria. Phase identification was conducted by means of powder x-ray diffractometry (XRD). The Seebeck coefficient and electrical resistivity were measured simultaneously using the DC four-probe method. Thermal conductivity was evaluated from thermal diffusivity and heat capacity measured by the laser-flash method and differential scanning calorimetry (DSC). To evaluate carrier concentration, Hall measurements were conducted at room temperature using van der Pauw method with magnetic field of 0.6 T.

RESULTS AND DISCUSSION
Phase stability and phase equilibria regarding (M^a, M^b)NiSn

Nearly single-phase half-Heusler alloys were fabricated using the OFZ-DS method; $(M^a_{0.5}, M^b_{0.5})$NiSn $(M^a, M^b$ = Hf, Zr, Ti) and (Ti_x, Zr_{1-x})NiSn $(x = 0$ to 1) alloys. Back scattered electron images of as-grown microstructures are shown in Fig. 1 for $(M^a_{0.5}, M^b_{0.5})$NiSn alloys compared with arc-melt counterparts. Arc-melt as-cast microstructures contain considerable amounts of coexisting impurity phases, such as Sn phase in the $(Ti_{0.5}, Zr_{0.5})$NiSn alloy, Ti_6Sn_5 phase in the $(Ti_{0.5}, Hf_{0.5})$NiSn alloy and $(Zr,Hf)_5Sn_3$ phase in the $(Zr_{0.5}, Hf_{0.5})$NiSn alloy. On the contrary, as-grown microstructures of OFZ-DS alloys are nearly single-phase of half-Heusler, particularly, the $(Zr_{0.5}, Hf_{0.5})$NiSn alloy is single-crystal. We have found a tendency of the phase separation between Ti-rich and Hf-rich half-Heusler (Ti, Hf)NiSn phases in the $(Ti_{0.5}, Hf_{0.5})$NiSn alloy through microstructure observation and quantitative chemical concentration measurement for constituent elements using EPMA. It is clearly observed in the arc-melt as-cast sample that dendrite microstructure has a concentration gradient as a result of the half-Heusler phase separation; core of half-Heusler dendrite is Ti-poor (Ti, Hf)NiSn while inter-dendrite region is Ti-rich (Ti, Hf)NiSn. The phase separation of half-Heusler is also observed in $(Ti_{0.5}, Zr_{0.5})$NiSn alloys, and Ti-poor (Ti, Zr)NiSn phase seems to be preferentially grown during OFZ-DS process. It suggests that considerable amount of Zr is rejected and enriched in an outer layer of an ingot while Ti-poor (Ti, Zr)NiSn phase grows preferably in the center of an ingot.

The phase separation between Ti-rich and Ti-poor half-Heusler phases can be confirmed using the powder XRD. Typical XRD profiles observed for (Ti_x, Zr_{1-x})NiSn $(x = 0$ to 1) alloys are represented as an example in Fig. 2. Splitting and broadening, or shift, of diffraction peaks of half-Heusler phases were observed depending on alloy compositions. This tendency is clearly seen in a blow up of (200) diffraction peak shown in (b) together with reported (200) peak positions of ternary TiNiSn and ZrNiSn [1]. The (200) peak position of $(Ti_{0.8}Zr_{0.2})$NiSn is close

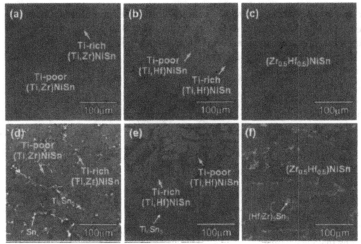

Figure 1. Back scattered electron images showing typical microstructures of (Ti$_{0.5}$, Zr$_{0.5}$)NiSn (a, d), (Ti$_{0.5}$, Hf$_{0.5}$)NiSn (b, e), and (Zr$_{0.5}$, Hf$_{0.5}$)NiSn (c, f), alloys. (a-c) directionally solidified as-grown samples and (d-f) arc-melt as-cast samples.

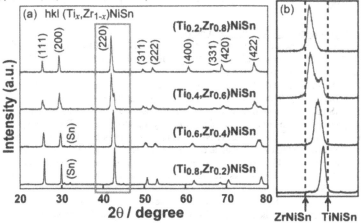

Figure 2. Typical profiles from powder XRD measurements conducted on (Ti$_x$, Zr$_{1-x}$)NiSn alloys.

to TiNiSn, and it shifts getting closer to ZrNiSn as the Ti:Zr ratio reaches to (Ti$_{0.2}$Zr$_{0.8}$)NiSn. It should be noted that similar tendency is observed in (Ti$_x$,Hf$_{1-x}$)NiSn alloys, however, diffraction peaks observed in (Zr$_x$,Hf$_{1-x}$)NiSn alloys are rather sharp since Zr and Hf are all-proportion miscible as forming a continuous solid solution of (Zr, Hf)NiSn.

Phase separation behavior in (Ti, Zr)NiSn and (Ti, Hf)NiSn systems was also evaluated using diffusion couples which consists of liquid Sn phase and solid TiNi phase containing 50 at.% Zr. A typical BEI and corresponding concentration profiles across the interface are shown in Fig. 3 for the case of Sn(L)/(Ti$_{0.5}$,Zr$_{0.5}$)Ni interface. The (Ti,Zr)NiSn phase forms in two different appearances at the interface; one is a layer consisting of two regions and the other is faceted grains growing toward Sn phase. The total growth thickness is around 40 μm. It is interesting that the (Ti,Zr)NiSn layer is composed of two layers. One is Zr-rich (Ti,Zr)NiSn phase (Ti:Zr = 1:2) formed on the (Ti$_{0.5}$, Zr$_{0.5}$)Ni side of the interface, and the other is Ti-rich (Ti,Zr)NiSn phase (Ti:Zr = 4:1) formed on the Sn side. Focusing on partitioning behavior of Zr between Ti-rich and Zr-rich phases, this result suggests the phase separation between TiNiSn and ZrNiSn. Composition ratio Ti:Zr indicates the solubility limit of Ti in ZrNiSn and Zr in TiNiSn vice versa. Note that the average chemical composition of faceted (Ti,Zr)NiSn grains is almost same as Ti-rich (Ti,Zr)NiSn.

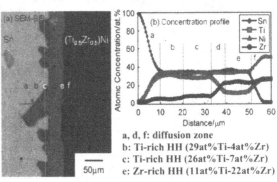

a, d, f: diffusion zone
b: Ti-rich HH (29at%Ti-4at%Zr)
c: Ti-rich HH (26at%Ti-7at%Zr)
e: Zr-rich HH (11at%Ti-22at%Zr)

Figure 3. A back scattered electron image (a) and corresponding chemical composition profiles (b) at the interface of diffusion couple (Ti,Zr)Ni/Sn[liquid] annealed at 1073 K for 1 h.

Thermoelectric properties of (M^a, M^b)NiSn alloys

Thermoelectric properties were evaluated for directionally solidified (M$^a_{0.5}$, M$^b_{0.5}$)NiSn (M = Hf, Zr, Ti) and (Ti$_{0.13}$, Zr$_{0.87}$)NiSn alloys. We pay much attention on the (Ti, Zr)NiSn system because it is desirable to develop ecological thermoelectric materials without using costly elements such as Hf in the case of the present work. First of all, the temperature dependence of (a) Seebeck coefficient, (b) electrical resistivity and (c) power factor is shown in Fig. 4. Ternary ZrNiSn and HfNiSn are represented together by dashed lines for a comparison. Note that data of TiNiSn single-phase alloy are not available since it is very hard to fabricate TiNiSn single-phase alloy even by the OFZ-DS method so far. HfNiSn has much higher Seebeck coefficient and higher electrical resistivity than ZrNiSn since carrier concentration of HfNiSn, 6.07 x 10^{24} m^{-3}, is lower than that of ZrNiSn, 9.48 x 10^{25} m^{-3}. Seebeck coefficient and electrical resistivity of (M$^a_{0.5}$, M$^b_{0.5}$)NiSn and (Ti$_{0.13}$, Zr$_{0.87}$)NiSn alloys are roughly within the range between ZrNiSn and HfNiSn. Values of Seebeck coefficient of (Ti$_{0.5}$, Zr$_{0.5}$)NiSn and (Ti$_{0.5}$, Hf$_{0.5}$)NiSn alloys are slightly lower than ZrNiSn at high temperature, and values of electrical resistivity of (Ti$_{0.5}$, Zr$_{0.5}$)NiSn alloy are lower than those of ZrNiSn. These results suggest that the substitution of Ti for Hf (and Zr) on the M site increases carrier concentration. According to Hall measurement at room temperature, the carrier concentration of (Ti$_{0.5}$, Zr$_{0.5}$)NiSn, 1.59 x 10^{25} m^{-3}, is almost comparable to that of (Zr$_{0.5}$, Hf$_{0.5}$)NiSn, 2.23 x 10^{25} m^{-3}. While the carrier concentration of (Ti$_{0.5}$, Hf$_{0.5}$)NiSn is 1.12 x 10^{26} m^{-3}, which is somehow about one order higher than the other alloys. Comparing two (Ti, Zr)NiSn alloys, it is interesting that (Ti$_{0.13}$, Zr$_{0.87}$)NiSn has much larger values of Seebeck coefficient than (Ti$_{0.5}$, Zr$_{0.5}$)NiSn while both alloys show almost comparable low values of electrical resitivity. As a consequent, (Ti$_{0.13}$, Zr$_{0.87}$)NiSn alloy exhibits the maximum power factor reaching to 4.5 mWm^{-1}K^{-2} at around 800 K. It is suggested that carrier concentration is smaller in (Ti$_{0.13}$, Zr$_{0.87}$)NiSn than in (Ti$_{0.5}$, Zr$_{0.5}$)NiSn. Since (Ti$_{0.5}$, Zr$_{0.5}$)NiSn

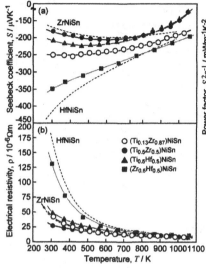

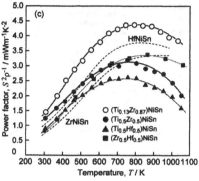

Figure 4. Temperature dependence of (a) the Seebeck coefficient and (b) the electrical resistivity, (c) electrical power factor measured for (Ti, Zr)NiSn alloys compared with (Ti, Hf)NiSn and (Zr, Hf)NiSn alloys

alloy has rather heterogeneous Ti-rich and Zr-rich two-phase half-Heusler microstructure while $(Ti_{0.13}, Zr_{0.87})NiSn$ alloy has single-phase microstructure, electrical properties may be deteriorated by these microstructural factors.

The main effect of substitution of M site elements is to reduce the lattice thermal conductivity via phonon scattering. Temperature dependence of thermal conductivity is shown in Fig. 5 (a) for $(M^a_{0.5}, M^b_{0.5})NiSn$ and $(Ti_{0.13}, Zr_{0.87})NiSn$ alloys. Comparing with ternary ZrNiSn and HfNiSn, thermal conductivity are reduced in these quaternary alloys. We previously reported that the carrier contribution of thermal conductivity κ_{car} is quite small in $(Zr_x, Hf_{1-x})NiSn$ (x = 0 to 1) single-phase alloys [6]. Therefore values of the lattice thermal conductivity κ_{lat} can be regarded close to total conductivity using the Wiedemann-Franz relationship, $\kappa_{total} = \kappa_{car} + \kappa_{lat}$ and $\kappa_{car} = (LT)/\rho$, where L is Lorenz number and ρ is electrical resistivity. Thereby, the lattice thermal conductivity should most effectively be reduced in $(Ti_{0.5}, Hf_{0.5})NiSn$ alloys probably by the solid solution effect since the differences in atomic mass and atomic radius are maximized in the case between Ti and Hf. It should be noted that the lattice thermal conduction can be reduced by microstructural factors, other than the solid solution effect, due to the phase separation. Thermal conductivity of $(Ti_{0.13}, Zr_{0.87})NiSn$ alloy is lower than that of $(Ti_{0.5}, Zr_{0.5})NiSn$ alloy partly because carrier concentration of the former is smaller than that of the latter. Additionally, the solid solution effect remarkably reduces the thermal conduction in the single-phase $(Ti_{0.13}, Zr_{0.87})NiSn$ alloy compared to the inhomogeneous two-phase $(Ti_{0.5}, Zr_{0.5})NiSn$ alloy.

To evaluate the potential of $(M^a_{0.5}, M^b_{0.5})NiSn$ and $(Ti_{0.13}, Zr_{0.87})NiSn$ alloys, the dimensionless thermoelectric figure of merit, ZT, were calculated from experimental results. Temperature dependence of ZT is shown in Fig. 5 (b). Among all the alloys, $(Ti_{0.13}, Zr_{0.87})NiSn$ alloy has the best power factor as mentioned above, and $(Zr_{0.5}, Hf_{0.5})NiSn$ alloy has the highest dimensionless figure of merit, ZT. Considering the good balance of power factor and ZT values, $(Ti_{0.13}, Zr_{0.87})NiSn$ with maximum ZT value of 0.65 at 800 K would be the most practical alloy.

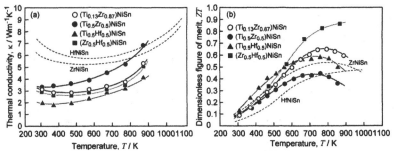

Figure 5. Temperature dependence of (a) the thermal conductivity measured by laser-flash method with DSC measurement and (b) dimensionless thermoelectric figure of merit evaluated for (Ti, Zr)NiSn alloys compared with (Ti, Hf)NiSn and (Zr, Hf)NiSn alloys.

CONCLUSIONS

Phase equilibria related to half-Heusler (M^a, M^b)NiSn $(M^a, M^b = Hf, Zr, Ti)$ systems were investigated with phase stability of these half-Heusler compounds. Moreover, thermoelectric properties were measured for directionally solidified $(M^a_{0.5}, M^b_{0.5})$NiSn and $(Ti_{0.13}, Zr_{0.87})$NiSn half-Heusler alloys fabricated using OFZ-DS method. Following concluding remarks can be drawn in the present work.

1. Phase separation of half-Heusler is observed between TiNiSn and ZrNiSn, and between TiNiSn and HfNiSn by means of microstructure observation, powder XRD and EPMA.
2. Partitioning behavior observed at the $Sn(L)/(Ti_{0.5}, Zr_{0.5})$Ni interface also indicates the phase separation of Ti-rich and Zr-rich (Ti,Zr)NiSn phases.
3. $(Ti_{0.13}, Zr_{0.87})$NiSn alloy has the best balance of maximum power factor of 4.5 mWm^{-1}K^{-2} and high ZT value of 0.65 at 800 K, as well as ecological alloy composition.

REFERENCES

1   F. G. Aliev, N. B. Brandt, V. V. Moshchalkov, V. V. Kozyrkov, R. V. Skolozdra and A. I. Belogorokhov, Z. Phys. B: Condens. Matter **75**, 167 (1989).
2. T. M. Tritt, S. Bhattacharya, Y. Xia, V. Ponnambalam, S. J. Poon and N. Thadhani, Appl. Phys. Lett. **81**, 43 (2002).
3   T. Katayama, S.-W. Kim, Y. Kimura and Y. Mishima: J. Electronic Mater., **32**, 1160 (2003).
4. Y. Kimura, T. Kuji, A. Zama, Y. Shibata, Y. Mishima: MRS Symp. Proc., **886**, 331 (2006).
5. Y. Kimura, A. Zama, Appl. Phys. Lett., **89**, 172110-1 (2006).
6. Y. Kimura, Y. Tamura and T. Kita, Appl. Phys. Lett. **92**, 012105-1, (2008).
7. C. Uher, J, Yang, S. Hu, D. T. Morelli and G. P. Meisner, Phys. Rev. B, **59**, 8615 (1999).
8. H. Hohl, A. Ramirez, C. Goldmann and G. Ernst, J. Phys.: Condens. Mater, **11**, 1697 (1999).
9. Q. Shen, L. Chen, T. Goto, T. Hirai, J. Yang, G. P. Meisner, C. Uher, Appl. Phys. Lett. **79**, 4165 (2001).
10. S. Katsuyama, H. Matsushima and M. Ito, J. Alloys and Compd., **385**, 232 (2004).
11. S. Sakurada and N. Shutoh, Appl. Phys. Lett. **85**, 1140 (2004).
12. K. Kurosaki, T. Maekawa, H. Muta, S. Yamanaka, J.Alloys Compd., **397**, 296 (2005).
13. B. Abeles, Phys. Rev. B, **29**, 1906 (1963).

Mater. Res. Soc. Symp. Proc. Vol. 1128 © 2009 Materials Research Society   1128-U01-10

Screening and Fabrication of Half-Heusler Phases for Thermoelectric Applications

Wilfried Wunderlich[1], Yuichiro Motoyama[1]
1) Tokai University, Fac. Eng, Materials Science Dept., Hiratsuka-shi, Japan

ABSTRACT

Half-Heusler phases have gained recently much interest as thermoelectric materials. Screening of possible systems was performed by drawing their stability region in a three-dimensional Pettifor map. The fabrication of Half-Heusler phases requires three steps, surface activation of the raw material by ball milling, arc-melting of pressed pellets and finally long-term annealing treatment in a vacuum furnace. On doped TiCoSb specimens, Seebeck coefficients of 0.1 mV/K, on NiNbSn 0.16 mV/K were measured, although the microstructure was not yet optimized.

INTRODUCTION

Thermoelectric materials (TE) are considered as clean energy sources helping to solve the severe CO_2- problem, but materials with higher efficiency need to be found. The figure-of-merit $ZT=S^2\sigma T/\kappa$ requires a high Seebeck coefficient S and electric conductivity σ and low thermal conductivity κ. For increasing ZT several concepts for materials design of thermoelectrics have been introduced [1-2], such as phonon-glass, electron-crystal, (PGEC), heavy rattling atoms as phonon absorbers, high density of states at the Fermi energy, differential temperature dependence of density of states, high effective electron mass [3], superlattice structures with their confined two-dimensional electron gas [4] and electron-phonon coupling [5,6]. In this study we focus on the search and fabrication of Half-Heusler (HH) structures, which have been found as successful thermoelectric materials, like NiTiSn [7,8]. The reason why HH [7-9], perovskite [3-5] and Skutterudite [2] are successful is sketched in fig. 1. The phonon wave pushes the electron waves through the crystal, when the electron-phonon coupling has suitable interaction energy [6]. This can be successful, when the electron waves have enough freedom to vibrate. These three crystal structures have vacant lattice positions or force atoms to sit in larger atomic distances than according to their atomic spheres. The empty space is one of the necessities for good thermoelectric materials.

In the first section the search for new Half-Heusler phases by three-dimensional Pettifor maps is described. The second and third sections describe the fabrication, experimentally obtained microstructures and the thermoelectric properties of different systems which were selected because they are possible candidates for HH alloys according to the 3- Pettifor maps.

THREE-DIMENSIONAL PETTIFOR MAPS

Drawing of Pettifor maps, see e.g. [10, 11] is the suitable method to display regions of element combinations, in which certain crystal structures are stable. To our best knowledge it is the first time to show such a map for ternary components XYZ. In HH-phases elements on each position of the three positions $X_1 Y Z$ come from different groups of the periodic table (fig. 2).

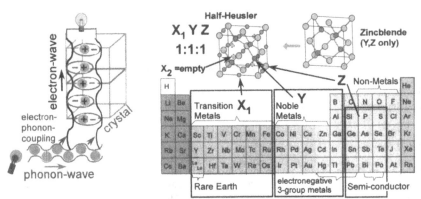

Figure 1. Sketch of the inter-action between phonon- and electron-waves in proper thermoelectric crystals

Figure 2. Half-Heusler crystal structure with X_1 Y Z positions occupied with elements from different groups in the periodic table. If X_1 is empty, Zincblende structure is obtained. If X_2 is occupied, Full-Heusler is obtained.

HH-phases are related to Full-Heusler (FH) phases X_1X_2YZ by leaving position X_2 empty and related to the Zincblende structure leaving position X_1 also empty (fig. 2). The crystallographic

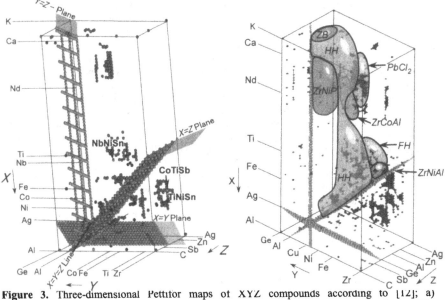

Figure 3. Three-dimensional Pettifor maps of XYZ compounds according to [12]; a) Half-Heusler structures only, b) regions of different crystal structures as marked (HH= Half-Heusler, FH=Full-Heusler, PbCl₂, ZrNiP, ZrCoAl, ZrNiAl).

22

Table 1. Half-Heusler structure and their competing crystal phases with their space group, reference name, a typical representative, and the number of representatives shown in fig. 3b.

Crystal structure	Space group (SG) #	referred in literature as	Represen- tative	appeared in Fig. 3 b n–times
HH	F −4 3m (216)	MgAgAs	TiCoSb	111
FH	F m −3 m (225)	VFe_2Si	$TiNi2Al$	26
ZB	F d −m 3 (216)	Zincblende	GaInSb	18
TiNiSi	P n a m (062)	TiNiSi ($PbCl_2$)	CrNiSb	97
ZrNiAl	P −6_2 m (189)	ordered Fe_2P	CrNiAs	53
FeSiV	P6/mmm (191)	CeNiSb	FeSiV	53
ZrNiP	P6_3/mmc (194)	ZrBeSi ($MgZn_2$)	ZrNiP	63
ZrCoAl	F d−3m S (227)	CoMnSb (CaF_2)	ZrCoAl	80
fcc–SS.	F m 3 m (225)	Ni	AgAuSi	–
bcc–SS.	I m 3 m (229)	Nb		–
hcp–SS	P6_3/mmc (194)	Cr		–
others				89

*) also available as anti-$PbCl_2$ structure; atomic positions strongly depend on the ionic radii.
Aberrations: HH=Half-Heusler, FH=Full-Heusler, ZB=Zincblende, SS=Solid Solution.

database [12] was searched for XYZ components and their crystallographic structures were put in the positions of sets of large tables, where x-, y- and z- axis correspond to each element on the crystallographic positions X_1, Y, Z. In Pettifor maps the elements are ordered according to the Mendeleev number [11], which starts with 1 (He), and ends at 103 (H). The regions for elements on X, Y, Z were restricted (fig. 3) because of the table's size. Symmetry planes with equal X=Y, Z=X and X=Y-elements are marked in fig. 3 with grey dots. Most of the crystal structure of XYZ compounds in the mentioned element range consists of those phases shown in table 1: HH-structures, in literature referred as MgAgAs, as shown in the third row together with the other XYZ-phases. These eight crystal structures cover more than 80% of all combinations in the 3-d Pettifor map (fig. 3b) and with ab-initio calculations [13] their stability and bond spectrum [14] was checked. HH structures cover a region much larger than from fig, 2 expected, limited by InLiAg, SbLiAg, PdFeTe and CuAgTe at each end, leaving many topics left for new research.

EXPERIMENTAL: FABRICATION AND MICROSTRUCTURE

Six alloy systems were selected in order to investigate whether HH phase exists as a stable phase. The metallic powders (μm-size, purity 99.999%, Fine Chemicals, Japan) with defined weight ratios were ball-milled for about 4h with 5mm Zirconia balls and then pressed into 8mm sized pellets. Subsequent 10kW arc-melting still leaves non-reacted Ti as black inclusions (fig. 4a) at TiCoSb, but during sintering of powder under Ar at 1023K for 10h 50μm sized HH-crystals were formed (fig. 5a) also confirmed by XRD diffraction, details see [15]. In the case of Mo its high melting point prevents sufficient diffusion but annealing under Ar at 1023K for 100h reduced the Mo particles remarkably from 15 μm to about 5 μm observed by SEM (Hitachi S-3200N, 30kV) equipped with EDS (Noran, Be-window). The microstructure after arc-melting is shown in fig. 4, a) TiCoSb, b) ZrCoSb, c) MoFeSb, d) NiCrSn, e) NbNiSn, f) NiTiSn. In all systems, the non-reacted transition metal elements Ti, Cr, Zr, Nb appear darker compared to Ni, Sn, or Sb. During solidification Sb (upper row) forms an intermetallic compound with Co or Fe and Sn (lower row) with Ni forming Ni_3Sn_2, Ni_3Sn_4 or NbNiSn. After annealing at 1023 K for 100h (fig. 5) diffusion drives microstructure homogeneous and binary particles smaller. In the TiCoSb system except the HH phase no other ternary phase was

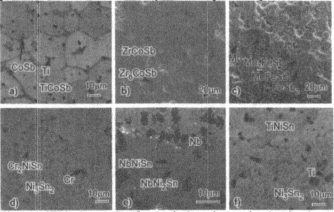

Figure 4. SEM micrographs of arc-melted specimens, showing the phases as marked; a) TiCoSb (HH), grey CoSb, dark Ti b) ZrCoSb, dark Zr_6CoSb, c) MoFeSb, dark Mo, bright FeSb, d) Cr_2NiSn, dark Cr, bright Ni_3Sn_2; e) NbNiSn, black Nb, dark $NbNi_2Sn$, bright NbNiSn or $Nb_3Ni_2Sn_2$, f) NiTiSn, dark Ti, bright Ni_3Sn_2.

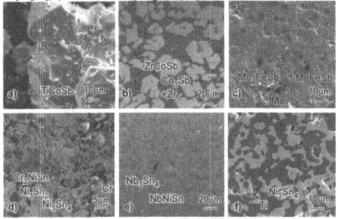

Figure 5. SEM micrographs after annealing (1023K 100h) of the phases, a) TiCoSb HH-crystals, b) ZrCoSb, dark Zr, bright CoSb, c) MoFeSb, dark Mo, grey Mo_2FeSb, d) Cr_2NiSn-Full-Heusler crystals, dark Cr, bright Nb_3Sn_2, Nb_3Sn_4; e) NbNiSn, dark Nb_3Sn_2, Nb_3Sn_4, or $Nb_3Ni_2Sn_2$, f) dark Ti and bright Ni_3Sn_4.

observed, in agreement with a study [16] on ball-milling, which aims a surface activation and hence higher reactivity. However, with increasing milling time, separation into two phases Ti and CoSb occurred, in agreement with our results on planar interface geometry [15]. In the case of ZrCoSb, HH is dominating, but also an eutectic $CoSb_3$-Zr microstructure remains due to non-stoichiometry. The MoFeSb system shows besides HH the Mo_2FeSb full-Heusler phase. According to the Pettifor map (fig. 3b) the CrNiSn system lies within the range of possible HH phases but it showed about 5µm large cubic-shaped Cr_2NiSn crystals, confirmed as full-Heusler phase. In the NbNiSn system after annealing hardly discernible NbNiSn, Nb_3Sn_2, Nb_3Sn_4 or $Nb_3Ni_2Sn_2$ phases appear. For the NiTiSn system decomposition into Ni_3Sn_4 and Ti occurred, indicating that the annealing temperature 1023K was higher than the temperature for which NiTiSn HH-phase is stable. Table 2 summarizes the results and indicates necessary optimization.

EXPERIMENTAL: SEEBECK- VOLTAGE MEASUREMENTS

The Seebeck voltage was measured in a self-built device made of refractories as explained in [17]. The pellet-shaped specimen lies with one end on a 15x15x2mm sized ceramics heater (Sakaguchi MS-1000) heated up to 1273K and the other end on a 30x10x10mm sized Cu heat sink. The Seebeck voltage U_S was measured with 1mm thick Ni-wires lying below and above the specimen with attached voltmeters [17] and were recorded together with temperature data in-situ in an attached computer. While Seebeck voltage measurements reported in literature are usually measured with small temperature gradients, for which theoretical conditions are valid for, in our case large temperature gradients of more than 600K are present. Such measurement conditions lead to somewhat higher values for the Seebeck voltages as literature data [9,17]. When heat flows steadily, the electric circuit of hot and cold end of the specimen was connected via resistances of R=1Ω, 100Ω, 10kΩ, or 1MΩ under closed circuit condition, and the electric current was measured and recorded.

The second and third row of table 2 show crystal structure calculation results explained elsewhere [14], the forth row summarizes the microstructure and the last two rows show Seebeck voltage at ΔT=600K and its maximum electric current. The maximum Seebeck voltage was measured in our apparatus on doped $NaTaO_3$ with Seebeck voltage -250mV and a maximum current of -330 µA [17], but the HH-specimens presented here, showed much smaller values. NbNiSn alloys show typically U=-100mV and I=-11mA, which yield to a Seebeck-coefficient

Table 2 Summary of calculation and experimental results

	Calculation Results		Experimental Results		
	Stable	Lattice		Seebeck-	
System	Lattice in	constant	Microstructure after annealing	Voltage at ΔT=600K	
	XYZ order	a [nm]	at 1023K for 100h	U [mV]	I_{max} [µA]
TiCoSb	TiCoSb	0.5873	HH; Ti+CoSb +3%Fe	−70	−3
ZrCoSb	ZrCoSb	0.5702	ZrCoSb, Zr_6CoSb	3	30
MoFeSb	MoFeSb	0.5588	Mo, MoSbFe, FeSb	7	11
NiCrSn	Cr_2NiSn	0.5902	Cr_2NiSn, Cr, Ni_2Sn	7	12
NbNiSn	NbNiSn	0.5987	HH, $Nb_5Ni_2Sn_2$, Ni_3Sn_2	−100	−11
TiNiSn	TiNiSn	0.5919	NiSnTi, $NiTi_3Sn$, $NiTiSn_3$ +3at%Fe	58	0.6

S=-100/600 mV/K=-0.16 mV/K. The Seebeck voltage strongly depends on the stochiometry and was highest for alloys with $Nb_{10}Ni_{10}Sn_8$ concentration. TiNiSn doped with 3at% Fe showed a positive Seebeck-voltage of 58mV. To conclude, further increase in Seebeck voltage is expected due to doping and microstructure optimization.

CONCLUSIONS

1) Half Heusler (HH) phases are promising candidates for thermoelectric materials approached from the metallic side due to one vacant lattice position. They are found to be stable phase over a wide range of element combinations as displayed here for the first time as three-dimensional Pettifor maps.
2) The best fabrication method for HH found in this study was, first ball-milling in order to activate the raw powder's surface, then arc melting and annealing.
3) Seebeck voltage measurements under large temperature gradient are able to judge the performance of thermoelectric candidate materials. Different behavior was found for some alloys when the voltage-current dependence was measured under closed electric circuit conditions, suggesting that the electric current measurements should also be included when comparing the performance of thermoelectrics.

ACKNOWLEDGEMENTS

Obtaining experimental data is acknowledged to Yoshikaji Aoki, Kosuke Nakatsuka, Kouta Okayama, Takayuki Nakagome, Kenji Uematsu, Souichiro Yoshimura.

REFERENCES

[1] M.S.Dresselhaus, MIT Boston (2001) http://web.mit.edu/afs/athena/course/6/6.732/www/new_part1.pdf
[2] G.S.Nolas, et.al., *MRS Bulletin* **31** 199-205 (2006); US Patent 6207888 (27.03.2001)
[3] W. Wunderlich, H. Ohta, and K.Koumoto, *arXiv/cond-mat*0510013 (to be published)
[4] K.H. Lee, Y. Muna, H. Ohta, and K. Koumoto, *Appl. Phys. Exp.* **1** 015007 (2008)
[5] M. Yamamoto, H. Ohta and K. Koumoto, *Appl.Phys.Lett.* **90** 072101 (2007)
[6] J. Sjakste, N. Vast, and V. Tyuterev, *Phys. Rev. Lett.* **99** 236405 (2007)
[7] S.R. Culp, S.J.Poon, T.M.Tritt, et al., *Appl. Phys. Lett.* **88** 042106 (2006)
[8] S. Sakurada, and N.Shutoh, *Appl. Phys. Lett.* **86** 082105 (2005)
[9] T.Sekimoto, K. Kurosaki, H. Muta, and S.Yamanaka, *J. All. Comp.*, **407** 326 (2006)
[10] D.G. Pettifor, and R.Podlucky, *Phys. Rev. Lett.* **53** 1080 (1984)
[11] S.Ranganathan, A. Inoue, *Acta Mat.* **54** 3647 (2006)
[12] FindIt, *Inorganic Crystallographic database Version 1.3.3*, NIST Gaithersburg / FIZ (2004)
[13] G. Kresse, and J. Hafner, *Phys. Rev. B* **49** 14251 (1994).
[14] W. Wunderlich, *Solid-State Electronics* **52** 1082 (2008); (and results to be published)
[15] Y. Aoki, and W. Wunderlich, *Proc. Jap. Inst. Met.* **141** 315 (9.2007); **142** 329 (3.2008)
[16] P. Amornpitoksuk, and S. Suwanboon, *J. Alloy Comp.* **462** 267 (2008)
[17] W. Wunderlich, *J. Nucl. Mat.* (in print, Doi: 10.1016/j.jnucmat.2009.01.007)

Mater. Res. Soc. Symp. Proc. Vol. 1128 © 2009 Materials Research Society

Mechanical and Thermal Properties of Single Crystals of Some Thermoelectric Clathrate Compounds

Norihiko L. Okamoto, Takahiro Nakano, Kyosuke Kishida, Katsushi Tanaka and Haruyuki Inui
Department of Materials Science and Engineering, Kyoto University, Sakyo-ku, Kyoto 606-8501, Japan

ABSTRACT

The mechanical and thermal properties of single crystals of the type-I clathrate compounds $Ba_8Ga_{16}Ge_{30}$ and $Sr_8Ga_{16}Ge_{30}$ have been investigated by measuring the elastic constants, coefficients of thermal expansion (CTE) and plastic deformation behavior in compression. The feasibility of these two clathrate compounds as a thermoelectric material in terms of mechanical stability under possible thermal stresses is evaluated by calculating thermal stresses that are expected to develop within these compounds when used as thermoelectric devises.

INTRODUCTION

Clathrate compounds have been rigorously investigated as promising candidates for practical thermoelectric materials because they exhibit relatively high electrical conductivity and very low lattice thermal conductivity [1-4]. These properties are considered to originate from their particular crystal structures consisting of polyhedral cages, which comprise tetrahedrally-bonded group IV/III atoms, and guest atoms, which are weakly bonded to the cages. Among a number of clathrate compounds with various structure types, the "type-I" clathrate compounds with the cubic space group $Pm\bar{3}n$ [5] have been investigated the most intensively [1-3]. In particular, the type-I clathrate compounds $Ba_8Ga_{16}Ge_{30}$ and $Sr_8Ga_{16}Ge_{30}$, in which Ba and Sr atoms are encapsulated by the $[Ga_{16}Ge_{30}]$ cage framework, have been reported to exhibit excellent thermoelectric properties at relatively high temperatures [1, 2, 6]. Knowledge on the mechanical properties of these type-I clathrate compounds is indispensable when their practical application in thermoelectric devices is considered, since these clathrate compounds will inherently be subjected to thermal stresses arising mostly from their own thermal expansion and contraction within thermoelectric devices [7]. To the best of our knowledge, however, almost nothing is known about the mechanical properties of these type-I clathrate compounds $Ba_8Ga_{16}Ge_{30}$ and $Sr_8Ga_{16}Ge_{30}$, except for low-temperature elastic constants for a few clathrate compounds [8]. In the present study, we investigate the elastic properties, thermal expansion and plastic deformation behavior of single crystals of $Ba_8Ga_{16}Ge_{30}$ and $Sr_8Ga_{16}Ge_{30}$ to elucidate their mechanical and thermal properties. Thermal stresses that are developed within these compounds when used as a thermoelectric material in thermoelectric devises are also evaluated with the values of elastic moduli and coefficients for thermal expansion determined in the present work adopting some simple models.

EXPERIMENTAL PROCEDURES

Single crystals of $Ba_8Ga_{16}Ge_{30}$ and $Sr_8Ga_{16}Ge_{30}$ were grown by the Czochralski method at a growth rate of 5 mm/h under an Ar gas flow. Specimens with a rectangular parallelepiped shape having three orthogonal faces parallel to {100} planes were cut from the single crystal by spark-machining for measurements of elastic constants and coefficients of thermal expansion (CTE). Measurements of elastic constants were carried out by the rectangular parallelepiped resonance (RPR) method [9] in the temperature range from room temperature to 873 K and 773 K for $Ba_8Ga_{16}Ge_{30}$ and $Sr_8Ga_{16}Ge_{30}$, respectively, while measurements of CTE were carried out with a push-rod type differential dilatometer (Shimadzu TMA-60) in the temperature range from 300 to 973 K at a heating rate of 5 K/min under an Ar gas flow. Compression tests were also carried out on an Instron-type testing machine in vacuum at a strain rate of 1×10^{-4} s^{-1} at 1123 and 973 K for $Ba_8Ga_{16}Ge_{30}$ and $Sr_8Ga_{16}Ge_{30}$, respectively. The compression axis was chosen to be [110] so that the operation of the most plausible slip system {001}<100> is facilitated.

RESULTS

Elastic constants

Values of single-crystal elastic constants, c_{ij}, of $Ba_8Ga_{16}Ge_{30}$ and $Sr_8Ga_{16}Ge_{30}$ are plotted in figures 1(a) and (b), respectively, as a function of temperature. Values of all the three independent elastic constants decrease monotonically with the increase in temperature for both $Ba_8Ga_{16}Ge_{30}$ and $Sr_8Ga_{16}Ge_{30}$. While the values of c_{11} and c_{44} for $Ba_8Ga_{16}Ge_{30}$ are larger than those for $Sr_8Ga_{16}Ge_{30}$, the value of c_{12} for $Ba_8Ga_{16}Ge_{30}$ is slightly smaller than that for $Sr_8Ga_{16}Ge_{30}$. The order of relative values of c_{12} and c_{44} is reversed between $Ba_8Ga_{16}Ge_{30}$ and $Sr_8Ga_{16}Ge_{30}$.

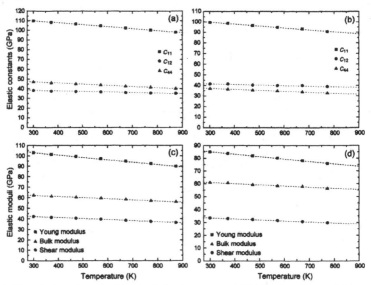

Figure 1. Three independent single-crystal elastic constants of (a) $Ba_8Ga_{16}Ge_{30}$ and (b) $Sr_8Ga_{16}Ge_{30}$ and polycrystalline bulk, shear and Young's moduli of (c) $Ba_8Ga_{16}Ge_{30}$ and (d) $Sr_8Ga_{16}Ge_{30}$ plotted as a function of temperature.

Polycrystalline elastic moduli [bulk (B), Young's (E) and shear (G) moduli] are evaluated from the values of single-crystal elastic constants by the Hill's method [10]. Values of polycrystalline elastic constants of $Ba_8Ga_{16}Ge_{30}$ and $Sr_8Ga_{16}Ge_{30}$ are plotted in figures 1(c) and (d), respectively, as a function of temperature. Values of all the three polycrystalline elastic constants decrease monotonically with the increase in temperature for both $Ba_8Ga_{16}Ge_{30}$ and $Sr_8Ga_{16}Ge_{30}$. Of importance to note is that the values of B for $Ba_8Ga_{16}Ge_{30}$ and $Sr_8Ga_{16}Ge_{30}$ are almost identical to each other whereas the values of Young's and shear moduli for $Ba_8Ga_{16}Ge_{30}$ are larger than those for $Sr_8Ga_{16}Ge_{30}$.

Thermal expansion

Relative elongation with respect to the original specimen length at room temperature for both $Ba_8Ga_{16}Ge_{30}$ and $Sr_8Ga_{16}Ge_{30}$ increases virtually linearly with increasing temperature. The thermal expansion behavior for the two clathrate compounds is almost identical with each other. The values of CTE averaged over the investigated temperature range are 14.2×10^{-6} and 14.1×10^{-6} K^{-1} for $Ba_8Ga_{16}Ge_{30}$ and $Sr_8Ga_{16}Ge_{30}$, respectively. These CTE values are comparable to those of typical metals but are considerably larger than that of pure germanium (5.92×10^{-6} K^{-1}) [11].

Compression tests

When referring to the crystal structure of type-I clathrate compounds, $\{001\}<100>$ is expected to be the most readily operating slip system. However, the shortest lattice translation vector along $<100>$ corresponding to the Burgers vector of slip dislocations is $a<100>$ (a: lattice constant), whose magnitude exceeds 1 nm [5]. Plastic deformation of type-I clathrate compounds is thus expected to be very difficult to occur. The temperatures at which compression tests were made correspond to as high as 0.90 and 0.93 the melting temperature for $Ba_8Ga_{16}Ge_{30}$ and $Sr_8Ga_{16}Ge_{30}$, respectively. Nevertheless, premature fracture occurs in the elastic region without showing any appreciable plastic strain for both $Ba_8Ga_{16}Ge_{30}$ and $Sr_8Ga_{16}Ge_{30}$. The stress values at which fracture occurs are 335 and 236 MPa for $Ba_8Ga_{16}Ge_{30}$ and $Sr_8Ga_{16}Ge_{30}$, respectively.

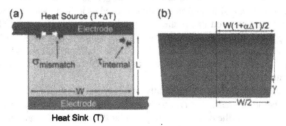

Figure 2. (a) Schematic illustration of thermal stresses developed in the thermoelectric material in operation; stresses ($\sigma_{mismatch}$) that are caused by thermal expansion mismatch between the thermoelectric material (with the length L and width W) and metal electrode, and stresses ($\tau_{internal}$) that are internally developed due to the temperature gradient ($\Delta T/L$) between the heat source and sink. (b) Schematic illustration of the generation of internal shear stresses generated in the direction perpendicular to the thermal gradient due to the different thermal expansion at heat source and sink positions.

DISCUSSION

The value of G is smaller for $Sr_8Ga_{16}Ge_{30}$ than for $Ba_8Ga_{16}Ge_3$. This may arise from the fact that the ionic radius of Sr is smaller than that of Ba, resulting in the larger open space for the Sr atom within the cage of the almost identical size [12]. When the crystal is sheared, the Sr atom can move more easily within the $[Ga_{16}Ge_{30}]$ cage so that the deformation of the cage framework is not disturbed seriously. This is consistent with the fact that the rattling motion is more significant for the Sr atom than for the Ba atom in the type-I clathrate compounds with the $[Ga_{16}Ge_{30}]$ cage framework [12].We evaluate possible thermal stresses arising in these clathrate compounds when used as a thermoelectric material in the thermoelectric devices with the values of elastic constants and CTE experimentally determined in the present study.

We now evaluate two different thermal stresses (figure 2); (i) stresses ($\sigma_{mismatch}$) that are caused by thermal expansion mismatch between the thermoelectric material and metal electrode [see figure 2(a)], and (ii) stresses ($\tau_{internal}$) that are internally developed due to the temperature gradient between the heat source and sink [see figure 2(b)].

As far as metals are used as the electrode, the CTE values of the clathrate compounds (14.2 and 14.1$\times 10^{-6}$ K^{-1} respectively for $Ba_8Ga_{16}Ge_{30}$ and $Sr_8Ga_{16}Ge_{30}$) do not differ so much from the corresponding value for usual metals (for example, 13.4$\times 10^{-6}$ K^{-1} for Ni [13]), and thus the values of the stress caused by thermal expansion mismatch between the thermoelectric material and metal electrode are expected to be small. The magnitude of the stress ($\sigma_{mismatch}$) generated at the interface between the thermoelectric material and metal electrode is evaluated with the following equations [14],

$$\sigma_{mismatch} = -E_{electrode} \cdot \varepsilon_{electrode} = E_{clathrate} \cdot \varepsilon_{clathrate} \quad (1)$$

$$\varepsilon_{total} = \varepsilon_{electrode} - \varepsilon_{clathrate} = (\alpha_{electrode} - \alpha_{clathrate}) \times \Delta t \quad (2)$$

where ε, α and Δt stand for strain, CTE value and the difference between the operation temperature and room (reference) temperature, respectively. As expected, the values of the stresses calculated with the equations (1) and (2) for $Ba_8Ga_{16}Ge_{30}$ and $Sr_8Ga_{16}Ge_{30}$ are as small as 38 and 29 MPa, respectively, even at a high operation temperature of 973 K. Since the CTE values for these clathrate compounds are a little larger than that for Ni, a small compressive stress is generated in them. If a ceramic material is used as a thermoelectric material, however, a large tensile stress is generated inside the material because of their small CTE values and large values of Young's moduli. For example, the $\sigma_{mismatch}$ value is calculated to be as large as 530 MPa for a typical ceramic material, Al_2O_3. Thus, these clathrate compounds seem favourable in avoiding development of stresses caused by thermal expansion mismatch between the thermoelectric material and metal electrode.

The stresses ($\tau_{internal}$) that are internally developed in the thermoelectric material inserted in a uniform temperature gradient, $\Delta T/L$, between the heat source and sink can be calculated as shear stresses generated in the direction perpendicular to the thermal gradient due to the different thermal expansion at heat source and sink positions [figure 2(b)]. Thus, the stress varies not only with the magnitude of temperature difference, ΔT, but also with the dimensions (length L and width W) of the thermoelectric material. This can be calculated with the following equation,

$$\tau_{internal} = \gamma G = \frac{\alpha \Delta T W}{2L} G = \frac{GW\alpha}{2} \times \frac{\Delta T}{L} \quad (3)$$

where γ and G stand respectively for the shear strain and the value of shear modulus of the thermoelectric material. The values of $\tau_{internal}$ are plotted in figure 3 as a function of temperature gradient ($\Delta T/L$) for various specimen widths, W, for $Ba_8Ga_{16}Ge_{30}$, $Sr_8Ga_{16}Ge_{30}$ and Al_2O_3. For any of the three materials, the magnitude of $\tau_{internal}$ calculated with the equation (3) increases with the increase in both the temperature gradient ($\Delta T/L$) and the specimen width (W). With the identical values of $\Delta T/L$ and W, the magnitude of $\tau_{internal}$ for $Sr_8Ga_{16}Ge_{30}$ is a little smaller than that for $Ba_8Ga_{16}Ge_{30}$, but the magnitude of $\tau_{internal}$ for Al_2O_3 is considerably larger than those for $Ba_8Ga_{16}Ge_{30}$ and $Sr_8Ga_{16}Ge_{30}$. The very large $\tau_{internal}$ value for Al_2O_3 obviously comes from the significantly large value of shear modulus. Horizontal lines in the figure correspond to the values of shear stresses ($\tau_{fracture}$) at which fracture occurs for the two clathrate compounds. When the value of W is 5 mm or less, the magnitudes of $\tau_{internal}$ developed in $Ba_8Ga_{16}Ge_{30}$ and $Sr_8Ga_{16}Ge_{30}$ are both smaller than those of $\tau_{fracture}$, avoiding fracture, even if the temperature gradient is increased to as large as 90 K/mm. However, if the value of W is increased to 10 mm or more, the magnitudes of $\tau_{internal}$ exceed those of $\tau_{fracture}$, even with the smaller temperature gradient. The critical temperature gradient at which the magnitude of $\tau_{internal}$ exceeds that of $\tau_{fracture}$ is 52 and 46 K/mm for W=10 mm, and 35 and 31 K/mm for W=15 mm for $Ba_8Ga_{16}Ge_{30}$ and $Sr_8Ga_{16}Ge_{30}$, respectively. In order to prevent fracture during the thermoelectric operation of the clathrate compounds, the temperature gradient must be smaller than these critical values.

CONCLUSIONS

In summary, the single-crystal elastic constants of $Ba_8Ga_{16}Ge_{30}$ and $Sr_8Ga_{16}Ge_{30}$ are determined in the temperature ranges from room temperature to 873 and from room temperature to 773 K, respectively. Coefficients of thermal expansion averaged over the temperature range from 300 to 973 K are determined to be 14.2×10^{-6} and 14.1×10^{-6} K^{-1} for $Ba_8Ga_{16}Ge_{30}$ and $Sr_8Ga_{16}Ge_{30}$, respectively. Single crystals of $Ba_8Ga_{16}Ge_{30}$ and $Sr_8Ga_{16}Ge_{30}$ fail in the elastic region without showing any plastic deformation in compression tests with the fracture stresses of 335 and 236 MPa, respectively. Thermal stresses developed in $Ba_8Ga_{16}Ge_{30}$ and $Sr_8Ga_{16}Ge_{30}$ are evaluated with the values of elastic moduli and CTE determined in the present work adopting some simple models. The thermal stresses developed due to CTE mismatch between the clathrate compounds and electrode is small when a usual metal such as Ni is used as the electrode. However, the internal shear stress developed due to temperature gradient between the heat source and sink can exceed the fracture stress when the temperature gradient is large (more than ~ 50 K/mm).

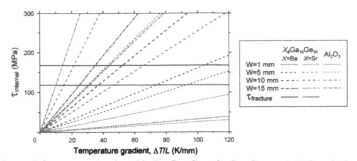

Figure 3. Internal shear stress developed in $Ba_8Ga_{16}Ge_{30}$, $Sr_8Ga_{16}Ge_{30}$ and Al_2O_3 calculated as a function of temperature gradient ($\Delta T/L$) for various specimen widths (W).

ACKNOWLEDGMENTS

This work was supported by Grant-in-Aid for Scientific Research (A) (No. 18206074) from the Ministry of Education, Culture, Sports, Science and Technology (MEXT), Japan, by a grant (Research for Promoting Technological Seeds, No. 06-014) from the Japan Science and Technology Agency and in part by the Foundation "Hattori-Hokokai."

REFERENCES

1. G. S. Nolas, J. L. Cohn, G. A. Slack and S. B. Schujman, *Appl. Phys. Lett.* **73**, 178 (1998).
2. A. Saramat, G. Svensson, A. E. C. Palmqvist, C. Stiewe, E. Mueller, D. Platzek, S. G. K. Williams, D. M. Rowe, J. D. Bryan and G. D. Stucky, *J. Appl. Phys.* **99**, 023708 (2006).
3. N. L. Okamoto, K. Kishida, K. Tanaka and H. Inui, *J. Appl. Phys.* **101**, 113525 (2007).
4. J. H. Kim, N. L. Okamoto, K. Kishida, K. Tanaka and H. Inui, *Acta Mater.* **54**, 2057 (2006).
5. B. Eisenmann, H. Schäfer and R. Zagler, *J. Less-Common Met.* **118**, 43 (1986).
6. N. L. Okamoto, K. Kishida, K. Tanaka and H. Inui, *J. Appl. Phys.* **100**, 073504 (2006).
7. B. C. Sales, B. C. Chakoumakos, R. Jin, J. R. Thompson and D. Mandrus, *Phys. Rev. B.* **63**, 245113 (2001).
8. K. Ueno, A. Yamamoto, T. Noguchi, T. Inoue, S. Sodeoka and H. Obara, *J. Alloys. Comp.* **388**, 118 (2005).
9. V. Keppens, M. A. McGuire, A. Teklu, C. Laermans, B. C. Sales, D. Mandrus, and B. C. Chakoumakos, *Physica B*, **316-317**, 95 (2002).
10. R. Hill, *Proc. Phys. Soc.* **A65**, 346 (1952).
11. K. Tanaka and M. Koiwa, *High Temp. Mater. Proc.* **18**, 323 (1999).
12. M. E. Straumanis and E. Z. Aka, *J. Appl. Phys.* **23**, 330 (1952).
13. Y. S. Touloukian, R. K. Kirby, R. E. Taylor and P. D. Desai, *Thermophysical Properties of Matter*, Vol. 12, *"Thermal Expansion"* (IFI/Plenum, New York, NY, 1975) p. 225.
14. N. L. Okamoto, M. Kusakari, K. Tanaka, H. Inui, M. Yamaguchi and S. Otani, *J. Appl. Phys.* **93**, 88 (2003).

Iron Aluminides—Physical Metallurgy, Processing and Properties

Mater. Res. Soc. Symp. Proc. Vol. 1128 © 2009 Materials Research Society   1128-U02-01

An Overview of the Mechanical Properties of FeAl

I. Baker
Thayer School of Engineering, Dartmouth College, Hanover, NH 03755

ABSTRACT

This paper presents an overview of the mechanical properties of the iron aluminide FeAl. Both the strength and ductility depend on a variety of parameters, including vacancy concentration, the environment (principally water vapor), alloy stoichiometry, temperature and grain size. The effects of alloying elements, particularly boron, are also briefly discussed. The review emphasizes how much of our current understanding of the strength and ductility of iron aluminides emanated from the discovery of the importance of both water vapor and thermal vacancies on these properties.

INTRODUCTION

Iron aluminides, i.e. Fe_3Al, FeAl, $FeAl_2$ Fe_2Al_5 and $FeAl_3$, have been of interest for over 75 years [1]. This review focuses on B2 or ordered body-centered cubic compound FeAl, see Figure 1. The compound is appealing because of its outstanding oxidation resistance, good corrosion resistance, good strength up to ~700K and low density (~6000 kg m^{-3}). Difficulties with this material finding commercial application include its lack of strength at high temperature, where it still shows excellent corrosion and oxidation resistance, and its poor low-temperature ductility, particularly with increasing aluminum content.

This short review outlines the effects of a number of parameters, i.e. vacancy concentration, the environment, alloy stoichiometry, temperature, grain size and alloying elements, on the mechanical properties of FeAl. Additional details can be found in a number of other reviews [2-14]. Our understanding of the effects of these different parameters was only possible once the rôles of the environment and of vacancies on ductility and strength were understood.

CRYSTAL STRUCTURE AND PHASE DIAGRAM

FeAl exists over a wide range of iron-rich compositions at low temperature, e.g. at 473K from ~36.5-49.5 at. % Al, whereas aluminum-rich compositions exist only at elevated temperature, see Figure 2. Hyperstoichiometric compositions of the contiguous $D0_3$-structured compound Fe_3Al disorder to the B2 structure at elevated temperatures that decrease with increasing aluminum content. Up to ~46 at. % Al, FeAl itself disorders to b.c.c. at increasing temperature with increasing aluminum content; binary alloys with greater aluminum concentrations appear to be ordered up to melting.

The Young's modulus, E, of polycrystalline FeAl is not strongly dependent on composition, with values of typically 250 GPa for 40-49 Al [15], and decreases linearly with increasing temperature. As measured by Zener's parameter, $A = 2 C_{44}/(C_{11}-C_{12})$, where C_{ij} are the elastic stiffness constants, FeAl is moderately anisotropic: A decreases slightly with increasing Al concentration from 4.4 at Fe-35Al to 3.8 for Fe-40Al [16].

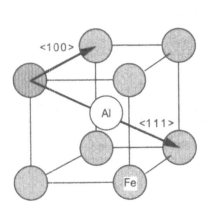

Figure 1. The B2 structure adopted by FeAl, showing the <111> and <100> slip vectors.

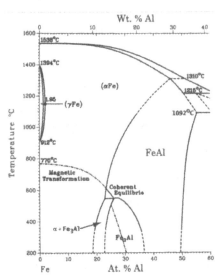

Figure 2. Iron-rich portion of the Fe-Al phase diagram. [17]

POINT DEFECTS

Anti-site atoms accommodate deviations from the stoichiometric composition in FeAl. In addition to anti-site Fe atoms, so-called 'triple defects' are also present, which consist of an Fe atom on the Al sublattice and two vacancies on the Fe sublattice. The concentration of these triple defects increases with increasing Al concentration.

The measured activation enthalpy for vacancy formation (E_f) is quite low, 91-74 kJ mol^{-1} for 38.5-47 at. % Al, and the activation entropy is $6k_B$, where k_B is Boltzmann's constant [18,19]. Thus, unusually high thermal-vacancy concentrations are present in FeAl at elevated temperatures, e.g., 2.5% in stoichiometric FeAl at 1000°C [20].

The vacancies that form at high temperatures are easily retained upon cooling due to the high activation enthalpy for vacancy migration (E_m) which ranges from 152-169 kJ mol^{-1} for 38.5-47 at. % Al [18,19]. In FeAl, therefore, $E_m \approx 2E_f$, unlike most metals where $E_m < E_f$. Note also that with increasing Al concentration, E_f decreases, i.e., the vacancy concentration increases), whereas E_m increases. Thus, it becomes increasingly difficult to anneal out the increasing numbers of thermal vacancies that are retained as the Al concentration increases.

DISLOCATIONS

Early data suggested that a transition from <111> to <100> slip occurs with increasing temperature [21,22]. However, later work showed that while <100> slip and, possibly, [110] slip, may occur at larger strains at elevated temperatures, yielding is due to the multiplication and glide of <111> dislocations, dissociated into anti-phase boundary (APB) coupled ½<111>

36

partials, at all temperatures [23,24]. The <100> dislocations form by the decomposition of <111> dislocations since they have the lowest self-energy [25].

Low temperature anneals (used to remove retained vacancies) produce: numerous faults; <111> dislocations, often as helices, after short anneals; and <100> dislocations loops after long anneals. These observations indicate that <111> dislocations are easy to nucleate by vacancy condensation but eventually decompose into the lower self-energy <100> dislocations [26].

STRENGTH

As outlined below, several factors influence the strength of FeAl. However, vacancies play a major impact on the strength, and without understanding their role, it is impossible to understand the effects of other parameters, a problem that bedeviled some early studies. Water vapor also affects on the strength, an effect that has been less studied than that of vacancies.

Effects of Vacancies

It is not possible to understand the effects of various parameters on the strength without removing the excess vacancies that are readily retained after elevated temperature anneals [27]. Even slow cooling rates retain excess thermal vacancies, a phenomenon arising because of the high E_m vacancy migration. In order to remove most of these vacancies, many workers have adopted the 5 day anneal at 673K used by Nagpal and Baker [27].

Chang et al. [20] showed that the microhardness of binary FeAl alloys, irrespective of aluminum concentration, is linearly related to the square root of the quenched-in thermal vacancy concentration, a relationship that can be rationalized by vacancy pinning of dislocations on the slip plane, which causes them to bow out. Later, Yang and Baker [28] showed that the room temperature yield strength increased with increasing thermal vacancy concentration for both polycrystals and single crystals of Fe-40Al at first rapidly and then more slowly as the concentration increased. Thermal vacancy concentrations in FeAl can be very large and a vacancy concentration of 9×10^{-3} increases the room temperature strength of Fe-40Al single crystals from ~125 MPa to ~550 MPa.

Effects of Environment

Liu, Lee, and McKamey [29] first noted that water vapor, by producing atomic hydrogen through the reaction $3H_2O + 2Al \rightarrow A_2O_3 + 6H$, has a profound effect on the mechanical properties of iron aluminides around room temperature. It has been shown to result in both a lower yield stress and lower subsequent flow stress in both FeAl single crystals and polycrystals when tested in air compared to tests performed in vacuum as long as the strain rate is less than $1 \, s^{-1}$ [30-32], see Figure 3. Conversely, at higher strain rates the effects of the water vapor/atomic hydrogen are not present, see Figure 3. Details of the mechanism for this strength reduction are not clear, but it is worth noting that most studies of the effects of grain size, alloy stoichiometry, alloying, etc on the strength of iron aluminides at room temperature have not considered this effect. Thus, this "hydrogen" or "water vapor" effect is probably convoluted with the effects of other parameters on the yield strength and subsequent flow strength.

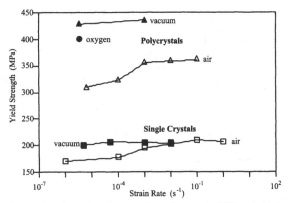

Figure 3. Yield strength as a function of strain rates for polycrystals [30] and single crystals [31] of FeAl in air and vacuum.

Effects of Alloy Stoichiometry

Ideally, in order to analyze the effects of alloy stoichiometry on the strength of FeAl, the critical resolved shear strength determined from single crystals of different compositions would be compared. However, while there have been mechanical tests on FeAl single crystals, these are insufficient to illustrate how the strength changes with aluminum content. Another way to establish the effect of aluminum content on the strength is to scrutinize the lattice resistance or lattice friction term, σ_o, in the Hall-Petch Relationship relating the yield strength, σ_y, to the grain size, d, i.e. $\sigma_y = \sigma_o + k_y d^{-1/2}$, as a function of aluminum content. (Testing very large grained polycrystals is another way, but care has to be taken that there are no textural differences between specimens.) σ_o shows a minimum at 45 at. % Al, and increases much more rapidly on the aluminum rich side of the minimum reflecting that hardening from anti-site atoms (iron rich side) is much less than hardening from vacancies (aluminum rich side) [10], see Figure 4.

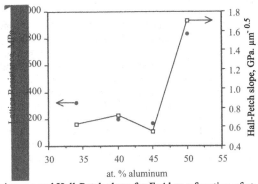

Figure 4. Lattice resistance and Hall-Petch slope for FeAl as a function of at. % Al [33].

38

This minimum in strength around Fe-45Al is observed at both room temperature and 77K, but the difference between the minimum at Fe-45Al and the maximum strength at the stoichiometric composition is less at 77K [8], see Figure 5.

In marked opposition to the low temperature behavior, the high-temperature creep strength of FeAl has been found to be independent of aluminum concentration [34].

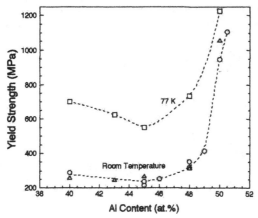

Figure 5. Yield strength as a function of Al concentration of FeAl annealed 5 days at 673K [8].

Effects of Grain Size

Grain size strengthening in FeAl, as measured by the k_y in the Hall-Petch relationship, is only weakly dependent on composition at room temperature for aluminum contents up to 45 at. % Al, but k_y for the stoichiometric FeAl is twice that of iron rich compositions, see Figure 4, indicating the greater difficulty of slip transmission at the stoichiometric composition. Boron has been shown to increase k_y for both Fe-40Al and Fe-45Al [35].

Effects of Temperature

B2 compounds typically show a yield strength that decreases rapidly with increasing temperature up to room temperature, followed by either a plateau or a peak (the so-called yield anomaly, YSA), before declining rapidly at temperatures above half the homologous melting point, see Figure 6. The YSA in many B2 compounds arises from dynamic strain ageing (DSA). However, while a YSA is observed at conventional strain rates (10^{-4}–10^{-3} s^{-1}) in iron-rich FeAl, the negative strain rate sensitivity is not seen [8,10]. For stoichiometric FeAl, the yield peak is only observed when straining is performed at higher rates (≥ 1 s^{-1}) [36], and becomes pronounced at very high strain rates (~2 x 10^3 s^{-1}) [37].

The YSA is obscured if quenched-in thermal vacancies are not eliminated prior to testing, see Figure 7 and disappears at very slow strain rates even in iron-rich FeAl [10].

A number of mechanisms have been proposed for the YSA: the <111> to <100> slip vector transition, cross-slip/pinning, APB relaxation, local dislocation climb-lock, dislocation decomposition, dislocation core dissociation, dislocation constriction/pinning and vacancy hardening (see, e.g., [10,12]). Of these only the vacancy hardening mechanism has been formulated as a theory [38]: the idea is that immediately below the yield strength peak vacancies are immobile and impede dislocation motion, while above the peak the vacancies are able to migrate and aid dislocation climb. Although individual vacancies cannot impede dislocation flow in a metal, in FeAl, a vacancy can cause the leading a/2<111> APB-coupled dislocation partial to climb. Since the trailing a/2<111> partial does not climb, APB created by this climbed segment is not removed by the trailing partial, causing a drag on the dislocation.

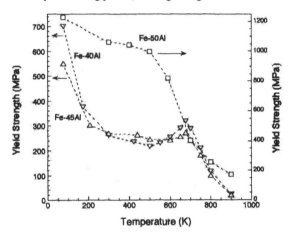

Figure 6. Yield strength versus temperature for large-grained, low-temperature annealed FeAl of various compositions strained under tension at a strain rate of 1×10^{-4} s^{-1} [8].

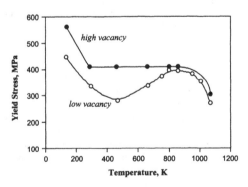

Figure 7. Yield Stress as a function of temperature for Fe-43Al-0.12B annealed for 1.25 hr at 1373 K and 2 hr at 973 K, with and without long term anneal of 120 hours at 673 K, which produce low and high vacancy concentrations respectively [39].

Using the approximation that the increase in strength from the vacancies is proportional to (vacancy concentration)$^{1/2}$ [20] and using the fact that the vacancy concentration increases exponentially with increasing temperature, the strength increase with increasing temperature up to the yield strength peak is proportional to $\exp[-E_f/2kT]$, where E_f is the activation enthalpy for vacancy formation, k is Boltzmann's constant and T is the absolute temperature [38]. Above the peak, the strength can be described by a diffusion-assisted deformation mechanism. Fitting experimental data for polycrystalline Fe-40Al to the model yields a value for E_f of 92 kJ/mol, which is close to the measured value. A reasonable value of the creep exponent, m (3.8), is obtained in the dislocation creep regime and the strength increase is reasonable for the vacancy concentrations present [38]. The model [38] allowed a number of predictions to be made, *viz.*,

- The yield strength below the peak is independent of strain rate whereas above the peak, in the dislocation creep regime, the yield strength is strain rate dependent. It follows that the yield strength peak should itself be strain rate dependent, the peak moving to higher stresses and higher temperatures with increasing strain rate and to lower stresses and lower temperatures with decreasing strain rate.
- Following the above argument, at very low strain rates, the yield peak should not be present.
- The model indicates that for short times, the time at the peak yield stress temperature prior to testing affects the magnitude of peak yield stress. This effect arises because vacancy formation is not instantaneous but requires time, with increasing times (which result in greater vacancy concentrations) leading to larger yield stresses.
- No orientation dependence of the yield peak is expected.

Not only have these predictions have been borne out by experiments [see references 8, and 40 for details], but also some of the experimental observations exclude dynamic dislocation-based mechanisms, e.g. climb decomposition, cross-slip pinning, glide decomposition, as possible causes for the YSA [13,41]. Morris and Morris-Muñoz [41], on examining available data, suggested that the vacancy-hardening model may also explain the YSA in Fe$_3$Al.

FeAl shows a yield stress tension/compression asymmetry at all temperatures, which may be related to the effect of hydrostatic pressure on the vacancy concentration or the effect of non-deviatoric stresses on the dislocation core structure. The latter effect could be the reason that the critical resolved shear stress of FeAl single crystals does not obey Schmid's law [10].

Alloying

As might be expected, both interstitials, such as boron, and substitutional elements increase the strength of FeAl both at room temperature and at elevated temperature with the substitutional strengthening being related to the misfit of the solute atoms, see reference 10 for details. Care has to be taken when comparing the effects of ternary additions since in addition to intrinsic strengthening ternary additions can slow the rate of vacancy removal during low temperature annealing so that vacancy hardening is actually being measured.

Auger electron spectroscopy studies have demonstrated that boron segregates to the grain boundaries and increases grain boundary strength [42]. However, some B remains in the lattice, inducing a lattice strain, and produces strengthening at room temperature that depends on both the aluminum concentration and the presence of vacancies [43]. Boron also shifts the yield strength peak to higher temperatures.

41

DUCTILITY

As for the strength, there are several factors that influence the ductility of FeAl. It should be noted though that the presence of water vapor has a marked effect on the ductility of FeAl, with vacancies playing a secondary deleterious role on the ductility. Without taking into account the role of these two parameters, the effects of other variables cannot really be understood.

Effects of Environment

The effects of the water vapor during tensile testing of iron-rich FeAl at slow to medium strain rates in air are so dominant that the true effects of other metallurgical variables are largely obscured. As noted above, Liu, Lee, and McKamey [29] first observed the effects of water vapor on FeAl and showed that polycrystalline Fe-36.5Al exhibited higher elongation in vacuum (5.5%) than in air (2%), and even greater elongation in dry oxygen (18%). Transgranular cleavage occurred in the specimens tested in air, while mixed mode failure occurred in vacuum and intergranular fracture occurred in oxygen. Their suggested mechanism of atomic hydrogen produced from water vapor diffusing to crack tips and aiding their advance was later supported by the work of Gleason et al. [44] who studied the reaction of water vapor at the surface of FeAl, and observed hydrogen formation occurring there together with oxidation of aluminum. Li and

Thus, factors, such as temperature, strain rate, and protective atmosphere, that affect the kinetics or energetics of the individual steps in the embrittlement of FeAl, such as the dissociative adsorption of H_2O, movement of H into the crack-tip region, etc. also affect its ductility.

The environmental and strain-rate dependence of the ductility in air has been observed by a number of workers on several different compositions of single crystals and polycrystals. Perhaps, the most dramatic effects were observed in tensile tests on single-slip-oriented Fe-43Al single crystals [32] where crystals strained at 2.5×10^{-3} s^{-1} exhibited 46% elongation in oxygen, but only 8-12% elongation in air, while only ~3% elongation was obtained at a slow strain rate of 4.4×10^{-6} s^{-1} in air. The fracture mode of the crystals tested in air changed from cleavage with numerous secondary cracks at the slow strain rate to a rougher fracture surface with no secondary cracks and some ductile dimples at high strain rate. The latter fracture mode was the same as that for crystals strained in oxygen.

It has been suggested [45] that atomic hydrogen segregation stabilizes the sessile <100> dislocations that form by the collision of gliding <111> dislocations [46] thereby promoting cleavage crack nucleation on {100}.

Effects of Vacancies

While the environment is the major cause of brittleness in iron-rich FeAl, vacancies also have a marked effect on ductility. Yang et al. [47] showed that the room-temperature fracture strengths of Fe-40Al single crystals in both air and vacuum are significantly increased due to vacancy hardening, but introducing a high vacancy concentration decreases elongations. The fracture mode was also found to be affected by the vacancy concentration: with a low vacancy concentration, fracture was by cleavage on both {100} and a variety of other planes, whereas at high vacancy concentration, cleavage occurs mainly on {100} and {110}. The latter fracture mode suggests that vacancies promote fracture along the slip planes by condensing on the {110} slip plane and forming crack nuclei.

Effects of Alloy Stoichiometry

There have been a number of studies of the effects of alloy stoichiometry on the ductility of FeAl. Probably the cleanest study is that by Liu and George [48] who performed tensile tests on polycrystalline binary FeAl under ultrahigh vacuum. The ductility depends strongly on Al content with high ductility and transgranular fracture at low Al contents, but a rapid drop in ductility and change to intergranular fracture with increasing Al, see Figure 8. The change to intergranular fracture as the aluminum concentration increases indicates the grain boundary weakening. In fact, room temperature intergranular fracture in near-stoichiometric compositions shows little evidence of plastic flow, failure occurring before yield under tension.

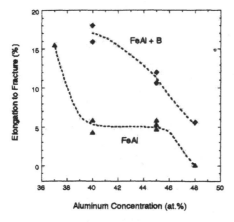

Figure 8. Effect of Al concentration and B doping (300 wppm) on the 'intrinsic' ductility and fracture behavior of FeAl in an ultrahigh vacuum of $\sim 10^{-10}$ torr [49].

Alloying

Schneibel et al. [50] studied the room temperature tensile elongation and fracture mode of Fe-45Al containing 5% of each of the first row transition metals, all of which decreased the ductility, irrespective of testing environment. Alloys with ternary additions whose atomic numbers are less than iron exhibited intergranular fracture, whereas additions whose atomic numbers are greater than iron exhibited transgranular fracture, probably reflecting the site occupancy of the ternary atoms. Liu and George [48] performed tensile tests on polycrystalline binary FeAl under ultrahigh vacuum with added B which segregates to the grain boundaries and suppresses intergranular fracture thereby shifting the ductile-brittle transition observed in unalloyed FeAl to higher Al levels, see Figure 8. Boron strengthening of grain boundaries in air is much less effective [13]. Several groups subsequently demonstrated boron strengthening of grain boundaries in FeAl.

Effects of Grain Size

The beneficial effects of refining the grain size have been observed during tensile testing of FeAl at high strain rate, ~1 s^{-1}, where the ductility is unaffected by the environment [30]. When tensile tested at ~1 s^{-1} in air at room temperature Fe-45Al shows an increase in elongation to fracture from ~6% to ~9% as the grain size is reduced from 80 μm to 28 μm, whereas at strain rates less than ~1 x 10^{-2} s^{-1} there is no difference in ductility for different grain sizes [51], see Figure 8. The effects of grain size are more dramatic in boron-doped material, where the elongation to failure increases from ~7% in large-grained Fe-45Al to ~15% in fine grained material tested in air at 1 s^{-1}, see Figure 9. Again, tests at slow strain rate show no differences.

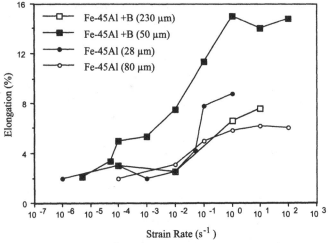

Figure 9. Elongation versus strain rate for both boron-doped and un-doped low-temperature-annealed Fe-45Al of different grain sizes tensile tested in air. After reference 50.

Effects of Temperature

For all binary compositions, the elongation of FeAl increases moderately from room temperature to 500 K, and then increases sharply with further increases in temperature, accompanied by considerable necking [52]. The reduction in area at the neck shows a sharp transition from <10% below 500 K, where deformation is homogeneous, to >80% above 600 K, with reductions in area >95% at 900 K. Compositions that at room temperature show wholly or partly intergranular fracture exhibit (more) cleavage fracture at elevated temperature. Small cleavage lips are observable on intergranular facets. At temperatures above 650 K, fracture is by ductile dimpled rupture, accompanied by grain boundary migration (above 800 K) and dynamic recrystallization (>900K). Excess quenched-in vacancies decrease the ductility at temperatures up to about 800K and solutes, such as boron, cause a dramatic ductility drop at high temperature, accompanied by intergranular fracture and grain boundary cavitation, see reference 10.

STRAIN-INDUCED FERROMAGNETISM

The phenomenon of strain-induced ferromagnetism in heavily-deformed intermetallic compounds that contain ferromagnetic elements and are fully disordered has been known for at least forty years and is well explained using the local environment model, which considers the number of like nearest-neighbor atoms around potentially ferromagnetic atoms in the compound. Intermetallic compounds, such as FeAl, that are subject to lighter deformation that does not completely disorder them can also show strain-induced ferromagnetism. This behavior has been modeled in FeAl by applying the local environment model to the disordered region inside APB tubes [53]. That the strain-induced ferromagnetism arises from APB tubes is consistent with the observations that:

- At temperatures where deformation produces strain-induced ferromagnetism, APB tubes are also present [54];
- APB tubes observed after room-temperature deformation can be annealed at temperatures of ~400 K, at which point the strain-induced ferromagnetism also disappears [55].

ACKNOWLEDGEMENTS

This research was supported by Award DE-FG02-07ER46392 from the Division of Materials Sciences, U.S. Department of Energy and grant DMR 0552380 from the National Science Foundation with Dartmouth College. Any opinions, findings, and conclusions or recommendations expressed in this material are those of the author(s) and do not necessarily reflect the views of the NSF, DOE or the U.S. Government.

REFERENCES

1. N. Ziegler, Trans AIME 100, 267 (1932).
2. I. Baker, P. Nagpal, in Processing and Fabrication of Advanced Materials for High Temperature Applications - II, ed.- T.S. Srivistan and V.A. Ravi (TMS, Warrendale, PA, 1993) p. 3-18.
3. I. Baker, P. Nagpal, in Structural Intermetallics, edited by R. Darolia et al. (TMS, Warrendale, PA, 1993) p. 463.
4. I. Baker, in Processing, Properties and Applications of Iron Aluminides, ed. - J.H. Schneibel and M.A. Crimp (TMS, Warrendale, PA, 1994) p. 101.
5. N.S. Stoloff, C.T. Liu, Intermetallics 2, 75 (1994).
6. D. Hardwick, G. Wallwork, Rev. High Temp. Mater. 4, 47 (1978).
7. X. Pierron, I. Baker, in Design Fundamentals of High Temp. Composites, Intermetallics and Metal-Ceramic Systems, ed. R.Y. Lin et al. (TMS, Warrendale, PA, 1996) p. 271.
8. E.P. George, I. Baker, in The Encyclopedia of Materials: Science and Technology, ed.- K.H.J. Buschow et al. (Elsevier Press, Pergamon, 2001) pp. 4201.
9. U. Prakash, R.A. Buckley, H. Jones, C.M. Sellars, ISIJ Int. 31, 1112 (1991).
10. I. Baker, P.R. Munroe, Int. Mat. Rev. 42, 181 (1997).
11. D.G. Morris, M.A. Morris-Muñoz, Intermetallics 7, 1121 (1999)
12. I. Baker, E.P. George, Proc. MRS 552, KK4.1.1 (1999).
13. C.T. Liu, E.P. George, P.J. Maziasz, J.H. Schneibel, Mater. Sci. Eng. A. 258, 84 (1998).

14. I. Baker, D. Wu, M. Wittmann, E.P. George, in Proc. 4th Pacific Rim Int. Conf. on Advanced Materials and Processing, Vol. I, (Jap. Inst. Mater., Sendai, Japan, 2001) p. 811.
15. M.R. Harmouche, A. Wolfenden, Mater. Sci. Eng. 84, 35 (1986).
16. H.J. Leamy, E.D. Gibson, F.X. Kayser, Acta Metall. 15, 1827 (1967).
17. Binary Alloy Phase Diagrams, ed - T.B. Massalski (ASM, Metals Park, OH, 1986), p. 112.
18. J.P. Riviere, J. Grihlé, Scripta Metall. 9, 967 (1975).
19. R. Wurschum, C. Grupp, H.-E. Schaefer, Phys. Rev. Letters, 75, 97 (1995).
20. Y.A. Chang, L.M. Pike, C.T. Liu, A.R. Bilbery, D.S. Stone, Intermetallics 1, 107 (1993).
21. M.G. Mendiratta, H.K. Kim, H.A. Lipsitt, Metall. Trans. A, 15A, 395 (1984).
22. I. Baker, D.J. Gaydosh, Mater. Sci. Eng. 96, 147 (1987).
23. B. Kad, J.A. Horton, Mater. Sci. Eng. 118, 239 (1997).
24. K. Yoshimi, S. Hanada, M.H. Yoo, Proc. MRS, 460, 313 (1997)
25. M.G. Mendiratta, C.C. Law, J. Mater. Sci. 22, 607 (1987).
26. M.A. Morris, O. George, D.G. Morris, Mater. Sci. Eng. A258, 99 (1998).
27. P. Nagpal, I. Baker, Metall. Trans. 21A, 2281 (1990).
28. Y. Yang, I. Baker, Intermetallics 6, 167 (1998).
29. C.T. Liu, E.H. Lee, C.G. McKamey, Scripta Metall. 23, 875 (1989).
30. L.M. Pike, C.T. Liu, Scripta Mater. 38, 1475 (1998).
31. D. Wu, I. Baker, Intermetallics 9, 57 (2001).
32. I. Baker, D. Wu, S.O. Kruijver, E.P. George, Mat. Sci. Eng. A. 329-331, 726 (2002).
33. I. Baker, P. Nagpal, F. Liu, P.R. Munroe, Acta Metall. Mat. 39, 1637 (1991).
34. J. D. Whittenberger: Mater. Sci. Eng. 57, 77 (1983); 77, 103 (1986).
35. L.M. Pike, C.T. Liu, Scripta Mater. 25, 2757 (1991).
36. I. Baker, Y. Yang, Mater. Sci. Eng. A239-240, 109 (1997).
37. Y. Yang, I. Baker, R.T. Gray, III, C. Cady, Scripta Mater. 40, 403 (1999).
38. E.P. George, I. Baker, Philos. Mag. 77, 737 (1998).
39. R. Carleton, E.P. George, R.H. Zee, Intermetallics 3, 433 (1995).
40. D. Wu, I. Baker, P.R. Munroe, E.P. George, Intermetallics 1, 103 (2007).
41. D.G. Morris, M.A. Morris-Muñoz, Intermetallics 13, 1269 (2005).
42. C.T. Liu, E. P. George, Scripta Metall. Mater. 24, 1285 (1990).
43. I. Baker, X. Li, H. Xiao, R. L. Carleton, E. P. George, Intermetallics 6, 177 (1998).
44. N.R. Gleason, C.A. Gerken, D.R. Strongin, Appl. Surf. Sci. 72, 215 (1993).
45. J.C.M. Li, C.T. Liu, Scripta Metall. Mater. 33, 661 (1995).
46. P.R. Munroe, I. Baker, Acta Metall. Mater. 39, 1011 (1991).
47. Y. Yang, I. Baker, E.P. George, Mater. Char. 42, 161 (1999).
48. C.T. Liu, E.P. George, Scripta Metall. Mater. 24, 1285 (1990).
49. J.W. Cohron, Y. Lin, R.H. Zee, E.P. George, Acta Mater. 46, 6245 (1998).
50. J.H. Schneibel, E.P. George, I.M. Anderson, Intermetallics 5, 185 (1997).
51. I. Baker, O. Klein, C. Nelson, E.P. George, Scripta Metall. Mater. 30, 863 (1994).
52. I. Baker, H. Xiao, O. Klein, C. Nelson, J.D. Whittenberger, Acta Metall. Mat. 43, 1723 (1995).
53. D. Wu, P.R. Munroe, I. Baker, Philos. Mag. 83, 295 (2003).
54. K. Yamashita, M. Imai, M. Matsuno, A. Sato, Philos. Mag. A 78, 285 (1998).
55. D. Wu, I. Baker, Mater. Sci. Eng., A, 329-331, 334 (2002).

Mater. Res. Soc. Symp. Proc. Vol. 1128 © 2009 Materials Research Society   1128-U02-02

Forging of Steam Turbine Blades With an Fe₃Al-Based Alloy

P. Janschek[1], K. Bauer-Partenheimer[1], R. Krein[2, 3], P. Hanus[2, 4] and M. Palm[2]

[1] Leistritz Turbinenkomponenten Remscheid GmbH, Lempstrasse 24, D-42859 Remscheid, Germany
[2] Max-Planck-Institut für Eisenforschung GmbH, Max-Planck-Strasse 1, D-40237 Düsseldorf, Germany
[3] now at: Salzgitter Mannesmann Forschung GmbH, Ehinger Str. 200, D-47259 Duisburg, Germany
[4] Technical University of Liberec, Studentska 2, Liberec 461 17, Czech Republic

ABSTRACT

The forging capabilities of two high-strength Fe₃Al-based alloys have been evaluated. Based on these results one alloy has been used for forging steam turbine blades. Blades of about 600 mm length were successfully forged by a standard procedure otherwise used for forging of 9-12 wt.% Cr steels. The forged steam turbine blades showed very good form filling, no pores and smooth surfaces. The blades were finished by cutting and grinding by standard procedures. The microstructure consists out of a Fe₃Al matrix with additional Laves phase which predominantly precipitated on grain boundaries. The large Fe₃Al grains in the cast precursors did only partially recrystallise during forging. This may be partially due to pinning of the grain boundaries by Laves phase precipitates. Cracks may form at those grain boundaries which are decorated with these brittle precipitates.

INTRODUCTION

Fe-Al-based alloys are light weight materials with favourable wear resistance and outstanding corrosion behaviour [1]. Recent advances in strengthening these alloys at high temperatures now make them suitable for applications in extreme environments such as steam turbines for power generation [2].

Forging has been applied to iron aluminides variously to consolidate powders and to produce billets and rods for subsequent rolling, see e.g. [3-5]. Though, little detailed information exists about the forgeability of these materials [6-8]. From the extensive work on iron aluminides carried out at Oak Ridge National Laboratory (ORNL) it is known that forging of Fe₃Al-based alloys is possible in the temperature range 850 to 1100 °C, however these forgings were conducted at a very slow speed [9]. Hot forging of FeAl-based alloys is also possible but necessitates heating of the dies [10]. For these FeAl-based alloys it was found that the forging window could be considerably enlarged if the microstructure was refined [10].

EXPERIMENTAL

Two alloys of nominal compositions Fe-25Al-20Ti-4Cr [11] and Fe-25Al-2Ta [12, 13] (all compositions in at.%) were produced by vacuum induction melting (VIM). In order to evaluate their forging behaviour these two alloys were forged on a 900 ton screw press. Eleven samples of 28 mm height and 25 mm diameter were forged at temperatures between 900 and 1150 °C. For forging the steam turbine blades, blanks of 660 mm length each weighing about 12 kg have been produced by investment casting. The blanks were heated up to 1100 °C and were then forged on a 630 kJ counterblow hammer with 10 blows each. The dies were standard tooling used for commercially produced steam turbine blades.

Samples for microstructural investigations were cut out off the forged parts by electrical discharge machining (EDM). For light optical microscopy (LOM) samples were ground up to 1200 grit and etched in a solution of 45 ml H_2O, 15 ml HCl, 30 ml HNO_3 and 10 ml HF. Scanning electron microscopy (SEM) and electron-probe microanalysis (EPMA) were performed on a JEOL 8100. Electron backscatter diffraction (EBSD) was performed with a TSL/EDAX system in a JEOL 6500 F field emission scanning electron microscope (FE-SEM) and in a JEOL 840 A SEM.

RESULTS AND DISCUSSION

Figs. 1 and 2 show the results of the preliminary forging tests on Fe-25Al-2Ta and Fe-25Al-20Ti-4Cr, respectively. After forging at temperatures between 900 and 1150 °C crack-free symmetric pancakes with smooth surfaces were obtained from Fe-25Al-2Ta (Fig.1). In contrast, Fe-25Al-20Ti-4Cr could not be deformed without substantial cracking even at 1150 °C (Fig. 2). Therefore Fe-25Al-2Ta was selected for forging of the steam turbine blades.

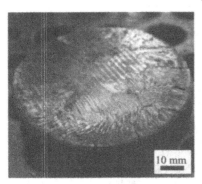 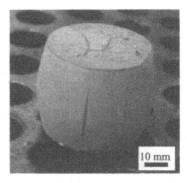

Figure 1. Fe-25Al-2Ta after forging at 900 °C.   **Figure 2.** Fe-25Al-20Ti-4Cr after forging at 1150 °C.

Figure 4. Investment cast blanks. Cracks marked by arrows formed during cooling. The blank shown in front was made by "tea pot" casting and shows no crack.

Figure 3. Segmented copper mould (centre) and tripod (left).

For the casting of the blanks for the steam turbine blades a copper mould was used which consists out of 6 segments held together in a tripod (Fig. 3). The first three castings were performed by pouring the melt from the top into the mould. After cooling it was realised that a large crack formed between the massive root and the blade during cooling (Fig. 4). Subsequent castings were realised by "tea pot" casting, i.e. by letting the melt rise from the ground of the mould. By this technique the formation of the above described cracks could be avoided (Fig. 4). All castings showed very good form filling with smooth surfaces and were free of pipes and pores (Fig. 5). Because no measures for grain refinement were undertaken columnar grains were found in the castings which made them rather brittle.

The forged steam turbine blades again showed very good form filling and smooth surfaces. Though, some of them split in two parts due to the severe crack between root and blade which was already present in the castings (Fig. 6). The blades were finished by cutting off the burrs on a band saw and afterwards grinding the edges by standard procedures. No heat treatment has been applied. Fig. 7 shows a forged blade after finishing.

Optical and scanning electron microscopy was performed on specimens taken from the root (R), the bottom (A), the middle (B) and the tip (C) of the airfoil (Fig. 7). In all parts of the blade the microstructure is characterized by large, elongated Fe_3Al grains (Fig. 8). By EPMA their composition has been determined as 73.4 at.% Fe, 25.3 at.% Al, 1.3 at.% Ta. EPMA line scans revealed that no chemical segregation occurs within the grains. The grain boundaries are

Figure 5. Investment cast blanks free of pipes (a) and pores (b) and with smooth surfaces (c).

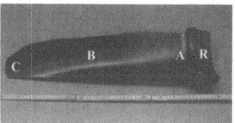

Figure 6. Steam turbine blades after forging. Some of the blades split into two parts due to a pre-existing crack in the casting.

Figure 7. Forged steam turbine blade after finishing. The letters denote the positions where samples for metallographic observations have been cut out of the blades.

partly decorated by fine precipitates (Fig. 8). In A some additional precipitates are also observed within the grains. Through EPMA the precipitates were identified as Laves phase (62 at.% Fe, 13.5 at.% Al, 24.5 at.% Ta). In addition a few, irregular-shaped Ta-rich particles (17 at.% Fe, 1 at.% Al, 82 at.% Ta) were detected (Fig. 8c). While the Laves phase apparently precipitated from the Fe-Al matrix which became supersaturated in Ta during cooling, the Ta-rich particles presumably stem from incomplete melting of the Ta during VIM. In some cases cracks propagated along grain boundaries which are decorated with the brittle Laves phase precipitates (Figs. 8c, d). Figs. 8e, f show the microstructure at the tip of the blade (C). Many "cracks" are

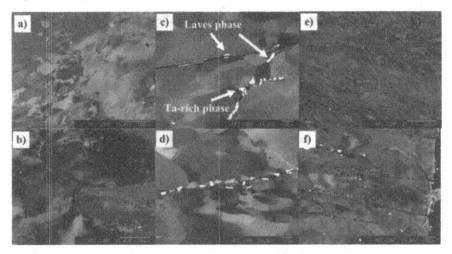

Figure 8. Scanning electron micrographs (back scatter electron (BSE) contrast) of parts R (a, b), A (c, d), and C (e, f). The compositions of the particles were determined by EPMA.

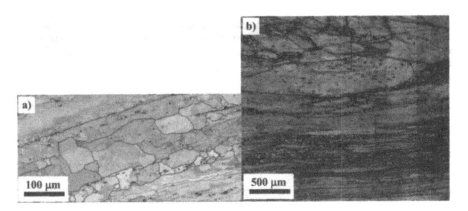

Figure 9. EBSD micrographs of part B (a) and part C (b). High angle grain boundaries (> 5°) are marked by black lines.

visible in this part of the blade which consist out of rows of pores (Fig. 8e). Apparently, they stem from cracks that formed during forging and which were partially closed by the forging process.

Through EBSD mapping the crystallographic orientations of the grains has been determined (Fig. 9). These mappings reveal that in all parts of the blade the large primary Fe_3Al grains which stem from the casting did only partially recrystallise. Fig. 9a shows an EBSD mapping of part B. While the grains in the central part of Fig. 9a are recrystallised, the large Fe_3Al grain in the upper left corner has not recrystallised at all. This behaviour becomes even more evident in part C (Fig. 9b). In this part at the tip of the blade significant grain refinement due to dynamic and post-dynamic recrystallisation is observed in some Fe_3Al grains (lower part of Fig. 9b) while others only show marked deformation but did not recrystallise (central part of Fig. 9b). The reason therefore may be pinning of the grain boundaries by the Laves phase precipitates.

SUMMARY AND CONCLUSIONS

The forging capabilities of two high-strength Fe_3Al-based alloys have been evaluated. While Fe-25Al-20Ti-4Cr did hardly deform even at the highest temperatures Fe-25Al-2Ta could easily be forged. Therefore the latter alloy has been successfully used for forging steam turbine blades by a standard procedure otherwise used for forging of 9-12 wt.% Cr steels. The forgings itself were free of any defects. After forging, the microstructure consists of large Fe_3Al grains and small precipitates of Laves phase which are predominantly found on the Fe_3Al grain boundaries. The large Fe_3Al grains formed during solidification of the cast blanks. They did not fully recrystallise during forging which may be partially due to pinning of the grain boundaries by the Laves phase precipitates. In addition, the decoration of the grain boundaries with these brittle precipitates enables the propagation of cracks. Nevertheless it can be concluded that forging of such advanced parts as a steam turbine blade is possible with a Fe_3Al-based alloy under standard industrial conditions. It is expected that refining the microstructure of the castings will result in a more recrystallised microstructure after forging. Due to the then smaller grain size also a more even distribution of the Laves phase is expected. Both factors will be beneficial for improving ductility.

ACKNOWLEDGMENTS

The authors wish to thank Frank Rütters and Jürgen Wichert for VIM and casting of the blanks, Gerd Bialkowski for EDM, Irina Wossack for SEM and EPMA, and Monika Nellessen for EBSD.

REFERENCES

1. C.G. McKamey, in: *Physical Metallurgy and Processing of Intermetallic Compounds*, eds. N.S. Stoloff and V.K. Sikka, Chapman & Hall, New York, 1996, pp. 351-391.
2. M. Palm, R. Krein, S. Milenkovic, G. Sauthoff, D. Risanti, C. Stallybrass and A. Schneider, *Mater. Res. Soc. Symp. Proc.*, **980**, II01-01 (2007).
3. J. Rodriguez, S.O. Moussa, J. Wall and K. Morsi, Scr. Mater., **48**, 707 (2003).
4. Y.D. Huang, W.Y. Yang and Z.Q. Sun, Mater. Sci. Eng., **A263**, 75 (1999).
5. Z.R. Zhang and W.L. Liu, Mater. Sci. Eng., **A423**, 343 (2006).
6. P. Zhao, D.G. Morris and M.A. Morris-Munoz, J. Mater. Res., **14**, 715 (1999).
7. X.Q. Yu and Y.S. Sun, Mater. Sci. Technol., **20**, 339 (2004).
8. R. Lyszkowski and J. Bystrzycki, Intermetallics, **14**, 1231 (2006).
9. V.K. Sikka, C.G. McKamey, C.R. Howell and R.H. Baldwin, *Fabrication and mechanical properties of Fe₃Al-based iron aluminides*, ORNL/TM-11456. Oak Ridge National Laboratory, 1990, pp. 1-76.
10. C.T. Liu, V.K. Sikka and C.G. McKamey, *Alloy development of FeAl aluminide alloys for structural use in corrosive environments*, ORNL/TM-12199. Oak Ridge National Laboratory, 1993, pp. 1-61.
11. R. Krein and M. Palm, Acta mater., **56**, 2400 (2008).
12. D.D. Risanti and G. Sauthoff, Intermetallics, **13**, 1313 (2005).
13. D.D. Risanti and G. Sauthoff, Mater. Sci. Forum, **475-479**, 865 (2005).

Mater. Res. Soc. Symp. Proc. Vol. 1128 © 2009 Materials Research Society   1128-U02-03

Grain Refinement for Strengthening in Fe₃Al-Based Alloys Through Thermomechanical Processing

Grain Refinement for Strengthening in Fe_3Al-Based Alloys Through Thermomechanical Processing

Satoru Kobayashi[1] Akira Takei[2] and Takayuki Takasugi[1,2]
[1]Osaka Center for Industrial Materials Research, Institute for Materials Research, Tohoku University, 1-1Gakuen-cho Naka-ku, Sakai, Osaka 599-8531, Japan
[2]Department of Materials Science, Graduate School of Engineering, Osaka Prefecture University, 1-1Gakuen-cho Naka-ku, Sakai, Osaka 599-8531, Japan

ABSTRACT

A thermomechanical process (TMP) for grain refinement was performed in bulk Fe_3Al-based alloys containing ~10% volume fraction of κ-Fe_3AlC precipitates. In the TMP, κ particles play an important role in reducing the inhomogeneity of recrystallization due to the matrix orientation. The grain size was refined to ~20 μm by optimizing the κ particle size. A fine-grained and pancake/recovered microstructure fabricated by the TMP showed more than 1200 MPa tensile strength and 8% tensile ductility at room temperature in air. The tensile strength of this material was higher than those of conventional wrought Fe_3Al alloys at temperatures between room temperature and 500 °C, and the specific tensile strength was as high as that of the Ti-6Al-4V alloy at temperatures above 400 °C.

INTRODUCTION

Fe_3Al-based intermetallic alloys have received attention so far as candidates for high temperature structural materials because of their excellent oxidation and sulphidation resistance at high temperatures, ubiquitous raw material and low density compared with conventional stainless steels and heat resistant alloys [1,2]. Research and development studies on these alloys were mainly focused on the following two issues: the improvement of (1) the limited room temperature ductility and strength by modification of the grain structure, the order state, surface conditions and alloying [3-6] and (2) poor high temperature properties by alloying through precipitation and/or solid solution strengthening [7-11].

Improvement of the room temperature ductility and strength was greatly achieved through a thermomechanical (warm rolling + annealing) process to produce pancake-shaped grains in which dislocations are well recovered [3-6]. Grain refinement (GR) is, on the other hand, a generally accepted fabrication process to improve strength and toughness in materials. GR was achieved in FeAl alloys through powder metallurgy (PM). PM manufactured Fe-40Al with very fine grain size (~1 μm) possesses an excellent combination of strength and ductility as well as fatigue properties [12]. However, GR method has not yet been established through wrought process in bulk Fe_3Al alloys.

We have investigated the texture and recrystallization of warm rolled Fe_3Al-based alloys with and without carbide particles [13-16], and have recently achieved to produce fairly homogeneous recrystallized structures with a grain size of ~20 μm in the Fe_3Al alloys containing κ-Fe_3AlC carbide particles [17]. The content of this paper is two-fold. The first is to present the effects of κ particles on the warm deformation and recrystallization of the Fe_3Al matrix phase. The second is to show the tensile properties for two types of fine-grained microstructures in a Fe_3Al alloy containing κ-Fe_3AlC particles.

RESULTS & DISCUSSION

The effects of κ particles on the deformation and recrystallization of the Fe₃Al matrix

It is difficult to obtain a homogeneously recrystallized grain structure in a particle-free Fe₃Al-based alloys due to the fact that the rotated cube oriented grains, {001}<110>, present in cast ingots, show a strong resistance against recrystallization [13]. The strong resistance against recrystallization of this orientation may be attributed to the symmetric activations of <111> slip systems during rolling deformation, forming less in-grain orientation gradients than other components.

The high stability of the {001}<110> orientation during rolling deformation can be destabilized by introducing hard particles into the matrix. As schematically shown in Figure 1, local shear zones will form around a hard particle under the global plane strain deformation, i.e. rolling deformation. The crystal may rotate in the shear zones, which can be the nucleation sites of recrystallization in a subsequent annealing process.

Figure 2 shows local shear zones formed around κ particles in a {001}<110> oriented grain in Fe-27Al-1.2C-1.0Mo (at. %) alloy warm rolled at 700 °C to a reduction of 10%. It can be clearly seen that the matrix crystal rotates near the particle with the rotation axis almost parallel to the transverse direction. The maximum rotation of the matrix was found to be more than 15°, which is necessary for the nucleation of recrystallization in a subsequent annealing. Figure 3 presents an annealed microstructure in the rotated cube oriented matrix. It can be recognized that recrystallized grains nucleated around κ particles.

It is generally accepted that the introduction of hard particles into a matrix phase promotes the nucleation of recrystallization of the matrix (particle stimulated nucleation of recrystallization, PSN), thereby reducing the grain size [18]. Fe₃Al alloys are no exception; in fact, finer averaged grain sizes were obtained in particle containing alloys [14-17]. It should be, however, emphasized that the important role of hard particles within the Fe₃Al matrix phase is to diminish the inhomogeneity of recrystallization originated from the matrix orientation.

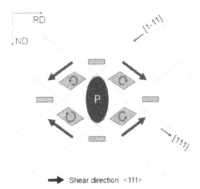

Shear direction <111>

Figure 1 The shear zones around the particle in the {001}<110> orientation under rolling deformation.

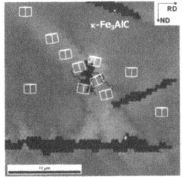

Figure 2 EBSD map of the alloy warm rolled, showing the crystal rotation around the particle.

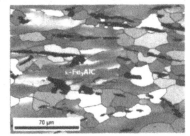

Figure 3 EBSD map of the alloy annealed at 900 °C for 2 min.

The optimization of κ particle size and distribution for grain refinement

Since the detailed experimental results and discussion for this subject were published in [17], a brief summary is presented here. A highest grain density is available in an optimized κ particle size (area) under the constant volume fraction of the particles (Fig. 4). The dependence of grain density on the particle size may be explained in terms of the following two facts: (1) The efficiency of the particle to generate nuclei, PSN efficiency, increases drastically in the small particle size range, but it remains almost constant in the large particle size range. (2) The density of the particles decreases monotonically with increasing particle size. The finest grain size obtained so far is ~20 μm in Fe₃Al-based alloys containing 10% fraction of κ particles.

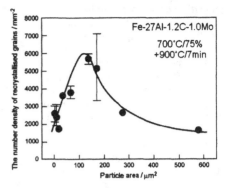

Figure 4 The relationship between the number density of recrystallized grains and averaged particle area

The tensile properties of fine-grained Fe₃Al alloys containing κ particles

Tensile properties were investigated for the two types of fine-grained micro-structures in Fe₃Al alloys containing ~10% volume fraction of κ particles; (1) recrystallized state and (2) pancake/recovered state. The second state was fabricated by warm rolling + recovery annealing preceded by a grain refinement process. Figure 5 shows EBSD maps of the microstructures. It can be seen that grains are more or less equiaxed in the first state (Fig. 5(a)). In the latter state, on the other hand, the grains are elongated along the rolling direction (Fig. 5(b)). The grain interiors contain misorientations due to subgrain structures. An averaged grain size of the two structures was 21 μm and 37 μm, respectively.

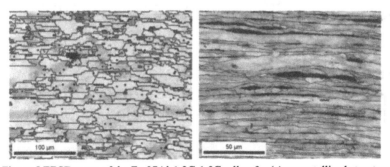

Figure 5 EBSD maps of the Fe-27Al-1.2C-1.0Cr alloy for (a) recrystallized structure fabricated by warm rolling at 650 °C for 92% + annealing at 900 °C for 7min, (b) pancake/recovered structure by warm rolling at 650 °C for 68% + annealing at 900 °C for 7 min, followed by warm rolling at 700 °C for 75% + annealing at 600 °C for 1 h. Boundary level is 15°.

Figure 6 shows the tensile properties of the two types of microstructures. The properties of a coarse-grained recrystallized structure are also included in the figure for comparison. It can be seen that grain refinement improves 0.2% proof stress, ultimate tensile stress (UTS) and elongation. However, the properties of the fine-grained and pancake/ recovered microstructure are even higher than those of the fine-grained recrystallized structure. Our microstructure and texture analyses indicate that the enhanced 0.2% proof stress in pancake/recovered microstructures is due to subgrain structures in grain interiors and the higher elongation is attributed to a smaller amount of <001> oriented grains with

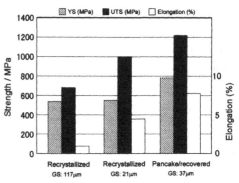

Figure 6 The room temperature tensile properties for different structures in the Fe-27Al-1.2C-1.0Cr alloy. Tensile tests were performed in air at a strain rate of 1.66×10^{-4} s^{-1} with the tensile axis parallel to the rolling direction.

respect to the tensile direction. The detailed results on the relationship between the microstructure, texture and tensile properties will be published in the proceeding of the same volume [19].

Figure 7 shows the high temperature 0.2% proof stress and tensile strength of a fine-grained and pancake/recovered microstructure, together with the data of comparative alloys [20]. FAS, FAL and FA-129 are Fe$_3$Al-based alloys fabricated by a warm rolling + recovery treatment [4]. The 0.2% proof stress of the fine-grained structure decreased by ~200 MPa at 200 °C but the stress remained up to 500 °C. Tensile strength of the sample gently decreases with increasing temperature below 400 °C but severely above 400 °C. It can be recognized that the 0.2% proof stress and tensile strength of the pancake/recovered microstructure are higher than those of comparatives between room temperature and 500 °C.

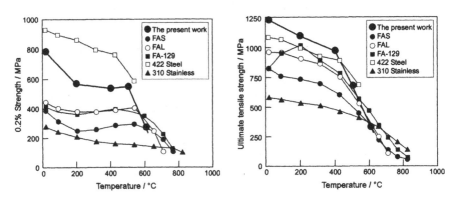

Figure 7 The high temperature tensile properties for a fine-grained and pancake/recovered microstructure in the Fe-27Al-1.2C-1.0Mo alloy. Tensile tests were performed in air at a strain rate of 1.66×10^{-4} s^{-1} with the tensile axis parallel to the rolling direction. The data for wrought Fe$_3$Al alloys, martensitic and austenitic steels are also included for comparison [20].

The specific tensile strength of the fine-grained and pancake microstructure is compared with those of Ti alloys and a wrought Ni base alloy [21, 22] in Figure 8. While the specific strength of the fine-grained Fe_3Al alloy is inferior to the Ti alloys at room temperature, it should be emphasized that the specific strength above 400 °C is as high as that of the Ti-6Al-4V alloy.

Commercial applications in a wide range of temperatures are considered for Fe_3Al-based alloys due to their excellent oxidation and sulphidation resistance, cheap material costs and low density [2, 4, 9, 23]. Improving high temperature properties above 600 °C is important for high temperature applications. The present study, however, indicates that fine-grained Fe_3Al alloys might be a ubiquitous high-specific-strength material to replace Ti alloys for medium high temperature applications around 400~500 °C. For this trial, fatigue tests and further improvement of the high temperature strength around 500 °C are currently on the way.

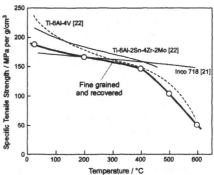

Figure 8 The specific tensile strength of a fine-grained and pancake/recovered structure in the Fe-27Al-1.2C-1.0Mo alloy. The density of 6.6 g/cm^3 was used for calculating the values.

SUMMARY

A TMP for grain refinement was performed in Fe_3Al-based alloys containing κ-Fe_3AlC precipitate particles, and the tensile properties of the fine-grained microstructures were investigated. The results obtained are summarized below:

(1) κ-Fe_3AlC particles play an important role in reducing the inhomogeneity of recrystallization due to the matrix orientation.

(2) Grain size was refined to ~20 μm by optimizing κ particle size.

(3) A fine-grained and pancake/recovered structure showed more than 1200 MPa tensile strength and 8% tensile ductility at room temperature in air.

(4) The tensile strength of this material was higher than that of conventional wrought Fe_3Al alloys at temperatures between room temperature and 500 °C, and the specific tensile strength was as high as that of a Ti-6Al-4V alloy at temperatures above 400 °C.

ACKNOWLEDGMENTS

This research was supported in part by the Grant-in-aid for Scientific Research from the Ministry of Education, Culture, Sports and Technology, JAPAN. The help in the use of FESEM-EBSD system by Dr. M. Demura at National Institute for Materials Science (NIMS), JAPAN is highly acknowledged. The authors also thank Prof. D. Raabe at Max-Planck-Institute fuer Eisenforschung GmbH, GERMANY for helpful suggestions.

REFERENCES

1. D.G. Morris, In: J.H. Schneibel et al, editor. Processing, Properties and Applications of Iron Aluminides, TMS, Warrendale, PA, 1994, p. 3.
2. N.S. Stoloff, Mater. Sci. Eng. A258, 1 (1998).
3. C.G. McKamey and D.H. Pierce, Scr. Metall. 28, 1173 (1993).
4. V.K. Sikka et al. US Patent No. 5084109.
5. Y.D. Huang, W.Y. Yang, Z.Q. Sun, Mater. Sci. Eng. A263, 75 (1999).
6. Y.D. Huang, W.Y. Yang, G.L. Chen, Z.Q. Sun, Intermetallics 9, 331 (2001).
7. C.G. McKamey, P.J. Maziasz, In: J.H. Schneibel et al, editor. Processing, Properties and Applications of Iron Aluminides, TMS, Warrendale, PA, 1994, p. 147.
8. A.Schneider, G. Sauthoff, Steel Res. Int. 75, 55 (2004).
9. D.G. Morris, M.A. Morris, C. Baudin, Acta Mater. 52, 2827 (2004).
10. M. Palm, Intermetallics 13, 1286 (2005).
11. P. Kratochvíl , P. Málek, M Cieslar, P. Hanus, J. Hakl, T. Vlasák, Intermetallics 15, 333 (2007).
12. Nano-DS Fe40Al Alloy Material Properties & Guidelines, project funded by the European Community under the 'Competitive and Sustainable Growth' Programme (1998-2002).
13. S. Kobayashi, S. Zaefferer, A. Schneider, D. Raabe, G. Frommeyer, Mater. Sci. Eng. A387-389, 950 (2004).
14. S. Kobayashi, S. Zaefferer, A. Schneider, D. Raabe, G. Frommeyer, Intermetallics 13, 1296 (2005).
15. S. Kobayashi, S. Zaefferer, D. Raabe, Mater. Sci. Forum 550, 345 (2007).
16. S. Kobayashi, S. Zaefferer, Mater. Sci. Forum 558-559, 235 (2007).
17. S. Kobayashi, T. Takasugi, Intermetallics 15, 1659 (2007).
18. F.J. Humphreys, M. Hatherly. Recrystallization and related annealing phenomina. 2nd ed. Pergamon; 2004, p. 285.
19. A. Takei, S. Kobayashi, T. Takasugi, submitted to proceedings of Mater. Res. Soc. Symp. (2008).
20. V.K. Sikka, International Symposium on Nickel and Iron Aluminides: Processing, Properties, and Applications, ASM, 1996, pp. 361-375.
21. J.D. Destefani, ASM Handbook, Tenth edition, ASM, Volume 2, p. 628, Fig. 30.
22. http://www.efunda.com/materials/alloys/titanium/titanium.cfm.
23. C.G. McKamey et al. US Patent No.4961903.

Mater. Res. Soc. Symp. Proc. Vol. 1128 © 2009 Materials Research Society   1128-U02-04

Ab Initio Study of Elastic Properties in Fe$_3$Al-Based Alloys

Martin Friák, Johannes Deges, Frank Stein, Martin Palm, Georg Frommeyer, and Jörg Neugebauer
Max-Planck-Institut für Eisenforschung GmbH, Max-Planck Str. 1, D-40237, Düsseldorf, Germany

ABSTRACT

Fe$_3$Al-based alloys constitute a very promising class of intermetallics with great potential for substituting austenitic- and martensitic steels at elevated temperatures. A wider use of these materials is partly hampered by their moderate ductility at ambient temperatures. Theoretical *ab initio* based calculations are becoming increasingly useful to materials scientists interested in designing new alloys. Such calculations are nowadays able to accurately predict basic material properties by needing only the atomic composition of the material. We have therefore employed this approach to explore (i) the relation between chemical composition and elastic constants, as well as (ii) the effect transition-metal substituents (Ti, W, V, Cr, Si) have on this relation. Using a scale-bridging approach we model the integral elastic response of Fe$_3$Al-based polycrystals employing a combination of (i) single crystal elastic stiffness data determined by parameter-free first-principles calculations in combination with (ii) Hershey's homogenization model. The *ab initio* calculations employ density-functional theory (DFT) and the generalized gradient approximation (GGA). The thus determined elastic constants have been used to calculate the ratio between the bulk B and shear G moduli as an indication of brittle/ductile behavior. Based on this approach we have explored chemical trends in order to tailor mechanical properties. Using this information we have cast a selected set of Fe$_3$Al-based ternary alloys, obtained for these the elastic constants by performing impulse excitation measurements at room as well as liquid nitrogen temperature and compared them with our theoretical results.

INTRODUCTION

The development of new lightweight materials is crucial for numerous energy-conversion applications in the automotive and aerospace industries. Low-cost and low-density materials operating at higher temperatures ensure a lower fuel consumption and environmentally cleaner and more efficiently produced electricity. Two basic options in materials design and/or functional optimization are the selection of an appropriate chemical composition and the processing of an optimized microstructure. Both characteristics are mutually interlinked and inherently multiscale in nature what make them challenging to study.

We address these fundamental aspects to a promising class of lightweight intermetallics: Fe$_3$Al-based alloys exhibiting an excellent high temperature oxidation and sulfidation resistance up to 1000 °C. Therefore, they possess a high potential as a low cost alternative to conventional ferritic, martensitic or austenitic steels [1]. Depending on the Al content Fe-Al alloys exhibit a disordered bcc-state, an ordered B2- or a D0$_3$-ordered superlattice structure. Most investigations deal with Fe$_3$Al-based alloys because of their lower density in comparison to conventional steels, the considerable room temperature ductility (that is quite unusual among intermetallics), as well as the soft magnetic properties.

For actual applications as structural materials, however, further improvements of these alloys e.g. via compositional optimization are considered. Specifically, these optimizations should lead to enhanced mechanical properties, such as improved room temperature ductility, or higher strength and creep resistance at temperatures above 600°C. As these alloys have a DO_3-ordered lattice, the deformation behaviour is rather complex exhibiting single, double, and fourfold dislocations which are separated by two types of anti-phase boundaries (APB) namely nearest-neighbour- and next-nearest-neighbour APB's. For the simulation of dislocation motion, or in general the deformation behaviour, the knowledge of the elastic properties is mandatory. Also it is obvious, that the plastic deformation should be facilitated if the APB energy is low, as less energy is stored in the APB's. The determination of APB energies by the measurement of the distance of separated superpartials in the TEM needs big efforts and may lead to divergent results. With the knowledge of the detailed site preferences and elastic constants APB energies can be calculated and the prediction of alloy compositions with low APB energies should be possible [2, 3]. For this, a detailed knowledge of the influence of alloying elements on the physical and mechanical properties is crucial but literature data are hardly available and the experimental determination is very time consuming. The use of new multidisciplinary approaches combining both experimental and theoretical methods is thus very desirable as it promises a remarkable reduction of the experimental effort and corresponding costs.

Theoretical *ab initio* calculations based on fundamental quantum-mechanical laws are nowadays increasingly used [4, 5] to accurately predict materials properties without any empirical input, solely from the knowledge of the atomic composition and/or structure. Basic mechanical and physical properties, like crystal structure and single crystal elastic constants, can be reliably calculated at zero temperature using density functional theory (DFT) [6, 7]. The calculated materials characteristics gathered at the atomistic level and our deeper understanding of the underlying processes can be effectively combined with advanced experimental techniques in order to design new materials with desired properties. Regarding specifically the iron-aluminides, (i) the site preference of the alloying elements and (ii) changes in the Young's modulus have been determined in this study. Computational results are then validated against experimental values of the Young's modulus in selected alloys.

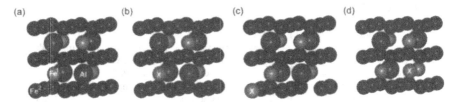

Figure 1. Schematic view (a) of the 16-atomic DO_3-based supercell which has been used in our study. The supercell of Fe_3Al that contains twelve Fe atoms distributed over two inequivalent Fe sublattices (Fe^2 depicted as small dark spheres and Fe^1 atoms shown as light spheres) and four Al atoms (the largest spheres). The substituent (marked X) can be located on either Fe^1 (b), Fe^2 (c), or Al-sublattice (d).

THEORETICAL CALCULATIONS

In order to analyse compositional trends of both thermodynamic and elastic properties of Fe_3Al-based materials, the supercell approach with a 16-atomic supercell derived from the DO_3 unit cell of Fe_3Al was employed (see Fig. 1a). The DO_3 unit cell consists of two iron sublattices, Fe^1 and Fe^2, the latter one having a twice higher number of the Fe atoms than the former one, and one Al sublattice (see Fig. 1). As the total number of atoms in the computational supercell is 16, the minimum concentration of the substituents when replacing a single Fe or Al atom is 1/16 or 6.25 at.%.

Site preference and elastic properties

The site preference of a substituent X replacing either the iron or aluminium atoms in Fe_3Al-based crystals will be addressed via calculations of the defect formation energy of ordered DO_3-based compounds with the element X (i) located at the Fe^1 sublattice (Fig. 1b), (ii) instead of one of the Fe^2 atoms (Fig. 1c), or as a substituent of one Al atom (Fig. 1d). The formation energy of a ternary Fe_iAl_jX described by a 16-atomic supercell ($i + j + 1 = 16$) with energy $E(Fe_iAl_jX)$ is then equal to $E_f(Fe_iAl_jX) = E(Fe_iAl_jX) - i\,\mu^{Fe} - j\mu^{Al}$, where μ^{Fe} and μ^{Al} are chemical potentials of ferromagnetic (FM), body-centered cubic (bcc) phase of Fe and non-magnetic (NM), face-centered cubic (fcc) of Al, respectively.

Single crystal elastic constants C_{ij} can nowadays be routinely predicted from *ab initio* calculations of the total energy changes as a function of specific lattice distortions applied to the undistorted ground state. For a cubic crystal three elastic constants are needed, i.e. three different distortions have to be studied. The first distortion is the isotropic volumetric change, i.e. the energy-volume dependence. The second derivative at the minimum of the energy-volume curve determines the bulk modulus B employing e.g. the Murnaghan equation of state [8]. The other two distortions are along the [001] and [111] directions (see e.g. Chen et al. [9]).

In a multi-physics approach, the single crystal elastic constants can be used to estimate a polycrystalline elastic modulus using various homogenization techniques as, e.g. the self-consistent continuum linear-elasticity Hershey's homogenization [10 - 13] which was used in this study. The homogenized polycrystalline elastic moduli can be further used to calculate essential engineering parameters, the bulk modulus B over the shear modulus G ratio as an approximative indicator of either ductile or brittle behavior. Despite of the fact that the *ab initio* calculations are performed for T=0K conditions, the qualitative validity of the predicted metallurgical trends can be extended also to finite temperature.

Parameters of the *ab initio* calculations

The *ab initio* calculations were performed using density functional theory in the generalized gradient approximation (GGA) [14] employing Projected Augmented Wave (PAW) potentials [15] as implemented in the Vienna Ab-initio Simulation Package (VASP) code [16, 17]. The plane wave cut-off energy has been 270 eV and an 8x8x8 Monkhorst-Pack mesh was used to sample the Brillouin zone of the supercells (Fig. 1). Both spin-polarized calculations and those without spin-polarization, i.e. non-magnetic, were performed and compared. The volume, shape, and internal parameters of the supercells were optimized with respect to the total energy.

EXPERIMENT

The theoretical research has been complemented with measurements of the Young's modulus at 77 K and room temperature. Both the calculations and measurements have been performed for a selected group of substituents (Ti, W, V, Cr, Si) that all have the solubility in Fe_3Al of at least 2 at.%. The Young's moduli have been determined by the impulse excitation technique. For this, samples with dimensions of 70 x 7 x 4 mm have been cut by spark erosion from vacuum induction melted ingots of several Fe_3Al-X alloys.

RESULTS

The site preference of the substituents has been studied by calculating the defect formation energy of ordered Fe_3Al-based compounds with a substituent X located on the three possible sublattices (see Figs. 1b-d). In agreement with previously published experimental data (see e.g. [18]) the obtained results clearly show a strong preference of W and V for the Fe^1 sites. The preference has been found weaker for Cr, making the Fe^2 substitution also possible especially at elevated temperatures, whereas Si atoms show a strong preference for the Al sites. The results are summarized in Tab. I. Importantly, the site preference can not be simply ordered with respect to the atomic radius r of the elements because $r^{Si} < r^{Fe} < r^{Al} < r^{Cr} < r^V < r^{Ti} < r^W$.

Table I. The differences of the defect-formation energy ΔE_f (eV/substituent) with respect to the energy obtained for the substitution that minimizes the defect formation-energy ΔE_f out of the three studied cases, i.e. a ternary element X substituting an atom on the sublattice Fe^1, Fe^2, or Al.

ΔE_f (eV/atom)	Ti	W	V	Cr	Si
Fe^1 site	0	0	0	0	0.070
Fe^2 site	0.876	1.034	0.750	0.150	0.592
Al site	0.826	0.659	0.643	0.456	0

Motivated by the use of Fe_3Al-based alloys as structural materials, the lattice constant change upon the substitution has been also theoretically determined. As the experimental measurements clearly show linear trends in the lattice parameter dependence as a function of the solute content [19], the values theoretically predicted for 6.25 at.% can be directly compared to the measurements for other concentrations. In a qualitative agreement with the experimental data [19, 20], the *ab initio* calculations predict an expansion of the lattice for Ti and W but a contraction in case of V, Cr, and Si.

In order to identify elements that may possibly increase the strength of the Fe_3Al-based materials, the change in the homogenized polycrystalline Young's modulus Y due to solutes has been determined. The calculations predict an increase in the moduli in case of all studied alloying elements. Assuming a linear increase in Y with solute atom concentration, the actual changes ΔY^{theo} obtained due to 6.25 at.% of the alloying element with respect to pure Fe_3Al are summarized in Tab. II.

Table II. The changes in the homogenized polycrystalline Young's modulus Y (in GPa per at.% with respect to Fe$_3$Al) in Fe$_3$Al-X (X=Ti, W, V, Cr, Si) as theoretically predicted (ΔY^{theo}) for 6.25 at.% of solutes and from a linear fit to the experimental data obtained for different solute contents (ΔY^{exp}). Also listed are theoretically predicted changes of the bulk modulus B, shear modulus G, and their ratio (B/G), all per atomic percent of the solute, with respect to Fe$_3$Al.

	Ti	W	V	Cr	Si
ΔY^{theo} (GPa/at. %)	4.7	9.3	6.8	3.1	5.5
ΔY^{exp} (GPa/at. %)	26.6	13.2	7.5	1.2	9.0
ΔB^{theo} (GPa/at. %)	-3.0	0.4	-1.4	-1.4	1.4
ΔG^{theo} (GPa/at. %)	2.0	4.2	3.2	1.5	2.3
$\Delta (B/G)^{theo}$ (1/at. %)	-0.10	-0.10	-0.10	-0.07	-0.05

To cross-check the validity of the *ab initio* prediction, the influence of the same five ternary elements on the Young's modulus has been investigated also experimentally. To perform measurements under conditions close to those of the *ab initio* calculations, the Young's moduli have been at first determined at 77 K. In agreement with the theoretical predictions, all the studied alloying elements increase the Young's modulus with respect to that of pure Fe$_3$Al in the concentration range above 1-2 at.% (see ΔY^{exp} Tab. II). A similar increase has been detected also at room temperature. The theoretical data (Tab. II) also show that all the studied elements reduce the B/G ratio from the value 2.42 that we obtained from the *ab initio* calculations for Fe$_3$Al (see Tab. II). If we consider the B/G ratio as an approximative indicator of either brittle ($B/G < 1.75$) or ductile ($B/G > 1.75$) behavior, then all the studied ternary elements are predicted to lower an inherent ductility of Fe$_3$Al but it should be noted that extrinsic effects (as e.g. environmental embrittlement) is not considered in our simulations.

A highly unexpected elastic effect in the Young's modulus compositional trends and significant deviation between the experimental data and theoretical predictions were observed in some Fe$_3$Al-Ti and Fe$_3$Al-W alloys. For small additions of the solute atoms, the expected linear trends, as reported, e.g., for iron in [21], have been not observed there. The Young's modulus has been found even to decrease in some cases in the concentration range below 1 at.% due to these solute atoms. As almost no experimental data are available in literature on the concentration dependence of the elastic properties for low solute concentrations in case of binary systems and even less for ternary systems, this interesting effect remains unexplained and will be a part of future investigations. Regarding the employed theoretical approach, special attention will be paid to the improvement of the homogenization technique in order to take into account (i) textures that were observed in some as-cast samples and which are not considered in Hershey's scheme, and (ii) possible dual-phase character of some Fe$_3$Al-Ti and Fe$_3$Al-W alloys.

CONCLUSIONS

The presented results show that very promising multi-disciplinary research combining both theoretical *ab initio* and experimental techniques can be effectively used for the development and optimization of new (here Fe$_3$Al-based) materials. The T = 0 K *ab initio* calculations of thermodynamic, structural, and elastic properties were complemented by Young's modulus measurements at both ambient and liquid-nitrogen temperatures.

The theoretical calculations of the solute formation energy were used to predict the site preference of different solute atoms. The DFT calculations further predict an expansion of the lattice due to the presence of Ti and W atoms but a contraction for V, Cr, and Si. Both thermodynamic and structural properties were found in an excellent qualitative agreement with measured data. Also determined were the integral elastic properties of the polycrystalline aggregates from a multi-physics combination of the *ab initio* calculations and Hershey's homogenization method. In agreement with the impulse excitation measurements performed at liquid-nitrogen and ambient temperatures, all studied ternary elements are predicted to increase the Young's modulus. The detected compositional trends in Young's modulus are unexpectedly complex in case of Fe_3Al-Ti and Fe_3Al-W showing deviations from a simple linear trend and a further methodological development is planned in order to explain this phenomenon.

ACKNOWLEDGMENTS

The research has been partly funded through the "Triple-M – Max-Planck Initiative on Multiscale Materials Modeling of Condensed Matter". We would also like to thank Prof. Dierk Raabe and Dr. Art William Counts (now at the Northwestern University, Evanston, IL, USA) from the Microstructure Physics and Metal Forming department at our institute for many valuable discussions.

REFERENCES

1. S. C. Deevi and V. K. Sikka, *Intermetallics* **4**, 357 (1996).
2. N. I. Medvedeva, Y. Gornostyrev, D. L. Novikov, O. N. Mryasov, and A. J. Freeman, *Acta Mater.* **46**, 3433 (1998).
3. F. Kral, P. Schwander, and G. Kostorz, *Acta Mater.* **45**, 675 (1997).
4. G. Ghosh, S. Delsante, G. Borzone, M. Asta, and R. Ferro, *Acta Mater.* **54**, 4977 (2006).
5. B. Ghosh, S. Vaynman , M. Asta, and M. E. Fine, *Intermetallics* **15**, 44 (2007).
6. P. Hohenberg and W. Kohn, *Phys. Rev.* **136**, B864 (1964).
7. W. Kohn and L. J. Sham, *Phys. Rev.* **140**, A1133 (1965).
8. F. D. Murnaghan, *Proc. Natl. Acad. Sci. USA* **30**, 244 (1944).
9. K. Chen, L. R. Zhao and J. S. Tse, *J. Appl. Phys.* **93**, 2414 (2003).
10. A. V. Hershey, *J. Appl. Mech.* **9**, 49 (1954).
11. W. A. Counts, M. Friák, D. Raabe and J. Neugebauer, *Acta Mater.* **57**, 69 (2009).
12. M. Friák, W. A. Counts, D. Raabe and J. Neugebauer, *Phys. Stat. Sol. b* **245**, 2636 (2008).
13. W. A. Counts, M. Friák, C. C. Battaile, D. Raabe and J. Neugebauer, *Phys. Stat. Sol. b* **245**, 2630 (2008).
14 J. P. Perdew, K. Burke, and M. Ernzerhof, *Phys. Rev. Lett.* **77**, 3865 (1996).
15. P. E. Blöchl, *Phys. Rev.* **B50**, 17953 (1994).
16. G. Kresse and J. Hafner, *Phys. Rev.* **B47**, 558 (1993).
17. G. Kresse and J. Furthmüller, *Phys. Rev.* **B54**, 11169 (1996).
18. S. Zuqing, Y. Wangyue, S. Lizhen, H. Yuanding, Z. Baisheng and Y. Jilian, *Mater. Sci. Eng.* **A258**, 69 (1998).
19. Y. Nishino, C. Kumada and S. Asano, *Scr. Mater.* **36**, 461 (1997).
20. L. Anthony and B. Fultz, *Acta Metall. Mater.* **43**, 3885 (1995).
21. G. R. Speich, A. J. Schwoebl, and W. C. Leslie, *Metall. Trans.* **3**, 2031 (1972).

Mater. Res. Soc. Symp. Proc. Vol. 1128 © 2009 Materials Research Society

Influence of Structure on the Magnetism of Fe$_{1-x}$Al$_x$ Alloys

E. Apiñaniz[1], D. Martín-Rodriguez[2,*], E. Legarra[2], J.J.S. Garitaonandia[2], F. Plazaola[2],
[1] Fisika Aplikatua I Saila, Ingenieritza Goi Eskola Teknikoa, UPV/EHU, Alameda Urquijo s/n, 48013 Bilbao
[2] Elektrizitate eta Elektronika Saila, Zientzia eta Teknologia Fakultatea, UPV/EHU, P.K. 644, 48080 Bilbao

* Currently at Bragg Institute, Australian Nuclear Science and Technology Organisation, PMB 1, Menai NSW 2234, Australia

ABSTRACT

With the aim of studying the relationship between structure and magnetism of Fe$_{1-x}$Al$_x$ alloys, annealed and as-crushed samples with $0.25 \leq x \leq 0.425$ been studied. X-rays and Mössbauer spectra show that the nearest neighborhood of each Fe atom plays a major role when determining the magnetic behavior of each Fe atom. In fact, these techniques allow us to determine in which surroundings these Fe atoms will behave paramagnetically. On the other hand, the influence of the lattice parameter has also been studied.

INTRODUCTION

One of the most prominent features of Fe rich FeAl alloys is that any structural change directly influences their magnetic behavior. That is to say, any slight mechanical deformation produced on the ordered alloy causes an abrupt increase of the ferromagnetic signal [1-8]. Even for alloys that show no magnetic signal in the ordered state, the mechanical deformation induces a clear magnetic signal. Moreover, small changes in Fe content of the alloy (less than 1 at.%) induce large changes in its magnetism. In order to explain this behavior the intimate relationship between microstructure and magnetism has to be taken into account; indeed, this is linked to the fact that these alloys show three main phases in the Fe-rich side of FeAl phase diagram: up to 19 at.% Al these alloys are reported to have A2 disordered structure [10]. When increasing the Al content D0$_3$ ordered structure appears (23-37 at. % Al). The third stable compound with a wide existence is B2 which exists for 37-50 at. % Al. [10]. These known structures and their relationship with the magnetism make Fe-rich FeAl intermetallic alloys very useful for testing theories and hypothesis of fundamental magnetism.

In this paper, we propose a preliminary model for explaining the magnetism of FeAl alloys with different Fe content, different treatments (annealed and mechanically crushed) and different structures. In addition, we will show the influence of addition of Si on the magnetism.

EXPERIMENT

In order to study the influence of the structure in the magnetism of these alloys, Fe$_{1-x}$Al$_x$ samples with $0.25 \leq x \leq 0.425$ have been analyzed. These samples were prepared by induction melting and cast into ingots in an Ar atmosphere; then powder was obtained by mechanical crushing. To obtain ordered samples, a long-time heat treatment was applied: powders were annealed at about 1200 K for 2 h and at 790 K for one week. XRD measurements were carried out at room temperature (RT) using Cu Kα radiation. 57Fe Mössbauer spectroscopy measurements were carried out at room temperature in transmission geometry using a conventional spectrometer with a 57Co-Rh source. The measured spectra were fitted using the NORMOS program developed by Brand et al. [11]

DISCUSSION

Figure 1 shows XRD patterns of annealed $Fe_{1-x}Al_x$ alloys. For x<0.325 the diffraction pattern shows superstructure peaks corresponding to $D0_3$ structure ((111), (311), (331)). Although not shown, X-ray diffraction patterns of as-crushed samples do not show superstructure peaks.

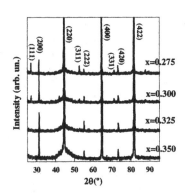

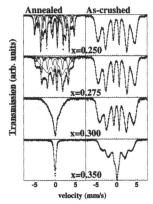

Figure 1. X-ray diffraction pattern of annealed $Fe_{1-x}Al_x$ samples.

Figure 2. Mössbauer spectra for different compositions of (left) annealed and (right) as-crushed samples.

As far as Mössbauer spectra are concerned, the annealed samples (see Figure 2 left) show narrow and well-defined sextets for x=0.25. Increasing the Al content, there is an overall decrease of the hyperfine field (B_{hf}), which clearly indicates that directing towards the B2 side of the phase diagram [10] the local magnetic contribution decreases. It could be due to the decrease of the Fe content or due to a structural change (Figure 1 shows that for x>0.30 the peaks corresponding to $D0_3$ structure disappear). For x>0.25 the sextets of the Mössbauer spectra widen, for instance at x=0.275 in order to fit the spectrum we used a B_{hf} distribution superposed to wide discrete sextets. For x=0.30 the discrete Mössbauer sextets disappear and a paramagnetic singlet appears, superposed to a magnetic B_{hf} distribution (see Figure 2 left). It indicates that besides the decrease of the magnetic contribution (less Fe atoms), the number of different local magnetic surroundings of Fe atoms is quite large, giving rise to a non-homogeneous B_{hf} distribution. Indeed, magnetic non-homogeneities or clustering processes have been proposed to explain the results obtained by small angle neutron diffraction experiments [12, 13]. For higher Al contents, the magnetic contribution in the annealed samples starts to disappear and for x=0.325 the spectrum shows only a paramagnetic singlet.

Although, as shown above, for annealed samples there is a clear magnetic evolution with composition, the mechanically deformed samples (see Figure 2 right) show very similar Mössbauer spectra for samples with x≤0.30, displaying only a very wide sextet. For higher x a paramagnetic singlet together with a magnetic contribution is observed. Thus, the magnetic contribution decreases with the increase in Al content.

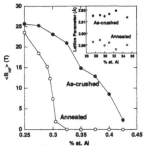

Figure 3. <B_{hf}> evolution with the composition.

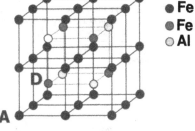

Figure 4. Fe-Al lattice [9].

Figure 3 shows the room temperature evolution of the mean hyperfine field (<B_{hf}>) as a function of composition in annealed and mechanically deformed samples. In annealed samples, <B_{hf}> decreases abruptly with the increase of Al content. At x=0.325, the X-ray diffraction patterns (Figure 1) show a structural transition where the peaks corresponding to D0$_3$ structure disappear, but the ones corresponding to B2 remains (in agreement with Massalski´s phase diagram [10]). This suggests a correlation between magnetic and crystallographic behaviors. Indeed, the magnetic transition occurs at the composition where the crystallographic one takes place.

The decrease of <B_{hf}> while D0$_3$ superstructure peaks are present can be explained as due to the lower Fe atom content. Moreover, as mentioned above, Figure 2 shows a clear evolution of the magnetic sub-spectra with the increase of the Al content in annealed alloys. However, the opposite is observed in Mössbauer spectra of as-crushed samples. In the latter, the shape of the spectra for x≤0.325 is identical (see Figure 2 right), even if <B_{hf}> decreases. In addition, Figure 3 shows a notable reinforcement of the magnetism in mechanically deformed alloys and it is worth noting that the ratio between as-crushed and annealed magnetic hyperfine fields increases with x.

Taking into account that <B_{hf}> at RT decreases with increasing Al content and that the main difference between B2 and D0$_3$ is the position of the atoms in BCC structure, we can discuss the results in annealed samples within the mean-field model of ferromagnetism [14]. In this model the Curie Temperature (T_c) is proportional to the interatomic exchange integral and to the coordination number of the strongly magnetic species around the transition metal site. Therefore, by looking at the local neighborhood of Fe atoms by means of Mössbauer spectroscopy we can determine the limiting number of nearest neighbor Al atoms around Fe atoms needed for the alloy to have T_c at RT in ordered alloys (i.e. a limiting number for a Fe atom to behave paramagnetically).

In Table I it can be read that in the B2 structure the probability for Fe atoms to sit on sites having rich Al environments is higher for higher Al contents, but even for high Al contents there is always a probability of finding Fe atoms surrounded by 8 Fe atoms. Indeed, for x=0.35, above

20% of the Fe atoms are surrounded by 8 Fe atoms. Nevertheless, the Mössbauer spectra of annealed samples with x>0.30 are singlets, which do not show any magnetic contribution at RT. As measurements have been performed at RT, the exchange energy is lower that the thermal energy, therefore we will assume that for x>0.30 all the Fe atoms are paramagnetic at RT. For this reason, for all the measured samples Fe atoms will show ferro- or paramagnetic behavior depending on the competition between net exchange interaction energy of Fe atoms with the surrounding and thermal energy, which is the same for all the samples.

Table I. Probabilities (%) for Fe atoms to sit on sites having 1, 2, 3, 4, 5, 6, 7 or 8 Fe atoms as nearest neighbors. D stands for Fe atoms at position D and A for Fe atoms at position A. A2 is a solid solution therefore all the positions are equivalent, i.e., only the probability for a Fe to be surrounded by 1-8 Fe atoms has to be taken into account.

$Fe_{1-x}Al_x$		x=0.250			x=0.275			x=0.300			x=0.350		
Position	Fe nearest neighbours	A2	DO_3	B2	A2	DO_3	B2	A2	DO_3	B2	A2	DO_3	B2
D	8		33,33	33,33		31,03	31,03		28,57	28,57		23,08	23,08
A	8	10,01		0,26	7,63		0,12	5,76		0,05	3,19		0,01
A	7	26,70		2,08	23,16		1,13	19,77		0,56	13,73		0,09
A	6	31,15		7,29	30,75		4,85	29,65		2,95	25,87		0,77
A	5	20,76		14,58	23,33		11,86	25,41		8,85	27,86		3,59
A	4	8,65	66,67	18,23	11,06	45,25	18,12	13,61	29,26	16,59	18,75	9,97	10,47
A	3	2,31		14,58	3,36	20,11	17,71	4,67	29,26	19,91	8,08	26,59	19,55
A	2	0,38		7,29	0,64	3,35	10,82	1,00	10,97	14,93	2,17	26,59	22,81
A	1	0,04		2,08	0,07	0,25	3,78	0,12	1,83	6,40	0,33	11,82	15,20
A	0	0,00		0,26	0,00	0,01	0,58	0,01	0,11	1,20	0,02	1,97	4,43

The stoichiometric B2 structure (x=0.50) shows two non-equivalent positions [9], position A is occupied by Fe atoms while all D positions are occupied by Al (see Figure 4). When the compositions are not stoichiometric (x<0.50) the exceeding Fe atoms sit in D positions, therefore there are some Fe (D positions) atoms surrounded by 8 Fe atoms and some with rich Al surroundings. The latter have very low magnetic moments as shown in ref. [2]. As mentioned above, even if there are Fe atoms surrounded by 8 Fe atoms no ferromagnetic contribution can be observed in the Mössbauer spectra for x>0.30 with B2 structure (see Figure 1). This behavior can be explained by the fact that if nearest neighbor Fe atoms are paramagnetic the atom surrounded by 8 Fe paramagnetic atoms will have a paramagnetic behavior.

From the pervious analysis of the B2 structure it cannot be concluded which is the minimum number of Al atoms for Fe to behave paramagnetically at RT. In order to determine this minimum number of atoms we will study the D0₃ structure where Fe atoms show a lower number of environments than the B2 one. Indeed, the stoichiometric D0₃ structure shows two non-equivalent positions [9]: position A is occupied by Fe atoms while position D is occupied evenly by Al and Fe (see Figure 4). When the compositions are not stoichiometric (x>0.25) the exceeding Al atoms sit in D positions, therefore there are still some Fe (D positions) atoms surrounded by 8 Fe atoms, and some Fe atoms located at position A surrounded by richer Al environment than the stoichiometric one. Table I shows that outside D0₃ stoichiometric composition there is a probability to find Fe atoms surrounded by 0, 1, 2 and 3 Fe atoms.

The Mössbauer spectrum for x=0.25 (D0₃ stoichiometry) shows well-defined sextets, which indicate that all the Fe atoms are ferromagnetic at RT, even the ones surrounded by 4 Al atoms. But for larger x the sextets are less defined, which is an indication of the presence of a distribution of hyperfine fields, that is, an increased number of different Fe environments. In addition, the area of the spectra decreases indicating that the number of ferromagnetic atoms is decreasing due to a larger amount of Al atoms and/or due to the decrease of T_C. For these compositions with higher x there is a probability of finding Fe atoms surrounded by 5 or more Al atoms (see Table I). Moreover, the spectra show that for these compositions there are less ferromagnetic atoms, therefore we can conclude that these Fe atoms (surrounded by 5 or more Al atoms) behave paramagnetically at RT.

When analyzing the B2 structure we established that Fe atoms surrounded by 8 Fe atoms will behave paramagnetically if these surrounding atoms are paramagnetic. On the other hand, by studying the DO_3 structure it has been concluded that the Fe atoms surrounded by 5 or more Al atoms behave paramagnetically at RT.

From table I the probability for a Fe atom in position A (i.e. rich Fe environment) to have 0, 1, 2, 3, 4, 5, 6, 7 and 8 paramagnetic atoms (i.e. Fe atoms with 5 or more Al atoms as nearest neighbors and so, with paramagnetic behavior at RT) can be calculated. It is clearly observed that the probability of finding a larger amount of Fe paramagnetic atoms is greater in B2 than in DO_3, within their range of existence. This could lead to the difference observed in the Mössbauer spectra. For low Al contents DO_3 shows well defined sextets (ferromagnetic Fe atoms), but increasing the Al content, some Fe atoms become paramagnetic. When the concentration of Fe paramagnetic atoms is high enough even Fe atoms surrounded by 8 Fe atoms (if these are paramagnetic) behave paramagnetically at RT. Therefore, these alloys show a singlet in the Mössbauer spectra. On the other hand, due to its structure and composition, B2 structure has a higher probability to have Fe atoms surrounded by paramagnetic atoms and so, it behaves paramagnetically at RT.

Up to now we have analyzed the influence of B2 and DO_3 crystallographic structures in the magnetic behavior at RT. In mechanically deformed samples no superstructure peaks appear in the X-ray patterns, which allows assuming, as a first approximation, that these samples are completely disordered, presenting the fundamental cubic A2 structure. In Figure 2 (right) Mössbauer data show that for x<0.35 all the contributions are ferromagnetic at RT and for x=0.35 a singlet superimposes the magnetic contribution. Table I (A2) shows that in x=0.325 there is a large probability to have Fe atoms surrounded by more than 5 Al atoms, which would imply the presence of a paramagnetic singlet in the Mössbauer spectrum. However, this is not the case and another effect has to be taken into account. Indeed, it has been proved [2, 5] that the lattice parameter (interatomic distance) has also a great influence on the magnetic properties. The inset of Figure 3 shows the mean lattice parameters of annealed and mechanically crushed samples for different compositions. Clearly, the crushing increases the lattice parameter in agreement with Nogues et al. [5], where by means of dichroism (XMCD) and ab-initio calculations they reported that, in the case of disordered $Fe_{0.6}Al_{0.4}$, experimentally the contribution of disorder and lattice expansion account for 65% and 35% of the magnetism of the alloy, respectively; while theoretically they account for 55% and 45%. They explain that with decreasing lattice parameter there is a strengthening of the hybridization, which gives rise to a charge transfer from the majority to the minority spin subbands, causing the decrease of the magnetic moment [see also ref. 2].

Finally, regarding the Si/Al substitution in $Fe_{0.6}Al_{0.4}$ and $Fe_{0.7}Al_{0.3}$ alloys, preliminary results indicate that in ordered $Fe_{0.6}Al_xSi_{0.4-x}$ alloy the Mössbauer spectra (not shown) can be explained assuming Si atoms (substituting Al) promote DO_3 structure. Moreover, Si atoms lead to a transition from B2 structure in x=0.4 to a mixture of B2 and DO_3 structures at x=0. In the case of $Fe_{0.7}Al_xSi_{0.3-x}$ the Mössbauer spectrum (see Figure 2 left) varies completely in comparison to that of $Fe_{0.7}Al_{0.3}$ giving rise the appearance of defined peaks in the Mössbauer spectra even with the minimum amount of Si (9 at.%Si) used in this work. This is an indication of the appearance of a more homogeneous distribution of hyperfine fields and therefore, compared to Al, more Si atoms in the nearest neighborhood of Fe are needed to make Fe atoms paramagnetic.

CONCLUSIONS

By means of magnetic measurements, X-ray diffraction and Mössbauer spectroscopy, a preliminary model that explains the relationship between the crystallographic structure and the magnetic behavior of FeAl alloys has been proposed. There are several factors that determine the magnetic properties of these alloys. On the one hand the local atomic environment of Fe atoms: we have concluded that at RT even the Fe atoms surrounded by 8 Fe atoms can behave paramagnetically if the latter are paramagnetic. In addition, we have established a limiting value for a Fe atom to be paramagnetic in ordered alloys at RT: Fe atoms surrounded by 5 or more Al atoms. On the other hand, at RT the increase of the interatomic distance enhances the magnetic moment of the Fe atoms due to charge transfer.

ACKNOWLEDGMENTS

We would like to aknowledge the Basque Government and the Spanish MEC for financial support under projects IT-382-07 and MAT-2006-12743.

REFERENCES

1. E. Apiñaniz, F. Plazaola, J. S. Garitaonandia, D. Martin, J. A. Jimenez, J. Appl. Phys. **93**, 7649 (2003)
2. E. Apiñaniz, F. Plazaola, J.S. Garitaonandia, Eur. Phys. J. B 31, 167 (2003)
3. E. Apiñaniz, F. Plazaola, J. S. Garitaonandia, J. Mag. Mag. Mat. **272-276**, 794-796 (2004)
4. D. Martín Rodríguez, E. Apiñaniz, F. Plazaola, J. S. Garitaonandia, J. A. Jiménez, D. S. Schmool and G. J. Cuello, Phys. Rev.B **71**, 212408 (2005)
5. J. Nogués, E. Apiñaniz, J. Sort, M. Amboage, M. d'Astuto, O. Mathon, R. Puzniak, I. Fita, J. S. Garitaonandia, S. Suriñach, J. S. Muñoz, M. D. Baró, F. Plazaola and F. Baudelet, Phys. Rev. B **74**, 024407 (2006)
6. A. Hernando, X. Amils, J. Nogues, S. Suriñach, M.D. Baró and M.R. Ibarra, Phys Rev B **58**, R11864 (1998)
7. X. Amils, J. Nogues, S. Suriñach, M.D. Baró, J.S. Muñoz, IEEE Trans. Magn. **34**, 1129 (1998)
8. E.P. Yelsukov, E.V. Voronina, V.A. Barinov, J. Magn. Magn. Mater. **115**, 271 (1992)
9. F. Schmid, K. Binder, J. Phys.: Cond. Matter 4, 3569 (1992).
10. T. B. Massalski, *Binary Alloy Phase Diagrams*, American Society for Metals, Metals Park, Ohio, 1986
11. R. A. Brand, J. Lauer, D.M. Herlach, J. Physics F: Metal Phys. **13**, 675 (1983)
12. D. Martín Rodríguez, F. Plazaola, J. J. del Val, J. S. Garitaonandia, G. J. Cuello, C. Dewhurst, J. Appl. Phys **99**, 08H502(2006)
13. W. Bao, S. Raymond, S. M. Shapiro, K. Motoya, B. Fåk, R. W. Erwin, Phys. Rev. Lett. **82**, 4711 - 4714 (1999)
14. L. C. Cullity, *Introduction to Magnetic Materials*, Addison-Wesley, Reading 1972

Mater. Res. Soc. Symp. Proc. Vol. 1128 © 2009 Materials Research Society   1128-U02-11

Ultrasonic Vacuum Chill Casting and Hot Rolling of FeAl-Based Alloys

Vladimír Šíma[1], Přemysl Málek[1], Petr Kozelský[2], Ivo Schindler[2], Petr Hána[3]
[1]Department of Physics of Materials, Faculty of Mathematics and Physics, Charles University, Ke Karlovu 5, CZ-121 16 Praha 2, Czech Republic
[2]Faculty of Metallurgy and Material Engineering, VŠB - Technical University of Ostrava, 17. listopadu 15, CZ-708 33 Ostrava-Poruba, Czech Republic
[3]Department of Physics, Technical University of Liberec, Studentská 2, CZ-461 17 Liberec, Czech Republic

ABSTRACT

An ultrasonic device was designed to fabricate relatively small vacuum chill castings of FeAl-based alloys with improved microstructure. A special hot-rolling procedure preventing thermal shocks was used for the thermomechanical treatment of cast alloys.

The efficiency of ultrasonic vacuum casting is manifested by improved microstructure of hot-rolled iron aluminides Fe – 40 at.% Al with addition of C or Zr and B or Zr and B with 1 wt.% of Y_2O_3 particles.

INTRODUCTION

Since about 1960 the iron aluminides based on the B2-ordered FeAl have been considered as promising materials for many applications. These expectations are not yet fully achieved due to various reasons and also limitations, coming, e.g., from the room-temperature brittleness of these materials.

Strong arguments, mentioned as the advantages of FeAl-based alloys, are the excellent chemical resistance against oxidation and sulphidation and low cost of raw materials. The only industrially-proved solution of the macroscopic brittleness can be seen in a very fine microstructure, with micrometric grain size. The intergranular brittleness of FeAl alloys may be suppressed by boron addition which strengthens grain boundaries [1–4]. The poor creep resistance at elevated temperatures can be improved by precipitation and solid solution hardening [5] or by dispersion of oxide particles [6]. The loss of strength above 600 °C is related to particle redissolution or coarsening [7]. At high temperatures above 1000 °C even a coarsening of the yttria (Y_2O_3) particles [8] (T_m= 2690 °C) is enabled by intrinsically high diffusion through the open bcc intermetallic matrix. The difficulties in conventional metallurgy processing (large grains and shrinkage cracks in castings, cracking of hot-rolled alloys) have focused many efforts on these materials to powder metallurgy [9,10]. Significant research has been devoted to the development of mechanically alloyed materials with addition of nanoscale Y_2O_3 particles preventing the onset of recrystallization in the consolidating powder [11].

In the present study we summarize our results on FeAl-based alloys prepared by conventional metallurgy processing, improved by using high-power ultrasound for solidifying the melt and a special hot-rolling procedure preventing thermal shocks. New results for Fe – 40 at.% Al with addition of Zr and B with 1 wt.% of Y_2O_3 particles are presented.

EXPERIMENTAL PROCEDURES

An ultrasonic device was designed to fabricate vacuum chill castings with dimensions of about 40 mm x 20 mm x 90 mm. During solidification of the melt, taking about 30 s, the ultrasonic vibrations were conducted to the ingot mould bottom. The output power of the ultrasound could be regulated in the range of about 50–1000 W. The generator and a piezoelectric transducer operated at a frequency of 40 kHz, the acoustic power was transferred to the ingot using a titanium sonotrode. The experimental arrangement is schematically shown in Figure 1. The alloys prepared in such a way will be designated as ultrasound-treated (UST) alloys.

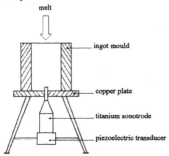

Figure 1. The experimental arrangement of the ingot mould and the piezoelectric transducer in the vacuum induction furnace.

If ultrasound was not applied during vacuum casting, the quality of castings was poor because of large grains and the inhomogeneous solidification microstructure with many shrinkage cavities or cracks. The subsequent thermomechanical treatment (hot rolling) of such an as-cast alloy was difficult, mainly because of fraying and cracking of the rolled alloy. The main reason for the poor quality of such hot-rolled products is the thermal shock arising at the surface of the hot casting, when the casting is coming into contact with the cold roller. A special stainless steel bandage of the casting rolled at 1200 °C in four passes (thickness reduction 12 %, 10 %, 10 %, 8 %) to plates of 12.5 mm thickness was a good solution to this problem [12].

The chemical composition of the alloys discussed in this study is summarized in Table I.

Table I. Chemical composition of alloys (at.%).

	Al	C	Zr	B	Fe
Fe–40Al	40	0.04(1)	-	-	balance
Fe–40Al–1C	40	1.0	-	-	balance
Fe–40Al–0.1Zr (UST)	40.7(4)	0.04(1)	0.08(2)	<0.01	balance
Fe–40Al–0.1Zr–0.03B	39.9(4)	0.04(1)	0.09(2)	0.04(1)	balance
Fe–40Al–0.1Zr–0.13B (UST)	39.7(4)	0.04(1)	0.09(2)	0.13(1)	balance
Fe–40Al–0.1Zr–0.13B +1 wt.% of Y_2O_3 (UST)	40.0(4)	0.04(1)	0.10(2)	0.13(1)	balance

The microstructure of the alloys was examined by light optical microscopy. Mechanical properties were measured using tensile tests (initial strain rate $1.0 \times 10^{-4} \, s^{-1}$) at a temperature of 600 °C. Further details are given elsewhere [12].

RESULTS AND DISCUSSION

The alloys (see Table I) can be divided into four basically different groups: Fe–40at.%Al, Fe–40at.%Al–1at.%C, Fe–40at.%Al–0.1at.%Zr with 0-0.13at.%B and Fe–40Al–0.1Zr–0.13B with addition of 1 wt.% of Y_2O_3 particles.

Microstructure of the UST Fe–40at.%Al [12]

The average grain size of the binary Fe–40at.%Al as-cast alloy, solidified without the UST, was 320 µm. The alloy of the same composition was then prepared using UST with two different intensity levels of ultrasonic sonication of the melt, namely with ultrasonic power of about 300 and 800 W. The average grain size of such samples was refined to 240 and 190 µm, respectively.

Microstructure of the UST Fe–40at.%Al–1at.%C [12]

There are three typical features in the microstructure of the ternary Fe–40at.%Al–1at.%C as-cast alloy solidified without the UST:

- chains of carbide particles Fe_3AlC with perovskite structure,
- isolated shrinkage cavities,
- cracks connected into asterisks.

Hot-rolling of such an alloy usually led to the formation of massive intercrystalline cracks, nucleated mainly on the casting defects. The microstructure of the same alloy, cast in the ultrasound with a power of 800 W, did not show any visible grain-refinement effect. On the other hand, the UST leads to a strong reduction of the shrinkage cavitation and practically prevents the formation of asterisk cracks in castings.

Microstructure and mechanical properties of the UST Fe–40at.%Al–0.1at.%Zr with 0-0.13at.%B [12]

The single-phase microstructure of hot-rolled castings of these alloys does not show any visible influence of the UST.

On the other hand, the improved quality of the UST alloy is obvious, if the tensile tests are evaluated. The values of the elongation to fracture A (tensile tests) (see Table II) and the time to rupture (in creep tests) [12], which are sensitive to a presence of casting defects in the alloy, are considerably higher for the UST alloy. The UST alloy showed a typical ductile failure, the failure of the alloy solidified without the UST was mostly brittle.

The yield stress R_p0.2 and ultimate tensile strength R_m data in Table II clearly demonstrate the expected strengthening effect of the boron addition at 600 °C.

Microstructure and mechanical properties of the UST Fe–40at.%Al–0.1at.%Zr–0.13at.%B with addition of 1 wt.% of Y_2O_3 particles

In [12] a further improvement of the UST of studied alloys was proposed: a grain refinement expected by ultrasonic treatment of the melt with addition of the Y_2O_3 powder. The yttria particles, used in powder metallurgy of FeAl-based materials for stabilization of the fine grain structure, should act in this case as additional nucleation centers activated by the UST of the melt. The yttria powder with particle size of 5-10 μm was used for preparation of such an UST alloy.

Table II. Data from tensile tests (yield stress (R_p0.2), ultimate tensile strength (R_m), elongation to fracture (A)) (initial strain rate 1.0 x 10^{-4} s^{-1}) at temperature of 600 °C.

alloy	R_p0.2 (MPa)	R_m (MPa)	A (%)
Fe–40Al–0.1Zr UST	238	275	13
Fe–40Al–0.1Zr–0.03B no UST	319	364	9
Fe–40Al–0.1Zr–0.03B UST	317	370	18
Fe–40Al–0.1Zr–0.13B UST	377	434	10
Fe–40Al–0.1Zr–0.13B + 1 wt.% of Y_2O_3 UST	424	455	21

a   b

Figure 2. Microstructure of the UST Fe–40Al–0.1Zr–0.13B alloy with 1 wt.% of Y_2O_3: (a) as-cast, (b) hot-rolled.

The microstructure of the as-cast and the hot-rolled alloy is shown in Figure 2. It can be seen that the role of the yttria particles did not come up to the expectations; the grain size is comparable to that for the alloy without yttria addition.

The microstructure of the as-cast alloy in Figure 2a shows a quite homogeneous distribution of yttria particles in relatively large grains with unstable wavy grain boundaries, the microstructure of the hot-rolled alloy in Figure 2b is partially recrystallized, a generation of some new grains is observed. The mechanical properties of the hot-rolled alloy were improved by the yttria addition, as can be seen from the data in Table II. The reason of this effect is not quite clear; an oxide dispersion strengthening (ODS) supposes much finer yttria particles (typically 10 nm) dispersed in the matrix.

CONCLUSIONS

The main results can be summarised as follows:

- the ultrasound application improves the casting microstructure and mainly the casting quality of FeAl-based alloys. This effect is supposed to be a result of the acoustic cavitation, which influences the solidification of the melt. The melt sonication improves the mold filling and prevents formation of shrinkage cavities in the cast piece [12].

- the use of UST during the solidification makes the subsequent hot-rolling process (thermomechanical treatment) easier and the quality of the resulting alloy is improved [12].

- the addition of micrometric yttria particles did not come up to the expectations of a grain refinement achieved with the UST via an enhanced nucleation. The idea was based on a concept of inoculation by oxide particles, which should act in this case as additional nucleation centres activated by the UST of the melt [13].

- the mechanical properties (yield stress, ultimate tensile strength, elongation to fracture) of the hot-rolled Fe–40Al–0.1Zr–0.13B alloy were improved by addition of micrometric yttria particles. The hardening mechanism in the UST alloy will be a subject of a further study.

ACKNOWLEDGMENTS

The authors would like to thank Grant Agency of the Czech Republic for the support of the grant 106/06/0019, the research was also a part of the projects MSM0021620834 and MSM 6198910015 financed by the Ministry of Education of the Czech Republic.

REFERENCES

1. S.C. Deevi and V.K. Sikka, *Intermetallics* **4**, 357 (1996).
2. C.G. McKamey, in *Physical Metallurgy and Processing of Intermetallic Compounds*, edited by N.S. Stoloff and V.K. Sikka, (Chapman and Hall, New York, 1996) pp.351.
3. P. Lejček and A. Fraczkiewicz, *Intermetallics* **11**, 1053 (2003).
4. H. Skoglund, M. Knutson and B. Karlsson, *Intermetallics* **12**, 977 (2004).

5. W.J. Zhang, R.S. Sundar and S.C. Deevi, *Intermetallics* **12**, 893 (2004).
6. E. Arzt, R. Behr, E. Göhring, P. Grahle and R.P. Mason, *Mater. Sci. Eng.* **A 234–236**, 22 (1997).
7. W.J. Zhang, R.S. Sundar and S.C. Deevi, *Intermetallics* **12**, 893 (2004).
8. C. Garcia Oca, M.A. Muñoz-Morris and D.G. Morris, *Intermetallics* **11**, 425 (2003).
9. D.G. Morris, J. Chao, C. Garcia Oca and M.A. Muñoz-Morris, *Mater. Sci. and Eng.* **A339**, 232 (2003).
10. G. Ji, T. Grosdidier, N. Bozzolo and S. Launois, *Intermetallics* **15**, 108 (2007).
11. R. Baccino, D. San Filippo, F. Moret, A. Lefort and G. Webb in *Proceedings of the PM'94, Powder Metallurgy World Congress, Paris, June 1994, Vol. II*, (Editions de Physique, Les Ulis, France, 1994) pp.1239.
12. V. Šíma, P. Kratochvíl, P. Kozelský, I. Schindler and P. Hána, *Int. J. Mater. Res.*, (in press)
13. G.I. Eskin, *Ultrasonic Treatment of Light Alloy Melts*, (Gordon and Breach Science Publishers, 1998).

Titanium Aluminides I—Physical Metallurgy, Processing and Properties

Mater. Res. Soc. Symp. Proc. Vol. 1128 © 2009 Materials Research Society   1128-U03-01

Solidification of TiAl-Based Alloys

Ulrike Hecht[1], D. Daloz[2], J. Lapin[3], A. Drevermann[1], V.T. Witusiewicz[1] and J. Zollinger[1]

[1]Access e.V., Intzestr. 5, 52072 Aachen, Germany (email: u.hecht@access.rwth-aachen.de)
[2]LSG2M, Ecole des Mines de Nancy, 54042 Nancy Cedex, France
[3]Institute of Materials and Machine Mechanics, Slovak Academy of Sciences, Racianska 75, 831 02 Bratislava 3, Slovak Republic

ABSTRACT

Titanium aluminides containing high niobium additions emerged as an attractive alloy family for automotive and aero-engine applications. Their processing by near net shape casting is rather demanding, not only due to easy contamination but also due to the fact that microstructure formation during solidification and subsequent solid state transformations sensitively depends on alloy composition and the applied processing conditions. The sequence of phase formation during solidification of the ternary alloy Ti-45Al-8Nb was analyzed based on non-equilibrium thermodynamic calculations and solidification experiments. This alloy solidifies completely via the β(Ti) phase, because the nucleation undercooling for the α(Ti) phase is high enough to prevent its formation. Thermodynamic calculations and experiments are shown to converge at last, with only minor improvements being necessary to correctly describe the metastable eutectic reaction "Liquid $\rightarrow \beta$(Ti) + γ-TiAl" that occurs at the end of the solidification path.

INTRODUCTION

Investment casting of TiAl-based alloys proved to be an attractive processing route for production of near net shape aero-engine and automotive parts. Worldwide efforts are being made to increase not only the reliability of casting processes and the quality of cast parts, but also the size of castings: large aero-engine blades and also industrial gas turbine blades are at the heart of today's developments. Within the European integrated project IMPRESS [1], the R&D activities related to TiAl processing equally push in this direction. One of the central research topics relates to microstructure formation during solidification. Of prime interest are the morphological features of the mushy zone, the segregation of alloying elements and the final grain size and texture of the as solidified phases.

When aiming to analyze phase formation and microsegregation during solidification of 3rd and 4th generation TiAl-based alloys, it is necessary to first examine the thermodynamic background on which to tackle solidification kinetics. The purpose of this paper is therefore focused on comparing thermodynamic calculations to solidification experiments for selected alloys from the ternary Ti-Al-Nb core system and to outline future research directions. The alloy composition range of interest is 44 to 46 at.% aluminum and around 8 at.% niobium.

THERMODYNAMIC BACKGROUND

For all aspects of solidification, the thermodynamic background is given by the phase equilibria that involve the liquid phase. From the recent thermodynamic description of phase equilibria in the Ti-Al-Nb system, proposed by Witusiewicz et al. [2], few things are taken up here in more detail. Compared to [2] minor modifications of the model parameters for the α(Ti) phase and the ordered β_0(Ti) phase were implemented, as to achieve an optimized description.

Figure 1 depicts the projection of the liquidus and solidus surfaces in the region of interest for TiAl-based alloys. These surfaces, and especially the course of the univariant peritectic reaction "Liquid + β(Ti) $\rightarrow$ α(Ti)", have long been at debate [3,4]. The thermodynamic description used throughout this paper, seems to be the most reliable in this respect. Recently, Shuleshova et al. [5] showed that the primary phase boundaries identified by synchrotron X-ray diffraction on levitated droplets well correspond to the calculated ones.

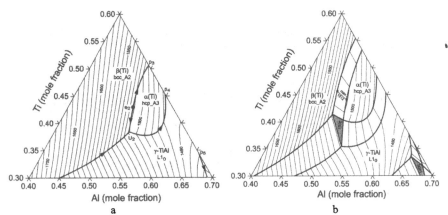

a   b

Figure 1. Projection of the liquidus (a) and the solidus(b) surfaces onto the Gibbs triangle of Al-Nb-Ti in the region of interest to TiAl-based alloys. Most relevant for TiAl-based alloys is the peritectic univariant reaction "Liquid + β(Ti) $\rightarrow$ α(Ti)" that extends from both sides of the quasi-binary eutectic point e_2 towards the peritectic nonvariant reaction of the binary Ti-Al system (p_3) and the quasi-peritectic or U-type reaction U_2, respectively.

Two features of the above thermodynamic description are important for phase formation during solidification: first, the cubic body centered β(Ti) phase is extended by niobium additions towards higher aluminum contents. In thermodynamic equilibrium conditions, alloys with 8 at % Nb and up to 46.4 at.% Al are consequently composed of single phase β(Ti) in the temperature range between solidus and beta-transus. Second, the peritectic reaction "Liquid + α(Ti) $\rightarrow$ γ-TiAl" from the binary system Ti-Al changes to the univariant eutectic reaction "Liquid $\rightarrow$ α(Ti) + γ-TiAl" for all Nb contents higher than 5 at.%. Experimental data on the solidification path of alloy Ti-54.4Al-8Nb confirm this and will be discussed elsewhere. The isopleth for 8 at.% Nb displayed in figure 2a compared to the binary Ti-Al phase diagram in figure 2b illustrate the above described features.

Both are relevant for phase formation during solidification in non-equilibrium conditions, since the liquid can reach compositions above 50 at.% Al due to microsegregation. Correspondingly the Nb content can decrease well below 5 at.%.

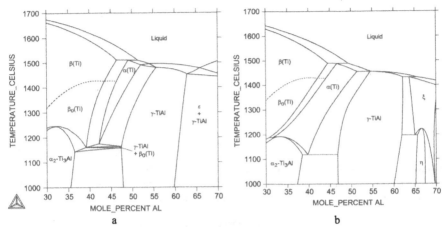

Figure 2. The calculated isopleth for Nb=8at.% (a) compared to the binary Ti-Al phase diagram (b) from [6] shows the β(Ti) and the ordered β₀(Ti) phase being stabilized by Nb. Accordingly the phase transformation from β(Ti) and β₀(Ti) to α(Ti) is shifted to higher Al-content. The isopleth section (a) also shows the univariant eutectic reaction "Liquid → α(Ti) + γ-TiAl"

Solidification paths, following either the lever rule or the Scheil-Gulliver model [7] were computed for alloys Ti-xAl-8Nb with x from 45 to 48at.% using the software ThermoCalc [8] and the above database for the ternary system Ti-Al-Nb, as shown in figure 3a. The results are explained for alloy Ti-45Al-8Nb only, being displayed in figure 3b: for lever rule conditions the alloy completes solidification via the β(Ti) phase within a narrow solidification interval that extends from 1581°C to 1531°C. For Scheil-Gulliver conditions, the peritectic reaction is met at around 1511°C, followed by growth of α(Ti) from the liquid. Even the second peritectic reaction, followed by growth of γ-TiAl is reached in the final stage of solidification. The solidification interval extends to 80 K, with solidus being as low as 1460°C. Also included in figure 3b is a Scheil calculation performed with the α(Ti) phase being suspended, a situation that could occur if the nucleation undercooling for the hexagonal close packed α(Ti) on the cubic body centered β(Ti) is high. In these conditions solidification continues with β(Ti) along the metastable path marked with an asterix and finishes with γ-TiAl forming in metastable eutectic reaction along with a small amount of β(Ti). Metastable path calculations must however be treated with special care. Quite often a fine-tuning of a thermodynamic description is required to yield reliable information on metastable equilibria.

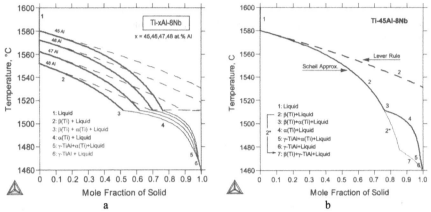

Figure 3. Evolution of fraction solid as function of temperature for different alloy compositions (a) from lever rule (dashed lines) and Scheil calculations (solid lines). For alloy Ti-45Al-8Nb (b) the Scheil calculation was also performed with α(Ti) suspended (thin line).

The evolution of fraction solid with temperature is associated to the evolution of phase composition, which reflects the different partition of the alloy components between the liquid and the growing β(Ti).Table I gives an overview of partition coefficients between β(Ti) and liquid in alloy Ti-45Al-8Nb that were extracted for selected temperatures from the Scheil calculation. The temperature of 1511°C corresponds to the onset of α(Ti) growth.

Table I. Partition coefficients of alloy components between β(Ti) and liquid in Ti-45Al-8Nb (at.%)

Temperature [°C]	Fraction β(Ti) [mole fraction]	Partition coefficients between β(Ti) and liquid		
		k_{Al}	k_{Nb}	k_{Ti}
1580.0	0.11	0.923	1.154	1.049
1530.0	0.67	0.916	1.170	1.067
1511.0	0.76	0.913	1.179	1.075

Figure 4 depicts the evolution of phase compositions in alloy Ti-45Al-8Nb for the alloying element Al as corresponding to the phase fraction evolution shown in figure 3b for the Scheil calculations with and without α(Ti) formation. In the absence of α(Ti) the β(Ti) phase grows as a metastable phase, marked as β(Ti)* within a composition range that extends above 46 at.% and up to 48 at.% Al. Correspondingly the Nb and Ti content in β(Ti)*, both with partition coefficients larger than unity would extend below 7.5 at.% Nb and below 46.4 at.% Ti, respectively. In the final stage of solidification metastable γ-TiAl is predicted to form in eutectic reaction.

Solidification experiments can be used to verify these calculations with respect to the sequence of phase formation and to the associated evolution of the chemical composition of the phases: solidification path analysis is commonly applied to this end.

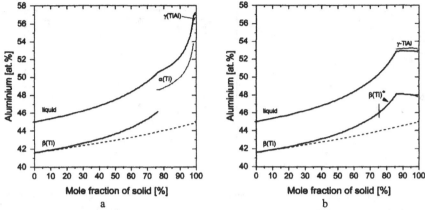

Figure 4. Evolution of the aluminium content in the different phases formed along a Scheil solidification path calculated for alloy Ti-45Al-8Nb with α(Ti) being present (a) and α(Ti) being suspended from the calculation. The dashed line corresponds to the lever rule calculation.

As will be discussed below, all solidification experiments with Ti-45Al-8Nb, but also Ti-46Al-8Nb [9], are completed without α(Ti) formation. For low cooling rates and hence for experimental conditions close to thermodynamic equilibrium this is no surprise. More interesting are experiments performed at high cooling rates: they should allow identifying the range of metastable β(Ti)* growth and hence estimating the nucleation undercooling for α(Ti).

UNIDIRECTIONAL SOLIDIFICATION EXPERIMENTS

Unidirectional solidification in Bridgman type furnaces is the method of choice, whenever attempting to analyze phase formation and microsegregation. Because growth occurs in a unidirectional temperature gradient imposed along the samples longitudinal axis, the mushy zone establishing in between the liquidus and solidus temperature will be localized in a specific region of the samples longitudinal axis. This mushy zone can be conserved for microstructure analysis by an efficient quenching operation at the end of a solidification experiment. We first describe the experimental challenges inherent to unidirectional solidification of TiAl-based alloys and then proceed to the results of microstructure and microsegregation analysis.

Alloy contamination is of concern whenever processing of TiAl-based alloys involves prolonged contact between the liquid and a ceramic crucible. A series of Bridgman experiments were carried out by Lapin et al. [9], aiming to investigate the contamination level that results from the contact between superheated liquid Ti-45.9Al-8Nb (at.%) alloy and densely sintered yttria crucibles supplied by Custom Technical Ceramics, Inc.

The oxygen content δ was measured by hot extraction in a LECO TC-436 N/O analyzer as function of the reaction time for different levels of superheat. The kinetics of oxygen pickup relative to the initial value of $δ_0 = 500$ wt.-ppm before processing was found to follow a power law with thermal activation. Figure 5 summarizes these results, while the mechanism behind the reaction will be discussed elsewhere.

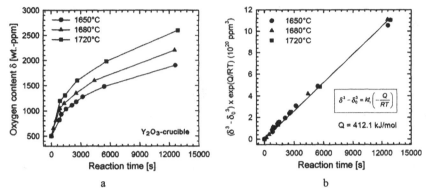

a   b

Figure 5. Dependence of the oxygen content on the reaction time between liquid Ti-46Al-8Nb (at.%) and yttria crucibles for different maximum temperatures of the liquid (a) The oxygen pickup relative to 500 wt.-ppm in the virgin material follows a thermally activated power law with the reaction constant $k=8.705 \times 10^{16}$.

The results show that yttria is not an inert containment for liquid Ti-Al-Nb alloys, but if contact durations are below 40 minutes, contamination levels below 1500 wt.-ppm can be achieved even when superheating amounts to 150 K.

A different set of experiments was performed at Access e.V. for microstructure and microsegregation analysis. Experiment durations were reduced as much as possible by choosing short samples, relatively high growth rates and short solidification lengths, just enough to reach steady state growth. The vertical Bridgman-Stockbarger furnace used for unidirectional solidification is schematically depicted in figure 6a. The furnace consists of a heating zone on top and a cooling zone with liquid metal inside a water cooled copper jacket at the bottom. The two zones are separated by a thermal insulation zone. The sample of 7.5 mm diameter and 100 mm length is mounted in the centre axis of the furnace, inside an yttria crucible. With heater temperatures set to1750°C and cooling bath temperatures around 30°C, the sample is molten during downward movement of the furnace and solidified during upward movement of the furnace. After a solidified length of about 40 mm the sample is quenched by a pneumatically activated movement of the furnace that brings the hot zone of the sample into the cooler. Three solidification experiments with an alloy of nominal composition Ti-45Al-8Nb were carried out at different furnace velocities, e.g. $v_1 = 33\ \mu ms^{-1}$, $v_2 = 46\ \mu ms^{-1}$, and $v_3 = 83\ \mu ms^{-1}$. The integral chemical composition of the samples was analyzed individually in the quenched part by Inductively Coupled Plasma Optical Emission Spectrometry (ICP) for the main elements and by hot extraction for oxygen. Aluminum contents ranging from 44.5 to 45.0 at.% and niobium contents ranging from 8.0 to 8.3 at.% confirmed the nominal values. Oxygen levels ranged from 1200 to 1565 wt.-ppm, being lowest for highest solidification velocity. We assume that the contamination effects on the microsegregation of the nominal alloying elements can be neglected at these contamination levels.

Figure 6b displays the temperature profile achieved inside a test sample in the temperature interval of interest. The temperature data were measured during a dedicated solidification experiment by a thermocouple placed inside the test sample, with solidification being performed at a velocity of v = 33 μms⁻¹. This temperature profile is assumed to vary only little with the solidification velocity in the range from 33 to 83 μms⁻¹. It was therefore used for all processed samples to estimate the experimentally observed solidus temperature using as an input the estimated temperature of dendrite tips and the measured length of the mushy zone. For the range of solidification velocities applied, dendrite tip undercooling values were estimated from marginal stability calculations to reach 6.5 to 9 K.

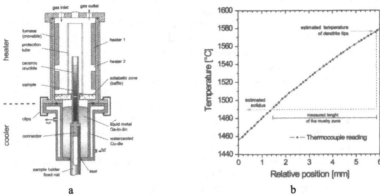

a   b

Figure 6. Scheme of the vertical Bridgman furnace with liquid metal cooling used for unidirectional solidification (a) and the temperature profile achieved along the longitudinal axis of a test sample in the range of the solidification interval (b).

Figure 7 shows the microstructure obtained in the sample Ti-44.7Al-8.2Nb (at.%) after unidirectional solidification at v = 46 μms⁻¹ in the temperature gradient G ≅ 2*10⁴ K/m. Similar microstructures were obtained at v = 33 μms⁻¹ and v = 83 μms⁻¹, respectively.

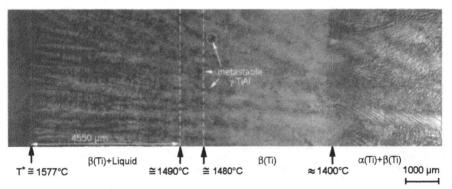

Figure 7. Micrograph of the mushy zone in Ti-44.7Al-8.2Nb (at.%) after unidirectional solidification at v = 46 μms⁻¹ and G ≅ 2*10⁴ K/m. The sharp brightness change around 1400°C is an artefact from stitching together a series of backscatter electron images.

85

The mushy zone was first characterized with regard to the solidification interval and the phases formed, the results being marked in figure 7. The dendrite tip temperature of 1577°C was estimated from the known liquidus temperature of the alloy (1585°C) and the estimated dendrite tip undercooling of 8 K. Solidus was identified to range in between 1490 and 1480°C and needs a special remark: The value of 1490°C corresponds to the length of 4550 μm and represents the position at which the aluminium content of the interdendritic, quenched liquid no longer increases. This position was determined from a series of line scans measured by Energy Dispersive Spectrometry (EDS) in the Scanning Electron Microscope (SEM) type Gemini 1550, the individual lines being set parallel to the solidification front at different positions in the mushy zone. However, solidification is completed only around 1480°C, with the metastable γ-TiAl growing from the last liquid. A small amount of metastable γ-TiAl, estimated to less than 5%, was identified by Electon Backscatter Diffraction (EBSD) even below 1480°C, but it quickly dissolves. In between 1490 and 1480°C the metastable γ-TiAl is the major phase obtained upon quenching the interdendritic liquid, this is why the composition of the quenched liquid remains virtually constant (compare fig. 4b).The average phase composition of the metastable γ-TiAl was measured by EDS to read 4.5 at.% Nb and 52.5 at.% Al. No α(Ti) phase was detected in the grown solid, the solidification path ending with mainly single β(Ti) and a small amount of metastable γ-TiAl.

All three factors, the absence of α(Ti), the presence of metastabile γ-TiAl and the large solidification interval, are indicative of solidification proceeding close to Scheil conditions, with α(Ti) formation being suppressed due to high nucleation undercooling.

To estimate the nucleation undercooling for α(Ti) from this experiment requires however additional information about the composition of the β(Ti) phase inside the mushy zone, specifically a prove that β(Ti) grows along its metastabile prolongation as β(Ti)*, as discussed in figure 4b. Microsegregation measurements are commonly used to analyze the evolution of phase composition along the mushy zone of unidirectional solidified samples. For TiAl samples in general this technique can be applied, but has to meet two challenges: First, as shown in figure 8, the bcc β(Ti) phase cannot be retained untransformed, even by efficient quenching.

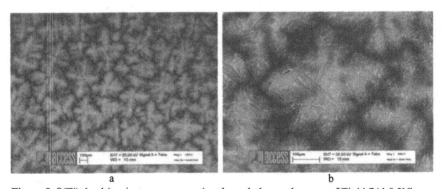

a   b

Figure 8. β(Ti) dendrites in transverse section through the mushy zone of Ti-44.7Al-8.2Nb (at.%) after unidirectional solidification at $v = 46$ μms^{-1} and $G \cong 2*10^4$ K/m (a) show a fine α(Ti) lath structure (b) formed by solid state transformation of β(Ti) to α(Ti) upon quenching.

During quenching the solid state transformation from β(Ti) to α(Ti) starts without being completed, leaving behind a network of fine α(Ti) laths. Recently Zollinger et al. [10] reported on how this substructure with its specific re-distribution of alloying elements introduces a significant scatter of composition profiles measured at the scale of the dendrites. An approach to re-calculate the composition profile of the β(Ti) phase before quenching is presently being developed based on knowledge of the tie-lines for β(Ti)-α(Ti) equilibria at temperatures below beta-transus. Results are however pending.

Second, a sluggish trace of macrosegregation was identified over the entire mushy zone, being attributed to thermo-solutal convection: From dendrite tips to the ground the average Al content decreases from 45.0 to 44.2 at.%, while Ti increases from 47.0 to 47.5 and Nb from 8.0 to 8.3 at%. These differences, though small, have to be accounted for during the detailed analysis of microsegregation, but are neglected in the following section for sake of clarity.

MICROSEGREGATION ANALYSIS

Microsegregation analysis was performed by a grid measurement technique that is more appropriate to capture microsegregation within a dendritic structure than line scans. Literature is rich with respect to the statistic analysis of the acquired data, to possible artefacts and to best ways of defining the measurement points, both their total number and position [11-14]. While the principle of the method is simple, the statistic theory behind it and the experimental fine-tuning, are less so.

Grid measurements were carried out in several transverse sections prepared from cutting the mushy zone at definite positions from tip to bottom. The local chemical composition data were acquired by EDS from a large number of individual spots equally spaced inside predefined grids. Grid sizes of 1000x1000 μm and a step size of 25 μm were used. One measurement is virtually an automated overnight job at the SEM. The EDS-spectra were quantified against an own Ti-44Al-8Nb-0.2C-0.2B standard, previously characterized by Elastic Recoil Detection Analysis (ERDA) at the Hahn-Meitner-Institute, Berlin. Measured datasets were subject to an adequate sorting procedure that reflects the expected segregation trend with increasing Al but decreasing Nb. For the TiAl-alloy discussed here, the datasets were sorted in ascending order following the difference between the Al and Nb content. Once in the sorted order, each point was assigned its cumulative fraction. Compositions potted against the cumulative fraction can easily be compared to thermodynamic calculations of phase composition as function of fraction solid.

Representative results are shown in figure 9 for two transverse sections that correspond to positions deep in the mushy zone, being quenched from 1510°C and 1500°C, respectively. The fraction of solid β(Ti) grown to the point of quenching was estimated to range around 85-90% at 1510°C and around 90-95% at 1500°C. For these conditions, e.g. below the peritectic temperature of 1511°C, the Scheil calculation with α(Ti) suspended predicts growth of the metastable β(Ti)*. Indeed the measured β(Ti) compositions yield Al-levels up to 47 at.% Al and down to 7 at.% Nb that well correspond to the calculated compositions of β(Ti)*. The differences between the calculated and the measured profiles in the region of high fractions of solid relate to the metastable γ-TiAl phase formed during quenching of the interdendritic liquid: the EDS-grids used were too coarse to correctly capture the fraction and composition of this phase. The thermodynamic description of the metastable γ-TiAl phase equally needs to be optimized. This is presently being done based on experimental data on the true, local composition of γ-TiAl. Improving both, the grid measurement and the database description will yield an excellent agreement between thermodynamic calculations and experiments in the region of high fraction of solid.

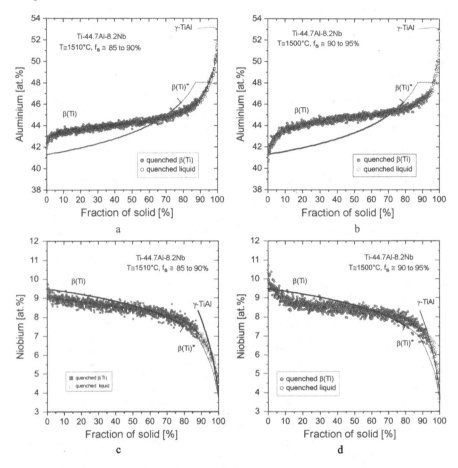

Figure 9. Experimental and calculated microsegregation profiles for the alloying components Al (a and b), and Nb (c and d) in the deep mushy zone of Ti-44.7Al-8.2Nb (at.%) solidified with 46 μms^{-1} under a temperature gradient of $2*10^4$ K/m. The calculated curves (solid lines) correspond to a Scheil calculation with α(Ti) being suspended. The transition from stable β(Ti) to metastable β(Ti)* is marked with a thin line, perpendicular to the curves.

In the region of low fractions of solid the measured composition profiles are indicative of significant back-diffusion for Al, but only limited back-diffusion for Nb. Future work is necessary to analyze the amount of back-diffusion, taking into account the above mentioned, small effect of thermo-solutal convection but also the effect of dendrite tip undercooling on the composition of the first solid formed. Irrespective of future refinements, the experimental results obtained so far shed some light on the nucleation undercooling for the α(Ti) phase: α(Ti) does not form along the solidification path of alloy Ti-44.7Al-8.2Nb, instead β(Ti) continues to grow as metastabile β(Ti)* well below the temperature of the peritectic reaction, e.g. well below 1511°C. Solidification ends with formation of metastable γ-TiAl from the last solidifying liquid, with solidus temperature as low as 1480°C. From these observations, the nucleation undercooling for α(Ti) on β(Ti) may be estimated to exceed 25 K. The high undercooling level implies that growth of α(Ti) rather than repeated nucleation should be observed in alloys that exhibit peritectic reaction and can explain why α(Ti) grains are often large and elongated [15], making it easy to achieve melt texturing. It may be speculated that the high nucleation undercooling for α(Ti) on β(Ti) relates to the Burgers orientation relationship[16] that must be established between the two phases and that nucleation of the peritectic α(Ti) at the interface between β(Ti) and liquid occurs in fact inside the solid β(Ti).

CONCLUSIONS

In this contribution we aimed to give an account of ongoing research activities related to solidification of TiAl-based alloys. The main purpose was to check, if the recently published thermodynamic database for the ternary system Ti-Al-Nb can serve as a solid background on which to tackle solidification kinetics. Solidification path analysis of unidirectionally solidified samples was selected as the method of choice for this purpose. Bridgman solidification of Ti-45Al-8Nb and Ti-46Al-8Nb in yttria crucibles was shown to yield reasonably low contamination levels, with room for further improvements.

First results from comparing experimental and calculated solidification paths for alloy Ti-45Al-8Nb are judged quite encouraging: thermodynamic calculations and solidification experiments converge at last, with some fine-tuning being necessary to achieve an excellent agreement. This fine-tuning encompasses the correct description of the metastable β(Ti) / γ-TiAl phase equilibria on the thermodynamic side and a refinement of microsegregation analysis on the experimental side. Once this is done, the thermodynamic database can serve as reliable input to dedicated microstructure simulation models.

Future directions of research may be outlined in three main areas:

(i)   Solidification experiments may be directed to alloys that form α(Ti) in the mushy zone, in order to understand more about the competition between nucleation and growth of α(Ti). Alloying additions that reduce the nucleation undercooling of α(Ti) on β(Ti) may be envisaged, but also boron additions known to promote nucleation of α(Ti) on borides rather than on β(Ti).

(ii)   Thermodynamic database development will continue from ternary to higher order systems, ideally leading to a comprehensive database like for steels or Ni-base superalloys.

(iii)   Finally, element diffusivities in the disordered β(Ti) phase and in the liquid phase are virtually unknown, but important input data for solidification models. Any progress in this field would be highly welcome.

ACKNOWLEDGEMENTS

This work was financially supported by the EC Integrated Project IMPRESS "Intermetallic Materials Processing in Relation to Earth and Space Solidification" under the contract No. NMP3-CT-2004-500635. For specific aspects of solidification path analysis we gratefully acknowledge the support by the Deutsche Forschungsgemeinschaft (DFG) within the SPP No. 1296 programme "Heterogeneous Nucleation and Microstructure Formation". Special thanks are due to Dr. Strub and Dr. Bohne from the Hahn-Meitner-Institute in Berlin, for having characterized the TiAl-standard used for quantifying EDS-spectra by ERDA.

REFERENCES

[1] D.J. Jarvis, D. Voss, *Mater. Sci. Eng. A* **413–414** (2005) 583–591.
[2] V. Witusiewicz, A. Bondar, U. Hecht, S. Rex, T. Velikanova, *J. Alloys Comp.* (in press), doi:10.1016/j.jallcom.2008.05.008
[3] C. Servant and I. Ansara, *Ber. Bunsenges. Phys. Chem.* **102** (1998), pp. 1189–1205.
[4] N. Saunders in: I. Ansara, A.T. Dinsdale and M.H. Rand, Editors, *COST 507: Thermochemical Database for Light Metal Alloys* **vol. 2**, European Communities, Brussels and Luxembourg (1998), pp. 342–345.
[5] O. Shuleshova, D. Holland-Moritz, W. Löser, A.Voss, H. Hartmann, U. Hecht, V. T. Witusiewicz, D. M. Herlach, B. Büchner, submitted to Acta Mat.
[6] V.T. Witusiewicz, A.A. Bondar, U. Hecht, S. Rex, T.Ya. Velikanova, *J. Alloys Comp.* **465** (2008) 64–77.
[7] E. Scheil, *Z. Metallkd.* **34** (1942) 70.
[8] B. Sundman, B. Jansson and J.-O. Andersson, *CALPHAD* **9** (1985) 153–190.
[9] J. Lapin, Tech. Rep., *IMPRESS* (2007).
[10] J. Zollinger, V. Witusiewicz, A. Drevermann, D. Daloz, U. Hecht, Int. *J. Cast. Met. Res.*, in press.
[11] M. N. Gungor, *Metall.Trans.* **20 A** (1989) 2529–2533.
[12] A. Hazotte, J. Lecomte, J. Lacaze, *Mater. Sci. Eng. A* **413–414** (2005) 223–228.
[13] M. Ganesan, D. Dye, P. Lee, *Metall. Mat. Trans.* **36 A** (2005) 2191–2203.
[14] A. Hazotte, J. Lacaze, *Trans. Indian. Inst. Met.* **60** (2007) 267–271.
[15] U. Hecht, V. Witusiewicz, A. Drevermann, J. Zollinger, *Intermetallics* **16** (2008) 969–978.
[16] W.G. Burgers, *Physica* **1** (1934) 561–86.

Mater. Res. Soc. Symp. Proc. Vol. 1128 © 2009 Materials Research Society   1128-U03-02

Grain Refinement of γ-TiAl Alloys by Inoculation

Daniel Gosslar[1], Robert Günther[1], Christian Hartig[1], Rüdiger Bormann[1], Julien Zollinger[2], Ingo Steinbach[3]

[1] Institute of Materials Science and Technology, Hamburg University of Technology, Eissendorfer Str. 42, D-21073 Hamburg, Germany

[2] RWTH-Aachen, Access e.V., Intzestr. 5, D-52072 Aachen, Germany

[3] Interdisciplinary Centre for Advanced Materials Simulation, Ruhr-University Bochum, Stiepeler Strasse 129, D-44780 Bochum, Germany

ABSTRACT

A significant grain refinement is achieved for a Ti-45Al (at.%) alloy via the addition of a novel Ti-Al-TiB$_2$ master alloy. This refinement is attributed to a low lattice mismatch of TiB$_2$ with the primary solidifying phase and its thermodynamic stability. Model calculations allow for the prediction of the grain size as a function of cooling rate, alloy constitution and particle size distribution and content. However, the model predictions are smaller than the experimentally observed grain size since the fraction of effective particles is overestimated.

INTRODUCTION

γ-TiAl alloys combine low specific weight and good strength properties at elevated temperatures up to 800 °C [1]. The use of γ-TiAl alloys instead of much heavier conventional high temperature materials in aircraft engines results in economical and ecological advantages [2]. Albeit the casting process, microstructure control during casting of γ-TiAl based alloys is essential to ensure homogeneous microstructures and to minimize chemical heterogeneities and casting texture. Indeed, the formation of large lamellar α$_2$(Ti$_3$Al) / γ(TiAl) grains is detrimental for ductility, strength properties and creep resistance [3].

For low boron additions the formation of TiB$_2$ and TiB (Bf) in the solid state leads to significantly refined and homogeneous microstructures in Ti-Al and Ti-Al-Nb alloys, respectively [4, 5]. This refinement is achieved via the nucleation of the α hcp phase on these borides during the bcc β to α transformation [4]. However, a direct refinement of the primary β phase by inoculation would also be beneficial since the morphology of the mushy zone and the associated feeding of solidification shrinkage could be controlled. Until now only little research has been carried out on the refinement of the β phase by inoculation of the melt.

In this work, grain refinement of a Ti-45Al alloy (all compositions are given in at.%) is investigated by inoculation. The first part of the paper is focussed on the determination of a potential boride inoculant particle by lattice mismatch considerations. The boride with the highest grain refining potential has been added to a Ti-45Al melt through an especially designed master alloy. In the second part, the casting microstructure is characterized in terms of inoculant particles and grain sizes. The measured grain size is compared to model predictions that account for cooling rate, alloy composition, particle fraction and particle size distribution.

INOCULANT SELECTION

The efficiency on the final grain size reduction reflects the potency of an inoculant particle [6]. A small crystallographic mismatch between the inoculant phase and the nucleating phase is a major criterion for this effectiveness beside the necessary stability in the melt. Lowering this mismatch reduces the contact angle and hereby increases the potency for heterogeneous nucleation which leads to a good grain refinement. The mismatch between two matching planes can be calculated simply by

$$\delta = \sum_{i=1}^{2} \frac{\left| d_S(u_i v_i w_i) - d_N(u_i v_i w_i) \right|}{d_N(u_i v_i w_i)} \tag{1}$$

where $d_N(u_i v_i w_i)$ is the spacing of atoms along a $u_i v_i w_i$ direction contained in one matching plane of the nucleus and $d_S(u_i v_i w_i)$ is the corresponding spacing for the substrate, the atom spacings are taken from two linear independent directions in each plane represented by the index i. If the symmetry of the matching planes is identical only one direction has to be considered. Table I summarizes the mismatch for the case of α or β nucleation on several titanium borides. For each case the possible minimum mismatch is given. To take the influence of aluminium concentration on the mismatch into account, the Vegaard's law has been applied to determine the atom spacing within the α and β phase for an average concentration of 45 at.% Al.

Table I. Minimum mismatch values for α- or β nucleation on titanium borides determined by equation (1).

Substrate	Space group [7]	Nucleus	Matching Planes, Substrate // Nucleus	Mismatch, %
TiB$_2$	P6/mmm (hexagonal)	α	{0001} // {0001}	7.0
		β	{10$\bar{1}$0} // {100}	4.6
TiB (B27)	Pnma (orthorhombic)	α	{100} // {10$\bar{1}$0}	4.0
		β	{100} // {100}	22.8
TiB (Bf)	Cmcm (orthorhombic)	α	{010} // {10$\bar{1}$0}	18.3
		β	{010} // {100}	5.2
Ti$_3$B$_4$	Immm (orthorhombic)	α	{010} // {10$\bar{1}$0}	17.8
		β	{010} // {100}	4.9

A mismatch below approximately 10 % is reported to indicate a good grain refining inoculant [8]. Thus, table I shows that the selected titanium borides should be either good grain refiners for α (TiB (B27)) or β (TiB$_2$, TiB (Bf)). TiB$_2$ is the only boride with a low mismatch for nucleation of both phases and fulfills the stability criterion; it is stable in a 45 at.% Al containing melt for boron contents higher than 1.5 at.% [9]. The chosen alloy composition in the following is therefore Ti-45Al with TiB$_2$ additions corresponding to an integral composition of Ti-45Al-2B. For this composition thermodynamic calculations predict a stable fraction of 1.0 wt.% TiB$_2$ in equilibrium with the first nucleated β phase.

EXPERIMENT

A Ti-45.4Al+10 wt.%TiB$_2$ master alloy has been prepared in three processing steps to achieve a composite with a homogenous boride distribution: 1. High energy ball milling of TiB$_2$ particles and pure titanium sponge granules in a Fritsch planetary ball mill (type 5). 2. Mixing of this powder blend and aluminium powder using a tumble mixer in a closed container. 3. Compacting the mixed powders to a tablet of 50 mm in diameter and 16 mm in height by a unidirectional press at room temperature. Milling and handling of the powder has been carried out under argon atmosphere in a glove box to avoid contamination by oxidation or humidity. After the 1st processing step, TiB$_2$ particles are distributed homogeneously in Ti-TiB$_2$ composite particles, which are embedded after the 3rd step in a compact aluminium matrix. This homogeneity is expected to avoid clustering of particles in the melt.

The casting experiment is carried out by induction melting of a Ti-45Al ingot together with the master alloy (Ti-45.4Al+10 wt.% TiB$_2$) and subsequent centrifugal casting into ceramic moulds of 25 mm in diameter. The melt was held at 1505°C (30°C superheat) for two minutes before being cast. The microstructure observation of the master alloy and as cast sample is performed by scanning electron microscopy in back scattered electron (BSE) mode. The boride particles are identified by energy dispersive X-ray spectroscopy (EDX). X-ray diffraction (XRD) measurements show that the present boride structure is TiB$_2$. The TiB$_2$ particle size distribution is determined software assisted, which assigns an equal circle diameter (ECD) to each particle.

MODELING

The initial β grain size is predicted according to the model introduced by Greer et al. [10] for aluminium alloys. The iterative numerical solution is carried out according to Guenther et al. [11]. The growth parameter S (defined in [10]) is approximated as a linear function of temperature for the nominal alloy composition. Table II summarizes the thermodynamic input parameters for the model calculations.

Table II. Input parameters for model calculations of Ti-45Al.

Quantity	Symbol	Units	Value	Reference
Enthalpy of fusion per unit volume	ΔH_V	J/m^3	1.30×10^9	[9]
Entropy of fusion per unit volume	ΔS_V	J/K m^3	7.25×10^5	[9]
Heat capacity of melt per unit volume	c_{PV}	J/K m^3	1.63×10^6	[9]
Solid-liquid interface energy	$\sigma_{S/L}$	J/m^2	0.1	[12]
Diffusivity in the melt	D_L	m^2/s	2.5×10^{-9}	[13]

RESULTS

Figure 1 shows the as cast microstructure. The microstructure is fully lamellar with a rather homogenous lamellar grain (colony) size. The BSE micrograph shows that every lamellar grain is surrounded by an aluminium enriched zone (dark contrast) which stems from enrichment of the liquid during the solidification. The observations also reveal that the initial equiaxed β grains (dendrites) have a similar grain size than the lamellar colony size (one lamellar orientation

93

per β dendrite). The measured grain size is in the range of 114 μm for an estimated cooling rate during solidification of 5 to 8 K/s.

The TiB$_2$ size distribution in the master alloy and after casting is shown in figure 2. The frequency of particles larger than 2 μm has increased and the size distribution has become broader. This difference is attributed to the change of volume fraction of borides during the casting process. Indeed the fraction of TiB$_2$ decreases upon heating and melting, leading to dissolution of the smallest particles; upon cooling and freezing, the fraction increases again with the growth of the remaining particles.

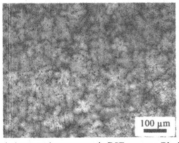

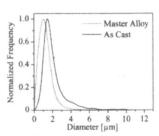

Figure 1. As cast microstructure in BSE contrast. Black dots were identified as TiB$_2$ particles by EDX.

Figure 2. TiB$_2$ particle size distribution in the master alloy (before casting) and after casting (as cast).

Concerning the influence on the model calculations by the different TiB$_2$ particle size distribution before and after casting (figure 2), no significant changes are predicted in the final grain size, or in the effective particle diameter contributing to grain initiation (figures 3 and 4). However, the predicted grain size is 25.8 μm at a cooling rate of 5 K/s; this value is more than 4 times smaller than the experimental one. The reasons leading to this discrepancy are discussed below.

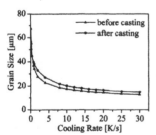

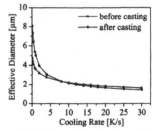

Figure 3. Model predictions of the β grain size as a function of cooling rate considering the TiB$_2$ particle size distribtion before and after casting.

Figure 4. Effective particle diameter contributing to grain initiation considering the TiB$_2$ particle size distribtion before and after casting.

DISCUSSION

The comparison of the measured grain size to literature data shows clear evidence of a grain refinement with TiB$_2$ additions compared to binary alloys. Imayev et al. obtained as cast structures with a grain size of 300 μm in a Ti-44Al alloy with a similar cooling rate as in the casting experiment presented here [14]. For a Ti-45Al alloy Zollinger measured grain sizes of 120 μm in ingots cast with significantly higher cooling rates (15 K/s) [15]. The obtained grain size of 114 μm measured here with a cooling rate of about 5 K/s thus can be considered as refined. However, the simulated grain size predicts a more than 4 times smaller size than the experimental one, whereas the model presented here is able to predict quantitatively the grain size in CP (commercial purity) aluminium [10] and magnesium [11] alloys. The reason for this discrepancy is caused by: 1. The high solute content in the present alloy (45 at.% Al) compared to the alloys of [10] and [11]. The model calculations neglect the solute diffusion prior to the attainment of free growth. The necessary solutal undercooling ΔT_S to drive this diffusion is strongly sensitive to the alloy constitution and cooling rate [16]. The available curvature undercooling ΔT_C is reduced and fewer particles contribute to grain initiation as predicted by the model. Consequently the final grain size is underestimated. 2. The centrifugal casting process causes a significant convection in the melt. Hereby the solute concentration profile at the solid (β nucleus) liquid interface is changed; the partition coefficient is influenced which leads to a local variation of the growth restriction parameter Q. Following from this a considerable amount of particles has a smaller value of Q causing an increase of the free growth velocity of initiated grains [10]. The final grain size is underestimated.

Another inhibition of grain size refinement in the present cast alloy is caused by the unstable fraction of TiB_2 particles in the melt. The fact that the borides grow upon cooling and solidification has an immediate consequence. It results in titanium depletion and thus aluminium enrichment around the growing boride particles. The local liquidus temperature in the vicinity of the particles is thus lower than the equilibrium one and prohibits β nucleation. However, the limiting factor for growth of the borides is the boron content. Since this element is in insufficient quantity in the melt the effects of the change in particle fraction on the local liquidus temperature are limited. Albeit their morphology, the growth of TiB_2 particles with sizes > 0.5 μm in Ti-Al melts results in large {10-10} and {0001} facets [17]. These planes are the most favourable to nucleation of both the α and β phase according to table I.

CONCLUSIONS

The work presented in this paper has lead to the following conclusions:
1. Based on lattice mismatch analysis and thermodynamic stability, TiB_2 is found to be a promising inoculant for both the β and α phase in a Ti-45Al alloy. A new Ti-45.4Al+10wt.% TiB_2 master alloy has been designed and prepared by planetary ball milling.
2. A significant grain refinement is achieved via the addition of this master alloy to a Ti-45Al melt compared to an unrefined casting.
3. The fraction of particles added to the melt changes slightly during the casting process; the effects on the size distribution are small, but need further investigations with respect to the refinement process.
4. The predicted grain size according to the present model calculation underestimates the experimentally measured grain size; it is found that the solutal undercooling associated to growth of nuclei has to be taken into account as well as the effect of convection in the melt in order to perform more quantitative predictions in Ti-45Al alloys.

ACKNOWLEDGEMENTS

This work is financially supported by the German Research Foundation (DFG) within the SPP No. 1296 programme. We thank U. Hecht for valuable discussions. Dr. F.-P. Schimansky has supplied generously titanium sponge granules.

REFERENCES

[1]   H. Kestler and H. Clemens in *Production, Processing and Application of g(TiAl)-Based Alloys, Titanium and Titanium Alloys*, edited by C. Leyens and M. Peters, (Wiley-VCH, Weinheim Germany, 2003) p. 35.
[2]   W. Smarsly and L. Singheiser, *J. Mat. Adv. Power Eng.*, Part II, 1731 (1994).
[3]   Y. Kim and D. Dimiduk, in *Structural Intermetallics*, ed. M. Nathal *et al.* (TMS), Warrendale, PA, 1997) p. 531.
[4]   U. Hecht et al., *Intermetallics* **16**, 969 (2008).
[5]   J. Zollinger et al., presented at the DFG Summer School within the SPP No. 1296 Programme, Herzogenrath, Germany, 2008 (unpublished).
[6]   M. Flemings, Solidification Processing, (McGraw-Hill, New York, 1974), p. 299.
[7]   U. Kitkamthorn et al., *Intermetallics* **14**, 759 (2006).
[8]   J. Reynolds and C. Tottle, *J. Inst. Metals* **80**, 1328 (1951).
[9]   V. Witusiewicz et al., *J. Alloy Comp.*, in Press (2008).
[10] L. Greer et al., *Acta. Mater.* **48**, 2823 (2000).
[11] R. Guenther et al., *Acta. Mater.* **54**, 5591 (2006).
[12] I. Egry et al., *Int. J. Thermophys.* **28**, 1026 (2008).
[13] V. Eremenko et al., *Fizika Khim. Mekh. Mater.* **14**, 3 (1978).
[14] R. Imayev et al., *Intermetallics* **15**, 451 (2007).
[15] J. Zollinger, PhD. Thesis, Institut National Polytechnique de Lorraine (NPL), (2008).
[16] T. Quested, PhD. Thesis, University of Cambridge, (2005).
[17] M. Hyman et al., *Metall. Trans.* **22A**, 1647 (1991).

Mater. Res. Soc. Symp. Proc. Vol. 1128 © 2009 Materials Research Society   1128-U03-03

Unidirectional Solidification and Single Crystal Growth of Al-Rich Ti-Al Alloys

Anne Drevermann[1], Georg J. Schmitz[1], Günther Behr[2], Elke Schaberger-Zimmermann[3], Christo Guguschev[2]
[1]ACCESS e.V., Intzestr. 5, 52072 Aachen, Germany
[2]Leibniz-Institut für Festkörper- und Werkstoffforschung Dresden, Helmholtzstr. 20, D-01069 Dresden, Germany
[3]Foundry Institute, Intzestr. 5, 52072 Aachen, Germany

ABSTRACT

To investigate the basic mechanical and thermomechanical properties of TiAl alloys with high Aluminium content sufficiently large single crystalline domains are required. To fabricate these samples undirectional solidification in Bridgman Stockbarger furnaces and optical floating zone devices were used. Focus of investigation were grain selection and impurity contamination. Both processes allow for growth of single crystal domains of some millimetres diameter and a few centimetres length. However in a Bridgman Stockbarger furnace the long contact times with the crucible proved detrimental to oxidation issues whereas in the optical floating zone device oxidation is negligible due to containerless processing.

INTRODUCTION

TiAl alloys with high aluminium content are considered as promising candidates for new materials and components. Compared to today's conventional gamma-Titaniumaluminides they offer increased oxidation resistance at high temperatures with simultaneously decreased weight [1].

In previous research single crystal material was produced with float zone technique [2] and travelling solvent / flux method [3] for TEM investigations of long-period superstructures in Al-rich TiAl. In this work unidirectional solidification in Bridgman-Stockbarger furnaces and in floating zone devices was used to generate single crystalline domains with diameters of a few millimetres diameter and a few centimetres length for mechanical and thermomechanical testing.

The objectives of the present research were (i) optimisation of the grain selection process, (ii) minimisation of stresses and strains leading to cracking of the samples and (iii) minimisation of impurities resulting from crucible materials.

Samples were processed in Bridgman-Stockbarger furnaces and floating zone devices. Parameters for the Bridgman-Stockbarger experiments comprised solidification velocity, temperature gradient, crucible material and the use of a self-seed [4]. Furthermore the processing of Al-rich TiAl in a Bridgman-Stockbarger process gives first insights into the problematics of processing in a crucible and provides cooling rates that are nearer to future industrial applications. Floating zone experiments complemented these investigations with a very high temperature gradient and processing without crucible.

EXPERIMENT

Several kilograms of master alloys were produced by Vacuum Arc Remelting. Precursor rods for the directional solidification and the floating zone experiments were produced by remelting these alloys in a cold-wall induction crucible and subsequent centrifugal casting into a permanent Niobium mould [5]. For the present experiments Ti - 60 at.% Al was used.

Goal of the experiments were single crystalline domains large enough for subsequent mechanical and thermomechanical testing [5]. Two phenomena leading to large crystalline domains have been investigated in detail:
  (i)   grain selection during unidirectional growth
  (ii)   seeded solidification in a temperature gradient using self seeds [4].

Bridgman experiments were performed to optimise the grain selection process while minimising oxidation and thermomechanical stress. The influences of solidification velocity, temperature gradient, crucible material, preoxidising treatment and use of a seed sample were investigated. The two Bridgman-Stockbarger furnaces have argon atmosphere and graphite heaters and provide different thermal gradients (Fig. 1a and b).

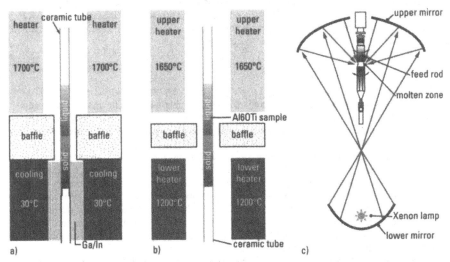

Figure 1. Schemata of the set-ups: **a)** high gradient furnace with a 40 mm baffle between heating and cooling zone, $G \approx 22$ K/mm; **b)** low gradient furnace with a 40 mm baffle between two heating zones, $G \approx 8$ K/mm; **c)** optical floating zone furnace, $G \approx 51$ K/mm.

In the high gradient furnace the solid-liquid interface is positioned in an isolation zone ("baffle") between the heating zone and a Ga/In liquid metal cooling [6]. In the low gradient furnace the baffle is located between two heating zones being operated at different temperatures. In both furnaces solidification proceeds by moving the furnace upwards while the sample is kept fixed. To ensure near steady-state conditions during the directional solidification only the middle part of a sample is directionally solidified while the lower part remains solid ensuring steady contact to the colder zone and the upper part remains in the hotter zone. The samples were 80 to 90 mm long and were processed in cylindrical, densely sintered yttria respectively alumina tubes.

An advanced method to create large single crystalline domains is based on the use of a seed crystal that has been prepared from a large domain resulting from a preceeding grain selection process. This seed sample was placed below an unprocessed sample. The unprocessed sample and part of the seed sample were molten before directional solidification started.

A further objective was the investigation of contamination (specifically oxidation) originating from the crucibles. Standard Alumina crucibles were compared to Yttria crucibles that already have been successfully used for short time exposure during casting of gamma TiAl alloys. In addition a preoxidation treatment was applied to some samples to create a protective oxide skin.

After directional solidification longitudinal sections of the samples were ground, polished and etched with Kroll solution. Thus different grains appear in different colours depending on their orientation when observed with polarised light. In addition SEM and EDX analysis were applied to non etched samples.

Eventually containerless zone melting experiments were performed in two different optical floating zone furnaces URN-Z2 (MPEI Moscow) and SFZ (IFW Dresden) [7, 8] (Fig. 1). Both devices are operating in argon atmosphere and are equipped with 5 kW Xenon lamps. The growth in the floating zone process was carried out without seeds.

Table I summarises the variation of all experimental parameters.

Table I. Variation of experimental parameters in the unidirectional solidification experiments

Sample	sol. velocity [μm/s]	crucible	diameter [mm]	temperature gradient [K/mm]	preoxidation	seed sample
YO_12.5	12.5	Y_2O_3	6.5	22 ± 2	–	–
YO_22.7	22.7	Y_2O_3	6.5	22 ± 2	–	–
YO_33.3	33.3	Y_2O_3	6.5	22 ± 2	–	–
YO_45.8	45.8	Y_2O_3	6.5	22 ± 2	–	–
AlO_45.8	45.8	Al_2O_3	7.5	22 ± 2	–	–
PreOx_45.8	45.8	Al_2O_3	7.5	22 ± 2	X	–
low_grad	45.8	Al_2O_3	7.5	8 ± 2	–	–
seed_50	50	Al_2O_3	7.5	22 ± 2	X	X
FZ_2.8	2.8	–	7.5	51	–	–

DISCUSSION / RESULTS

A series of Bridgman experiments with different solidification velocities show that grain selection improves with increasing velocity (Fig. 2). Respective samples display severe cracking due to the brittleness of the material. To reduce thermomechanical stress being associated with the high temperature gradient one sample was directionally solidified in another furnace with a much lower temperature gradient. This lead to a less efficient grain selection but reduced cracking susceptibility (Fig. 3).

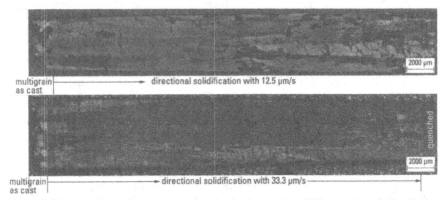

multigrain as cast — directional solidification with 12.5 µm/s

multigrain as cast — directional solidification with 33.3 µm/s

Figure 2. Longitunal cuts showing the grain selection improving with increasing velocity. Both samples show severe cracking due to thermomechnical stresses and as a result of the metallographical preparation.

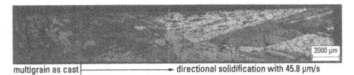

multigrain as cast — directional solidification with 45.8 µm/s

Figure 3. Less efficient grain selection resulting from a lower temperature gradient (Table I).

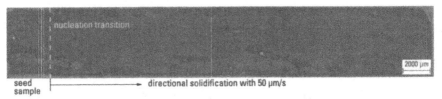

seed sample — directional solidification with 50 µm/s

Figure 4. Sample being nucleated on a previously directionally solidified seed sample. The orientation transfer from the seed sample to the seeded sample has been successful. Large single grain domains of several millimetres width and several centimetres length have been produced.

Densely sintered yttria and alumina crucibles were compared regarding their inertness against the strongly reducing melt of the TiAl alloy. Figure 5 shows the rim of the samples just above the maximum melting position.

In addition to this strongly oxidised rim, Y_2O_3 and Al_2O_3 particles are observed in the bulk material which increase in size and abundancy along the length of the solidified sample, i.e. proportional to the contact time of the melt with the crucible (Fig. 6).

 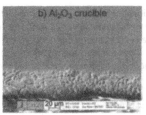 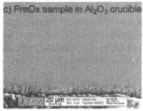

Figure 5. Rim oxidation of the samples near the maximum melting position where the contact time between melt and crucible is minimal and identical for all the samples; **a)** ~ 10 μm oxidised rim; **b)** up to 50 μm oxidised rim; **c)** PreOx sample in Al$_2$O$_3$ crucible, ~ 10 μm oxidised rim, exposed to oxygen atmosphere at 500°C for 24 hours prior to directional solidification.

 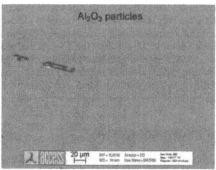

Figure 6. SEM-images revealing oxide formation in the upper regime of the samples. Al$_2$O$_3$ particles are usually smaller than the Y$_2$O$_3$ particles. Moreover processing in both crucible types leads to oxide precipitates on a submicrometre scale.

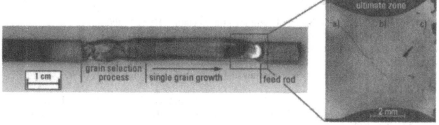

Figure 7. Ti-60 at.%Al single crystal. The grain selection from small to large grains takes place during the first few millimetres of growth. After a length of about 20 mm solidification continues as single crystal. Due to segregation during crystal growth the final growth zone is enriched in Al. The image of the ultimate zone shows **a)** single crystal, **b)** precipitates in the frozen zone, **c)** recrystallization microstructure of the unmolded feed rod displaying orientation contrasts between different grains.

In the containerless directional solidification using the optical float zone devices grain selection towards forming a single crystal takes about 20 mm of solidification length (Fig. 7). The selection process is enhanced by a strong convex curvature of the solid liquid interface. Optical microscopy in polarised light and scanning electron microscopy have verified the single crystallinity with high perfection and with negligible traces of secondary phases in the grown crystal.

CONCLUSIONS

Relatively large domains of single grain material of Ti-60at.%Al can be produced by unidirectional solidification in Bridgman-Stockbarger furnaces. Nucleation on seeding material proved to be beneficial and successful with respect to enhanced grain selection. The highly reducing character of the TiAl melt and a very low solubility of the γ-TiAl phase for oxygen [9] lead to oxidation problems in all crucible materials being investigated including the preoxidised samples. This is due to the long exposure times typical for crystal growth experiments. For possible future applications, e.g. castings of polycrystalline parts and components, with shorter exposure times to the crucibles the investigated crucible materials may be sufficient.

With respect to the fundamental investigations requiring large single crystalline domains however, grain selection during the floating zone process is more effective and leads to high quality single crystals extending over the entire sample diameter and several centimetres length. Oxidation effects in these samples are negligible due to the containerless processing.

ACKNOWLEDGEMENTS

We gratefully acknowledge financial support from the Deutsche Forschungsgemeinschaft (DFG) for funding this work under DFG grant number SCHM323/7-1 within the group research project Al-rich Ti-Al alloys (PAK 19).

REFERENCES

1. M. Palm, L.C. Zhang, F. Stein, G. Sauthoff, Intermetallics **10**, 523 (2002).
2. T. Nakano, K. Hayashi, Y. Umakoshi, Phil. Mag. A, 763, (2002).
3. D. R. Johnson, H. Inui, M. Yamaguchi, Intermetallics **6**, 547 (1998).
4. G.J. Schmitz, H. Weiss, Ch. Wolters, U. Hashagen, B. Zeimetz, Proceedings of the 2nd European Conference on Advanced Materials and Processes, Cambridge, ed. by T.W. Clyne, 89 (1991).
5. M. Paninski, A. Drevermann, G.J. Schmitz, M. Palm, F. Stein, M. Heilmaier, N. Engberding, H. Saage, D. Sturm, in Ti-2007 Science and Technology, ed. by M. Niinomi (Japan Institute of Metals) 1059.
6. A. Drevermann, C. Pickmann, L. Sturz, G. Zimmermann, Proceedings of 2004 IEEE International Ultrasonics, Ferroelectrics and Frequency Control Joint 50th Anniversary Conference, Montreal, 537.
7. A.M. Balbashov, S.K. Egorov, J. Crystal Growth **52**, 498 (1981).
8. D. Souptel, W. Löser and G. Behr, J. Crystal Growth **300**, 538 (2007).
9. A. Menand, H. Zapolsky-Tatarenko. A. Nérac-Partaix, Mater. Sci. Eng. **A250**, 55 (1998).

Mater. Res. Soc. Symp. Proc. Vol. 1128 © 2009 Materials Research Society   1128-U03-04

Development of G4 TiAl Alloys by Spark Plasma Sintering

Houria Jabbar[1, 2], Jean-Philippe Monchoux[1, 2], Marc Thomas[3] and Alain Couret[1, 2]

1 : CNRS ; CEMES (Centre d'Elaboration de Matériaux et d'Etudes Structurales) ; BP 94347, 29 rue J. Marvig, F-31055 Toulouse, France
2 : Université de Toulouse ; UPS ; F-31055 Toulouse, France
3 : DMMP/ONERA, 29 Avenue de le Division Leclerc, BP , 92322 Châtillon Cedex, France

ABSTRACT

G4 alloys ($Ti_{51}Al_{47}Re_1W_1Si_{0.2}$) are developed by Spark Plasma Sintering (SPS) with the aim to improve the creep resistance of SPS materials. The microstructure is analyzed by Scanning and Transmission Electron Microscopies (SEM and TEM). The mechanical properties at low and high temperatures are measured. The addition of heavy elements does not lead to an improvement of the mechanical strength.

INTRODUCTION

Large scattering in the mechanical properties is identified as a main drawback for the use of TiAl alloys in blades of aircraft engines [1]. This is particularly true for casting alloys because of the presence of some structural inhomogeneities which are inherent to the process [2]. On the other hand, powder metallurgy allows a reduction of solidification heterogeneities and a better control of the mechanical properties. In this context, the present work is aimed at exploring the Spark Plasma Sintering (SPS) technique which is a fast and cost effective PM route. It consists of sintering a powder by the simultaneous application of a pulsed direct current (DC) of high intensity and of uniaxial pressure [3, 4].

The first SPS compactions of TiAl alloys were performed in the 90's and they have demonstrated the feasibility of sintering (for a more detailed review see Ref. [5]). These studies led to well compacted alloys which generally suffer of inhomogeneous microstructures. Investigations of mechanical properties have been unfortunately scarcely performed. More recently, a very high strength has been achieved in compression through the fabrication of nanostructured alloys [6]. Our group has recently sintered two prealloyed TiAl powders with the following compositions: a reference GE alloy $Ti_{49}Al_{47}Cr_2Nb_2$ and a boron containing alloy $Ti_{51}Al_{44}Cr_2Nb_2B_1$ [5]. The as-produced microstructures are: $\alpha_2+\gamma$ two phase and duplex, respectively. The $\alpha_2+\gamma$ two phase microstructure is mainly formed by γ grains due to a sintering temperature situated in the $\alpha + \gamma$ two phase field. The formation of the duplex microstructure is attributed to the crossing of the $\alpha + \beta$ two phase field, as a result of the decreasing transformation temperature by lowering the aluminium content. Concerning the mechanical properties, the room temperature behaviour appears to be excellent whereas the creep resistance is limited.

The improvement of the creep resistance is our purpose in the present paper, which leads to sinter a G4 TiAl alloy containing heavy elements: Ti(47-49)Al-1Re-1W-0.2Si. Indeed, a better creep resistance was obtained for G4 alloys than for GE alloys elaborated by both cast and PM routes [7, 8]. For the cast alloy, this effect was attributed to a secondary β precipitation at lamellar interfaces and to a Re enrichment of the non-lamellar interdendritic areas. Additionally, a sub-transus heat treatment leading to a duplex microstructure provides the better creep resistance for the PM alloy. It is worth noting that prealloyed niobium rich powders (Ti45Al8.5Nb(BWY)) have been recently sintered by SPS [9, 10]. In this last work, single phase and lamellar microstructures containing β precipitates were formed depending on the sintering temperature. Compression tests have shown that the near lamellar microstructure exhibits a high strength at high temperatures with respect to the current literature.

EXPERIMENTAL DETAILS

A prealloyed $Ti_{51}Al_{47}Re_1W_1Si_{0.2}$ powder fabricated by gas atomisation process at GKSS was used in the present work. Experiments were carried out in a Sumitomo 2080 apparatus (Sumitomo Coal Ming Co. Japan) implemented in the "Plateforme Nationale de Frittage Flash/CNRS" in Toulouse, France. Figure 1 shows a sample heated at 950°C in the SPS set-up. This moderate temperature was selected with the aim to study the microstructure of powder particles before a complete sintering step. Accordingly, interparticle boundaries are clearly visible due to an incomplete sintering. Some of these particles exhibit a pronounced dendritic morphology. The dendritic cores are Ti rich and contain β precipitates. The interdendritic areas are Al rich. The characteristic dimension of this dendritic structure is about 10 µm. No major difference is found between the centre and the periphery of every particle. It is considered that the powder particles are with this morphology when the sintering process starts in the SPS machine.

The powder was compacted by SPS at 1125°C under an uniaxial pressure of 100MPa following the experimental procedure described in details in Ref. [5]. The powder was filled in a graphite die and between graphite punches. The heating rate was programmed to be 100°C/min and a holding time of only 2 min was selected. The chamber is maintained under vacuum (a few Pa). The diameter and the thickness of the billet are 36 mm and 6 mm, respectively. By the activation of the α transformation, it has been shown that the difference between the programmed temperature and the sample temperature is about 140°C for TiAl samples of 36 mm in diameter; in other words the temperature of the material is the SPS temperature + 140°C [5].

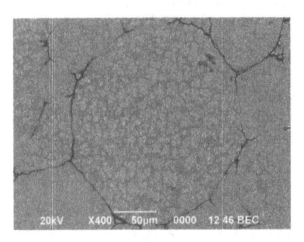

Figure 1. Dendritic morphology of a G4 powder particle.

MICROSTRUCTURES

Figures 2 & 3 display the microstructure for the alloy sintered at 1125°C. The SEM micrograph (figure 2a) shows a slight heterogeneity in the microstructure depending on the location. In some areas, the initial dendritic morphology is clearly visible (see for example the place marked DM). Dendrites containing β precipitates in bright contrast are surrounded by interdendritic channels free of these precipitates at this observation scale (figure 2b). At some other places (marked H), the microstructure is formed by a more homogeneous γ matrix containing β precipitates. Observations along the radius of the billet have shown that this microstructure is conserved in the entire surface. The TEM micrograph (figure 3) shows that the microstructure is formed by γ grains which contain fine intragranular β precipitates. The γ grain size is about a few micrometers. β intergranular precipitates with a triangular shape are also observed at the γ/γ grains boundaries.

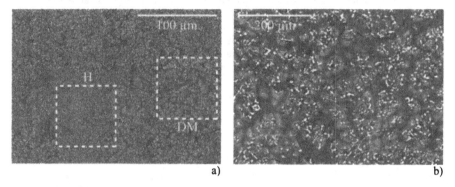

a)   b)

Figure 2. SEM micrographs of the microstructure of the G4 alloys sintered at 1125°C at various magnifications.

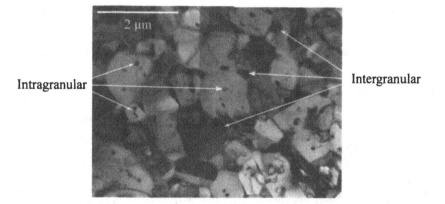

Figure 3. TEM micrograph of the microstructure of the G4 alloys sintered at 1125°C.

MECHANICAL PROPERTIES and DEFORMATION MICROSTRUCTURES

The tensile properties of the G4 SPS alloy are compared with those of the reference and boron containing SPS alloys that were presented elsewhere [5] (figure 4). Tensile tests were performed at room temperature at a $10^{-4}s^{-1}$ nominal strain rate. The YS of the G4 alloy is higher than that of the reference alloy with a $\gamma+\alpha_2$ two phase microstructure and lower than that of the boron alloy with a duplex microstructure. The tensile elongation of the G4 alloy (1.35%) is lower than those of the other alloys. As for the reference alloy with the $\gamma+\alpha_2$ two phase microstructure, the YS is followed by a plateau-like region, characteristic of a single phase alloy. This is consistent with the high similarity of the two microstructures, which are near γ microstructures containing a secondary phase (α_2 for the reference alloy and β for the G4 alloy).

The creep properties have been evaluated at 700°C, under 300 MPa. Since compression tests in the reference and boron alloys and tensile tests in the G4 alloy were performed, we will compare only the minimum creep rates and not the primary stage. Indeed, a possible slight misalignment of the compression setup at the beginning of the test can affect the values characterizing the primary stage. It is worth pointing out that the tensile primary stage of the G4 alloy reaches only 0.15% of strain. Table I compares the minimum creep rate of the three SPS alloys. The G4 alloy has a similar creep resistance than the two other alloys.

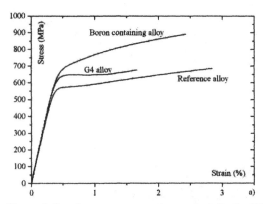

Figure 4. Tensile curves at room temperature for the G4, Reference and Boron containing alloys.

Table I. Minimum creep rate for the three alloys (700°C, 300 MPa).

Alloy	Minimum creep rate
G4	$1\ 10^{-7}\ s^{-1}$
Reference alloy	$2\ 10^{-7}\ s^{-1}$
Boron containing alloy	$1\ 10^{-7}\ s^{-1}$

Figure 5 shows typical deformation microstructures of G4 (a) and reference (b) alloys. Thin foils were cut in samples strained till rupture at room temperature. In both cases, the deformation is due to ordinary dislocations which are elongated along their screw orientation and which are anchored at many pinning points. Such a configuration is usually observed after deformation at room temperature in conventional TiAl alloys [11-14]. These pinning points have been found to result from a very fine chemical segregation [15]. A few superlattice dislocations and twins have also been observed in the deformed samples.

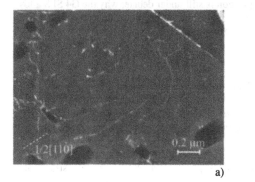
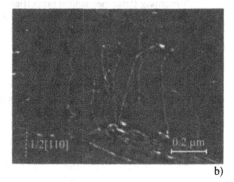

a)   b)

Figure 5. Deformation microstructures in the G4 (a) and reference (b) alloys

DISCUSSION

Due to the 140°C of difference between the SPS and sample temperatures, the dendritic γ microstructure presented in figures 2 & 3 has been achieved for an actual temperature of 1275°C. According to the phase diagram describing the G4 alloy which was proposed in Ref. [7], this temperature is situated in a three phase field ($\alpha + \gamma + \beta$). A discrepancy appears between the SPS microstructure and that observed in the Hipped alloy subsequently treated at 1275°C during 4h and air cooled [16], for which the microstructure contains α_2 grains with a grain size identical to that of γ grains and which covers about 20% of the surface fraction. On the reverse, γ equiaxed areas of the standard cast alloy (Hipping : 1275°C/172 MPa/4h + solution annealing 1275°C/20h + argon furnace cooling) exhibit a γ dendritic microstructure similar to that of the SPS alloy (see figure 5(e) of Ref. [7]). The retention of the dendritic morphology and the absence of formation of α_2 or lamellar grains is probably the result of the shortness of the SPS process.
Concerning the microstructure of SPS G4 alloys, it should be underlined the coexistence of areas in which the initial dendritic morphology was retained and areas in which it is not so noticeable. At this stage, it is postulated that the latter areas could be those strongly deformed during the sintering process. Work is in progress to verify this point. Nevertheless, it is worth pointing out that β precipitates are retained in the entire microstructure.

The tensile properties at low temperatures and the creep strength at high temperature of the G4 SPS alloy are similar to those of the reference and boron containing alloys, despite the addition of refractory elements. In particular, they are very close to those of the $\gamma + \alpha_2$ two phase microstructure (reference alloy). This results from the activation of the same elementary deformation mechanisms in the same γ phase, as evidenced in the present paper by *post mortem* TEM analyses of samples deformed at room temperature (figure 5). Fort the case of creep, that is also probably due to the segregation of refractory elements in β precipitates in the entire microstructure, which are in turn not present in the γ matrix to reduce the mobility of climbing dislocations. Consistently, β precipitates are too much distant to participate to the strength of the alloy. One way of improvement would be to reduce the volume fraction of the β phase which is not considered in the literature as creep resistant. Accordingly, a number of post heat treatments should help homogenise refractory elements through the matrix. Concerning these mechanical properties a question is still under debate: why the G4 SPS alloy is less ductile than the two other alloys? An attempt is currently being made to correlate this lack of ductility to the heterogeneity of the microstructure.

CONCLUSIONS

A G4 TiAl alloy containing heavy elements has been elaborated by Spark Plasma Sintering at 1125°C. Fully sintered compacts have been obtained in less than 30 min. The microstructure consists of γ grains containing β precipitates. It has been found that the dendritic morphology of the initial powder and the primary β precipitation are retained in the sintered alloy due to fast SPS processing.

The addition of refractory elements fails to improve the mechanical properties, probably because of the retention of the dendritic morphology which leads to some structural heterogeneity and of this primary β precipitation which segregates these elements.

ACKNOWLEDGMENTS

The authors are thankful for the SPS facilities at the PNF2 (Plateforme Nationale de Frittage Flash/CNRS" in Toulouse, France).

This study was developed in the framework of the SPALSMAP project which is supported by the French regions Midi-Pyrénées and Aquitaine and by the TURBOMECA (Groupe SNECMA) and LIEBHERR companies. All of them are strongly acknowledged.

REFERENCES

1. A. Lasalmonie, *Intermetallics* **14**,1123 (2006).
2. M. Thomas and S. Naka, *Matériaux et Techniques* **1-2**,13 (2004).
3. G. Molénat, M. Thomas, J. Galy and A. Couret, *Adv. Eng. Mat.* **9**, 667 (2007).
4. Z.A. Munir, U. Anselmi-Tamburini and M. Ohyanag, *J. Mater. Sci.* **44**, 763 (2006).
5. A. Couret, G. Molénat, J. Galy and M. Thomas, *Intermetallics* **16**, 1134 (2008).
6. H.A. Calderon, V. Garibay-Febles, M. Umemoto and M. Yamaguchi, *Mater. Sci. and Eng. A* **329-331**, 196 (2002).
7. M. Grange, J.L. Raviart and M. Thomas, *Metal. Trans. A* **35A**, 2087 (2004).
8. M. Thomas, J.L. Raviart and F. Popoff, *Intermetallics* **13**, 944 (2005).
9. Y.H. Wang, J.P. Lin, Y.H. He, Y.L.Wang and G.L. Chen, *Intermetallics* **16**, 215 (2008).
10. Y.H. Wang, J.P. Lin, Y.H. He, Y.L.Wang and G.L., *Mater. Sci. and Eng. A* **489**, 55 (2008).
11. A. Couret, *Phil. Mag. A* **79**, 1977 (1999).
12. A. Couret, *Intermetallics* **9**, 899 (2001).
13. S. Sriram, DM. Dimiduk, PM. Hazzledine and V.K. Vasudevan, *Phil. Mag A* **76**, 965 (1997).
14. B. Viguier, KJ. Hemker, J. Bonneville, F. Louchet and J.L. Martin, *Phil. Mag. A* **71**, 1295 (1995).
15. S. Zghal, A. Menand and A. Couret, *Acta Maert.* **46**, 5899 (1998).
16. O. Berteaux and M. Thomas, *unpublished results* (2008).

Mater. Res. Soc. Symp. Proc. Vol. 1128 © 2009 Materials Research Society   1128-U03-05

Hot-Die Forging of a β-Stabilized γ-TiAl Based Alloy

Wilfried Wallgram[1], Helmut Clemens[2], Sascha Kremmer[1], Andreas Otto[3], Volker Güther[3]
[1]Bohler Schmiedetechnik GmbH & CoKG, Kapfenberg, Austria
[2]Department of Physical Metallurgy and Materials Testing, Montanuniversität Leoben, Leoben, Austria
[3]GfE Metalle und Materialien GmbH, Nuremberg, Germany

ABSTRACT

Because of the small "deformation window" hot-working of γ-TiAl alloys is a complex and difficult task and, therefore, isothermal forming processes are favoured. In order to increase the deformation window a novel Nb and Mo containing γ-TiAl based alloy (TNM™ alloy) was developed. Due to a high volume fraction of β-phase at elevated temperatures the alloy can be hot-die forged under near conventional conditions, which means that conventional forging equipment with minor and inexpensive modifications can be used. With subsequent heat-treatments balanced mechanical properties can be achieved. This paper summarizes our progress in establishing a "near conventional" forging route for the fabrication of γ-TiAl components. The results of lab scale compression tests and forging trials on an industrial scale are included. In addition, the mechanical properties of forged and heat-treated TNM™ material are presented.

INTRODUCTION

The strong need for higher efficiency, reduced CO_2 emissions and weight reduction in aircraft engines leads to a demand of innovative light-weight high-temperature resistant materials which can partly substitute the materials currently employed. Presently, Ni-base alloys are used for turbine blades in aero-engines which satisfy the high mechanical and thermal requirements, like creep resistance and toughness as well as thermal stability of the microstructure. A disadvantage of Ni-base alloys, however, is their high density of about 8 g/cm^3. Intermetallic titanium aluminides, which exhibit a low density of about 4 g/cm^3, are certainly among the most promising candidates to meet the required thermal and mechanical specifications. After a long period of fundamental research and technological studies the production of engine blades made of γ-TiAl alloys is at the verge of realization. Today there are mainly two industrial approaches for the fabrication of γ-TiAl turbine components. As reported recently, the production via the investment casting route has been established for serial production [1]. However, it is well-known that the mechanical properties of γ-TiAl alloys in cast (and heat-treated) condition are lower than those of wrought processed γ-TiAl alloys. During the last decade several "wrought" γ-TiAl alloys with complex alloy compositions have been developed [2]. These alloys exhibit excellent mechanical properties, but show a narrow processing window [3]. Therefore, these alloys can only be forged under isothermal conditions. However, isothermal forging of γ-TiAl must be performed at high temperatures, requiring special dies and environmental conditions which increase manufacturing costs. Therefore, Nb and Mo containing γ-TiAl alloys have been developed which are appropriate to hot-die forging under near conventional conditions. These so-called TNM™ alloys possess a high amount of unordered β-phase with bcc lattice at elevated temperatures which improves hot workability. This paper elucidates the hot-die forging route developed for TNM™ alloys.

EXPERIMENT

The investigated γ-TiAl alloy has a nominal composition of Ti-43Al-4Nb-1Mo-0.1B (in atomic percent). The ingot was produced via a triple vacuum arc remelting process to achieve good chemical homogeneity [4]. Then the ingot with 200 mm in diameter was steel canned and hot-extruded to a diameter of 30 to 55 mm, depending on application. Subsequently, a stress relieve heat-treatment was conducted. The phase transition temperatures of the alloy, which are the basis of hot-working and heat-treatments, were determined by differential scanning calorimetry (DSC) using a Setaram Setsys Evolution [5]. Lab scale compression tests were performed on a DIL805 A/D supplied by Bähr Thermoanalyse GmbH. The industrial hot-die forging process was carried out on a conventional hydraulic press with a maximum force of 1000 metric tons. The press is specially automated to guarantee a low deformation speed which is essential due to the high strain rate sensitivity of γ-TiAl alloys [6]. To prevent heat loss during the forging process the surface of the billet was covered with a special heat shielding and the dies were pre-heated to a temperature of approximately 400°C to 800°C below the billet temperature. Prior to forging the billets were heated in an electric furnace in argon atmosphere and held for a defined time. Subsequently, the forged material was subjected to different heat-treatments in order to obtain balanced mechanical properties. Tensile and creep tests were conducted according to ASTM-E8 and ASTM-E139.

RESULTS AND DISCUSSION

In order to study the deformation behavior of the investigated alloy lab scale compression tests were carried out. Samples with a diameter of 4 mm and a length of 10 mm were heated in helium atmosphere up to temperatures in the range of 1240°C to 1300°C and deformed at strain rates between 0.05 and 0.5 s^{-1}. Figure 1 presents schematically the flow stress as a function of temperature and strain rate. As expected, a strong strain rate dependency is observed. This behavior can be explained by thermally activated dislocation motion [6]. Of particular interest is the distinct maximum of the flow stress close to the α-transus temperature (T_α). The temperature of the peak maximum corresponds with that temperature where thermodynamic calculations predict the minimum in ductile β-phase as well as in γ-phase fraction [7]. At elevated temperatures both β and γ provide good deformation behavior which is attributed to a sufficient number of slip systems [6,7]. Close to T_α, however, the hexagonal α-phase is the dominant microstructural constituent showing a lower deformability.

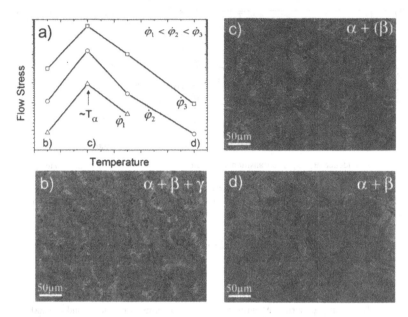

Figure 1. (a) Dependence of the flow stress of a TNM™ alloy on temperature and strain rate. Due to confidentiality reasons the numerical values of flow stress and temperatures have been omitted; (b) – (d) show SEM images of hot-extruded samples which were deformed under compression at different temperatures (see (a)) using identical strain rates. After deformation to a certain strain the samples were cooled with He gas. (b) Deformation was conducted within the ($\alpha+\beta+\gamma$) phase region below T_α. (c) compression test at the maximum of the flow stress, i.e. where the β-phase shows its minimum (see Figure 2b). (d) Deformation above T_α in the $\alpha+\beta$ phase-field region. All SEM images were taken in back-scattered electron mode.

To gain a better understanding of the existing phase fields and the dependence of the transition temperatures on the Al content DSC measurements were performed. Figure 2a shows the phase diagram according to DSC measurements conducted with a heating rate of 40 K/min. It should be noted that the influence of the heating rate on the transition temperatures is rather small [8]. From Figure 2a it is evident that in the range of 43 to 44 at% Al a two-phase-field ($\alpha + \gamma$) and a three-phase-field ($\alpha + \gamma + \beta$) exist below T_α. Above the single α-phase region a two-phase-field region ($\alpha + \beta$) occurs. However, as described in another contribution of these proceedings [5] the existence of a α-phase region is related to the dissolution behavior of the β-phase. It was found that the single α-phase field occurs for longer annealing times only. For short-term annealing treatments within the temperature range of the single α-phase region a minimum in β-phase volume fraction is observed (Figure 2b). The temperature where the minimum in β-phase volume fraction occurs critically depends on the Al-content of the alloy.

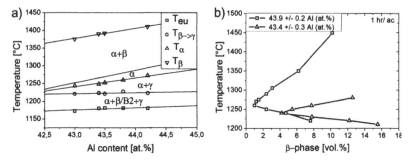

Figure 2. (a) Dependence of the phase transition temperatures on Al-content. The transition temperatures were determined by DSC measurements employing a heating rate of 40 K/min (see text). (b) Dependence of β-phase fraction on temperature for two Ti-xAl-4Nb-1Mo-0.1B alloys with different Al content is shown. The samples were annealed for 1 hour, followed by air cooling. The minimum of the β-phase fraction correlates with the maximum in flow stress (Figure 1a).

With this knowledge it was possible to establish a safe process window where hot-die forging can be performed on an industrial scale. In order to prove the feasibility of a blade forging process where upsetting (pressing along extrusion direction) and side-pressing (pressing perpendicular to extrusion direction) of the material is necessary, the experimental setup included upsetting experiments using cylinders with a diameter of 40 mm and a length of 40 mm as well as side pressing experiments with cylinders of 40 mm in diameter and 40-260 mm in length. In Figure 3 representative results from various forging experiments are shown together with two different blade forgings. In addition, all forging experiments were supported by finite element simulations using the software packages Deform 2D and 3D. From the simulation it was possible to extract information on temperature changes, local strain rates and on the stress distribution during forging.

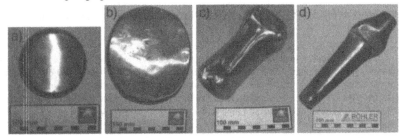

Figure 3. TNM™ billets after (a) upsetting to an overall strain of 0.7 and (b) after side pressing to an overall strain of 1.3. Photographs (c) and (d) show different TNM™ blade pre-forms manufactured by upsetting both sides of a cylindrical bar and subsequent side pressing to final shape (3-step forging process). After forging the surface of the depicted examples was ground.

112

Finally, samples were taken from forged pancakes (e.g. see Figure 3, pancake c) and subjected to heat-treatments in order to reduce the portion of β-phase and to adjust a microstructure capable of providing the required mechanical properties. The forged material was subjected to the following two-step heat treatments:

- (A) 1230°C/ 1 hour/ air cooling + 850°C/ 6 hours/ furnace cooling
- (B) 1230°C/ 1 hour/ air cooling + 950°C/ 6 hours/ furnace cooling.

The first step, conducted within the (α + γ + β) phase field region, adjusts the size of the α-grains which then transform to (α₂ + γ) lamellar colonies upon air cooling. However, due to the high cooling rate only few and very thin γ-lamellae have been formed. In addition, the volume fraction of the β-phase is reduced because the chosen temperature is close to the temperature where the minimum in β-phase fraction is expected (Figure 2b). Because the annealing step is performed in the (α + γ + β) phase region, a certain amount of small equiaxed γ-grains is present after cooling to room temperature. The second step of the heat-treatment, annealing at 850°C or 950°C for 6 hrs followed by furnace cooling, is a so-called "stabilization" treatment. Here, a further removal of β-phase takes place and the constituting phases approach a condition closer to thermodynamic equilibrium which also can be seen in the complete formation of the (α₂ + γ) lamellar colonies [9]. For detailed information on the evolution of the microstructure during heat-treatments the reader is referred to another paper of these proceedings [10]. Figures 4a, b show the results of tensile and creep tests conducted on alloy Ti-43Al-4Nb-1Mo-0.1B after heat treatments (A) and (B). The higher yield strength of the material after heat-treatment (A) can be connected to the smaller lamellar spacing within the (α₂ + γ) lamellar colonies. For test temperatures higher than 650°C a significant increase in tensile elongation is observed. At room temperature the material subjected to heat-treatment (B) shows a higher fracture strain than the material heat-treated according to (A). The differences in yield strength and fracture strain can be explained by the differences in the morphology of the microstructure as well as differences in the occurring phase fractions. Results of creep tests conducted at 800°C and 300 MPa show similar results. Here, heat-treatment (A) leads to a higher creep strength which also is attributed to the smaller lamellar spacing within the (α₂ + γ) lamellar colonies [10]. It can be summarized that heat treatment (A) leads to a low fracture strain, but shows the best creep resistance, whereas (B) increases fracture strain, but decreases creep resistance.

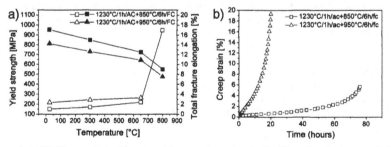

Figure 4. (a) Yield strength (■, ▲) and total fracture elongation (□,Δ) of forged and subsequently heat-treated Ti-43Al-4Nb-1Mo-0.1B material. The heat-treatments are indicated in the diagram. (b) Results of creep tests conducted in air at 800°C and 300 MPa. Note: the relatively high creep stress of 300 MPa at 800°C was selected to discern the effect of microstructural differences within short test times.

SUMMARY

TNM™ alloys are multi-phase TiAl alloys which contain a balanced concentration of the β-stabilizers Nb and Mo. These alloys exhibit a high volume fraction of ductile bcc β-phase at temperatures where hot-deformation processes, such as forging, rolling, etc., are conducted in order to facilitate plastic deformation. This novel β-solidifying γ-TiAl based alloys with adjustable β-phase fraction show a wide forging window with regard to billet temperature, die temperature and ram speed, providing the foundation for a robust industrial forging process. Due to a high volume fraction of the β-phase at hot-working temperature the alloy can be forged under near conventional conditions. The technical feasibility of manufacturing a turbine blade in a three-step forging process has been demonstrated. Furthermore, it is possible to forge this alloy above T_α, where extensive grain growth is suppressed due to the presence of a $(\alpha+\beta)$ phase region. With subsequent heat-treatments a defined and significant reduction of the β-phase is achieved, allowing the adjustment of optimized microstructures which leads to balanced mechanical properties.

REFERENCES

1. M. J. Weimer, T. J. Kelly, Paper presented at the 3rd international Workshop on γ-TiAl Technologies, Bamberg, Germany, (March 28 – 31, 2006).
2. F. Appel, J. D. H. Paul, M. Oehring, C. Buque, in: Gamma Titanium Aluminides 2003, eds. Y.-W. Kim, H. Clemens, A. H. Rosenberger (The Minerals, Metals and Materials Society (TMS), Warrendale, 2003), pp. 139-151.
3. S. Kremmer, H. F. Chladil, H. Clemens, A. Otto, and V. Güther, in Ti-2007 Science and Technology, Vol. 2, The Japan Institute of Technology, Japan 2008, 989.
4. V. Güther, A. Otto, J. Klose, C. Rothe, H. Clemens, W. Kachler, S. Winter, S. Kremmer, in: Structural Aluminides for Elevated Temperature Applications, eds. Y.-W. Kim, D. Morris, R. Yang, C. Leyens, (The Minerals, Metals and Materials Society (TMS), Warrendale, 2008), pp. 249-256.
5. H. Clemens, B. Boeck, W. Wallgram, T. Schmoelzer, L. M. Droessler, G. A. Zickler, H. Leitner, A. Otto, these proceedings.
6. F. Appel, H. Kestler and H. Clemens, in: Intermetallic Compounds – Principles and Practice, Volume 3, Progress, eds.: J. H. Westbrook and R. L. Fleischer (Chicester, UK: John Wiley Publishers, 2002), 617-642.
7. H. Clemens, W. Wallgram, S. Kremmer, V. Güther, A. Otto, A. Bartels, *Adv. Eng. Mater.* **10**, 707 (2008).
8. H. F. Chladil, Ph.D. thesis, Montanuniversität, Leoben, Austria 2007.
9. H. Clemens, H.F. Chladil, W. Wallgram, G.A. Zickler, R. Gerling, K.-D. Liss, S. Kremmer, V. Güther, W. Smarsly, Intermetallics **16,** 827 (2008).
10. L.M. Droessler, T. Schmoelzer, W. Wallgram, L. Cha, G. Das, H. Clemens, these proceedings.

Mater. Res. Soc. Symp. Proc. Vol. 1128 © 2009 Materials Research Society 1128-U03-06

Experimental Studies and Thermodynamic Simulations of Phase Transformations in Ti-(41-45)Al-4Nb-1Mo-0.1B Alloys

Helmut Clemens[1], Barbara Boeck[1], Wilfried Wallgram[2], Thomas Schmoelzer[1], Laura M. Droessler[1], Gerald A. Zickler[3], Harald Leitner[1,3], Andreas Otto[4]
[1]Department of Physical Metallurgy and Materials Testing, Montanuniversität Leoben, A-8700, Leoben, Austria
[2]Bohler Schmiedetechnik GmbH&CoKG, A-8605, Kapfenberg, Austria
[3]Christian Doppler Laboratory for Early Stages of Precipitation, A-8700, Leoben, Austria
[4]GfE Metalle und Materialien GmbH, D-90431, Nürnberg, Germany

ABSTRACT

TNM™ alloys are novel γ-TiAl based alloys which exhibit a high concentration of β-stabilizing elements such as Nb and Mo. Due to the high volume fraction of disordered β-phase these alloys can be hot-die forged under near conventional conditions. In this study, solid-state phase transformations and phase transition temperatures in Ti-(41-45)Al-4Nb-1Mo-0.1B (in at%) alloys were analyzed experimentally and compared to thermodynamic calculations. Results from scanning electron microscopy, conventional and high-energy X-ray diffraction as well as differential scanning calorimetry were used for the characterization of the prevailing phases and phase transformations. For the prediction of phase stabilities and phase transition temperatures thermodynamic calculations were conducted. ThermoCalc® was applied using a commercially available TiAl database. Combining all results a stable as well as a metastable phase diagram for Ti-(41-45)Al-1Mo-0.1B alloys is proposed.

INTRODUCTION

The continuous demand for weight reduction and higher engine efficiencies in automotive, aerospace and energy industries pushes the materials applied today towards their limits. Therefore, these industries have a strong need for the development of novel light-weight materials which can withstand temperatures up to 800°C, while maintaining acceptable mechanical properties. Intermetallic γ-TiAl based alloys are certainly among the most promising candidates to possess the required thermal and mechanical properties [1]. In order to increase the economic feasibility of wrought processing for manufacturing of γ-TiAl components, alloys are needed which can be processed "near conventionally", which means a conventional forging equipment with minor and inexpensive modifications can be used. Thus, a fine-grained casting microstructure is favorable to both, ingot breakdown and secondary forming operations [2,3]. Cast TiAl alloys based on Ti-(41-45)Al with additions of Nb and Mo solidify via the β-phase, exhibiting a fine grained and texture-free microstructure with modest micro-segregations [2-4], whereas peritectic alloys (solidification via the α-phase) show anisotropic microstructures as well as significant texture and segregation [5]. Because Nb and Mo represent the most relevant alloying elements, this TiAl alloy family has been named "TNM™ alloys" in order to distinguish them from the well-known TNB alloys, which rely on a high Nb concentration only [6]. TNM alloys show an *adjustable* volume fraction of disordered β-phase at elevated temperatures which acts as ductile phase [2-4], because the bcc β-lattice provides a sufficient number of independent slip systems. In previous studies it has been shown that TNM alloys exhibit an improved deformability at elevated temperature, where, for example, processes such as rolling and forging are performed [2,3]. However, at service temperature, which is in the range of 700°C to 800°C, the volume fraction of the β-phase, which then shows an ordered B2 structure, should be small in order not to deteriorate creep properties. For TNM alloy Ti-43Al-4Nb-1Mo-0.1B it was demonstrated that the β/B2 volume fraction can effectively be reduced by subsequent two-step heat treatments, leading to balanced mechanical properties, i.e. good creep strength and sufficient plastic fracture strain at room temperature [7]. For thermo-mechanical processing and subsequent heat treatments the phase diagram of TNM alloys is of particular importance. Therefore, in the present work phase transformations in Ti-(41-45)Al-4Nb-1Mo-0.1B (in at%) alloys were analyzed experimentally and compared to thermodynamic calculations. Combining all results both a metastable and a stable phase diagram are proposed for TNM alloys, describing the response of the material to short-term and long-term treatments at elevated temperatures.

EXPERIMENTAL/THERMODYNAMIC CALCULATIONS

In this work the solid-state phase transformations in Ti-xAl-4Nb-1Mo-0.1B alloys as a function of the Al content x were studied on arc-melted alloy buttons [8]. Buttons of about 150 g weight were melted in a laboratory arc-melting furnace on a water-cooled copper plate under Argon atmosphere. The investigated materials included six Ti-Al-4Nb-1Mo-0.1B alloys with Al concentrations in the range of 41 - 45 at%. As starting materials high-purity metals and Mo-Al, Al-B and NbTiAl master alloys were used. The buttons were melted twice to ensure sufficient homogeneity. The nominal chemical composition of the investigated alloys and the results of chemical analysis are summarized in Table I. For all alloys the total amount of interstitial impurities (O,N,C,H) was smaller than 1000 wt.-ppm.

Table I. Nominal composition of the investigated TNM alloys. In parenthesis are the results of chemical analysis. All values in atomic percent.

Element	Alloy 1	Alloy 2	Alloy 3	Alloy 4	Alloy 5	Alloy 6
Al	41 (40.9)	42 (n.d.)	43 (42.8)	43.5 (43.6)	44 (44.1)	45 (44.4)
Nb	4 (3.97)	4 (n.d.)	4 (3.87)	4 (3.96)	4 (3.97)	4 (3.85)
Mo	1 (0.97)	1 (n.d.)	1(0.83)	1 (0.96)	1 (0.96)	1 (0.89)
B	0.1 (0.12)	0.1 (n.d.)	0.1 (0.12)	0.1 (0.12)	0.1 (0.89)	0.1 (0.11)

n.d.: not determined.

The sequence of the occurring phase transformations was studied by differential scanning calorimetry (DSC) using a Setaram Setsys Evolution, a sample carrier with S-type thermocouples and alumina crucibles with a cover. The measurements were performed in a dynamic helium atmosphere with a flow rate of 20 ml/min. DSC experiments were carried out from room temperature up to 1480°C at heating rates of 10, 20 and 40 K min^{-1}. The sample material used for DSC measurements was first annealed at 1150°C for 6 hrs in order to avoid DSC signals stemming from a strong thermodynamic imbalance. The distribution and constitution of the phases within the different alloys was determined using a Zeiss Evo 50 scanning electron microscope (SEM) equipped with an energy dispersive X-ray analysis (EDX) unit. Conventional X-ray diffraction (XRD) experiments were conducted on a Siemens D500 in Bragg-Brentano mode using CuK_α radiation. High-energy X-ray diffraction (HEXRD) experiments were carried out at the synchrotron beamline ID15B at ESRF in Grenoble, France.

116

The benefit of X-rays with a energy of around 100 keV is their high penetration power, which allows to investigate polycrystalline bulk materials [9]. Here, *in-situ* observations were conducted during continuous heating experiments up to 1400°C. More experimental details are reported by Clemens et al. [4].

For the prediction of phase stabilities and phase transition temperatures thermodynamic calculations were conducted. ThermoCalc® was applied using a commercially available TiAl database [10]. Details regarding modeling can be found in the reviews of Ansara [11] and Saunders and Miodownik [12].

RESULTS AND DISCUSSION

Figure 1a shows a section of the Ti-xAl-4Nb-1Mo phase diagram calculated with ThermoCalc. The phase diagram shows that solidification via the β-phase ends for an Al content of about 44 at% and no single α-phase region can be expected for Al concentrations lower than ~45 at%. The addition of Nb and Mo to the binary TiAl alloy changes the eutectoid line into a four-phase-field area ($\alpha_2 + \alpha + \gamma + \beta$), thus the former eutectoid temperature T_{eu} changes into a starting and finishing temperature. In previous studies, however, the thermodynamic database used was found to poorly describe the transition temperatures and phase proportions in TiAl alloys exhibiting high concentrations of β-stabilizing alloying elements (especially with regard to the eutectoid temperature T_{eu}) [13].

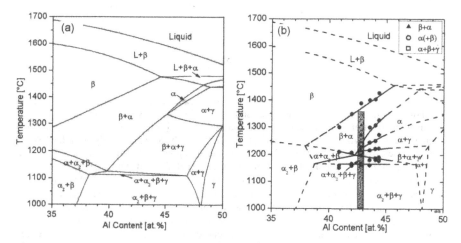

Figure 1. (a) Section of the phase diagram Ti-(35-50)Al-4Nb-1Mo including 450 wt.-ppm oxygen calculated with ThermoCalc; (b) proposed phase diagram according to the results of DSC measurements (●), heat treatments (symbols ▲,○,□ depict data for alloy 3, the analyzed phases are stated in the inset; see also Figure 3) and *in-situ* HEXRD experiments (the grey contrasted rectangle indicates the scanned temperature range where the following phases were detected: low temperatures, range 1: $\alpha_2 + \gamma + \beta$; high temperatures, range 2: β + α).

The phase transformation temperatures of the six alloys (Table I) were investigated by means of DSC measurements conducted at different heating rates. In order to achieve a non-dynamic transition temperature, the DSC data were extrapolated by a linear approach as described in [13]. For example, Figure 2 shows a section of the DSC curves for alloy 2, 3 and 6 for a heating rate of 40 K min[-1]. For better readability the baseline is subtracted and the appearing reactions are indicated by arrows. The onset of the corresponding peak was taken for the evaluation of the eutectoid starting temperature, whereas the maximum of the peak was used to determine the α- and β-transus temperatures [8]. For alloy 6 a shift of the eutectoid reaction peak to a higher temperature was observed. This shift might be attributed to local chemical fluctuations within the analyzed sample volume. According to Figure 1a the composition of alloy 2 and alloy 3 is in the vicinity of the calculated eutectoid point (~ 40 at% Al), whereas alloy 6 almost comes in contact with single α-phase field (at ~1325°C). Figure 1b shows the phase diagram according to the obtained DSC data.

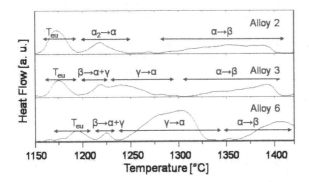

Figure 2. DSC curves of alloy 2, 3 and 6 (see Table I) measured at a heating rate of 40 K min[-1]. For better readability the baseline is subtracted and the appearing reactions are indicated by arrows.

Obviously, there are some significant differences to the calculated phase diagram. Firstly, the level of the eutectoid temperature is underestimated by the thermodynamic calculation. This relatively high difference between predicted and experimental values was also found for other high Nb bearing TiAl alloys and is reported in [13]. Secondly, a detailed analysis of the DSC curves has clearly shown that the position of the eutectoid point is shifted to a higher Al concentration. Finally, and most important for processing at elevated temperatures, the experimental data give evidence for the existence of a single α-phase field in TNM alloys with Al contents higher than 42 at%. However, if the single α-phase region is rather small it can be passed without significant growth of the α-grains, provided the dissolution rate of the (metastable) β-phase is slow. For verification of the extrapolated DSC transformation temperatures different annealing treatments were chosen, i.e. samples of all alloy variants were annealed at different temperatures and subsequently water quenched [8]. For example, Figure 3 presents the microstructure and the corresponding XRD spectra of alloy 3 after annealing at 1205°C, 1240°C, 1275°C, and 1310°C for 1 hour followed by water quenching. From the appearing phases and their volume fraction it is evident that the annealing treatments were conducted in different phase fields (see Figure 1b). Due to the high volume fraction of α and β-phase present at 1275°C and 1310°C, grain growth during annealing was suppressed effectively. It should be noted that according to Figure 1b the annealing treatment at 1240°C was conducted

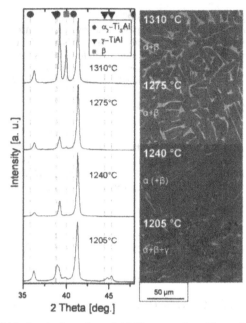

Figure 3. Microstructures and corresponding XRD spectra for alloy 3 after annealing at 1205°C, 1240°C, 1275°C, and 1310°C for 1 hour following water quenching (see text). 2θ is the scattering angle. The SEM images were taken in back-scattered electron mode.

within the single α-phase field. However, from Figure 3 it is evident that after annealing for 1 hour a small amount of β-phase is still present. This supports the idea of its rather sluggish dissolution behavior which can be attributed to the presence of Nb and Mo which exhibit a low diffusibility. However, if the annealing time at 1240°C is extended to 10 hours a complete dissolution of the metastable β-phase was observed, along with significant coarsening of the microstructure [8]. A similar behavior was observed for all alloy variants heat-treated in the single α-phase field. Because dissolution is a diffusion-controlled process, the time necessary to remove the β-phase decreases with increasing temperature. The results of HEXRD investigations support the findings derived from DSC and heat treatments [14]. Exemplarily, the phases present in alloy 3 at selected temperatures are included in Figure 1b. The phases were determined from Debye-Scherrer rings which were recorded *in-situ* during continuous heating experiments conducted at a heating rate of 2 Kmin[-1].

From the obtained results it is evident that the database used does not yield good predictions (Figure 1a). The calculated phase diagram reflects rather the behavior of Ti-(41-45)Al-4Nb-1Mo-0.1B alloys for short-term heat treatments, where no single α-phase field occurs (note that the calculation leads to lower T_{eu}, higher T_β and a wrong position of the eutectoid point). For longer heat-treatments, however, the system approaches equilibrium conditions which are reflected in the phase diagram shown in Figure 1b. For a comprehensive description of the phase diagram, including all results derived from HEXRD studies as well as from neutron diffraction experiments on the ordering temperature of the β-phase, the reader is referred to a forthcoming publication [14].

SUMMARY

TNM alloys are novel γ-TiAl based alloys with adjustable β/B2-phase fraction and excellent hot workability. For an understanding of the solid-state phase reactions taking place during heat-treatments as well as for successful thermo-mechanical processing it is essential to know the phases present at a chosen annealing/processing temperature. Therefore, solid-state phase transformations and phase transition temperatures in Ti-(41-45)Al-4Nb-1Mo-0.1B alloys were analyzed experimentally and compared to thermodynamic calculations which were found to differ strongly from the experimental results. Combining all experimental results a phase diagram has been established for Ti-(41-45)Al-4Nb-1Mo-0.1B alloys.

ACKNOWLEDGMENTS

The authors thank Arno Bartels, Andreas Stark, Volker Güther, Thomas Buslaps, and Klaus-Dieter Liss for fruitful cooperation. Likewise, we are grateful for the beam-time and overall support delivered at the ESRF in Grenoble, France. Part of this study was supported by the Austrian Fonds zur Föderung der wissenschaftlichen Forschung FWF (P20709-N20).

REFERENCES

[1] *Structural Aluminides for Elevated Temperatures*, edited by Y-W. Kim, D. Morris, R. Yang, and C. Leyens (TMS, Warrendale PA, 2008).
[2] H. Clemens, H.F. Chladil, W. Wallgram, B. Böck, S. Kremmer, A. Otto, V. Güther, A. Bartels, in [1], p. 217.
[3] H. Clemens, W. Wallgram, S. Kremmer, V. Güther, A. Otto, A. Bartels, *Adv. Eng. Mater.* **10**, 707 (2008).
[4] H. Clemens, H.F. Chladil, W. Wallgram, G.A. Zickler, R. Gerling, K.-D. Liss, S. Kremmer, V. Güther, W. Smarsly, *Intermetallics* **16**, 827 (2008).
[5] V. Küstner, M. Oehring, A. Chatterjee, V. Güther, H. Clemens, F. Appel, in *Gamma Titanium Aluminides 2003*, edited by Y-W. Kim, H. Clemens and A. H. Rosenberger (Warrendale, PA: TMS, 2003), p. 89.
[6] F. Appel, M. Oehring, R. Wagner, *Intermetallics* **8**, 1283 (2000).
[7] L.M. Droessler, T. Schmoelzer, W. Wallgram, L. Cha, G. Das, H. Clemens, these proceedings.
[8] B. Boeck, Diploma thesis, Montanuniversität, Leoben, Austria (2008).
[9] W. Reimers, A.R. Pyzalla, A. Schreyer, H. Clemens (Eds.), Neutrons and Synchrotron Radiation in Engineering Materials Science (WILEY-VCH, Weinheim, Germany, 2008).
[10] N. Saunders, in *Gamma Titanium Aluminides 1999*, edited by Y-W. Kim, D.M. Dimiduk and M.H. Loretto (TMS, Warrendale PA, 1999), p. 183.
[11] I. Ansara, *Int. Met. Reviews* **22**, 20 (1979).
[12] N. Saunders, A.P. Miodownik, *CALPHAD - A Comprehensive Guide* (Elsevier Science, New York, 1998).
[13] H.F. Chladil, H. Clemens, G.A. Zickler, M. Takeyama, E. Kozeschnik, A. Bartels, T. Bulaps, R. Gerling, S. Kremmer, L. Yeoh, K.-D. Liss, *Int. J. Mat. Res.* **98**, 1131 (2007).
[14] W. Wallgram, H. Clemens, B. Böck, T. Schmölzer, G.A. Zickler, A. Otto, *Intermetallics* (in preparation).

Microstructure and Tensile Ductility of a Ti-43Al-4Nb-1Mo-0.1B Alloy

Laura M. Droessler[1], Thomas Schmoelzer[1], Wilfried Wallgram[2], Limei Cha[1], Gopal Das[3], Helmut Clemens[1]

[1]Department of Physical Metallurgy and Materials Testing, Montanuniversität Leoben, Franz-Josef-Str. 18, A-8700 Leoben, Austria

[2]Bohler Schmiedetechnik GmbH&CoKG, Mariazeller Str. 25, A-8605 Kapfenberg, Austria

[3]Pratt & Whitney, 400 Main Street M/S 114-43, East Hartford, CT 06108, USA

ABSTRACT

The microstructural development of a forged Ti-43Al-4Nb-1Mo-0.1B (in at%) alloy during two-step heat-treatments was investigated and its impact on the tensile ductility at room temperature was analyzed. The investigated material, a so-called TNM™ gamma alloy, solidifies via the β-route, exhibits an adjustable β/B2-phase volume fraction and can be forged under near conventional conditions. Post-forging heat-treatments can be applied to achieve moderate to near zero volume fractions of β/B2-phase allowing for a controlled adjustment of the mechanical properties. The first step of the heat-treatment minimizes the β/B2-phase and adjusts the size of the α-grains, which are a precursor to the lamellar γ/α₂-colonies. However, due to air cooling after the first annealing step, the resulting microstructure is far from thermodynamic equilibrium. Therefore, a second heat-treatment step is conducted below the eutectoid temperature which brings the microstructural constituents closer to thermodynamic equilibrium. It was found that temperature and duration of the second heat-treatment step critically affect the solid-state phase transformations and, thus, control the plastic fracture strain at room temperature. Scanning and transmission electron microscopy studies as well as hardness tests have been conducted to characterize the multi-phase microstructure and to study its correlation to the observed room temperature ductility.

INTRODUCTION

In the last decades, intermetallic γ-TiAl based alloys have been developed which can replace Ni-base alloys up to temperatures of 800°C, while maintaining acceptable mechanical properties. Their low density, high specific stiffness, high yield strength, and good creep resistance have allowed this class of materials to be used as turbocharger wheels and valves in combustion engines [1]. Drawbacks which have delayed and are still delaying their widespread use are difficult hot-workability and limited ductility at room temperature. Hot workability can be improved by stabilizing the ductile disordered bcc β-phase through additions of β-stabilizers such as Nb and Mo. These Nb and Mo bearing alloys, which have Al contents in the range of 42 to 44 at%, are so-called TNM™ alloys. Specific information on TNM™ alloys can be found in [2,3]. At elevated temperatures, a high volume fraction of β-phase can be obtained which allows forging under nearly conventional conditions [4]. Following forging the material was subjected to different heat treatments to control the microstructure. Tensile tests at room temperature on a TNM™ gamma alloy after an optimized heat treatment show a remarkably high plastic fracture strain with no degradation in hardness.

EXPERIMENT

In this paper, a TNM™ alloy with a nominal composition of Ti-43Al-4Nb-1Mo-0.1B (in at%) was investigated. The ingot was produced by vacuum arc remelting (VAR) and remelted three times to assure adequate chemical homogeneity [5]. The as-cast ingot was hot-extruded from a diameter of 200 mm to a diameter of 54 mm with an extrusion ratio of about 14:1. A subsequent stress relieve heat-treatment was performed at 980°C for 4 h under argon atmosphere. The extruded rod was turned to a diameter of 40 mm and was then side-pressed to a thickness of 20 mm in a single step under near-conventional forging conditions [4]. Subsequently, the forged material was subjected to different two-step heat-treatments along with Vickers hardness tests as summarized in Table I. Note that heat-treatment #1 represents the material after the first heat-treatment (HT) step, i.e. after annealing in the ($\alpha+\beta+\gamma$)-phase region and subsequent air cooling. The second heat-treatment step (#2...4) is conducted below the eutectoid temperature and can be regarded as "stabilization" treatment [2].

Table I: Heat-treatments performed on the TNM-alloy and resulting HV5 hardness values (AC: air cooling, FC: furnace cooling).

HT	1st heat treatment step				2nd heat treatment step			hardness HV5
#1	1230°C	1h	AC		-	-	-	598
#2	1230°C	1h	AC	+	850°C	6h	FC	646
#3	1230°C	1h	AC	+	900°C	6h	FC	650
#4	1230°C	1h	AC	+	950°C	6h	FC	647

Scanning electron microscopy (SEM) was used to conduct microstructural characterization on specimens prepared by standard metallographic methods [6]. All SEM images were obtained with a Zeiss EVO 50 in back-scattered electron (BSE) mode. Transmission electron microscopy (TEM) investigations were conducted with a Philips CM-12 and a FEI CM200 with acceleration voltages of 120 kV and 200 kV, respectively. X-ray diffraction (XRD) experiments were carried out on a Siemens D500 using Bragg-Brentano mode and CuK$_\alpha$ radiation. Tensile tests were conducted with round specimens at room temperature and at a strain rate of $1 \cdot 10^{-4}$ s^{-1}. The dimensions of gauge length and width were 25.4 mm and 4.7 mm, respectively.

RESULTS AND DISCUSSION

From Table I it is evident that the second heat-treatment step, conducted within the ($\alpha_2+\beta$/B2+γ)-phase region, increases the hardness of the material considerably compared to HT #1. It is interesting to note that hardness values remain more or less the same for the second heat treatments conducted from 850°C to 950°C. It appears that the temperature range at which the second heat treatment step is carried out does not influence the hardness. The results of the tensile tests are shown in Table II.

Table II: Plastic tensile strain to fracture at room temperature for specimens subjected to heat-treatments #2 and #4 (average values of two tested specimens).

Heat treatment	Plastic fracture strain [%]
# 2	1.2
# 4	3.1

It is interesting to note, that the plastic fracture strain is more than twice the value for specimens subjected to heat-treatment #4 than for those subjected to heat-treatment #2, although similar hardness values were determined. The microstructures corresponding to the heat-treatments of Table I are displayed in Figure 1.

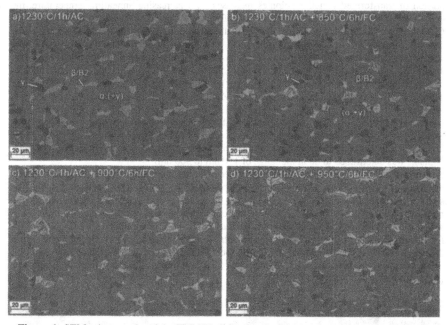

Figure 1: SEM micrographs of the TNM™ alloy after the four heat-treatments (see Table I). After the second heat-treatment step (b to c), decomposition of β/B2-grains is observed as well as the occurrence of a cellular (discontinuous) reaction in the α_2/γ-lamellar grains. All images were taken in the BSE mode.

In the SEM images three different microstructural constituents can be detected. Globular β/B2- and γ-grains are located at grain boundaries and triple points of $(\alpha_2+\gamma)$-colonies. The β/B2-grains appear as brightest phase, the γ-grains show a dark contrast and the $(\alpha_2+\gamma)$-colonies show a light grey contrast. Figure 1a shows the microstructure of the sample subjected only to the first part of the heat-treatment (1230°C for 1 h, followed by air cooling). The average size of the α_2-grains was determined to be 20 μm. It should be noted that air cooling hampers the transformation of the α-grains to lamellar α_2/γ-colonies. Since HT #1 is conducted near to the α-transus temperature, the microstructure contains a small fraction of γ-phase. For more information regarding the phase diagram of TNM™ alloys, the reader is referred to another paper of these proceedings [7].

During the second heat-treatment step, the formation of lens-shaped γ-platelets within the β/B2-grains is observed (Figures 1b-d). An orientation relationship has been established between these γ-platelets and the β/B2-matrix [6,8]. It has been found, that the close-packed planes of γ and β/B2 are parallel to each other, which leads to the following crystallographic relationship: $\{111\}_\gamma \| \{110\}_{\beta/B2}$. At temperatures at which the second heat-treatment is conducted, a fraction of the lamellar α_2/γ-colonies undergoes a so-called cellular (discontinuous) reaction [2,9]. In Figure 1b (HT #2) the start of the cellular reaction can be observed at α_2/γ-colony boundaries. As the portion of transformed lamellar colonies increases with increasing annealing temperature, the occurrence of the cellular reaction can be identified more clearly in Figures 1c (HT #3) and 1d (HT #4). From Figure 1d it is evident that the cellular reaction leads to a partial breakdown of the lamellar colonies and thus to a local refinement of the microstructure. The driving force of the cellular reaction stems from the strong chemical nonequilibrium after the first heat-treatment step and from the high interface energy provided by the formation of an ultrafine lamellar structure during the second heat-treatment step (see below). Further details will be given in a forthcoming paper [10]. The volume fraction of β/B2- and γ-grains determined by quantitative phase analysis was found to vary only slightly for the four different heat-treatments. TEM investigations conducted on all heat-treated samples revealed the presence of ω-phase within the β/B2-grains. Since ω shows a $B8_2$ structure and, therefore, exhibits lower symmetry than the β/B2-phase [10,11,12], its formation might affect the ductility of the material. The average size of the ω-domains was determined to be in the range of 150 nm to 300 nm.

The most crucial microstructural development occurring during the second heat-treatment step is the formation of fine γ-lamellae from the supersaturated α_2-grains. This precipitation reaction starts at a temperature of 720°C, as determined by means of *in-situ* heating experiments using high energy X-ray diffraction (HEXRD). In this experiment the heating rate was 20 K/min. TEM studies on lamellae formation in supersaturated α_2-grains have been reported by Cha et al. [13] for similar TiAl alloys. Figure 2 shows representative TEM images of the lamellar α_2/γ-colonies formed during heat-treatments #1, #2 and #4 (see Table I). The crystallographic orientation of the γ-laths with respect to α_2 is described by the well-known Blackburn relationship [14]: $(111)_\gamma \| (0001)_{\alpha_2}$ and $[-101]_\gamma \| [11\text{-}20]_{\alpha_2}$.

Figure 2: TEM images of α_2/γ-lamellae after heat-treatment #1 (a), #2 (b) and #4 (c). The images were obtained in the $<110]_\gamma \| [2\text{-}1\text{-}10]_{\alpha_2}$ zone axis.

From Figure 1a it is evident, that only a few γ-lamellae are present after air cooling from the (α+β+γ)-phase region. Figures 2b and 2c show that the mean interface spacing, including α_2/γ- and γ/γ-interfaces, depends on the temperature at which the second heat-treatment is conducted.

Holding at 850°C for 6 hrs (HT #2) has led to a mean interface spacing of 13 nm while for HT #4 (950°C/6h/FC) an average interface spacing of 27.3 nm was determined.

Since the volume fraction of the globular β/B2- and γ-grains obtained after the three different second heat-treatment steps are essentially identical (compare Figures 1b to d), differences in morphology or the phase fraction within the predominant α_2/γ-colonies have to be responsible for the observed differences in the plastic fracture strain at room temperature (see Table II). To analyze the change of the γ-phase fraction during the applied heat-treatments X-ray diffraction measurements were conducted. The patterns obtained are shown in Figure 3.

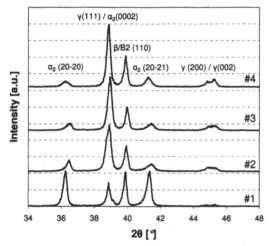

Figure 3: XRD patterns of the four differently heat-treated samples (see Table I). The intensity of the γ-peaks increases with increasing temperature of the second heat-treatment step, while the intensity of the α_2-peaks decreases.

After the first heat-treatment step (HT #1) the intensity of the γ-peaks is low, whereas a high intensity for the α_2-peaks is observed. If a second heat-treatment is performed (HT #2-4) the γ-phase fraction increases significantly, while the α_2-phase decreases. With increasing temperature of the second heat-treatment step, the γ-phase fraction increases at the expense of α_2.

During the second heat-treatment the formation of γ-laths in supersaturated α_2-grains as well as in β/B2-grains is observed (Figures 1 and 2). The increase in hardness during the second heat-treatment step is primarily caused by the formation of fine γ-lamellae within the α_2-grains.

With increasing temperature of the second heat-treatment step an increase in tensile ductility is observed (Table II), whereas the hardness seems to be unaffected by the temperature at which the second heat-treatment step is conducted. From the XRD investigations (Figure 3) and additional TEM studies it is evident, that the volume fraction of γ-phase increases in the lamellar colonies with increasing temperature. Simultaneously, the volume fraction of α_2-phase decreases. It is well known, that the plastic deformation behavior in ($\alpha_2+\gamma$)-TiAl alloys is dominated by the γ-phase. Here, dislocation movement is easier achieved than in α_2. Additionally, a cellular reaction takes place, leading to a local refinement of the microstructure. With increasing annealing temperature this effect increasingly dominates the prevailing microstructure. Both effects, i.e. the increase of γ-phase within the lamellar α_2/γ-colonies and the refinement due to a cellular

125

reaction, increase the plastic fracture strain at room temperature considerably. Presently, a study is conducted to separate the effect of both mechanisms.

SUMMARY

Two-step heat-treatments have been performed on a TNM™ alloy with the nominal composition of Ti-43Al-4Nb-1Mo-0.1B (in at%). After the first heat-treatment step, the microstructure consists of a small volume fraction of globular β/B2- and γ-grains as well as supersaturated α_2-grains with an average size of 20 µm. The second heat-treatment increases the materials hardness considerably. Additionally, the temperature of the second heat-treatment step has a strong influence on the plastic fracture strain at room temperature. Microstructural investigations have been conducted by SEM, TEM and XRD. It has been shown that the observed increase in hardness can be attributed to the formation of γ-lamellae within the supersaturated α_2-grains. This process is accompanied by a significant increase of the γ-phase fraction. In addition, with increasing temperature of the second heat-treatment step, the microstructure is partly refined due to a cellular reaction. Both effects lead to an increase of the plastic fracture strain at room temperature for which values up to 3.1 % have been measured.

REFERENCES

[1] H. Kestler, H. Clemens, in Titanium and Titanium Alloys, edited by C. Leyens, M. Peters (WILEY-VCH, Weinheim, 2003), pp. 351-392.

[2] H. Clemens, H.F. Chladil, W. Wallgram, B. Böck, S. Kremmer, A. Otto, V. Güther, A. Bartels, in Structural Aluminides for Elevated Temperature Applications, edited by Y.W. Kim, D. Morris, R. Yang, C. Leyens, (The Minerals, Metals and Materials Society (TMS), Warrendale, 2008), pp. 217-228.

[3] H. Clemens, W. Wallgram, S. Kremmer, V. Güther, A. Otto, A. Bartels, Adv. Eng. Mater. **10**, 707 (2008).

[4] W. Wallgram, S. Kremmer, H. Clemens, A. Otto, V. Güther, these proceedings.

[5] V. Güther, A. Otto, J. Klose, C. Rothe, H. Clemens, W. Kachler, S. Winter, S. Kremmer, in Structural Aluminides for Elevated Temperature Applications, edited by Y.W. Kim, D. Morris, R. Yang, C. Leyens, (The Minerals, Metals and Materials Society (TMS), Warrendale, 2008), pp. 249-256.

[6] L.M. Droessler, Characterization of β-Solidifying γ-TiAl Alloy Variants Using Advanced In- and Ex-Situ Investigation Methods, (Diploma thesis, Montanuniversität Leoben, 2008).

[7] H. Clemens, B. Boeck, W. Wallgram, T. Schmoelzer, L.M. Droessler, G.A. Zickler, H. Leitner, A. Otto, these proceedings.

[8] Z.W. Huang, Acta Mater. **56**, 1689 (2008).

[9] S. Mitao, L.A. Bendersky, Acta Mater. **45**, 4475 (1997).

[10] H. Clemens, W. Wallgram, T. Schmoelzer, L. Droessler, L. Cha, G. Das, Intermetallics (in preparation).

[11] L.A. Bendersky, W.J. Boettinger, B.P. Burton, and F.S. Biancaniello, and C.B. Shoemaker, Acta Mater. **38**, 931 (1990).

[12] G.Das and D.B. Snow, Unpublished results, (2008).

[13] L. Cha, C. Scheu, H. Clemens, H.F. Chladil, G. Dehm, R. Gerling, A. Bartels, Intermetallics **16**, 868 (2008).

[14] M.J. Blackburn, in The Science, Technology and Application of Titanium Alloys, edited by R.I. Jaffee, N.E. Promisel, (Plenum Press, New York, 1970), pp. 633.

Mater. Res. Soc. Symp. Proc. Vol. 1128 © 2009 Materials Research Society 1128-U03-11

An Overview of the ESA-ESRF-ILL Collaboration in the Framework of the IMPRESS Integrated Project

G. Reinhart[1,2,3], G.N. Iles[1,2,3], G.J. McIntyre[2], A.N. Fitch[3], D.J. Jarvis[1]

[1] European Space Agency, Keplerlaan 1, 2201 AZ, Noordwijk, The Netherlands
[2] Institut Laue-Langevin, 6 rue Jules Horowitz, BP 156, 38042 Grenoble cedex 9, France
[3] European Synchrotron Radiation Facility, 6 rue Jules Horowitz, BP 220, 38043 Grenoble cedex 9, France

ABSTRACT

Three intergovernmental research organizations from the EIROforum collaboration: the European Space Agency (ESA), the European Synchrotron Radiation Facility (ESRF) and the Institut Laue-Langevin (ILL), are cooperating to perform advanced experimental characterization in the field of materials science within the framework of the IMPRESS Integrated Project. This project aims to develop and test two distinct prototype-based intermetallic materials: (i) γ-TiAl turbine blades for aero-engines and stationary gas turbines, and (ii) Raney-type Ni-Al catalytic powder for use in hydrogen fuel cell electrodes and hydrogenation reactions. The opportunity to carry out investigations combining the use of both synchrotron radiation at the ESRF and neutrons at the ILL provides unique experimental data to complement other benchmark experiments performed on the ground and in microgravity. We present an overview of the different synchrotron X-ray and neutron characterization techniques implemented at ESRF and ILL to study the solidification and subsequent processes leading to the final products for these two materials.

INTRODUCTION

EIROforum is a partnership grouping Europe's seven largest intergovernmental research organizations. Among them, the European Synchrotron Radiation Facility (ESRF) and the Institut Laue Langevin (ILL) both located in Grenoble France, and the European Space Agency (ESA) have combined resources in order to perform key materials science experiments for the benefit of the IMPRESS Integrated Project [1,2]. IMPRESS is an acronym for Intermetallic Materials Processing in Relation to Earth and Space Solidification [3,4]. This project is coordinated by ESA within the European Commission's 6th Framework Programme (FP6). The main scientific objective of IMPRESS is to gain a better understanding of the strategic links between material processing routes, material structure and final properties of intermetallic alloys. Technically, the project aims to develop and test two distinct prototype-based intermetallic materials (i) TiAl and (ii) NiAl.

In the case of (i) structural turbine blade applications, titanium aluminides have been investigated for many years and most effort is now concentrated on γ-TiAl with various alloying additions. These alloys offer excellent mechanical properties at high temperature up to 750 °C, combined with low density and non-burn. They have the potential to replace nickel superalloy turbine components, in turn offering substantial weight reduction for gas turbine components.

In the case of (ii) catalytic applications, intermetallic materials such as Raney-type nickel-aluminide alloys have been used since the 1920s in the chemical industry [5]. Conventionally, Raney-type catalytic powders are produced by a cast-and-crush method followed by leaching with aqueous sodium hydroxide in order to remove aluminum atoms randomly at the unit-cell level revealing a highly porous skeletal structure of activated nickel. In IMPRESS, manufacturing routes such as gas-atomization and nano-particle synthesis are being explored as new ways of producing powders with improved activity and selectivity.

One of the unique parts of IMPRESS is the ability to perform benchmark experiments in space thereby removing the effects of gravity. Space experiments on MAXUS sounding rockets are in preparation to study solidification phenomena affected by gravity, for example, the settling or flotation of grains/inclusions in the melt. It is equally important for the development of reliable computer models of solidification to have accurate measurements of thermophysical properties of liquid metals. In this area, use is being made of a containerless melt processing technique known as electromagnetic levitation (EML) regularly used in microgravity on the ESA/DLR (German Aerospace Center) parabolic flight campaigns and TEXUS sounding rockets. The final class of experiments being carried out in microgravity involves the production of Ni-based nanoparticles from the vapour phase. Under weightless conditions and without convective effects, the agglomeration kinetics of the nanoparticles can be studied, including the magnetic interactions between particles.

Carrying out extensive investigations combining the use of the intense neutron source available at the ILL and the powerful synchrotron X-ray source of the ESRF to complement the experiments performed by the IMPRESS partners on the ground and in microgravity is an exceptional opportunity to create a research capability unparalleled in the world. This paper describes a number of different characterization techniques applied so far in the IMPRESS Project. Details of the individual instruments named below can be found in [6,7].

CHARACTERIZATION TECHNIQUES

Synchrotron X-ray diffraction using a 2D-type detector

Hot Isostatic Pressing (HIPping) is a densification technique widely used in the metallurgical industry, for example, to remove the internal porosity generated during casting. In recent work [8] it has been found that the pressures used during such a process can alter the equilibrium proportion and thus the properties of phases in a Ti-Al-Nb alloy. Precise measurements of the lattice parameters of the various phases present at room temperature and at the HIPping temperature were performed on beamline ID15B of the ESRF using an induction furnace and a 2D detector to calculate the corresponding variation in atomic volume and thus determine the change in volume fraction. Furthermore, the 2D image and integrated diffraction pattern contain more information than traditional powder patterns by enabling the rapid viewing of a large solid angle of detection within a single collection frame as shown in figure 1.

(a)   (b)

Figure 1. 2D diffraction patterns of a Ti-Al-Nb alloy at (a) room temperature and (b) T = 680 °C

Grazing incidence synchrotron X-ray diffraction (XRD)

A pending issue of Ti-Al based alloys for turbine blades is the degradation of various mechanical properties, especially room-temperature ductility, after exposure to air at temperatures higher than 700 °C [9]. The application of a protective coating or alloying are ways to minimize the influence of oxidation [10], however, the origin of the resulting embrittlement of these alloys remains unclear. One assumption is that gas diffusion into the surface modifies the initial lattice. The surface is then put into stress leading to the formation of cracks [11]. Surface investigations were performed on the high-resolution diffraction beamline ID31 of the ESRF. The highly monochromatic beam available was used at different wavelengths to vary the penetration depth and precise $sin^2\Psi$ measurements were recorded, where Ψ is the angle between the scattering vector and the sample's normal direction. The underlying stress in the sample can be found from the gradient of the graph of d-spacing plotted against $sin^2\Psi$.

In situ and real-time energy dispersive X-ray diffraction

Despite the two different application areas of the IMPRESS project, the two intermetallic compounds investigated have a common feature; their respective production process are primarily governed by out-of-equilibrium solidification behaviour. The ID15 beamline has been used to perform high-energy-dispersive diffraction on electromagnetically levitated melts. This allowed an *in-situ*, real-time study of solidification from the undercooled melt to be made. Such experiments enabled the determination of phase selection and nucleation undercooling values in various Ti-Al and Ni-Al Alloys [12,13].

High resolution synchrotron X-ray and neutron powder diffraction

Raney-type Ni is obtained by leaching away the aluminum of Ni-Al powders with aqueous sodium hydroxide. The leaching conditions as well as the characteristics of the initial powder have strong effects on the properties of the final catalyst [14]. Gas-atomization is a promising way to obtain a more efficient catalyst by rapidly solidifying fine liquid droplets. Previous experiments combining X-ray and neutron diffraction of atomized powders reported variations in phase fraction with grain size of the droplet as well as the occurrence of phases which could not be identified [15]. Powder diffraction patterns of Ni-Al powders with selected grain sizes and composition, before and after leaching, were recorded on the high resolution beamlines ID31 of the ESRF and D20 of the ILL. The very high angular resolution available on ID31 can separate peaks of unidentified phases which are generally small and overlap the reflections of other phases. Neutron diffraction patterns can show strong, well-defined diffraction peaks even at high angles and thus provide precise information on the atomic structure and phase fraction through Rietveld refinement analysis.

Synchrotron X-ray microtomography

Analyses by neutron and X-ray powder diffraction are efficient methods to determine average phase content, phase composition and phase fraction but provide limited morphological information on the microstructure, especially in the case of complex dendritic structures. Local investigations at the grain scale were carried out by performing synchrotron microtomography with the best pixel resolution available on the ID19 beamline of the ESRF (0.17 µm/pix.). 3D reconstructions of an atomized and a Raney-type droplet are shown in figure 2. Phase-fraction measurements can be compared to those obtained from the Rietveld refinement of diffraction patterns. Morphological analysis of the skeletal structure observed after leaching will provide significant parameters such as surface area, porosity and tortuosity which are critical properties affecting the activity of the final catalyst.

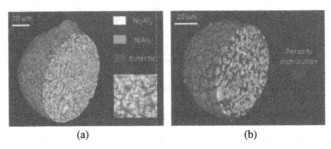

(a)   (b)

Figure 2. 3D reconstructions of (a) Ni-Al atomized droplet and (b) passivated Raney-type Ni droplet.

In situ and real-time neutron diffraction

Previous studies have shown that the activation process of Raney nickel is very fast in its initial stage (of the order of a few seconds) [14]. In order to predict catalytic activity, the transformation of Ni-Al alloys to metallic nickel requires characterization of the starting alloy and the phase content of the powders directly during the leaching process. Thus, dynamic *in situ* neutron diffraction was performed on the D20 instrument at the ILL whilst Ni-Al powders of selected composition and size were leached with aqueous sodium deuteroxide of various concentrations. The large and intense neutron beam available on D20 illuminates a sufficient amount of product during the reaction and provides sequences of diffraction patterns that display the evolution of the different species with time, as shown in figure 3. The amount of Ni_2Al_3 decreases while Raney nickel forms, inducing a decrease in intensity of the peaks related to Ni_2Al_3 and the development of a broad peak related to Ni, accompanied by the appearance of small-angle scattering.

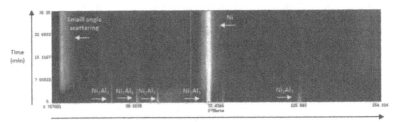

Figure 3. Typical sequence of diffraction patterns showing a decrease of Ni_2Al_3 and formation of Raney nickel.

Small-angle neutron scattering

Strong scattering at small angles is often observed in the powder diffraction patterns of leached Raney nickel. The cause of this can be the presence of very small particles (10 to 200 Å) which cannot be characterized by X-ray or neutron diffraction or by tomography at the micron scale. For this reason, experiments to determine the mean particle size and shape were performed on the small-angle neutron scattering instrument D11 of the ILL. Additional information will be obtained on the surface area of all the particles as well as on the porosity of powders.

MICROGRAVITY RESEARCH

Nickel and its alloys with high specific surface area are interesting catalysts for anodic/cathodic reactions in fuel cells. Evaporation from the melt in an inert gas environment is one of the options to produce such highly porous catalyst metals. In this process nanoscale particles grow and agglomerate via Brownian motion of the nanoparticles and cluster-cluster aggregation via van der Waals, electrostatic or magnetic interactions. Since evaporation takes place at high temperature, convective flow of the hot aerosol occurs. In addition larger agglomerates tend to sediment out. Both processes dominate particularly at low flow rates, i.e., at extended agglomeration times. Therefore microgravity is employed effectively to suppress both convective motion and sedimentation of the larger agglomerates [16]. A custom-built, microgravity-compatible inert-gas condensation source has been flown onboard an Airbus A300 during three ESA parabolic flight campaigns. Samples of Ni agglomerates have been grown during the microgravity phase of the parabola from an aerosol uninhibited by the effects of gravity [17]. Aircraft parabolic flights provide repetitively periods of up to 20s of reduced gravity during ballistic flight manoeuvres and are used to conduct, amongst other things, short microgravity investigations in Physical and Life Sciences and Technology. Their use is complementary to other microgravity platforms, such as drop towers, sounding rockets, low-earth orbit (LEO) capsules and the International Space Station (ISS) [18].
Samples made in microgravity have been measured at the ESRF using magneto-optical Kerr effect (MOKE) and XRD whilst experiments are planned to perform X-ray magnetic circular dichroism (XMCD) on the samples in order to determine the contribution of spin and orbital moment to the total magnetic moment.

CONCLUSION AND PERSPECTIVES

Many investigations in several scientific research areas are being performed at the ILL and the ESRF in the framework of the IMPRESS Integrated Project. Precise measurements and valuable information are being collected which will be used to validate multi-scale numerical models. Further analysis will improve the general understanding of the different phenomena occurring during the manufacture of Ti-Al turbine blades and Raney-type Ni catalysts. The utilization of novel processing routes such as gas atomization, sophisticated EML techniques and benchmark experimentation in space puts IMPRESS at the forefront of materials development. Combined with the rich environment at Grenoble, the most intense continuous neutron source in the world and one of the highest resolution synchrotron powder X-ray beamlines, this collaboration will continue to generate results at the cutting edge of science.

131

ACKNOWLEDGMENTS

The authors would like to thank all the beamline scientists and engineers of ILL and ESRF for their generous help and support, as well as the European partners involved : IRC University of Birmingham (UK), Krakow University of Mining and Metallurgy (PL), Leiden University (NL), Leibniz Institut for Solid State and Material Research Dresden IFWD (DE), German Aerospace Center DLR (DE), University of Rouen (FR), Armines-CEMEF (FR), ACCESS e.V (DE), Fraunhofer Institute for Materials IFAM (DE), British Ceramic Research Ltd (UK) and Novespace (FR). The IMPRESS Integrated Project (Contract NMP3-CT-2004-500635) is co-funded by the European Commission, the European Space Agency and the Swiss Government.

REFERENCES

1. D.J. Jarvis, General Workplan ESA-ESRF-ILL (2006).
2. ESA-ESRF-ILL Memorandum of Understanding (2008).
3. D.J. Jarvis and D. Voss, *Mat. Sci. Eng.* **A583**, 413 (2005).
4. M. Heppener, O. Minster and D.J. Jarvis, *Act. Astr.* **63**, 20 (2008).
5. M. Raney, US Patent No 1 563 787 (1925).
6. http://www.esrf.fr/UsersAndScience/Experiments
7. http://www.ill.eu/instruments-support/instruments-groups/yellowbook/
8. A. Huang, D. Hu, M.H. Loretto, J. Mei and X. Wu, *Script. Mater.* **56**, 253 (2007).
9. R. Pather, W.A. Mitten, P. Holdway, H.S. Ubhi, A. Wisbey and J.W. Brooks, *Intermetallics* **11**, 1015 (2003).
10. E. Godlewska, M. Mitoraj, F. Devred and B. E. Nieuwenhuys, *J. Thermal Anal.Cal.* **88**, 225 (2007).
11. X.Wu, A. Huang, D. Hu and M. Loretto, *Intermetallics* (submitted).
12. O. Shuleshova, D. Holland-Moritz, H.-G. Lindenkreuz, W. Löser and B. Büchner, proceedings of the 13th International Conference on Rapidly Quenched & Metastable Materials (RQ13), Dresden, Germany (2008) (submitted).
13. O. Shuleshova, D. Holland-Moritz, H.-G. Lindenkreuz, W. Löser and B. Büchner, proceedings of the 2nd International Conference on Advances in Solidification Processes (ICASP2), Graz, Austria (2008) (submitted).
14. F. Devred, B.W. Hoffer, W.G. Sloof, P.J. Kooyman, A.D. van Langeveld and H.W. Zandbergen, *Appl. Catal. A: General* **244**, 291 (2003).
15. C. Bao, U. Dahlborg, N. Adkins, and M. Calvo-Dahlborg, proceedings of the Annual Meeting of the French Society for Metallurgy and Materials, Paris, France (2008) (submitted).
16. S. Lösch, B. Günther, G.N. Iles, (unpublished).
17. S. Lösch, B. Günther, G.N. Iles, U. Buengener, (unpublished).
18. V. Pletser, A. Pacros, O. Minster, *Microgravity Sci. Technol.* **20**, 177 (2008).

Titanium Aluminides II—Structure, Properties and Coatings

Mater. Res. Soc. Symp. Proc. Vol. 1128 © 2009 Materials Research Society   1128-U04-02

In Situ TEM Observation of Precipitation Reactions in $Ti_{40}Al_{60}$ and $Ti_{38}Al_{62}$ Alloys and Symmetry Relations of the Phases Involved

Klemens Kelm[1,2], Julio Aguilar[3], Anne Drevermann[3], Georg J. Schmitz[3], Martin Palm[4], Frank Stein[4], Nico Engberding[4] and Stephan Irsen[2]
[1] Center for Materials Analysis, Technische Fakultät, Christian-Albrechts-Universität Kiel, Kaiserstraße 2, D-24143 Kiel, Germany
[2] Research Center Caesar, Ludwig-Erhard-Allee 2, D-53175 Bonn, Germany
[3] ACCESS e.V., Intzestraße 5, D-52072 Aachen, Germany
[4] Max-Planck-Institut für Eisenforschung, Max-Planck-Str. 1, D-40237 Düsseldorf, Germany

ABSTRACT

The phase evolution in two alloys, $Ti_{40}Al_{60}$ and $Ti_{38}Al_{62}$ (in at.%) was investigated by *in situ* heating experiments in a transmission electron microscope (TEM). In $Ti_{40}Al_{60}$, metastable Ti_3Al_5 precipitates have been found at room temperature. These transformed to TiAl by heating up to 900 °C. Subsequent cooling to 800 °C results in the reappearance of Ti_3Al_5 accompanied by precipitates of h-TiAl$_2$. Both precipitate from TiAl which becomes supersaturated in Al during cooling. In $Ti_{38}Al_{62}$, TiAl and h-TiAl$_2$ are preserved during the heat treatment, while Ti_3Al_5 transforms to TiAl when heated to 900 °C and precipitates when cooled below 800 °C. The crystallographic interrelations of the phases involved are analyzed using the formalism of group-subgroup relationships. This allows classification the phase transitions crystallographically and helps in understanding the different transformation kinetics.

INTRODUCTION

Today, TiAl alloys rich in titanium are used for valves and turbocharger rotors for automotive applications as well as for low pressure turbine blades in aerospace industry [1]. Otherwise, TiAl alloys rich in aluminum have an even lower density and show better oxidation resistance [2] compared to their Ti-rich counterparts. This makes Al-rich titanium alloys promising candidates for light-weight, high performance materials at elevated temperatures.

During a preliminary study on Al-rich TiAl alloys it was found that duplex or lamellar microstructures of TiAl + r-TiAl$_2$ can be obtained by appropriate heat treatments [3]. However, in as-cast alloys the metastable phases h-TiAl$_2$ and Ti_3Al_5 are usually observed as precipitates in the TiAl matrix [4]. In the present work we carried out *in situ* heating experiments in a TEM to analyze the temperature dependent phase transitions in $Ti_{40}Al_{60}$ and $Ti_{38}Al_{62}$ (all compositions are given in at.%). To get some insight in the phase transition kinetics we analyze the crystallographic interrelations of TiAl with those of the metastable phases Ti_3Al_5 and h-TiAl$_2$ using the group-subgroup relationships.

Two alloys were used for the TEM investigations. According to the Ti-Al phase diagram $Al_{60}Ti_{40}$ and $Al_{62}Ti_{38}$ should become single-phase TiAl above about 1050 °C and 1150 °C, respectively [5]. Below these temperatures they consist of TiAl$_2$ – either stable r-TiAl$_2$ or metastable h-TiAl$_2$ – and a TiAl matrix. Furthermore, they may contain metastable Ti_3Al_5, whose stability is not entirely clear by now, but which may dissolve between 800 °C and 900 °C [6].

EXPERIMENTAL DETAILS

Rods of about 8 mm in diameter of the two alloys were produced by centrifugal casting using the induction skull melting technique [7]. From these rods, small cylinders of 14 mm length and 3 mm diameter were cut by spark erosion. These cylinders were sliced in discs again using spark erosion and subsequently ground on SiC paper to a thickness of about 100 µm. Final thinning to electron transparency was carried out by electropolishing at -20 °C using a mixture of methanol, n-butanol and perchloric acid (36:20:3 by volume) as electrolyte.

The transmission electron microscope investigations were carried out on a LEO 922A TEM with an in-column filter operated at 200 keV using a Gatan 652 heating holder. During the heating experiments, most images were zero-loss filtered to suppress thermal diffuse scattering.

RESULTS AND DISCUSSION

In $Al_{62}Ti_{38}$, TiAl, h-$TiAl_2$ and Ti_3Al_5 are present at room temperature (RT) as is depicted in the selected area electron diffraction (SAED) pattern in Fig. 1a. By imaging the domains of h-$TiAl_2$ using dark field microscopy (Fig. 1b) it can be seen that h-$TiAl_2$ persists up to 1000 °C. The metastable Ti_3Al_5 disappears between 800 °C and 900 °C as proved by the disappearance of the corresponding reflections in the SAED pattern (Fig. 1c). Slow cooling to room temperature retains the microstructure and Ti_3Al_5 reappears (Fig. 1d). The orientation relationships between the phases are $[001]_{TiAl}\|[001]_{h\text{-}TiAl2}$, $[100]_{TiAl}\|[100]_{h\text{-}TiAl2}$ and $[100]_{TiAl}\|[010]_{h\text{-}TiAl2}$ (two orientation domains of h-$TiAl_2$) and $[001]_{TiAl}\|[001]_{Ti3Al5}$, $[110]_{TiAl}\|[100]_{Ti3Al5}$. These orientation relationships are in agreement with previous observations [8]. The setting of TiAl corresponds to setting 1 in Table I, the space groups and cell parameters for the other phases are also listed there.

The observations on $Al_{60}Ti_{40}$ differ from the above observations in that only TiAl and Ti_3Al_5 are present in the as-cast state (Fig. 2a). The Ti_3Al_5 is finely dispersed in the TiAl matrix, as it was first observed by Loiseau et al. [9]. Upon heat treating the sample in the TEM at 900 °C for 30 min. the reflections of Ti_3Al_5 vanish. This state can be conserved to RT by rapid cooling (Fig. 2b). Upon reheating the sample to 780 °C and holding for 30 min., Ti_3Al_5 precipitates again. Furthermore, the reflections of h-$TiAl_2$ appear in the SAED pattern (Fig. 2c). By heating to 900 °C both superstructure phases dissolve again, leaving single-phase TiAl. In contrast to this rather short and dynamic experiments, heating the $Al_{60}Ti_{40}$ alloy in sealed quartz ampoules for 10 h at 900 °C results in the formation of r-$TiAl_2$ precipitates (Fig. 3).

Table I. Lattice constants (in pm) and space groups of TiAl, h-$TiAl_2$, r-$TiAl_2$ and Ti_3Al_5.

Phase	TiAl [10] (setting 1)	TiAl [10] (setting 2)	h-$TiAl_2$ [11] (ZrGa$_2$-type)	r-$TiAl_2$ [12] (HfGa$_2$-type)	Ti_3Al_5 [13]
Lattice parameters (in pm)	$a = 400.4$ $c = 407.1$	$a = 283.2$ $c = 407.1$	$a = 1209.44$ $b = 395.91$ $c = 403.15$	$a = 397.11$ $c = 2431.31$	$a = 1128.61$ $c = 403.11$
Space group	$P\,4/mmm$	$P\,4/mmm$	$C\,mmm$	$I\,4_1/amd$	$P\,4/mbm$

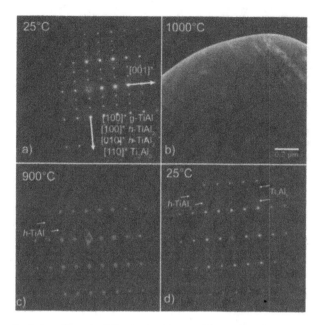

Figure 1. $Al_{62}Ti_{38}$ a) SAED pattern of $Al_{62}Ti_{38}$ of the area shown in (b) at RT showing the phases TiAl, h-TiAl$_2$ (two orientations) and Ti$_3$Al$_5$. b) Dark field image at 1000 °C using the (301)-reflection of h-TiAl$_2$. c) SAED pattern of the area shown in (b) at 900 °C exhibiting no Ti$_3$Al$_5$ superstructure reflections. 1d) SAED pattern of the region in (b) at RT after cooling from 900 °C where Ti$_3$Al$_5$ reflections are present again. Indexing of TiAl refers to setting 1 in Table I.

According to the results of Stein et al. [6], Ti$_3$Al$_5$ is metastable and dissolves near 900 °C in Al$_{60}$Ti$_{40}$ and in Al$_{62}$Ti$_{38}$ at a slightly lower temperature. This result formerly obtained by differential thermal analysis [6] has now been verified by *in situ* TEM. When the temperature is lowered from 1000 °C below the dissolution limit at about 900 °C, Ti$_3$Al$_5$ precipitates again from TiAl which becomes supersaturated in Al during cooling. This precipitation and dissolution was also observed for h-TiAl$_2$ in Al$_{60}$Ti$_{40}$. In Al$_{62}$Ti$_{38}$ h-TiAl$_2$ persisted up to 1000 °C. Thus, this alloy composition is within the two-phase field TiAl + TiAl$_2$ within the entire investigated temperature range, which is in accordance with the phase diagram [5].

At this point the question arises why the two metastable phases h-TiAl$_2$ and Ti$_3$Al$_5$ precipitate first within minutes while the precipitation of the thermodynamically stable r-TiAl$_2$ takes several hours. With respect to the atomic arrangement, topological aspects [14], the arrangements of atoms in the (002)-plane of the aluminides [8], or the interfacial energy [6] have been used to explain this observation. While interfacial energies affect the stability of phases only for very small particles, which can be the case for the very small precipitates of Ti$_3$Al$_5$, topological arrangements are usually governed by radii quotients and electronegativity. The description of structures as layers is often misleading because of the arbitrary nature of the

137

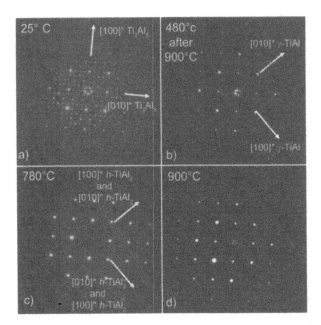

Figure 2. $Al_{60}Ti_{40}$ a) SAED pattern of $Al_{60}Ti_{40}$ recorded at RT where the phases TiAl and Ti_3Al_5 are present. b) SAED pattern at 480 °C after fast cooling from 900 °C where only TiAl reflections are visible. c) SAED pattern taken after 30 min. at 780 °C showing TiAl, h-$TiAl_2$ (two orientations) and Ti_3Al_5. d) SAED pattern after reheating to 900 °C showing only TiAl reflections. Indexing of TiAl refers to setting 1, Table I.

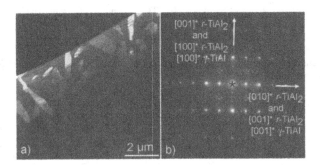

Figure 3. a) Dark field image of $Al_{60}Ti_{40}$ heat treated in sealed quartz ampoules for 200 h at 900 °C. Phases present are TiAl and r-$TiAl_2$. b) SAED pattern of area shown in (a), showing the two orientation variations of r-$TiAl_2$. Indexing of TiAl refers to setting 1, Table I.

partitioning, but this idea can be extended systematically by analyzing the group-subgroup-relationships concerning the structures involved. The results of the analysis of the group-subgroup relationships for Ti$_3$Al$_5$ are visualized as a Bärnighausen tree in Figure 4a.

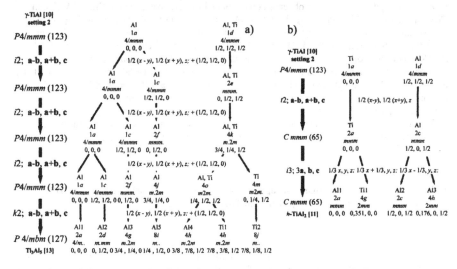

Figure 4. Bärnighausen tree visualizing the group-subgroup relationships of TiAl and Ti$_3$Al$_5$ (a) and TiAl and h-TiAl$_2$ (b). It is noted that the start settings for TiAl correspond to setting 2 in Table I, while the indexing of the SAED patterns in Figs. 1 - 3 refers to setting 1.

First, one should note that the starting coordinates for TiAl refer to setting 2 in Table I, where the coordinates for setting 1 correspond to the first *isomorphic* transformation of order 2 (*i*2). Two additional subsequent *i*2 transitions partly remove the degeneracy of the mixed positions while the final splitting of the 4o site releases the statistical occupation delivering the atomic coordinates for Ti$_3$Al$_5$ [15] (*klassengleiche* transition of order 2, *k*2). The group-subgroup relation for TiAl and h-TiAl$_2$ is visualized in Fig. 4b. Again, one has to start with setting 2 (Table I) for TiAl. The initial symmetry reduction from P4/*mmm* to C*mmm* (*translationengleiche* transition of order 2, *t*2) does not release any statistical occupation. This is achieved by the *isomorphic* transition tripling the cell and thus producing the well-known crystallographic parameters of h-TiAl$_2$. Both transitions can be regarded as Peierls distortions since they remove the degeneracy of atomic positions by symmetry reduction.

For the transition of TiAl to r-TiAl$_2$ no group-subgroup relationship can be established, since the symmetry relations demand a ratio of the c-axis of 1:12 for a phase of space group P4/*mmm* and a group-subgroup related phase of space group I4$_1$/*amd* [16, 17]. Furthermore, no group-subgroup relationship exists between the space groups C*mmm* and I4$_1$/*amd*. Thus, for the transformation of h-TiAl$_2$ to r-TiAl$_2$ or the precipitation of r-TiAl$_2$ from TiAl extensive rearrangement of the lattice is necessary, resulting in the long transition times being observed.

CONCLUSIONS

For the first time, we were able to follow the phase evolution in Al-rich TiAl alloys *in situ*. The precipitation sequence observed on quenched samples has been confirmed. While in $Ti_{40}Al_{60}$ the metastable phases Ti_3Al_5 and h-$TiAl_2$ disappear above 900 °C, in $Ti_{38}Al_{62}$ h-$TiAl_2$ persists up to 1000 °C after short times. No formation of r-$TiAl_2$ occurred in the samples during short-time annealing in the TEM. This can be understood by the fact that the structures of h-$TiAl_2$, as well as of Ti_3Al_5, are related to the structure of TiAl by a group-subgroup relationship. Thus, these phases form by a Peierls distortion of the crystal structure of TiAl. This, together with the fact that the transformations observed require a diffusion length only of the dimension of their cell parameters explains the high speed of ordering in contrast to the precipitation of r-$TiAl_2$. For the formation of the latter phase from TiAl a substantial rearrangement of the atomic array in three dimensions is necessary, needing more time than the small distortions leading to the formation of the observed metastable phases h-$TiAl_2$ and Ti_3Al_5.

ACKNOWLEDGMENTS
We gratefully acknowledge financial support from the Deutsche Forschungsgemeinschaft (DFG) for funding the group research project Al-rich Ti-Al alloys (PAK 19).

REFERENCES

1. C. Leyens, M. Peters (eds.), *Titanium and Titanium Alloys*, Wiley-VCH, Weinheim, 2003.
2. Z. Liu, G.Wang, *Mater. Sci. Eng.* **A397**, 50 (2005).
3. M. Palm, F. Stein, in: *Gamma Titanium Aluminides 1999*, edited by Y.W. Kim, D.M. Dimiduk and M.H., TMS, Warrendale, (1999) pp. 161-168.
4. L.C. Zhang, M. Palm, F. Stein, G. Sauthoff, *Intermetallics* **9**, 229 (2001).
5. J.C. Schuster, M. Palm, *J. Phase Equil.* **27**, 255 (2006).
6. F. Stein, L.C. Zhang, G. Sauthoff, M. Palm, *Acta Mater.* **49** 2919 (2001).
7. M. Paninski, A. Drevermann, G.J. Schmitz, M. Palm, F. Stein, M. Heilmaier, N. Engberding, H. Saage, D. Sturm, in: *Ti-2007 Science and Technology*, edited by M. Ninomi et al., Vol. 2, Japan Institute of Metals, Sendai, (2007) pp. 1059-1062
8. T. Nakano, A. Negeshi, K. Hayashi, Y. Umakoshi, *Acta Mater.* **47** 1091 (1999).
9. A. Loiseau, A. Lasalmonie, G. van Tendeloo, J. van Landuyt, S. Amelinckx, *Acta Cryst.* **B41**, 411 (1985).
10. P. Duwez, J.L. Taylor, *Trans. AIME* **194**, 70 (1952).
11. J.C. Schuster, H. Ipser, *Z. Metallkd.* **81**, 389 (1990).
12. J. Braun, M. Ellner, *J. Alloys Compd.* **309**, 118 (2000).
13. R. Miida, S. Hashimoto, D. Watanabe, *Jap. J. Appl. Phys.* **21**, L59 (1982).
14. U.D. Kulkarni, *Acta Mater.* **46**, 1193 (1998).
15. G. Gosh, M. Asta, *Acta Mater.* **53**, 3225 (2005).
16. Th. Hahn (ed.), *International Tables for Crystallography*. Vol. A, 5th edition, Kluwer Academic Publishers, Dordrecht, 2002.
17. H. Wondratschek, U. Müller (eds.), *International Tables for Crystallography*, Vol. A1, Kluwer Academic Publishers, Dordrecht, 2004.

Mater. Res. Soc. Symp. Proc. Vol. 1128 © 2009 Materials Research Society   1128-U04-04

Observation of <2c+a> Dislocation Glide in Duplex Ti-48at.% Al After Room Temperature Tensile Deformation

Jörg M. K. Wiezorek and Andreas K. Kulovits
Department of Mechanical Engineering and Materials Science, Swanson School of Engineering, University of Pittsburgh, 848 Benedum Hall, 3700 O'Hara Street, Pittsburgh, PA 15261, USA

ABSTRACT

In this study we investigated the deformation behavior of the hexagonal ordered phase α_2-Ti$_3$Al in Duplex TiAl under tensile loading. Transmission electron microscopy (TEM) revealed that the orientation relation ships (OR) between α_2-Ti$_3$Al and the L1$_0$ ordered γ-TiAl phase are very different as compared to the OR common in fully lamellar PST TiAl. We observed deformation related <2c+a> dislocation activity on pyramidal slip systems in the α_2-phase during post situ TEM analyses. We rationalize this observation by the possible build up of pile up stresses in γ-TiAl due to the different OR with the α_2-Ti$_3$Al phase that can possibly lead to the activation of <2c+a> dislocation activity on pyramidal slip systems with similarly resolved stresses in the α_2-Ti$_3$Al phase.

INTRODUCTION

Two phase TiAl based engineering alloys offer attractive combinations of density-specific properties rendering them attractive for selection in structural components of advanced transportation systems to realize energy savings during operation [1]. Slightly substoichiometric TiAl based alloys typically comprise at least two phases, namely the tetragonal L1$_0$-ordered phase γ-TiAl and a minority fraction of the hexagonal DO$_{19}$-ordered phase α_2-Ti$_3$Al. Generally two different microconstituents are distinguished, a fully lamellar polytwinned aggregate, PST TiAl (synthetically polytwinned TiAl) and TiAl with equiaxed microstructures, Duplex TiAl. During general shape changes at room temperature both constituent phases must accommodate plastic deformation. The deformation behavior of the α_2-Ti$_3$Al phase has important implications for the mechanical behavior of TiAl based alloys [2]. α_2-Ti$_3$Al exhibits a yield stress anomaly for the <2c+a> dislocation slip systems, the activation of which is required for general shape change [3]. Furthermore, transmission electron microscopy (TEM) after deformation of single crystals of Ti$_3$Al revealed an interesting tension-compression anomaly for the glide activation of <2c+a> dislocation [4,5]. At room temperature <2c+a> dislocations glide occurs on $\{11\bar{2}1\}$ type pyramidal planes under compression loading and on $\{2\bar{2}01\}$ type pyramidal planes under tensile loading [4,5]. In addition the Yield stress required to activate <2c+a> slip on pyramidal planes is about 2-3 times as high as compared to <a> dislocation slip on basal and prismatic planes [3]. Hence, commonly applied tensile stresses are insufficient to activate <2c+a> dislocation activity. We previously reported the choice of pyramidal glide plane critically influences details of the transfer of plastic deformation induced shears associated with dislocation glide and mechanical twinning activity from the γ-phase to the α_2 phase in lamellar grains of TiAl alloys during room temperature compression [6,7].

Here we used TEM to investigate the dislocation glide activity in the α_2-phase constituent in polycrystalline two-phase Ti-48at.% Al with a duplex microstructure after room temperature tensile deformation to failure. The results, activity of <2c+a> dislocation on pyramidal slip planes during tensile testing are discussed in relation to the mechanical behavior of α_2-Ti$_3$Al, including the choice of <2c+a> dislocation glide plane, intraphase interface shear transfer and mechanical properties of duplex two-phase TiAl.

EXPERIMENTAL PROCEDURE

Round bar tensile test specimens have been machined by surface grinding from heat-treated sections of a hot extruded Ti-48at.% Al alloy with duplex microstructure as described previously in [6]. Tensile tests have been performed in air to failure (failure strain ε_F = 2.5%) at room temperature using a strain rate of 1 x 10^{-4}s^{-1}. Thin discs have been obtained by sectioning perpendicular to the loading axis from areas adjacent to the fracture surface. TEM specimens were prepared by twin jet electropolishing using a solution of 6 vol.% perchloric acid and 34 vol.% butoxyethanol in methanol at a temperature of about -35°C. A Philips CM200 transmission electron microscope operating at 200 kV has been used to analyze microstructure. We used the CaRIne v3.1 version to simulate stereographic projections.

RESULTS AND DISCUSSION

Figure 1a shows a typical overview of the equiaxed microstructure of Duplex TiAl used for the investigation. Figure 1b on the right shows a lamellar like α_2-phase grain that has been used for TEM analysis after the tensile testing.

Figure 1: a) TEM bright field overview image of duplex TiAl used for the investigation; b) TEM bright field image of the area of α_2-phase used for the investigation

In fully lamellar PST TiAl commonly the closed packed (111) planes of the γ-TiAl phase are parallel to the closed packed (0001) planes of the α_2 – Ti$_3$Al phase and the closed packed $[\bar{1}10]$ directions of the γ-TiAl phase are parallel to the closed packed $[11\bar{2}0]$ directions of the α_2 – Ti$_3$Al phase [2]. In this study we used nano-beam diffraction with a probe size of about 50nm in the TEM to analyze the orientation relationships (OR's) between the γ-TiAl phase and the α_2–Ti$_3$Al (Figure 2).

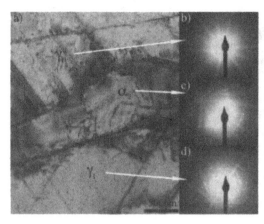

Figure 2: a) bright field TEM micrograph of the area of interest; nano-beam diffraction pattern images of the b) γ_{II}-phase grain, c) the α_2-phase grain and d) the γ_I-phase grain

The analysis yielded the following orientation relationships (OR's) between the γ_{II}-phase and the α_2-phase grain and between the α_2-phase and the γ_I-phase grain:

$$\gamma_{II} - phase \approx [\overline{8}43](13\,\overline{1}) \parallel \alpha_2 - phase \approx [11\overline{2}0](\overline{2}20\overline{3})$$
$$\gamma_I - phase \approx [06\overline{7}](200) \parallel \alpha_2 - phase \approx [11\overline{2}0](\overline{2}20\overline{3})$$

These OR's deviate significantly from the usual OR commonly observed in fully lamellar PST TiAl, where respective atomically close-packed planes of both phases and close-packed directions in the closed packed planes are parallel [2].

In order to determine the Burgers vector of the dislocations observed in the area a $\vec{g} \bullet \vec{b}$ analysis has been carried out. In figure 3 two bright field images, that were taken using different imaging conditions, are displayed. For the bright field image in figure 3a $\vec{g} = 0002$ has been used for imaging. Dislocation line segments are clearly visible. Due to the invisibility criterion $\vec{g} \bullet \vec{b} = 0$, these dislocation line segments must contain a c-component in the burgers vector, as $\vec{g} = 0002$ has been used for imaging. In image 3b the same dislocation line segments disappear, when imaged with the $\vec{g} = 20\overline{2}0$. A Burgers vector of the type <2c+a> like $\vec{b} = \frac{1}{3}[1\overline{2}16]$ would fulfill the $\vec{g} \bullet \vec{b} = 0$ invisibility criterion for both imaging conditions. Hence, the $\vec{g} \bullet \vec{b}$ analysis shows that <2c+a > dislocations are present after tensile test deformation in Duplex TiAl. Previously <2c+a> type dislocations have not been observed under tensile testing in TiAl [6,7]. Commonly only <a> slip is observed [6,7] as the yield stresses required to activate <a> slip is only about half the yield stress required to activate <2c+a> slip on pyramidal planes [3]. In addition the externally applied stresses are insufficient to activate <2c+a> slip on pyramidal planes.

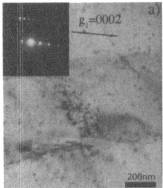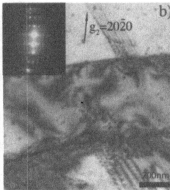

Figure 3:
Bright field
images for
the $\vec{g} \cdot \vec{b}$
analysis
taken using
different
imaging
conditions:
a) $\vec{g} = 0002$
and b)
$\vec{g} = 20\bar{2}0$

Under the influence of the electron beam over a period of 10-20 minutes we observed dislocation motion, which occurred in cascades of rapid rearrangements followed by long periods (one to several minutes) of stability of the dislocation configurations (Figure 4).

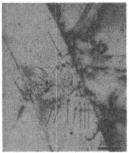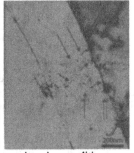

Figure 4: Bright field images of a time sequence using the same imaging condition showing dislocation motion under the influence of the electron beam (differently colored arrows point out unstable changing dislocation segments)

Figure 4 shows a sequence of bright field images that were taken using the same imaging conditions at different times. Arrows point out different dislocation segments that change during the time sequence. This indicates that the observed dislocations are unstable. The activation energy provided by the electron beam was sufficient to allow "unstable" dislocations to move and relax into lower energy configurations. Hence, these dislocations must be a product of the stress applied during the tensile testing. A possible explanation for the activity of <2c+a> dislocation slip on pyramidal glide planes might be found in the very different orientation relationships (OR's) between the α_2-Ti$_3$Al phase and the γ-TiAl phase we determined here, which differ from that typically observed in lamellar or PST TiAl [2].

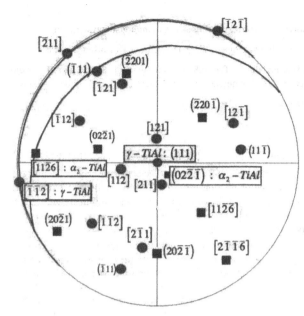

Figure 5: Stereographic projection constructed using [111] as the projection normal direction and the orientation relationship between the γ_{II} – TiAl and the α_2 – Ti$_3$Al grains. Indicated in red is the trace of the (111) plane highlighting the $[\bar{1}\,\bar{1}2]$ direction of the γ_{II} – TiAl grain; in blue is the trace of the $(02\bar{2}\,\bar{1})$ plane and highlighting the $[11\bar{2}6]$ direction of the α_2 – Ti$_3$Al grain.

We hypothesize, that possibly pile up stresses, originating from active twin dislocation systems i.e. systems with $\vec{b} = \frac{1}{6}\langle11\bar{2}\rangle$ on (111) planes in the γ-TiAl phase, could superimpose with externally applied stresses to produce locally sufficiently large and suitable stress states in the α_2-Ti$_3$Al phase to activate <2c+a> dislocation glide on pyramidal planes that therefore experience same sign and significant magnitude resolved shear stresses as compared to the twin dislocation systems in the γ-TiAl phase. Both the bright field image in figure 3b and the bright field images in figure 4 show characteristics of twin dislocation activity in the in the γ-TiAl phase at the interface with the in 5he α_2-Ti$_3$Al phase in the vicinity of the <2c+a> dislocation activity. To evaluate our hypothesis we simulated stereographic projections using the experimentally obtained local OR's in order to determine whether <2c+a> pyramidal slip systems with similar resolved shear stresses as compared to the twinning slip system in the γ-TiAl phase are available. Figure 5 shows an example of such an analysis. The traces of the (111) plane of the γ-TiAl phase and the $(02\bar{2}\,\bar{1})$ plane of the α_2-Ti$_3$Al phase have the same inclination relative to the projection normal. Additionally the $[\bar{1}\,\bar{1}2]$ direction of the γ-TiAl phase is almost parallel to the $[11\bar{2}6]$ direction of the α_2-Ti$_3$Al phase. That suggests that the pyramidal slip system with

$$\vec{b} = \frac{1}{3}[11\bar{2}6] \text{ on } (02\bar{2}\,\bar{1})$$

would have a similar resolved shear stress acting as the twinning slip system with $b = \frac{1}{6}\left[\bar{1}\,\bar{1}\,2\right]$ on the (111) octahedral plane in the γ-TiAl phase. This observation is consistent with previous findings that in Ti48at.%Al under tension <2c+a> slip occurs on pyramidal planes of the type $\{2\bar{2}01\}$, whereas under compression <2c+a> glide occurs on pyramidal planes of the type $\{11\bar{2}1\}$ [4,5].

SUMMARY

We observed dislocation activity of <2c+a> dislocations, possibly on $\{2\bar{2}01\}$ type pyramidal slip planes in the α_2-phase in the Duplex Ti-48at.% Al after tensile loading. We also find very different orientation relationships (OR's) between the α_2-Ti$_3$Al and the γ-TiAl phase. We propose that pile up stresses caused by the twinning slip systems in the γ-TiAl phase in addition with externally applied stresses lead to the activation of <2c+a> dislocations on $\{2\bar{2}01\}$ type pyramidal planes with similar resolved shear stresses. This work shows that under the appropriate crystallographic and loading conditions on pyramidal planes c – component slip in the α_2 – phase can be accomplished under tensile loading, for instance for the α_2 - , γ - phase orientation relationship we observed. Other orientation relationships for the two phases may exist for which the resolved shear stress resulting from combined effects of local pile up stresses at the hetero - interface and the external tensile loading suffices for activation of pyramidal plane c – component dislocation slip systems. However based on the limited fields of view we have investigated to date, it appears that the observation we report is unique, implying that the majority of the α_2 – phase fraction contained in the Ti-48at.% Al does not exhibit deformation behavior required for a general shape change and improved ductility.

ACKNOWLEDGEMENTS

We gratefully acknowledge provision of PST material by Dr. Dimiduk of the Wright Patterson Air Force Base and helpful discussion with Profs. H.L. Fraser and M.J. Mills. JW gratefully acknowledges support from the national Science Foundation for part of this work while working at the Ohio State University under the guidance of Prof. H.L. Fraser.

REFERENCES

[1] Clemens, H, and Kestler, H., Adv. Eng. Mater. **2** (2000) 551.
[2] Fujiwara, T, et al., Phil. Mag. A, 1990, **61**, 591.
[3] Y. Umakoshi et al. Acta Met. et Mat. 41 (1993) 1149.
[4] Legros, M., Minonishi, Y., and Caillard, D., Phil. Mag. A, **76** (1997) 995.
[5] Legros, M., Minonishi, Y., and Caillard, D., Phil. Mag. A, **76** (1997) 1012.
[6] J. M. K. Wiezorek, P. M. DeLuca, M. J. Mills, H. L. Fraser, Phil. Mag. Lett. **75** (1997) 271.
[7] J.M.K. Wiezorek, X.D. Zhang, W.A.T. Clark, H. L. Fraser, Phil. Mag. A **78** (1998) 217.

Mater. Res. Soc. Symp. Proc. Vol. 1128 © 2009 Materials Research Society   1128-U04-05

TEM of C-Component Dislocations Associated With Pyramidal Slip Activity in Hexagonal α2-Ti3Al

Jörg M. K. Wiezorek and Andreas K. Kulovits
Department of Mechanical Engineering and Materials Science, Swanson School of Engineering, University of Pittsburgh, 848 Benedum Hall, 37000 O'Hara Street, Pittsburgh, PA 15261, USA

ABSTRACT

While only a minor phase constituent, the deformation behavior of the hexagonal α_2-Ti$_3$Al phase, significantly affects the mechanical properties of two-phase TiAl based alloys. We have used conventional and high-resolution transmission electron microscopy to investigate the fine structure of pyramidal plane glide dislocations, with Burgers vectors of the type $\mathbf{b}=<2\mathbf{c}+\mathbf{a}>$, in α_2-Ti$_3$Al after room temperature compression of binary polysynthetically twinned TiAl normal to the lamellar interfaces to nominal plastic strains of 1%-7%. We report atomic resolution observations of non-co-planar dislocation core configurations for $<2\mathbf{c}+\mathbf{a}>$-dislocations and show that translamellar deformation twins active in the majority γ-TiAl phase play an important role in facilitating pyramidal plane slip in α_2-Ti$_3$Al in the lamellar two-phase alloys.

INTRODUCTION

The hexagonal ordered phase Ti$_3$Al is a minor constituent in two-phase TiAl-based alloys, which offer density specific property combinations attractive for high temperature structural applications in advanced airborne and ground transportation system [1]. The morphologically lamellar grains in fully-lamellar and duplex microstructures of polycrystalline TiAl-based alloys consist of thin slabs of the tetragonal ordered γ-TiAl and α$_2$-Ti$_3$Al, conjugated across their respective atomically close-packed planes and close-packed directions [2]. Lamellar grains exhibit anisotropic mechanical properties and require plasticity of the minority α$_2$-Ti$_3$Al phase for deformation normal to the lamellar interfaces [2]. Single crystals of Ti$_3$Al are also mechanically strongly anisotropic [3]. The $\{0001\}<11\bar{2}0>$ and $\{1\bar{1}00\}<11\bar{2}0>$ slip systems, involving dislocations with Burgers vectors of the type $\mathbf{b}=1/3<11\bar{2}0> = <\mathbf{a}>$, can be activated with relative ease in Ti$_3$Al, while c-component dislocation slip, involving dislocations with Burgers vectors $\mathbf{b}=1/3<11\bar{2}6> = <2\mathbf{c}+\mathbf{a}>$ gliding on first-order pyramidal planes, $\{2\bar{2}01\}<11\bar{2}6>$, and/or second-order pyramidal planes, $\{11\bar{2}1\}<11\bar{2}6>$, operates only for loading close to the c-axis [3-6]. The pyramidal plane $<2\mathbf{c}+\mathbf{a}>$-slip systems are associated with anomalous yielding [3]. A compression-tension anisotropy has been reported for $<2\mathbf{c}+\mathbf{a}>$-dislocation glide at room temperature in α$_2$-Ti$_3$Al with preference for the first-order pyramidal plane slip systems, $\{2\bar{2}01\}<11\bar{2}6>$, for tensile loading and for the second-order pyramidal planes, $\{11\bar{2}1\}<11\bar{2}6>$, for compressive loading [4,5]. Previous detailed investigations using in-situ tensile straining experiments and weak-beam dark field (WBDF) imaging in the transmission electron microscope (TEM) documented a complex nature of the pyramidal plane c-component dislocation glide processes in α$_2$-Ti$_3$Al [4,5]. Thus, large amounts of debris, including dislocation $<\mathbf{a}/2>$- and $[\mathbf{c}]$-type dislocation loops, are characteristically associated with pyramidal plane slip in Ti$_3$Al and non-planar configurations c-component dislocations have been observed. The respective dissociation and decomposition reactions of the $<2\mathbf{c}+\mathbf{a}>$-dislocations can involve length scales that are at or beyond the imaging resolution limit of WBDF TEM imaging (~2-3 nm). The anisotropic deformation behavior of the minority phase strongly affects the mechanical properties and behavior of the morphologically lamellar grains in two-phase TiAl alloys and therefore requires detailed understanding to support further improvement and development of this class of intermetallic-based engineering materials [6-8]. We previously reported on the role of translamellar deformation modes in the majority γ-TiAl phase on the activation of slip in the minority α_2-Ti$_3$Al phase [7,8]. In an extension of this prior work, we used here binary polysynthetically twinned (PST) Ti-48at.%Al model alloys to study the fine structure of c-component dislocations in the minority α_2-Ti$_3$Al phase activated during room temperature compressive loading by post-mortem conventional and high-resolution transmission electron microscopy (TEM and HREM).

EXPERIMENTAL PROCEDURES

Rods of PST-TiAl, with nominal binary composition of Ti-48at.%Al, 12.5 mm in diameter and up to 175 mm in length have been obtained from the Materials Directorate, Wright Laboratories, Wright-Patterson Air Force Base. Details of PST-TiAl synthesis, compositional analysis, and procedures for compression coupon preparation have been reported previously [7]. Rectangular shape compression coupons with square cross sections (4.0x4.0 mm^2) and 6.0 mm high have been deformed at room temperature at a constant rate of 5.6×10^{-4} s^{-1} to total strains between 0.01 and 0.07 with the compression axis oriented normal to the lamellar interfaces in the PST-TiAl, i.e., parallel to [111] in the γ-phase and [0001]=[c] in the α_2-phase. Twin-jet electropolishing using a solution of 6 vol.% perchloric acid and 34 vol.% butoxyethanol in methanol at a temperature of about -35° C and ion-beam milling with a liquid nitrogen cooled stage have been employed to obtain thin foils for the TEM investigation. A Philips CM200 and a Phillips CM300FEG TEM instrument have been used at 200 kV and 300 kV.

RESULTS AND DISCUSSION

At all levels of plastic straining normal to the lamellar interfaces the γ-phase lamellae accommodated the majority of the strain in the PST-TiAl compression coupons by activation of dislocation glide and mechanical twinning systems experiencing high resolved shear stresses, which are associated with Burgers and twinning shears with components normal to the interfaces, while the α_2-lamealle appeared relatively defect free in comparison. Observations related to the γ-phase deformation have been reported previously [9]. In concert with previous work <a>-dislocations with b=<a> have not been activated during N-orientation loading of PST-TiAl [2,8]. After 1% of nominal plastic compressive strain the TEM observations indicate that the α_2-lamellea remain essentially defect free, with very limited and localized deformation involving short segments of hooked shaped c-component dislocations, typically in the vicinity of stress concentrations associated with translamellar mechanical twins in the adjacent g-lamellae (e.g. labels A, B in Fig. 1). Attempts to determine the actual Burgers vector of these dislocations, which near deformation twins (labels T1 and T2, Fig. 1) in the adjacent γ-lamellae, by diffraction contrast tilting experiments has proven unsuccessful to date, due to the short line length of these dislocations and the proximity of the lamellar (γ/α_2)-interface. However, the contrast behavior of the dislocations A and B is consistent with <c+a/2>-dislocations (Fig. 1) contained n (0-22±1) and (11-2±1) respectively (Fig. 1). Thus, the defects near B in Fig. 1(b) exhibit a straight line morphology when viewed along a [-1100], whereas at least the defect above the label A appeared

148

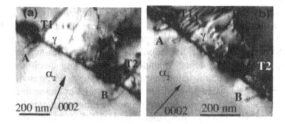

Figure 1: Early stages of pyramidal plane <2c+a> dislocation activity in the α_2-lamellae after 1% of strain; bright field TEM with **g**=0002 for (a) BD ~ <11-20> and (b) BD~ <01-10>.

to have significant curvature in this foil geometry. Interestingly, when viewed along [-2110] essentially the reverse is the case (Fig. 1a). We propose that these short <c+a/2>-dislocations are generated by the local stress concentrations of the translamellar deformation twins, involving the activation of interfacial sources (e.g. near B, Fig. 1) and possibly a direct twin shear transfer mechanism (e.g. near A, Fig.1) [8], and fail to contribute significantly to strain accommodation. In contrast, after compression to 0.03 strain <c+a/2>-dislocations are activated in pyramidal slip bands and the α_2-phase exhibits plasticity (Fig. 2). We observed two types of slip bands inclined to the lamellar interfaces (label D, Fig. 2) in an opposite sense (labels A, B and C, Fig. 2). This morphology was typical of the deformation microstructure in the α_2-lamellae after 0.03 strain. The defect analysis for the α_2-lamellae involved a series of tilting experiments, the results of which are not all shown here, since they were practically identical to those reported in [8]. Thus the <2c+a> dislocations in the planes with traces marked A and C have Burgers vectors parallel to [1-21-6] and [2-1-16] and glide on second-order pyramidal planes (-12-1-1) and (2-1-1-1) respectively. The contrast behavior of the defects associated with the slip-band marked B in Fig. 2(a) is consistent with <2c+a> dislocation activity in the first-order pyramidal plane (02-2-1).

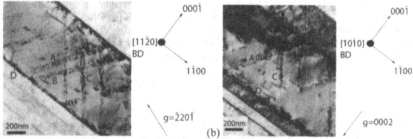

Figure 2: Morphology of pyramidal plane <2c+a> dislocation activity after 3% of strain; Bright field TEM for different beam directions (BD) and imaging vectors, g, as indicated; the traces of the planes labeled A to D have been identified from stereographic analyses (see text for details).

Fringe contrast (Fig. 2) and copious amounts of debris were associated with the slip activity of the <2c+a>-dislocations, consistent with previous studies [4, 5, 8]. Qualitative agreeing with a model proposed for the debris genesis [4,5] most of the debris visible for **g**=2-201 is invisible for **g**=0002 (Fig. 2), indicating that the majority of the small loop defects have Burgers vectors contained in the basal plane (0001). The <c+a/2>-dislocations generally exhibited convoluted line morphologies with numerous pinning points between bowed-out segments (Fig. 3). WBDF TEM details that at least some of these pinning points appear to be associated with local decompositions and/or dissociations of the individual <c+a/2>-dislocations as indicated by the small arrows in Fig 3(a) and along defect A. Under certain imaging conditions the dissociation of the <c+a/2>-superpartial dislocations into two partial dislocations has been detected in WBDF images (label B in Fig. 3(a)). A line profile from the boxed area of the central region along the line length of the defect B clearly demonstrates the possible dissociation with an image peak separation of about (2.67±0.17) nm in this projection (Fig3(b)). It has proven impossible to resolve this type of dissociation consistently in WB images produced with reflections **g** other than 0002. Similar difficulties associated with the diffraction contrast studies of the fine structure of <c+a/2>-superpartials in deformed α_2-Ti$_3$Al has been encountered in earlier studies [4,5].

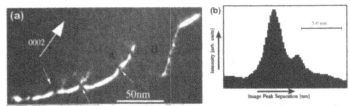

Figure 3: Weak beam dark field, (g,5.3g) in (a); <c+a/2> superpartials after 3% strain exhibiting dissociation into partials, local decomposition and complex line morphology with pinning points; (b) line profile from region marked B in (a) showing image peak separation of ~ 2.5 nm.

The HREM micrographs of Fig. 4 have been obtained for a beam direction, **BD**=[11-20]. Fig. 4(a) shows the fine structure of two <c+a/2>-dislocations, labels #1 and #2, contained in one of the pyramidal plane slip bands, which has been determined to be associated with (2-201) from additional diffraction contrast experiments (not shown here). A schematic of the overall configuration of this non-planar superpartial dipole is depicted in Fig. 4(b). Completing Burgers circuits around these two dislocations identified them as superpartials with a total Burgers vector consistent with **b**=1/6[2-1-1-6], which would be glissile in (2-201). Thus, the start (S) and end points (F) together with the Burgers circuit closure failure, **b**=1/6[2-1-1-6], and the spacing of the basal, $d_{0001}{\approx}4.7$ Å, and prism, $d_{1{\sim}100}{\approx}5.0$ Å, planes are marked in Fig. 4(c). The core structure of each of the individual superpartials #1 and #2 was quite complex. Thus, the terminations of the extra-half planes associated with the [**c**]-component and that associated with the <**a**>-component of the total Burgers vector were displaced with respect to one another by about 2.75 nm along the [1-100]-direction (Fig. 4(c). The [**c**]-component dislocation core is contained in (1-100), while the <**a**/2>-core is contained in (0001). These observations are consistent with a decomposition of the <c+a/2>-superpartials into two dislocations with orthogonal Burgers vectors and the formation of a non-planar, locked core-configuration:

$$1/6[2\text{-}1\text{-}1\text{-}6](2\text{-}201) \quad \Rightarrow \quad [000\text{-}1](1\text{-}100) + 1/6[2\text{-}1\text{-}10](0001) \tag{1}$$

or more generally

$$<c+a/2>\{2\text{-}201\} \quad \Rightarrow \quad [c]\{1\text{-}100\} + <a/2>(0001) \tag{2}$$

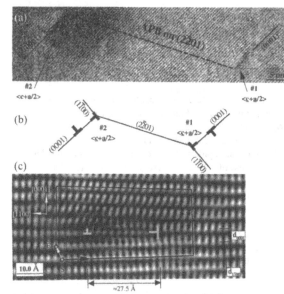

Figure 4: HREM of a dipole of <c+a/2> dislocations for BD=[11-20] after 3% strain. (a) Overview, the dipole related superpartials labeled #1 and #2; (b) Schematic of Z-shape dipole including the traces of basal, prism and pyramidal planes; (c) Wiener-filtered image of the decomposed non-planar core of superpartial #1 (a) and (b) with Burgers circuit.

The glide of a single superpartial with $b=<a/2>$ on the basal plane would generate an APB in Ti$_3$Al. The displacement associated with $b=<a/2>=1/6[2\text{-}1\text{-}10]$ shifts the prism planes (1-100) by half the prism plane spacing, $1/2d_{1\text{-}100} \approx 2.5$Å, across the basal plane APB in the [11-20]-projection. The frequencies due to the individual (2-200) superlattice planes with a spacing of $d_{2\text{-}200} \approx 2.5$ Å, i.e., half of $d_{1\text{-}100}$, are resolved. These superlattice planes are continuous across the apparent basal plane fault, which is consistent with the presence of such a basal plane APB. Precluding non-conservative processes, decompositions as described in (1) can occur only along line segments of the corresponding <c+a/2>-dislocations parallel to [11-20], the common trace of (2-201), (1-100) and (0001). With line direction $u=[11\text{-}20]$ the original <c+a/2>-dislocation, $b=1/6[2\text{-}1\text{-}1\text{-}6]$, would be of near edge-character, $\approx76°$ between u and b, the <a/2>-dislocation would be of 60°-character, and the [c]-dislocation would be of pure edge-character. The locked configuration for the overall <2c+a>-superdislocation forming as a result of the decompositions of each superpartial core according to (1) is depicted schematically in Fig 4(b).

The pyramidal <c+a/2>-dislocation slip systems are the only deformation modes available to the α_2-lamellae which can facilitate the accommodation of strains associated with the applied stress loading normal to the lamellar interfaces in the PST-TiAl. Furthermore, general ductility, or the α_2-phase cannot be provided without the activation of the pyramidal <c+a/2>-dislocation slip systems. The complex dislocation fine structures of the <c+a/2>-dislocations are in concert with the high τ_{CRSS}-values of these pyramidal slip systems [10]. Observations virtually identical to the two-peak WBDF images of Fig. 3 have been reported previously for defects in the first-order pyramidal planes, {2-201}, of monolithic Ti$_3$Al with stoichiometric composition [4, 5]. To be consistent with Fig. 3 the partial dislocations from the dissociation of the <c+a/2>-dislocation must have Burgers vectors with significant components parallel to [c]. In analogy to α-Ti, two possible suitable dissociation schemes can be proposed for the <c+a/2>-superpartials [11, 12]:

$$1/6[1\text{-}21\text{-}6] \quad \Rightarrow \quad 1/6[0\text{-}11\text{-}3] + 1/6[1\text{-}10\text{-}3], \tag{3}$$

$$1/6[1\text{-}21\text{-}6]_{(2\text{-}201)} \quad \Rightarrow \quad 1/24[1\text{-}54\text{ -}12]_{(2\text{-}201)} + 1/8[1\text{-}10\text{-}4]_{(2\text{-}201)} . \tag{4}$$

Reaction (3) involves partial dislocations with Burgers vector components normal to a macroscopic (2-201)-slip plane. Reaction (4) is planar and has been proposed as the more likely dissociation scheme for the first-order pyramidal plane [4,5]. Here <c+a/2>-slip has been observed more frequently on the second-order pyramidal planes, {11-21} and the Burgers vectors involved in the dissociation reaction (3) are contained in the appropriate second-order pyramidal plane, i.e. (1-211). The decomposition of <2c+a>-superdislocations in Ti₃Al into dislocations with orthogonal Burgers vectors similar to that identified here experimentally by HREM for the <c+a/2>-superpartials (Fig. 4) has been proposed previously based purely on theoretical considerations [13]. The non-planar core of the <c+a/2>-dislocations extended over a distance of ≈2.75nm on the basal plane, The decomposition reactions (1) and (2) result in the formation of very effective locks along line segments of the <c+a/2>-dislocations parallel to the appropriate <a>-direction and could produce pinning points at irregular intervals along the dislocation line (Fig. 3(a)). Furthermore, the decomposition of a <c+a/2>-dislocation according to (1) is neutral regarding the elastic line energy.

CONCLUSIONS

Our findings are consistent with interpretations put forward in [4, 5, 8] regarding the APB-dragging and the debris formation associated with pyramidal plane <c+a/2>-slip. Here we report the formation of sessile non-planar cores according to the type of reactions (2), representing a new micro-mechanism contributing to the high τ_{CRSS} of pyramidal <c+a/2>-slip in the α₂-phase.

ACKNOWLEDGEMENTS

We gratefully acknowledge provision of PST material by Dr. Dimiduk, Wright Patterson Air Force Base, support for JW from NSF for part of this work while at the Ohio State University, and helpful discussion with Profs. H.L. Fraser and M.J. Mills.

REFERENCES

[1] Clemens, H, and Kestler, H., Adv. Eng. Mater. 2 (2000) 551.
[2] Fujiwara, T, et al., Phil. Mag. A, 1990, 61, 591.
[3] Umakoshi, Y. , et al., Acta Metall. Mater. 41 (1993) 1149.
[4] Legros, M., Minonishi, Y., and Caillard, D., Phil. Mag. A, 76 (1997) 995.
[5] Legros, M., Minonishi, Y., and Caillard, D., Phil. Mag. A, 76 (1997) 1012.
[6] Umakoshi, Y., and Nakano, T., Acta Metall., 41 (1993) 1155.
[7] Wiezorek, J.M.K., et al., Phil. Mag. Let. 75 (1997) 271.
[8] Wiezorek, J.M.K., et al., Phil. Mag. A 78 (1998) 217.
[9] Wiezorek, J.M.K., Zhang, X.D., and Fraser, H.L., MRS Symp. Proc. Vol. 819 (2004) N4.5.1.
[10] Inui, H., et. al, Phil. Mag. A, 1993, 69, 1161
[11] Jones, I.P., and Hutchinson, W.B., Acta metall., 1981, 29, 951
[12] Bacon, D.J., and Liang, M.H., Phil. Mag. A, 1986, 53, 163
[13] Loretto, M.H., Phil. Mag. A, 1992, 65, 1095.

Mater. Res. Soc. Symp. Proc. Vol. 1128 © 2009 Materials Research Society   1128-U04-07

Initial Stages of Lamellae Formation in High Nb Containing γ-TiAl Based Alloys

Limei Cha[1,2], Christina Scheu[2,3], Gerhard Dehm[4], Ronald Schnitzer[2], and Helmut Clemens[2]
1. Materials Center Leoben Forschung GmbH, Leoben, Austria
2. Department of Physical Metallurgy and Materials Testing, Montanuniversität Leoben, Austria
3. Department of Chemistry and Biochemistry Ludwig-Maximilians-University Munich, Germany
4. Department of Materials Physics, Montanuniversität Leoben, and Erich-Schmid Institute of Materials Science, Austrian Academy of Sciences, Leoben, Austria

ABSTRACT

In this study a two-step heat treatment was applied to high Nb containing γ-TiAl based alloys in order to investigate the initial stage of the lamellae formation in ordered α_2-grains as well as in massively transformed γ_m-grains. The first heat treatment step, conducted in the single α-phase field followed by oil quenching, leads to a microstructure consisting of supersaturated α_2-grains and a small volume fraction of γ_m-grains. The second step of the heat treatment was performed below the eutectoid temperature, i.e. within the ($\alpha_2+\gamma$)-phase field region and was again followed by oil quenching. There, the formation of ultra-fine γ-lamellae takes place in the α_2-grains and (some) fine α_2-lamellae are formed in the γ_m-grains. In both cases the lamellae show a Blackburn orientation relationship with the matrix grain. It was found that the precipitation of γ-laths in the supersaturated α_2-grains is faster than the formation of α_2-laths in γ_m-grains. The characteristics of the initial stage of formation were investigated by transmission electron microscopy.

INTRODUCTION

Intermetallic titanium aluminides are innovative materials for application in aero-engines and combustion-engines [1]. In two-phase γ-TiAl based alloys the mechanical properties are often determined by the presence of a lamellar microstructure which can be substantially varied by modifying the length scale between the internal γ-TiAl/α_2-Ti$_3$Al heterophase as well as γ-TiAl/γ-TiAl homophase boundaries [1-3]. To control the microstructure and thus to adjust the desired mechanical properties, it is important to understand the formation and growth processes, as well as the morphology of the (α_2-Ti$_3$Al+γ-TiAl) lamellar structure.

In γ-TiAl based alloys, a lamellar microstructure can be obtained from the unordered α-phase or the ordered α_2-phase by different routes [4]. Route I is a slowly cooling from the single α-phase field region, i.e. the following reactions are taking place: α→(α+γ)$_{Lamellar}$→($\alpha_2+\gamma$)$_{Lamellar}$. Route II uses the transformation from α→α_2→($\alpha_2+\gamma$)$_{Lamellar}$ by rapid cooling from the α-phase field region and subsequent aging below the eutectic temperature. A lamellar microstructure can also be produced via a third route which involves the formation of the γ_m-phase by massive transformation and a subsequent aging treatment (route III), i.e. α→γ_m→($\alpha_2+\gamma$)$_{Lamellar}$ [4,6]. In all three cases the Blackburn orientation relationship (BBOR), i.e. $(0001)_{\alpha2}[2\text{-}1\text{-}10]_{\alpha2}//\{111\}_\gamma<110>_\gamma$ exists between the α_2 and γ laths. It was found that γ precipitated in α or α_2 by dislocation decomposition (from <2-1-10>/3 to <10-10>/3) accompanied by stacking faults and interdiffusion [7-9]. On the other side α_2 is formed in γ_m matrix when the following dislocation decomposition takes place: <110]/2 to <2-1-1]/6 [10, 11]. The growth of γ as well as α₂ lath is performed by a terrace-ledge-kink mechanism [4-9]. Depending on the cooling condition from the single α-phase field, the microstructure prevailing at room temperature can consist of supersaturated α₂-grains and massively transformed γₘ-grains [3-9]. Ageing of such a microstructure leads to lamellae formation both in α₂- and γₘ-grains. The formation of a lamellar structure from α, α₂ and γₘ grains (route I, II and III, respectively) has been the topic of former studies [4-9]. However, the lamellar formation in a mixed α₂- and γₘ- microstructure has rarely been analyzed in massive transformed γ-TiAl based alloys. It was found by transmission electron microscopy (TEM) that a few (α₂+γ)-lamellae were located next to α₂/γₘ interfaces and in α₂-grains [10, 11]. Diffraction experiments revealed that between these α₂- and γₘ-grains the BBOR exists [10, 11].

In this work the early stage of lamellar formation in a microstructure consisting of α₂ and γₘ grains is investigated. The difference of the lamellae formation, i.e. in α₂ and γₘ grains, is analyzed and discussed. The main investigation tools are scanning electron microscopy (SEM) and TEM.

EXPERIMENT

For the present investigation a high Nb containing Ti-45Al-7.5Nb (composition in atomic percent) alloy was used. A two-step heat treatment was applied to study the initial stage of lamellae formation. At first, the samples were heated above the α-transus temperature (~1295°C), held for 10 minutes in the single α-phase region and then oil quenched (OQ) to room temperature (sample type A). In a second step, the samples were continuously heated with a rate of 20°C/min to a temperature of 790°C and then immediately followed by OQ (sample type B). The microstructure of the samples (sample type A and B) was studied by SEM and TEM. SEM investigations were performed in back-scattered electron (BSE) mode in a Zeiss EVO 50. Conventional TEM studies were carried out using a Philips CM12. The deTiAls for sample preparation are described in reference [2].

RESULTS AND DISCUSSION

a): Morphology of γₘ- and α₂-grains

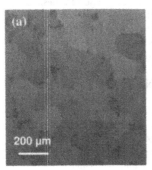 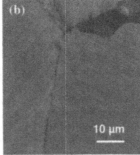

Figure 1. SEM images of Ti-45Al-7.5Nb: (a) after the first oil quenching (sample A). The large grains are supersaturated α₂ phase and the dark appearing phase at the α₂ grain boundary is γₘ. (b) shows the microstructure of the alloy after the two step heat treatment (sample B) with higher magnification. The image shows lamellar colonies and some remaining γ grains at colonies boundaries.

154

SEM images taken in BSE mode from samples A and B are shown in Fig. 1. The α_2 grain size ranges between 100 and 150 μm. From Fig. 1a it is evident that a small amount (~ 10%) of irregular shaped γ_m-grains exist at grain boundaries (GB) and triple junction of α_2 grains, and no lamellar structures could be detected by SEM. For the sample B an almost fully lamellar microstructures occurred within the former α_2-grains and a small number of remained γ-grains is present at α_2 grain boundaries (Fig. 1b). Due to the limited spatial resolution of the SEM, TEM technique was employed to study the microstructures mainly at areas close to α_2/γ_m interfaces. In the following the different types of γ modification are distinguished by subscripts, e.g. γ_m indicates massively transformed γ, and γ_L denotes a γ-lath. The superscripts indicate the orientation of grains, for example, α_2^1 and α_2^2 exhibit different orientations and γ^1 and γ^{VI} are variants in one single $(\alpha_2+\gamma)_L$ colony (see [4]). An α_2 grain, which has a BBOR with both γ^1 and γ^{VI}, is denoted as α_2^1. Thus, a lamellar structure can consist of the following variants: $(\alpha_2^1+\gamma_L^1+\gamma_L^{VI})_L$, $(\alpha_2^1+\gamma_L^1)_L$ and/or $(\alpha_2^1+\gamma_L^{VI})_L$.

b): Early stage of lamellar formation in α_2-grains adjacent to an α_2/γ_m interface

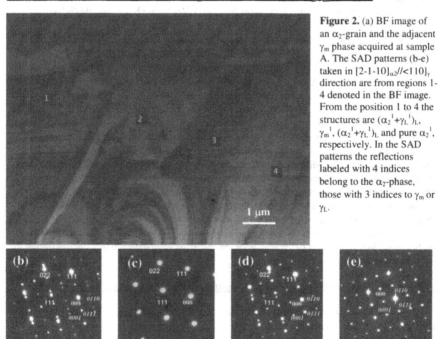

Figure 2. (a) BF image of an α_2-grain and the adjacent γ_m phase acquired at sample A. The SAD patterns (b-e) taken in $[2\text{-}1\text{-}10]_{\alpha2}//<110>_\gamma$ direction are from regions 1-4 denoted in the BF image. From the position 1 to 4 the structures are $(\alpha_2^1+\gamma_L^1)_L$, γ_m^1, $(\alpha_2^1+\gamma_L^1)_L$ and pure α_2^1, respectively. In the SAD patterns the reflections labeled with 4 indices belong to the α_2-phase, those with 3 indices to γ_m or γ_L.

155

Fig. 2a is a TEM bright-field (BF) image taken from sample A. It displays an α_2^1 matrix grain and an adjacent γ_m^1 grain with irregular shape. Combining the BF image and the selected area diffraction (SAD) patterns taken in the vicinity of the α_2/γ_m interface, it can be concluded that some fine γ laths (γ_L^1) occur within the α_2^1 grain (Fig. 2a), which show a BBOR with α_2^1 (Figs. 2b-2e). It should be noted that, another α_2 (α_2^2) grain showing no BBOR with γ_m^1 is also adjacent to this γ_m^1 grain, but it is not included in Fig.2a. The point of origin of the γ_L^1 precipitated in the α_2 grain is always connected with the α_2^1/γ_m^1 interface. However, there is no obvious grain boundary between γ_L^1 and γ_m^1. In Fig. 2 all the diffraction patterns were taken in $[2\text{-}1\text{-}10]_{\alpha 2}//<1\text{-}10]_\gamma$ direction.

Zhang et al. [11] studied massive transformation in Ti-49Al, Ti-48Al-2Mn-2Nb, Ti-55Al-25Ta, and Ti-50Al-20Ta alloys by applying two-step heat treatments similar to our study. They detected a $(\alpha_2^1+\gamma_L^1)/\gamma_m^1/\alpha_2^2$ structure in their samples and suggested that γ_L^1 lamellae, which had a BBOR with α_2^1 but not with α_2^2 grain, is the site of origin of the γ_m^1 grain which then grows into α_2^2. It was concluded that the γ_L^1 laths precipitate from α_2^1/α_2^2 grain boundary and grew firstly into α_2^1 grain. Then the γ_m^1 grain is formed from the same α_2^1/α_2^2 grain boundary but spread into the α_2^2 grain showing no BBOR. This seems to fit to the expected phase formation sequence, i.e. γ_m forms at a lower temperature and always consumes the adjacent α_2 grain with which it possesses no BBOR.

However, when a γ_L^1 lath precipitates from an α_2^1/α_2^2 grain boundary, this γ_L^1 and its variant (γ_L^{V1}) should have the same probability to form within the α_2^1 grain. Therefore, a $(\alpha_2^1+\gamma_L^1+\gamma_L^{V1})$ lamellar structure should be the preferred microstructure if γ_L^1 lath forms before γ_m. However, the diffraction patterns in our study and the work of Zhang et al. (see Fig. 3 of Ref. [11]) reveal only one type of γ, i.e. the $(\alpha_2+\gamma_L^1)$ lamellar structure which is adjacent to γ_m and no γ_L^{V1} was detected. Especially in Fig. 2a, only the $(\alpha_2^1+\gamma_L^1)$ lamellar structure is observed at different edges of the same γ_m grain. The absence of the γ_L^{V1} variant might be due to the sequence of the phase formation, i.e. γ_m^1 formed at first, then γ_L^1 expanded into the adjacent α_2^1 to generate $(\alpha_2^1+\gamma_L^1)$ lamellae. In this assumption, γ_L^{V1} has a much lower chance to be produced due to the additional interface energies of γ_m^1/γ_L^{V1}, and no γ_L^{V1} were found after the first heat treatment step. Therefore, it is more reasonable that γ_m^1 is the point of origin. The described sequence of phase formation is supported by adiabatic consideration where the latent heat of transformation from α into γ_m indicates a local temperature raise sufficient to promote the $(\alpha_2+\gamma)_L$ lamellar formation locally. This point will be analyzed in more detail in a future study. In summary, during the first heat treatment, the γ_m^1-phase forms firstly, then γ_L^1 lamellae start to expand from the α_2^1/γ_m^1 interface. In this condition, the γ_L^1 laths displayed the same orientation with the parent γ_m^1 grain and thus also have BBOR with the α_2^1 grain. However, no γ_L^{V1} was formed. Compared to the number of γ_L^1 laths formed in α_2^1 grains, only few α_2^1 laths were detected in γ_m^1 grains during the first heat treatment step.

C). Early stage of lamellar formation in γ_m matrix adjacent to an α_2/γ_m interface

The SEM images of sample B after the second heat treatment (Fig. 1b) clearly reveal that lamellae were formed within α_2 grains. During the second annealing step γ_m becomes thermodynamically instable, thus facilitating a reaction $\gamma_m \rightarrow \alpha_2 + \gamma$. In order to investigate this process and how it is influenced by the adjacent α_2 grains, the following part focuses on the area around the former γ_m grains.

The TEM bright-field image in Fig. 3a shows a microstructure of γ_m^1/α_2^1, where the γ_m grain (in the following labeled as γ_m^1) has a BBOR with α_2^1 grain. The corresponding dark-field image (Fig. 3b), which was taken by using the (0002) reflection of α_2^1, reveals that a few α_2 laths have formed at the γ_m^1/α_2^1 interface and extend into the γ_m^1 grain. SAD patterns (Fig. 3c-3f) were taken from different regions across the γ_m^1/α_2^1 interface (see positions 1-4 in Fig. 3a). Position "1" is the inner part of γ_m^1 grain, "2" an area in the γ_m^1 grain close to the γ_m^1/α_2^1 grain boundary, "3" is the region in the α_2^1 grain near to the γ_m^1/α_2^1 grain boundary, and "4" is the interior of α_2^1 grain. The convergent electron beam diffraction pattern taken from region 1 (Fig. 3c) shows that the inner part is pure γ_m^1-phase. The SAD recorded from region 2 (Fig. 3d) indicates a $(\alpha_2^1+\gamma_m^1)_L$ lamellar structure formed at the left side of the γ_m^1/α_2^1 interface. In the area close to the γ_m^1/α_2^1 interface, but in the former α_2^1 grain (region 3), two γ variants were detected (Fig. 3e), i.e. a $(\alpha_2^1+\gamma_L^1+\gamma_L^{VI})_L$ structure occurs. In the interior of the α_2^1 grain (Fig. 3f), $(\alpha_2^1+\gamma_L^{VI})_L$ is the predominant structure, although some γ_L^1 can still be revealed. Therefore, it can be concluded that, starting from the γ_m^1/α_2^1 interface, γ_L^1 laths and α_2^1 laths extend into the α_2^1 and γ_m^1 grain, respectively. In the former γ_m^1 grain a few α_2^1 laths were formed, while in the former α_2^1 grain a $(\alpha_2^1+\gamma_L^1+\gamma_L^{VI})$ lamellar structure was generated. Certainly, for the α_2 grains without adjacent γ_m grains, a $(\alpha_2^1+\gamma_L^1+\gamma_L^{VI})$ lamellar structure is the common feature [2].

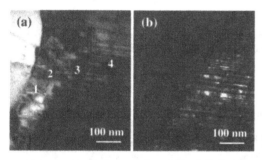

Figure 3. (a) BF and (b) DF images of the α_2/γ_m grain boundary region in sample B. All diffraction patterns were taken in $[2\text{-}1\text{-}10]_{\alpha 2}//<110>_\gamma$ directions. Due to the small width of the γ_m grain, convergent electron beam diffraction (c) was performed at position 1. The SAD patterns in (d-f) are corresponding to the positions of 2-4. The structures of region 1-4 are γ_m^1, $(\alpha_2^1+\gamma_L^1)_L$, $(\alpha_2^1+\gamma_L^1+\gamma_L^{VI})_L$ and $(\alpha_2^1+\gamma_L^{VI})_L$, respectively.

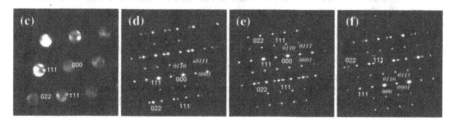

Previous studies [3,9,12] reported the existence of crossed α_2 laths in γ_m grains. Since there are four equally crystallographic {111} planes in γ_m, which are inclined 70.5° relative to each other, the (0002) planes of α_2 precipitates have four possibilities to match the (111) planes of the γ_m parent phase, when the aging time is long enough to bring the system close to thermal equilibrium conditions. Thus, crossed α_2 laths might be observed in a γ_m grain in <011> imaging direction. However, in our experiments the samples were continuously heated to 790°C, and then immediately quenched to room temperature. For this reason, the α_2 laths had very limited time to precipitate in the γ_m-grains. Therefore, α_2^1 laths showing the same orientation as the adjacent α_2^1 grain have formed preferentially in the γ_m^1 grain, resulting in the $(\alpha_2^1 + \gamma^1)_L$ lamellar structure observed in our sample and only few crossed α_2-laths were found. It might be speculated that a higher temperature and/or an extended aging time leads to the formation of crossed α_2-laths in γ_m grains. In-situ heating TEM experiments are planned to confirm this assumption.

CONCLUSIONS

The initial stages of lamellae formation in a Ti-45Al-7.5Nb alloy were investigated. Using a two-step heat treatment, $(\alpha_2 + \gamma)$ lamellar structures can be formed in supersaturated α_2-grains as well as in γ_m-grains both displaying a BBOR. After the first quenching, some γ_L^1 lamellae occur in the α_2^1 matrix, exhibiting a BBOR. These γ_L^1 laths have the same orientation as the adjacent γ_m^1 grain. When the specimen is subjected to the second heat treatment step, a $(\alpha_2^1 + \gamma_L^1 + \gamma_L^{VI})_L$ lamellar structure is formed in the former α_2^1 grain. In contrast, in the former γ_m^1 grain, which has a BBOR with the adjacent α_2^1 grain, only a few α_2^1 laths were formed.

ACKNOWLEDGEMENT

Part of this work was financially supported by the Austrian Federal Government and the Styrian Provincial Government under the frame of the Austrian COMET Competence Centre Program.

REFERENCES

1. H. Kestler, H. Clemens, in Titanium and Titanium Alloys, (Eds. C. Leyens, M. Peters), WILEY-VCH, Weinheim 2003, pp. 351.
2. L. Cha, C. Scheu, H. Clemens, H. Chladil, G. Dehm, R. Gerling, A. Bartels, Intermetallics **16**, 868 (2008).
3. H. Clemens, A. Bartels, S. Bystrzanowski, H. Chladil, H. Leitner, G. Dehm, Intermetallics **14**, 1380 (2006).
4. F. Appel, R.Wagner, Mater. Sci. Eng. R **22**, 187 (1998).
5. H. Park, S. Nam, N. Kim, S. Hwang, Scripta Mater. **41**, 1197 (1999).
6. R. Pond, P. Shang, T. Cheng, M. Aindow, Acta Mater. **48**, 1047 (2000).
7. G. Qin, S. Hao, X. Sun, Scripta Mater. **39**, 289 (1998).
8. A. Denquin, S. Naka, Acta Mater. **44**, 343 (1996).
9. T. Kumagai, E. Abe, T. Kimura, M. Nakamura, Scripta Mater. **34**, 235 (1995).
10. P. Wang, D. Veeraraghavan, M. Kumar, VK. Vasudevan, Metall. Mater. Trans. **33**, 2353 (2002).
11. X. Zhang, S. Godfrey, M. Weaver, M. Strangwood, P.Threadgill, MJ. Kaufman, M. Loretto, Acta Mater. **44**, 3723 (1996).
12. R. Schnitzer, H. Chladil, C. Scheu, H. Clemens, S. Bystrzanowski, A. Bartels, S. Kremmer, Prakt. Metallogr. **44(9)**, 430 (2007).

Mater. Res. Soc. Symp. Proc. Vol. 1128 © 2009 Materials Research Society 1128-U04-09

Surface Treatment of TiAl With Fluorine for Improved Performance at Elevated Temperatures

A. Donchev, P.J. Masset, M. Schütze
DECHEMA e.V. Karl-Winnacker-Institut, Theodor-Heuss-Allee 25, D-60486 Frankfurt/ Germany

ABSTRACT

Alloys based on aluminium and titanium are possible materials for several high temperature applications. The use of TiAl would increase the efficiency of e.g. aero turbines, automotive engines and others due to their properties, among others low specific weight and good high temperature strength. The oxidation resistance is low at temperatures above approximately 800°C so that no long term use of TiAl-components is possible without improvement of the oxidation behaviour. Small amounts of halogens in the surface zone of TiAl-samples lead to a dramatic improvement of the oxidation resistance at temperatures up to 1100°C for more than 8000 hours in air. In this paper results of the work on the halogen effect over the last years are presented. The results of thermogravimetric measurements, thermocyclic oxidation tests of small coupons and thermodynamic calculations for different atmospheres (e.g. air, H_2O, SO_2) are shown and the halogen effect mechanism is discussed. The postulated mechanism is in good agreement with the results of the oxidation tests. The limits of the halogen effect will also be mentioned. Predictions for the halogenation of TiAl-components can be given so that the processing can be planned in advance.

INTRODUCTION

Intermetallic titanium aluminides are potential light weight materials for several high temperature applications [1]. Their high temperature strength would allow the use at temperatures above 800°C but their oxidation resistance is very poor in this temperature range so it has to be improved. The specific weight of TiAl-alloys is about 4 g/cm³ and thus only half of the normally used Ni-based alloys or steels. This weight reduction may be a way to increase the efficiency of e.g. automotive or jet engines and to reduce fuel consumption and noise. One possibility to improve the oxidation behaviour drastically is the halogen treatment [2]. The oxidation kinetics is reduced by several orders of magnitude and the formation of a non protective mixed scale ($TiO_2/TiN/Al_2O_3$) during oxidation in air is suppressed instead an adherent alumina layer is formed even under thermocyclic conditions which protects the base material [3]. Especially fluorine has proven to be the best doping element [4]. In this paper results of untreated and fluorine treated TiAl-samples oxidized isothermally and thermocyclically in dry and wet air and air plus SO_2 are presented. The calculated limits for the fluorine effect are shown and the effect of SO_2 in the oxidation is discussed thermodynamically. The established halogen effect model for the oxidation in air is extended to SO_2 containing environments. In this paper results of thermodynamic calculations and isothermal and thermocyclic oxidation tests at elevated temperatures are presented.

EXPERIMENTAL

Coupons of the technical TiAl-alloys TNB (Ti-40Al-10Nb) and TNB V5 (Ti-45Al-5Nb-0.2B-0.2C) were cut from bars, ground to 1200 and 2400 grid, respectively, with SiC paper, rinsed with distilled water, cleaned ultrasonically in ethanol and dried in air. One set of such prepared samples was oxidized in dry and wet air and air plus SO_2 (0.1 vol.%) without further treatment to investigate the oxidation behavior of the pure material. Other samples were treated with fluorine before oxidation. Different fluorine treatments were used. Fluorine was applied to the surface by beam line ion implantation (BLI[2]), plasma immersion ion implantation (PI[3]), gas phase processes, spraying, painting or dipping. Isothermal thermogravimetric measurements were performed in a vertical tube furnace under a steady flow of synthetic air. Thermocyclic or quasi isothermal oxidation tests were executed in horizontal tube furnaces again with a steady flow of synthetic air moistened with 10% H_2O by passing the air through a thermostat. Tests with SO_2 were performed by adding SO_2 into the air stream. Post experimental investigations included metallography, light microscopy (LM), scanning electron microscopy (SEM) and energy dispersive X-ray (EDX).

RESULTS AND DISCUSSION

The halogen effect is based on the gas phase transport model proposed by Donchev et al. [5]. The boundaries of metal halide partial pressures of the chemical cycle (Equations 1 and 2) where a positive effect is expected [5] are shown in figure 1.

$$2\,Al(s) + X_2 \longrightarrow 2\,AlX(g) \tag{1}$$

$$4\,AlX(g) + 3O_2(g) \longrightarrow 2\,Al_2O_3 + 2\,X_2 \tag{2}$$

In equations 1 and 2 X represents the halogen species (fluorine, chlorine, bromine or iodine). Thermodynamic calculations were performed to assess the conditions for the effectiveness of the halogen effect in the case of fluorine. The corridor situated between $p(X)_{min.}$ and $p(X)_{max.}$, where a positive halogen effect is expected [6], is a function of the alloy composition, the oxidation temperature and the halogen used. The halogen pressure $p(X)$ represents the sum of the volatile halogen species which might be involved in the halogen effect. It was shown that the fluorine partial pressures for a positive effect are similar for the alloys $Ti_{0.6}Al_{0.4}$ to $Ti_{0.5}Al_{0.5}$ which corresponds to the composition range of most of the industrial alloys (Figure 1). The minimum and maximum halide partial pressures as well as the width of the halogen partial pressure area characterizing the halogen effect increase with temperature [6]. A positive effect is given by the formation of a protective alumina scale. The thermodynamic calculations show the exclusive formation of gaseous Al-fluorides if the fluorine partial pressure stays with in the fluorine window which is defined by the minimum amount of Al-fluorides for the formation of the alumina scale and the maximum where the partial pressure Ti-fluorides exceed the value of the Al-fluorides. The results also give no hind of a negative effect of additional elements like Nb or others. The partial pressures of their fluorides are several orders of magnitude smaller than those of the Al-fluorides [5]. The thermodynamic calculations were limited to binary alloys as the ternary system Al-Ti-Nb system is completely resolved and still under active investigations [7, 8]. In addition, the quaternary system Ti-Al-Nb-O is partially described and no thermodynamic database is available [9].

160

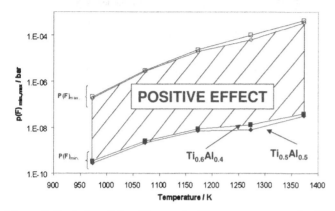

Figure 1: Evolution of the $p(F)_{max}$ and $p(F)_{min}$ vs. temperature for different alloy compositions. $p(F)_{max}$: (∇) $Ti_{0.5}Al_{0.5}$, (M), $p(F)_{min}$: (!) $Ti_{0.5}Al_{0.5}$, (Λ) $Ti_{0.6}Al_{0.4}$ [6].

The system Ti-Al-O-S was investigated to evaluate the conditions of formation of sulfides during the oxidation process. Figure 2 shows the calculated stability diagram of Ti and Al versus the O_2 and SO_2 partial pressures at 1050 °C which represents approx. the upper temperature of use of TiAl alloys, i.e. in turbochargers. Under these conditions, the lower and upper limit for the O_2 partial pressure is 10^{-33} bar (O_2 equilibrium partial pressure at the oxide/metal interface) and 0.8 bar (air), respectively. The SO_2 partial pressure varies from 10^{-30} bar (SO_2 equilibrium partial pressure for the metal sulfide formation) to 10^{-3} (SO_2 partial pressure in synthetic air for the experiments presented). It shows that both Al_2S_3 and TiS_2 are stable for oxygen partial pressures lower than 10^{-11} bar at 1050 °C (green dotted line). For lower temperatures the stability domains of $TiO_2 + Al_2O_{12}S_3$ extend to lower O_2 and SO_2 partial pressures. According to these thermodynamic predictions some sulfides should form especially in the vicinity of the oxide/metal interface where the oxygen partial pressure is the lowest in the case the oxide scale would not be dense enough and allows the ingress of SO_2. At 1050 °C, the compound $Al_2O_{12}S_3$ does not form in the partial pressure range investigated. In addition, the results of these thermodynamic calculations showed that the boundaries for the halogen effect do not change in the presence of SO_2 and are similar to what has been determined previously [6]. It means that the fluorination parameters may be kept constant even for such environments.

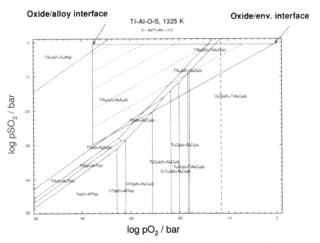

Figure 2: Calculated stability diagram at 1050 °C of Ti and Al as a function of oxygen and sulfur dioxide partial pressures.

In Figure 3 results of thermocyclic 25h-cycle tests in wet and dry air and metallographic cross sections are shown. These observations showed that the fluorination improves significantly the oxidation behavior of TiAl-alloys in wet and dry air [7]. Nb leads to a reduction of the mass gain of TiAl-alloys which is explained by several mechanisms [3] but does not promote the formation of a pure alumina scale. This is only found on TiAl-alloys after halogen treatment.

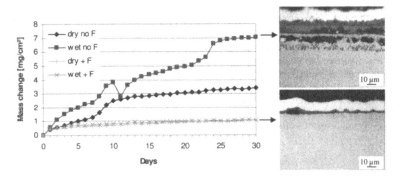

Figure 3: Mass change behavior of the alloy TNB during a 25h-cycle test in wet and dry air at 900°C (24h on temperature/1h cooling)

The mass gain of the untreated alloy is always higher than after F-treatment. In wet environments the oxidation rate is increased so that the mass gain is higher [8]. The mass loss in wet air after 10 days was caused by spallation of oxide flakes. The mixed oxide scale got too

162

thick so that during cooling tensions due to the thermal expansion mismatch could only be released by cracking and finally spallation. The fluorine treatment works also in wet environment. The alumina grow is not influenced by the water vapor so that the curves are almost equal. The results shown in Figure 3 were achieved with the beamline implantation technique. This method can not be used for complex components so that different fluorination processes were developed. In Figure 4 the mass gain behavior of the third generation TiAl-alloy TNB V5 with and without F-treatment during a 25h-cycle test and SEM-micrographs of the surface are presented.

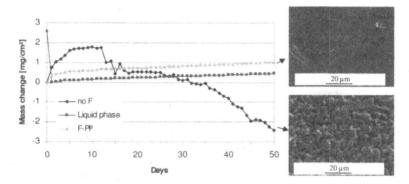

Figure 4: Mass change behavior and SEM-micrographs of the surface of the alloy TNB V5 during a 25h-cycle test in dry air at 900°C (24h on temperature/1h cooling)

The untreated alloy is very susceptible to spallation so that it can not be used without oxidation protection and the surface is covered with coarse TiO_2-crystals. The fluorine effect works under these conditions. The mass loss after the first day of the liquid phase treated sample is due to evaporation of the residues of the F-compound. The surface is much smoother. The results of oxidation in air plus SO_2 are shown in Figure 5.

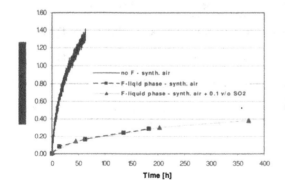

Figure 5: Comparison of the oxidation curves at 900 °C TNBV5 alloy. (_) untreated and oxidized under synthetic air, (■) F-treated and oxidized under synthetic air, (▲) F-treated and oxidized under synthetic air + 0.1 v/o SO_2.

163

After 50 h of oxidation at 900 °C, the mass up-take for the untreated specimen is close to 1.4 mg/cm^2 whereas it is lower than 0.2 mg/cm^2 for the F-treated specimen. Under SO$_2$ containing environment the mass variation versus time for the fluorinated sample was found to be the same as under synthetic air (Figure 5). It shows that the halogen effect is still effective even in the presence of SO$_2$ which agrees with the thermodynamic predictions. The oxidation curves follow parabolic laws and it leads to kinetic constants of 2 x 10^{-13} g^2.cm^{-4}.s^{-1} for F-treated specimens (in synthetic air with and without SO$_2$) whereas it is close to 5 x 10^{-11} g^2/cm^4.s^1 for untreated specimens.

CONCLUSIONS

This work shows that the fluorine effect works under thermocyclic conditions even in wet and SO$_2$ containing environments. The fluorine treatment is a very promising way to improve the oxidation resistance of TiAl alloys for high temperature applications up to at least 900°C.

ACKNOWLEDGEMENTS

The work was funded by the German Ministry of Economics and Technology via the German Federation of Industrial Research Associations (AiF) under the contract numbers 176 and 262 ZBG which is gratefully acknowledged by the authors. The work is a cooperation with the Research Center Dresden where the implantation work was carried out which is also acknowledged. Many thanks go to Rolls-Royce Germany and Tital for providing the materials in this work.

REFERENCES

1. F. Appel and M. Oehring, "γ-Titanium Aluminide Alloys: Alloy Design and Properties," *Titanium and Titanium Alloys*, ed. C. Leyens and M. Peters (Wiley-VCH, 2003) pp. 89-152.
2. M. Schütze and M. Hald, Mat. Sci. Eng. **239-240**, 847-854 (1997).
3. C. Leyens, "Oxidation and Protection of Titanium Alloys and Titanium Aluminides," *Titanium and Titanium Alloys*, ed. C. Leyens and M. Peters (Wiley-VCH, 2003) pp. 187-230.
4. A. Donchev and M. Schütze, Mat. Sci. Forum **261-264**, 447-454 (2004).
5. A. Donchev, B. Gleeson and M. Schütze, Intermetallics **11**, 387-398 (2003).
6. P.J. Masset and M. Schütze, Adv. Eng. Mater. **10** 666-674 (2008).
7. V.T. Witusiewicz, A.A. Bondar, U. Hecht, T.Ya. Velikanova , J. Alloys Comp. **472(1-2)**, 133-161 (2009).
8. O. Rios, S. Goyel, M.S. Kesler, D.M. Cupid, H.J. Seifert, F. Ebrahimi, Scripta Mater. **60(3)** 156-159 (2009).
9. M. Allouard, Y. Bienvenu, I. Nazé, C.B. Bracho-Traconis, J. Physique IV, Coll. C9, suppl. to J. Phys. III, **3**, 419-428 (1993).
10. A. Donchev, E. Richter, M. Schütze and R. Yankov, Intermetallics **14**, 1168-1174 (2006).
11. A. Zeller, F. Dettenwanger and M. Schütze, Intermetallics **10**, 59-72 (2002).

Mater. Res. Soc. Symp. Proc. Vol. 1128 © 2009 Materials Research Society   1128-U04-10

The Role of Fundamental Material Parameters for the Fluorine Effect in the Oxidation Protection of Titanium Aluminides

Hans-Eberhard Zschau and Michael Schütze

DECHEMA e.V., Karl-Winnacker-Institut, Theodor-Heuss-Allee 25, D-60486 Frankfurt am Main, Germany

ABSTRACT

The increasing interest in Gamma-TiAl based alloys is motivated by their excellent specific strength at high temperatures which offers a high application potential in aerospace and automotive industries. To improve the insufficient oxidation resistance at temperatures above 750 °C the fluorine effect leading to the formation of a protective alumina scale offers an innovative way. After F implantation of TiAl and oxidation at 900 °C/1000 °C the fluorine maximum determined by using PIGE is found to be at the metal/oxide interface. The time dependent behaviour of the fluorine content is characterized by a fast decrease of the fluorine concentration during heating followed by a moderate decrease. As parameter for the stability c_F^{max} was defined. The time behaviour can be described by an exponential decay function with a constant part of $0.3 - 1.2$ at.-% offering a stable oxidation protection by the fluorine effect and its possible technical application. A theoretical modelling performed by using the diffusion coefficient of F in TiAl at 900 °C fits the experimental data with a good agreement.

INTRODUCTION

Due to their high specific strength at high temperatures the γ-TiAl based alloys are believed to have a high application potential in high temperature technologies, especially in aerospace and automotive industries. In contrast to the presently used Ni-based superalloys their specific weight is about 50 % lower leading to a lower moment of inertia for rotating parts like turbocharger rotors, exhaust valves or turbine blades. Thus lower mechanical stresses and a reduced fuel consumption can be expected. However the oxidation resistance of γ-TiAl inhibits an industrial use at temperatures above 750 °C [1]. To reach higher service temperatures the surface modification is expected to avoid any detrimental influence on the excellent mechanical properties of the material. The oxidation resistance can be improved by using the so-called "halogen-effect". In [2-7] a dense alumina scale protecting the material against corrosion was formed on the surface after the application of small amounts of chlorine. However the alumina scale failes during thermocyclic loading. After fluorine ion implantation [8] or liquid phase treatment with diluted HF [9] in combination with oxidation at 900 °C/air a dense alumina scale was achieved showing pronounced adherence even during cyclic oxidation. The halogen effect can be explained by a thermodynamic model assuming the preferred formation and transport of volatile Al-halides through pores and microcracks whithin the metal/oxide interface and their conversion into alumina forming a protective oxide scale on the surface [4]. This model was proven by the F-depth profiles obtained by PIGE (Proton Induced Gamma-ray Emission) [8,9] showing a distinct maximum located at the metal/oxide interface [8-11]. In contrast to the initial F-concentration a pronounced fluorine decrease was found within the first hours of oxidation at 900 °C. In [10,11] it was shown that significant fluorine loss occurs during the sample heating.

The remaining fluorine content of less than 5 at.-% is essential for a stable fluorine effect. However technical use is only possible if the fluorine effect can be stabilized for a time of at least 1000 hours. This work is dedicated to the role of fundamental material parameters related to the stability of the fluorine effect.

EXPERIMENTAL

Cast γ-TiAl (Ti-50 at.-% Al) and the technical γ-Met (46.6 at.-% Al) manufactured by powder metallurgy were prepared as coupons of size $8*8*1$ mm^3 and polished with SiC paper down to 4000 and 1200 grit resp. Microstructural investigations showed minor amounts of the α_2-Ti$_3$Al phase (lamellar structure) within the γ-TiAl phase. For standard implantation a fluence of 2×10^{17} F cm^{-2} at 20 keV was used [8]. The mean projected range of the F-ions is about 34 nm [12]. All beam line implantations were carried out at the 60 kV implanter of the Institute of Nuclear Physics (IKF) of the Johann Wolfgang Goethe - University in Frankfurt/Main using CF$_4$ in the gas source. The samples were oxidized isothermally and thermocyclically (cycles of 1 h and 24 h) in a furnace at 900 °C and 1000 °C in lab air. The fluorine concentration depth profiles were determined by using the non-destructive PIGE-technique [13]. The PIGE measurements were performed at the 2.5 MV Van de Graaff accelerator of the IKF using the nuclear reaction ^{19}F(p, $\alpha\gamma$)^{16}O at a resonance energy of proton energies E$_p$=340 keV and 484 keV, resp., and detecting the high energetic γ-rays (5-6 MeV) with a 10 inch NaI-detector. The information depth of the PIGE depth profiling is within $1.4 - 1.5$ µm, whereas the depth resolution near the surface is about 25 nm. Finally all samples were inspected by metallographic methods and SEM.

RESULTS AND DISCUSSION

Dependence of c_F^{max} from the oxidation time

The F-depth profiles resulting from beam line implantation in TiAl can be calculated by using the Monte Carlo simulation software T-DYN [14]. The F-depth profiles were measured before and after oxidation by using PIGE. In Fig. 1 implantation profiles of 1.23×10^{17} F cm^{-2} and 2×10^{17} F cm^{-2} at 20 keV are presented showing relatively high F conentrations within the near surface region. For the fluence 2×10^{17} F cm^{-2}/20 keV the F maximum of nearly 50 at.-% is located in a depth of about $35 - 40$ nm. In contrast to this the F-profiles show a significantly different behaviour already after 1 hour of oxidation. The F content is reduced mostly by evaporation of TiF$_4$ during the heating process as reported in [10,11]. The F-profile can be divided into 3 different regions as described in [8-10] and illustrated in Fig.2. A negligible F content occurs within the formed initial alumina scale of about 200 nm thickness. This alumina scale – once formed – acts as a diffusion barrier and prevents further F loss from the surface. The F maximum of about 3 at.-% is located at the metal/oxide interface, followed by a region which is characterized by a slight diffusion of F into the metal.

For longer oxidation times the F maximum drops to 1-2 at.-% after 500 h/900 °C (see Fig. 1). These results suggest that the F-concentration at the metal/oxide interface is an important parameter for the stability of the fluorine effect. We define this parameter as c_F^{max}. One focus of this work is on the time behaviour of c_F^{max}. In the case of isothermal oxidation at 900 °C first results have been reported in [15]. In the present work the time for isothermal oxidation has been extended to 4000 h showing a nearly constant time behaviour of c_F^{max} as depicted in Tab. 1. The

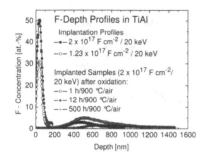

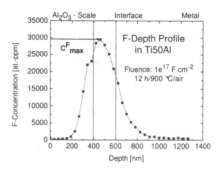

Figure 1: Time behaviour of the F-depth profiles after implantation and after isothermal oxidation (900 °C/air).

Figure 2: F-depth profile of an implanted TiAl-sample (10^{17} F cm^{-2} / 20 keV) after oxidation (12 h/900 °C/air).

Treatment/Oxidation	Parameters		
	A	C_0	t_1
F-Implant. (900 °C/isoth.)	2.53	1.19	137.18
F-Implant. (900 °C/cycl.)	3.18	1.06	65.77
F-Implant. (1000 °C/isoth.)	4.18	0.27	97.27
F-Implant. (1000 °C/cycl.)	3.69	0.77	80.02

Table 1: Paramters of the decay function describing the c^F_{max} – dependence after single F implantation and oxidation.

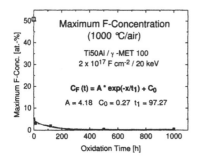

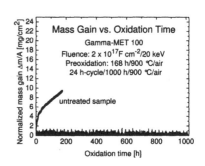

Figure 3: An exponential decay function with a constant term fits the time behaviour of c^F_{max} (2×10^{17} F cm^{-2} /20 keV) during isothermal oxidation (1000 h/1000 °C/air).

Figure 4: Mass gain after oxidation (24 h – cycle/1000 h/900 °C/ air). The Gamma-Met 100 sample was implanted (2×10^{17} F cm^{-2} / 20 keV) and preoxidized (168 h/900 °C/air).

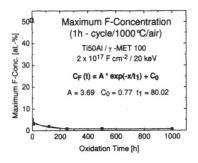

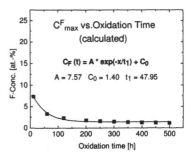

Figure 5: An exponential decay function describes the time behaviour of c^F_{max} (2×10^{17} F cm^{-2} /20 keV) during cyclic oxidation (1000 h/ 1000 °C/ air).

Figure 6: Modelling of the F-maximum c_F^{max} at the metal/oxide – interface during oxidation at 900 °C.

SEM inspection showed a dense alumina scale on the sample surface revealing that the F-effect is still working. The time dependence of c_F^{max} after passing the heating phase can be expressed by an exponential decay function starting at t=1 hour of the following type (1) where t denotes the time, t_1 the time constant, A the variable F amount at t=0 and c_0 the asymptotic F amount.

$$c_F^{\;max}(t) = c_0 + A \; \exp\left(-\frac{t}{t_1}\right) \qquad (1)$$

The equation 1 suggests that after formation of a dense alumina scale the c_F^{max} decreases only slightly approaching a constant value of c_0. The fit parameters summarized in Tab. 1 allow the description of the F reservoir essential for the oxidation protection reaching a constant value of about 1 at.-%. By increasing the oxidation temperature to 1000 °C the kinetics of c_F^{max} corresponds also to eq. 1 as illustrated in Fig. 3. The parameters in Tab. 1 indicate that a fluorine content of 0.27 at-% stabilizes the F effect even at 1000 °C at least within 1000 h. However components in aerospace and automotive environments undergo service conditions which are characterized by thermocyclic loading between 900 °C and ambient temperature. Therefore cyclic oxidation (1 h-cycles/900 °C/air) was performed to study the fluorine content up to an oxidation time of 2592 h. Before starting the cyclic process a preoxidation was done to establish a dense alumina scale on the surface. The time behaviour of the F maximum is also fitted by eq. 1, whereas the parameters in Tab. 1 reveal the presence of about 1 at.-% fluorine at the metal/oxide interface. The mass gain depicted in Fig. 4 shows the existence of an alumina kinetics at cyclic oxidation (24 h – cycle/1000 h/900 °C/air). The results for the cyclic loading at 1000 h/1000 °C (Fig. 5) indicate the stability of the F-content also under these extreme conditions as summarized in Tab. 1. The F-maximum reaches values of about 0.7 at.-%. In agreement with the metallographic inspections it was demonstrated that an excellent oxidation protection can be obtained already after single F implantation. A detailed study for the case of fluorine double implantation leads to an increasing F-content c_0 for the oxidation parameters discussed above [16].

Diffusion coefficient of Fluorine and the modelling of c_F^{max}

After the formation of an alumina scale the stabilitiy of c_F^{max} depends on the inward diffusion of F into the metal. In [15] the F diffusion coefficient D in TiAl at 900 °C was determined. By using the mean value of 1.56 x 10^{-15} cm²/s a theoretical modelling of the dependence of c_F^{max} from the oxidation time at 900 °C was performed. The c_F^{max} was desribed by the solution of the diffusion equation (2) where Q denotes the number of fluorine atoms in at./cm²

$$c(x,t) = \frac{Q}{2\sqrt{Dt\pi}}\exp\left(-\frac{x^2}{4Dt}\right) \qquad (2)$$

deposited at the metal/oxide interface, x the distance from the metal/oxide interface, t the time and D the diffusion coefficient. The result in Fig. 6 confirms the experimental behaviour by showing a nearly constant F amount of 1 at.-% for longer oxidation time.

CONCLUSIONS

The F concentration at the metal/oxide interface c_F^{max} was identified to be a fundamental material parameter for the stability of the fluorine effect. Isothermal and thermocyclic exposures were performed at 900 °C and 1000 °C. After single F implantation and F loss during heating the time behaviour of c_F^{max} corresponds to an exponential decay function with a constant term reaching values between 0.27 and 1.2 at.-%. From the results a stable F amount at the metal/oxide interface can be concluded already after single fluorine implantation leading to a stable oxidation protection at temperatures of 900 °-1000 °C. The diffusion coefficient of F into TiAl at high temperatures was identified to be a second fundamental material parameter to establish a stable F effect.

ACKNOWLEDGEMENT

The work has been funded to a large extent by Deutsche Forschungsgemeinschaft (DFG) under contract no. SCHU 729/15-1, which is gratefully acknowledged by the authors. The authors thank also Dr. L. Schmidt and Dipl.-Phys. Ing. S. Neve (IKF) for the support at the implanter and the accelerator group of Dr. K. Stiebing (IKF).

REFERENCES

1. A. Rahmel, W. J. Quadakkers, M. Schütze, *Mat. and Corr.* **46,** 271 (1995).
2. M. Kumagai, K. Shibue, K. Mok-Soon, Y. Makoto, Intermetallics, **4,** 557 (1996).
3. M. Schütze, M. Hald, *Mat. Sci. Eng.*, **A239-240,** 847 (1997).
4. A. Donchev, B. Gleeson, M. Schütze, *Intermetallics* **11,** 387 (2003).
5. G. Schumacher, F. Dettenwanger, M. Schütze, U. Hornauer, E.Richter, E. Wieser, W. Möller, *Intermetallics*, **7,** 1113 (1999).
6. G. Schumacher, C. Lang, M. Schütze, U. Hornauer, E. Richter, E.Wieser, W. Möller, *Mat. and Corr.*, **50,** 162 (1999).
7. U. Hornauer, E. Richter, E. Wieser, W. Möller, G. Schumacher, C Lang, M. Schütze, *Nucl. Instr. & Meth. in Phys. Res.*, **B 148,** 858 (1999).
8. H.-E. Zschau, V. Gauthier, M. Schütze, H. Baumann, K. Bethge, *Proc. International Symposium Turbomat*, June 17-19, 2002, Bonn, ed. W. Kaysser, 210 (2002).
9. H.-E. Zschau, V. Gauthier, G. Schumacher, F. Dettenwanger, M. Schütze, H. Baumann, K. Bethge, M. Graham, *Oxid. Met.*, **59,** 183 (2003).
10. H.-E. Zschau, M. Schütze, H. Baumann and K. Bethge, *Nucl. Instr. & Meth. in Phys. Res.*, **B 240,** 137 (2005).
11. H.-E. Zschau, M. Schütze, H. Baumann and K. Bethge, Mat. Sc. For. **461-464,** 505 (2004).
12. J. Ziegler, J. Biersack, U. Littmark, *The stopping and range of ions in solids*, Version 95, 1995, Pergamon Press, New York.
13. J. R. Tesmer, M. Nastasi (Eds.), *Handbook of Modern Ion Beam Materials Analysis*, Materials Research Society, Pittsburgh 1995, PA, USA
14. J. Biersack, *Nucl. Instr. & Meth. in Phys. Res.* **B 153,** 398 (1999).
15. H.-E. Zschau, M. Schütze, H. Baumann and K. Bethge, *Nucl. Instr. & Meth. in Phys. Res.*, **B 257,** 383 (2007).
16. H.-E. Zschau, S. Neve, P. J. Masset, M. Schütze, H. Baumann, K. Bethge, Submitted to *Nucl. Instr. & Meth. in Phys. Res.*

Poster Session:
Iron Aluminides, Titanium Aluminides,
Nickel Aluminides and Silicides

Mater. Res. Soc. Symp. Proc. Vol. 1128 © 2009 Materials Research Society

1128-U05-01

Crystal Structure and Thermoelectric Properties of Mn-Substituted Ru$_2$Si$_3$ With the Chimney-Ladder Structure

Tatsuya Koyama, Norihiko L. Okamoto, Kyosuke Kishida, Katsushi Tanaka and Haruyuki Inui
Department of Materials Science and Engineering, Kyoto University, Sakyo-ku, Kyoto 606-8501, Japan

ABSTRACT

Phase relationships of manganese-substituted ruthenium sesquisilicide alloys have been investigated by using scanning and transmission electron microscopy. A series of chimney-ladder phases Ru$_{1-x}$Mn$_x$Si$_y$ are formed over a wide compositional range between Ru$_2$Si$_3$ and Mn$_4$Si$_7$. The thermoelectric properties of the directionally solidified alloys with the nominal compositions of Ru$_{1-x}$Mn$_x$Si$_y$ ($0.55 \leq x \leq 0.90$) have been investigated as a function of the Mn content and temperature. The thermoelectric power factor and dimensionless figure-of-merit (ZT) for the alloys with high Mn contents ($x \geq 0.75$) increase with the increase in the Mn content. The ZT value for the crystal with $x = 0.90$ is as high as 0.76 at 874 K.

INTRODUCTION

The high-temperature (HT) phase of ruthenium sesquisilicide (Ru$_2$Si$_3$) possesses the tetragonal Ru$_2$Sn$_3$-type structure, which is known as one of the chimney-ladder structures [1,2]. Some of the family of compounds with the chimney-ladder structures are known to exhibit a high Seebeck coefficient and low thermal conductivity simultaneously so that they have been investigated as a candidate for thermoelectric materials [3-5] because the thermoelectric performance is evaluated with the dimensionless figure-of-merit, $ZT = \alpha^2 T/(\rho \cdot \lambda)$, where α, ρ, λ and T stand for Seebeck coefficient, electrical resistivity, thermal conductivity and temperature, respectively. The chimney-ladder compounds expressed with the general chemical formula of M$_n$X$_{2n-m}$ (M: transition metal element, X: group 13 or 14 element, n, m: integers) possess a particular tetragonal crystal structure, in which the unit cell consists of M (Ru) subcell with the atomic arrangement of the β-Sn type (chimney) and X (Si) subcell with the atomic arrangement of a coupled helices (ladders) with both the chimney and ladder being aligned along the c-axis of the tetragonal unit cell [1,6,7]. We have recently found out that substitutions of Ru (group 8) in Ru$_2$Si$_3$ with Re (group 7) stabilize the HT phase with the chimney-ladder structure to appear at low temperatures so that a series of chimney-ladder phases, Ru$_{1-x}$Re$_x$Si$_y$ are formed over a wide composition range so that the valence electron count (VEC) per M atom in the compounds is maintained (VEC=14) [8,9]. Of interest to note is that the X/M values for these chimney-ladder phases deviate from the ideal values expected from the VEC=14 rule and also that semiconducting behavior of the chimney-ladder phases depends on the extent of compositional deviation from the ideal VEC=14 values [10].

In the present study, we investigate the phase relationships and crystal structures in Mn-substituted ruthenium sesquisilicide alloys by scanning electron microscopy (SEM) and transmission electron microscopy (TEM) to clarify how the composition range of the chimney-ladder phases in the Ru-Mn-Si system deviates from the ideal one that satisfies the VEC=14 rule. We also investigate the thermoelectric properties of directionally solidified crystals of the chimney-ladder compounds in the Ru-Mn-Si system as a function of the Mn content and temperature with an expectation of further enhancing the thermoelectric properties.

EXPERIMENTAL PROCEDURES

Polycrystalline specimens with nominal compositions of $Ru_{1-x}Mn_xSi_{1.5}$ and $Ru_{1-x}Mn_xSi_{1.75}$ ($x = 0.20$, 0.40, 0.60, and 0.80) were prepared by arc-melting elemental Ru, Mn, Si under an Ar gas flow. Microstructures of the as-grown samples were examined by SEM and TEM, in which chemical compositions of interested phases were estimated by energy dispersive x-ray spectroscopy (EDS). Directional solidification were made for some $Ru_{1-x}Mn_xSi_y$ alloys ($x = 0.50$, 0.55, 0.75, 0.85 and 0.90) in order to measure the thermoelectric properties. Seebeck coefficient and electrical resistivity were measured by the static dc and four-probe methods, respectively, while thermal conductivity was estimated from the values of thermal diffusivity and specific heat measured by the laser flash method.

RESULTS

Phase equilibria and structural variation in the chimney-ladder phases

Figures 1(a) and (b) show SEM backscattered electron images (BEIs) of the as-arcmelted samples with the nominal alloy composition of $Ru_{0.40}Mn_{0.60}Si_{1.5}$ and $Ru_{0.40}Mn_{0.60}Si_{1.75}$, respectively. Two phases are identified for both the samples with the major phase being the $(Ru,Mn)Si_y$ sesquisilicide phase. While the second phase observed for the $Ru_{1-x}Mn_xSi_{1.5}$ alloy is the $(Ru,Mn)Si$ monosilicide phase, it is the Si phase for the $Ru_{1-x}Mn_xSi_{1.75}$ alloy (as indicated by white and black arrows in figures 1(a) and (b), respectively). The volume fraction of the $(Ru,Mn)Si$ monosilicide second phase increases as the Mn content in $Ru_{1-x}Mn_xSi_{1.5}$ alloys increases, while the opposite is true for the Si second phase in $Ru_{1-x}Mn_xSi_{1.75}$ alloys, indicating that the Si content in the sesquisilicide phase increases as the Mn content increases. The major $(Ru,Mn)Si_y$ sesquisilicide phase is always accompanied by contrast variation within the region.

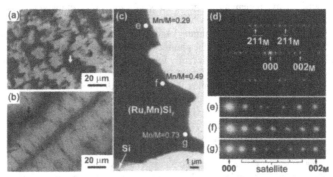

Figure 1. SEM-BEIs of as-arcmelted samples with nominal alloy compositions of (a) $Ru_{0.40}Mn_{0.60}Si_{1.5}$ and (b) $Ru_{0.40}Mn_{0.60}Si_{1.75}$. (c) TEM bright-field image of $Ru_{0.40}Mn_{0.60}Si_{1.75}$. (d) SAED pattern taken from position 'e' in (c). (e)-(g) Enlarged $00l$ systematic rows in the SAED patterns taken from positions denoted 'e', 'f' and 'g' in (c), respectively.

The extent of the compositional variation in the $(Ru,Mn)Si_y$ sesquisilicide phase is hardly changed significantly by annealing at 1273 K for 96 h.

Figure 1(c) shows a TEM bright-field image of a sample with the nominal composition of $Ru_{0.40}Mn_{0.60}Si_{1.75}$. A grain of the $(Ru,Mn)Si_y$ sesquisilicide phase is located next to a grain of the Si phase. TEM-EDS analysis revealed that the Mn content in the sesquisilicide phase decreases as the distance from the Si grain increases. Selected-area electron diffraction (SAED) patterns taken from the positions in the sesquisilicide phase denoted 'e', 'f' and 'g' in figure 1(c) all exhibit characteristics of the chimney-ladder structure consisting of diffraction spots from the M subcell and those from the Si subcell as shown in figure 1(d). This indicates that the HT-phase of Ru_2Si_3 with the chimney-ladder structure is stabilized by the substitution of Ru with Mn. All the SAED patterns are indexed as those with the $[120]_M$ incidence when referred to the M subcell but the spacing of satellite spots from the Si subcell in the 00l systematic row increases as the Mn content decreases (figures (e)-(g)). The Si/M atomic ratios (y) in the corresponding chimney-ladder phases which were determined from the ratio of the spacing of satellite spots from the Si subcell to that from the M subcell [8] are plotted in figure 2(a) against the Mn content determined by TEM-EDS analysis. The values of the Si/M ratios increase as the Mn content increases and those for most chimney-ladder phases ($0.14 \leq x \leq 0.87$) are larger than those expected from the VEC=14 rule. Chimney-ladder phases are thus confirmed to exist in a wide composition range of $0.14 \leq x \leq 0.97$ and $1.584 \leq y \leq 1.741$ when expressed in the form of $Ru_{1-x}Mn_xSi_y$. Phase relationships among sesquisilicide, monosilicide and Si phases are described in figure 2(b).

Microstructure and thermoelectric properties of directionally solidified alloys

Five different nominal compositions were selected for crystal growth by directional solidification based on the result described in the previous section. Contrary to the case of chimney-ladder compounds in the Ru-Re-Si system [10], it was impossible to obtain single crystals of the chimney-ladder compounds in the Ru-Mn-Si system. However, crystal orientation analysis by the X-ray back reflection Laue method has revealed that crystals grow preferentially along the c-axis of the tetragonal chimney-ladder phases. The sesquisilicide phase in directionally solidified alloys also exhibits compositional variation due to the formation of many different chimney-ladder phases, as observed in the as-arcmelted samples (figures 1(a) and (b)).

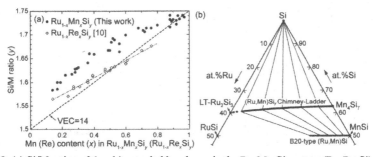

Figure 2. (a) Si/M ratios of the chimney-ladder phases in the Ru-Mn-Si system (Ru-Re-Si) plotted against the Mn (Re) content [10]. (b) Phase relationships among silicon, $(Ru,Mn)Si_y$ sesquisilicide and $(Ru,Mn)Si$ monosilicide phases.

Thermoelectric properties of the directionally solidified $Ru_{1-x}Mn_xSi_y$ alloys were evaluated along the growth direction. Values of Seebeck coefficient and electrical resistivity for these alloys are plotted in figures 3(a) and (b), respectively, as a function of temperature. The alloys with $x = 0.50$ and 0.55 exhibit negative values in the whole temperature range investigated whereas those with $x = 0.75$, 0.85 and 0.90 exhibit positive values. While the values of electrical resistivity for the alloy with $x = 0.75$ decrease with the increase in temperature, those for the alloys with $x = 0.50$, 0.85 and 0.90 increase with increasing temperature. Values of thermal conductivity for alloys with $x = 0.85$ and 0.90 are plotted in figure 4(c) as a function of temperature. The values of thermal conductivity for both alloys increase with the increase in temperature. Values of dimensionless figure of merit (ZT) calculated for alloys with $x = 0.85$ and 0.90 are plotted in figure 4(d) as a function of temperature, together with the similar plots for the $Ru_{0.40}Re_{0.60}Si_{1.663}$ alloy, which is the best alloy in the Ru-Re-Si ternary system [10]. The highest ZT value for the alloy with $x = 0.90$ is as high as 0.76 at 874 K, which is much higher than that obtained for the $Ru_{0.40}Re_{0.60}Si_{1.663}$ alloy (ZT=0.56 at 973 K) [10].

DISCUSSION

As described above, chimney-ladder compounds in the Ru-Mn-Si system also behave as electronic compounds, following the VEC=14 rule [6]. However, these chimney-ladder compounds in the Ru-Mn-Si system do not precisely obey the VEC=14 rule, as shown in figure 4(a) where the VEC values of the experimentally observed chimney-ladder phases are plotted as a function of the Mn content. The VEC values for the chimney-ladder phases with compositions close to Mn_4Si_7 ($\sim 0.9 \leq x \leq 0.97$) are slightly smaller than 14. This is reasonable in view of the fact that the chimney-ladder phase in the Mn-Si binary system is reported to have a solid

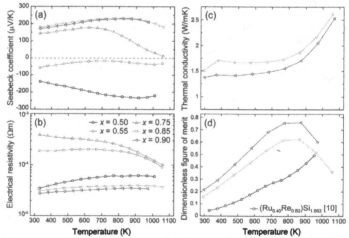

Figure 3. Temperature dependence of thermoelectric properties of directionally solidified alloys $Ru_{1-x}Mn_xSi_y$. (a) Seebeck coefficient, (b) electrical resistivity, (c) thermal conductivity, (d) dimensionless figure of merit (ZT). The highest value of ZT obtained among the Ru-Re-Si chimney-ladder compounds are also indicated in (d) [10].

176

solubility range from Mn_4Si_7 (Si/M=1.75) to $Mn_{11}Si_{19}$ (Si/M=1.727) so as to form a series of compounds with VEC<14. On the other hand, the VEC values for the chimney-ladder compounds with compositions of $0.14 \leq x \leq \sim0.9$ are larger than 14. Upon alloying Mn_4Si_7 with Ru, the value of the Si/M ratio should decrease with the increase in the Ru content so as to maintain VEC=14. The larger VEC values of these chimney-ladder compounds indicate that Si atoms exist in their lattices in excess of what is expected from the VEC rule. The amount of excess Si atoms (i.e., the VEC value) in $(Ru,Mn)Si_y$ compounds increases with the increase in the Ru content up to $x = \sim0.5$ (figures 2(a) and 4(a)). Of interest to note is that the cell volume of the M subcell v_M ($=a_M^2 \times c_M$) increases with the increase in the Ru content also up to $x = \sim0.5$. These indicate that these excess Si atoms are introduced to maintain the atomic packing (defined as the ratio of the volume of atoms located in the unit metal subcell to the volume of the unit metal subcell) of the chimney-ladder structure, as observed in the Ru-Re-Si system [10]. The values of the atomic packing factor calculated for actual compositions of the experimentally observed chimney-ladder phases are plotted as a function of the Mn content, together with those estimated for the corresponding VEC=14 compositions. Indeed, the values of the atomic packing factor for the experimentally determined compositions vary with the composition less significantly than those estimated for the corresponding VEC=14 compositions, indicating that the chimney-ladder phases are stabilized by maintaining the atomic packing factor in a certain range. Thus, the VEC value of $(Ru,Mn)Si_y$ chimney-ladder compounds increases with the increase in the Ru content (up to $x = \sim0.5$) due to the increased amount of excess Si atoms, which are introduced in the lattice to maintain the atomic packing factor in a certain range in the circumstance of the increase in the cell volume of the M subcell. Upon further alloying Mn_4Si_7 with Ru ($0.14 \leq x \leq \sim0.5$), the VEC values do not significantly change with the Ru content. In the alloy composition range, the change in the cell volume of the M subcell is not so significant when compared to that in the alloy composition range of $\sim0.5 \leq x < 1$) (figure 4(a)). This indicates that the amount of excess Si atoms to be introduced to maintain the atomic packing factor may not be so large in the alloy composition range. These may be due to the fact that the deviation of VEC values of these chimney-ladder compounds from the ideal value (VEC=14) are in their upper limit and that the introduction of more excess Si atoms in their lattices may destabilize the chimney-ladder crystal structure.

For the chimney-ladder compounds ($Ru_{1-x}Re_xSi_y$) in the Ru-Re-Si system, the Si/M ratio

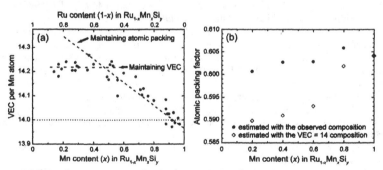

Figure 4 (a) VEC values per M atom for the $Ru_{1-x}Mn_xSi_y$ chimney-ladder phases plotted against the Mn content. (b) Atomic packing factors of the $Ru_{1-x}Mn_xSi_{1.5}$ samples ($x = 0.20, 0.40, 0.60,$ and 0.80) and Mn_4Si_7.

also deviates from that expected from the VEC=14 rules, as shown in figure 2(a) [8,10]. The alloys with the Si/M values larger than those expected from the VEC=14 rule exhibit n-type conduction and vice versa. The considerably small values and of Seebeck coefficient for the $Ru_{1-x}Mn_xSi_y$ alloy with $x = 0.55$ and 0.75 are considered to be due to the cancelling out of the contributions from the n- (Mn-poor) and p-type (Mn-rich) phases. On the other hand, once an alloy consists of either of n- or p-type compounds, the values of Seebeck coefficient become high as observed in alloys with $x = 0.50$ (n-type) and $x = 0.85$ and 0.90 (p-type). The ZT value (0.76 at 874 K) for the alloy with $x = 0.90$ is higher than the highest value (0.49 at 950 K) obtained for the chimney-ladder compound in the Ru-Re-Si system ($Ru_{0.40}Re_{0.60}Si_{1.663}$) [10].

CONCLUSIONS

In summary, the $HT-Ru_2Si_3$ phase with the chimney-ladder structure is stabilized by the substitution of Ru with Mn to form a series of chimney-ladder compounds over a wide compositional range between Ru_2Si_3 and Mn_4Si_7. The compositional region of the chimney-ladder compounds, $Ru_{1-x}Mn_xSi_y$ is slightly deviated from the compositional line connecting Ru_2Si_3 and Mn_4Si_7. The compositional deviation is explained in terms of the VEC rule and atomic packing. The thermoelectric properties of the directionally solidified alloys of $Ru_{1-x}Mn_xSi_y$ have been investigated as a function of the Mn content and temperature. The ZT value for the alloy with $x = 0.90$ is as high as 0.76 at 874 K, which is higher than that obtained for the ternary alloys in the Ru-Re-Si system.

ACKNOWLEDGMENTS

This work was supported by Grant-in-Aid for Scientific Research (A) (No. 18206074) and Young Scientists (Start-up) (No. 20960049) from the Ministry of Education, Culture, Sports, Science and Technology (MEXT), Japan and in part by the Global COE (Center of Excellence) Program of International Center for Integrated Research and Advanced Education in Materials Science from the MEXT, Japan.

REFERENCES

1.   D. J. Poutcharovsky, K. Yvon and E. Parthé, *J. Less-Common Met.* **40**, 139 (1975).
2.   C. P. Susz, J. Muller, K. Yvon and E. Parthé, *J. Less-Common Met.* **71**, P1 (1980).
3.   B. A. Simkin, Y. Hayashi and H. Inui, *Intermetallics*, **13**, 1225 (2005).
4.   L. Ivanenko, A. Filonov, B. Shaposhnikov, G. Behr, D. Souptel, J. Schumann, H. Vinzelberg, A. Plotnikov and V. Borisenko, *Microelectronic Eng.* **70**, 209-214 (2003).
5.   V. K. Zaitsev, *Thermoelectric properties of anisotropic MnSi$_{1.75}$*, in CRC handbook on thermoelectrics, edited by D. M. Rowe (CRC Press, New York, 1994), pp. 299-309.
6.   H. Nowotny, *The chemistry of extended defects in non-metallic solids*, edited by E. R. Eyring and M. O'Keefe (Amsterdam, North-Holland, 1970), p223.
7.   H. Q. Ye and S. Amelinckx, *J. Solid State Chem.* **61**, 8 (1986).
8.   B. A. Simkin, A. Ishida, N. Okamoto, K. Kishida, K. Tanaka and H. Inui, *Acta Mater.* **54**, 2857 (2006).
9.   A. Ishida, N. L. Okamoto, K. Kishida, K. Tanaka and H. Inui, *Mater. Res. Soc. Symp. Proc.* **980**, II05-37 (2007).
10. K. Kishida, A. Ishida, T. Koyama, S. Harada, N. L. Okamoto, K. Tanaka and H. Inui, *Acta Mater.* [submitted].

Mater. Res. Soc. Symp. Proc. Vol. 1128 © 2009 Materials Research Society

Diffusion Paths for the Formation of
Half-Heusler Type Thermoelectric Compound TiNiSn

Chihiro Asami[1], Yoshisato Kimura[1], Takuji Kita[2] and Yoshinao Mishima[1]
[1]Tokyo Institute of Technology, Materials Science and Engineering, 4259-G3-23 Nagatsuta, Midori-ku, Yokohama 226-8502, Japan.
[2]Toyota Motor Corporation, Higashifuji Technical Center, 1200 Mishuku, Susono, Shizuoka 410-1193, Japan.

ABSTRACT

Diffusion paths for the formation of the Half-Heusler phase TiNiSn were determined using solid/liquid diffusion couples. The most interesting result is that a single-phase TiNiSn layer forms at the TiNi/Sn(L) interface during annealing at 1073 K for only 1 h. Moreover, faceted grains of TiNiSn single-crystals grow at the interface towards the liquid Sn presumably by the solidification during isothermal holding at 1073 K. The TiNiSn phase layer which forms on the TiNi side of the interface consists of very fine sub-micron grains, and faceted TiNiSn single-crystal grains which form on the other side of the interface all have the same crystallographic orientation. While only TiNiSn forms at the TiNi/Sn(L) interface, the Heusler phase TiNi$_2$Sn also forms with TiNiSn at the TiNi$_3$/Sn(L) interface and Ti$_6$Sn$_5$ is observed at the Ti$_2$Ni/Sn(L) interface.

INTRODUCTION

The Half-Heusler compound TiNiSn is one of the most promising candidates for thermoelectric materials which can be used to directly convert waste heat to clean electric energy at high temperatures (around 1000 K) [1-3]. Half-Heusler compounds have the cubic C1$_b$ type ordered structure denoted as ABX which consists of four interpenetrating fcc sub-lattices of the elements A, B, X and vacancies [4]. Of these compounds TiNiSn is the most attractive one because not only it has excellent thermoelectric properties but also it consists of eco-friendly elements which are neither toxic nor costly. However, for TiNiSn the problem that fabrication of single-phase TiNiSn alloys is quite difficult since the melting temperature of Sn, 505 K, is much lower than that of the other constituent elements. Therefore it is hard to avoid heterogeneous solidification and hence the formation of coexisting impurity phases which tend to degrade the thermoelectric properties of the TiNiSn alloy. The isothermal section at 770 K reported by Stadnyk et al. is the only available phase diagram of the ternary Ti-Ni-Sn system [5]. There exist two ternary compounds, Half-Heusler TiNiSn and Heusler TiNi$_2$Sn, which are in equilibrium with each other. Both are nearly line compounds having quite restricted off-stoichiometric composition ranges. Existing phase diagram information is insufficient to understand the formation of TiNiSn. We fabricated nominally stoichiometric TiNiSn alloys using arc melting and subsequent annealing at 1073 K for 2 weeks [2], though, considerable amounts of metallic Ti$_6$Sn$_5$ and Sn remain as impurity phases. Since TiNiSn single-phase alloy fabrication requires long annealing times and a tremendous amount of energy, we believe it is necessary to develop new fabrication processes.

The present work aims at establishing the basis for new fabrication processes for TiNiSn alloys without involving non-equilibrium solidification. Therefore the diffusion paths for the TiNiSn phase formation were investigated using diffusion couples composed of liquid Sn and solid Ti-Ni binary phases. The relatively high diffusion rate in the liquid phase can be advantageous for an ecological fabrication process if a target material can be formed in a short process time.

EXPERIMENT

To determine diffusion paths for the formation of TiNiSn diffusion couples consisting of liquid Sn and a solid Ti-Ni binary compound were assembled as depicted in Fig. 1. A lump of Sn was embedded in a hole of 3.6 mm diameter in a piece of TiNi alloy to avoid the molten Sn being spilt over. TiNi, $TiNi_3$ and Ti_2Ni were selected as solid phases used in the liquid(Sn)/solid diffusion couples. Nominal compositions of 66.7at.%Ti-33.3at.%Ni, 49.5at.%Ti-50.5at.%Ni and 25at.%Ti-75at.%Ni were chosen to prepare single-phase alloys of Ti_2Ni, TiNi and $TiNi_3$, respectively. The diffusion couples were wrapped in a sheet of Ta foil and heat-treated at temperatures from 773 K to 1273 K for 1 h in flowing argon gas atmosphere, and then cooled in air. The alloys with nominal compositions shown in Table 1 were prepared by arc melting using high purity raw materials of Ti (>4N), Ni (>4N) and Sn (>4N) in argon gas atmosphere to investigate phase equilibria in the Ti-Ni-Sn ternary system. Several alloys were annealed at 1073 K for 336 h in a flowing argon gas atmosphere. Microstructures were observed by means of scanning electron microscopy (SEM) using a back scattered electron compositional image (BEI) on an electron probe micro-analyzer (EPMA), JEOL JXA-8900R. Quantitative chemical analyses were performed by EPMA operated at acceleration voltage of 20 kV. The spot size of electron probe was about 1 μm and absorbed current was controlled to be about 20 nA. Additionally, information on the crystallographic orientations was obtained using electron back scattered diffraction (EBSD) attached to a field emission (FE-) SEM, JEOL JSM-7001F.

RESULTS AND DISCUSSION

Formation behavior of TiNiSn phase at the solid/liquid(Sn) interface

Typical microstructures at the interface of three diffusion couples, (a) TiNi/Sn(L), (b) $TiNi_3$/Sn(L) and (c) Ti_2Ni/Sn(L), annealed at 1073 K for 1 h are shown by back scattered electron images (BEIs) in Fig. 2. Concentration profiles of Ti, Ni and Sn which were measured across the interface are shown in Fig. 2(d-f). Two interesting microstructural features are commonly observed at all three interfaces; one is a relatively thick TiNiSn layer consisting of two regions with or without coexisting phases, and the other is large faceted TiNiSn crystals which are mostly surrounded by the Sn melt. These faceted TiNiSn crystals are supposed to grow towards the Sn liquid phase regions during isothermal holding at 1073 K. Voids (or pores) and impurity phases such as Ni_3Sn_4 are formed during solidification of the Sn liquid upon cooling from 1073 K because of thermal shrinkage and of compositional changes in Sn, respectively. It should be noted that the TiNiSn layer includes a lot of small grains which can be distinguished by their light gray contrast shown in Figs. 2 (a) and (b). Nevertheless, chemical composition profiles indicate the formation of almost stoichiometric TiNiSn phase layer without having any

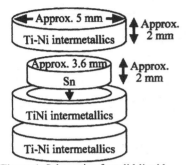

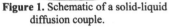

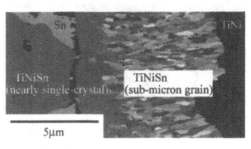

Figure 1. Schematic of a solid-liquid diffusion couple.

Figure 3. An inverse pole figure map of TiNi/Sn(L) annealed at 1073 K for 1 h.

other coexisting phases.

To confirm whether the layer is single-phase TiNiSn or not, X-ray mappings by energy dispersive spectroscopy (EDS) and EBSD measurements were conducted using a FE-SEM with higher spatial resolution. A typical result of crystallographic orientation mapping based on inverse pole figures measured by EBSD for the TiNi/Sn(L) interface is shown in Fig. 3. It should be noted that grains having different orientations are distinguished by the gray scale contrast. It can be summarized here that the TiNiSn layer is indeed composed of two different regions; one on the TiNi side of the interface consists of very fine sub-micron TiNiSn grains while the other on the Sn(L) side is composed of coarse TiNiSn grains. The light gray contrasts observed in BEIs of Figs. 2 (a) and (b) are attributed to the electron-channeling contrasts of TiNiSn grains which have the different crystallographic orientation with each other, since the contrast slightly changes with the acceleration voltage. Note that the light gray seam observed in the same figure is also supposed to be due to the difference of electron-channeling contrast. The boundary between two types of TiNiSn layers, consisting of very fine grains and of coarse grains, is supposed to be the initial interface. The x-ray mapping of EDS also supports these results that the formed layers at the interface are TiNiSn single-phase.

Moreover, the large faceted TiNiSn crystals which are mostly surrounded by Sn melt are single-crystals and most of them have the same crystallographic orientation, i.e. their <001> direction is perpendicular to the interface. It suggests that the possible growth mechanism of faceted TiNiSn crystals is the solidification during isothermal holding at 1073 K, since <001> is a preferred growth direction of typical fcc crystals. Large faceted morphology indicates the slow growth rate and a certain amount of accumulative time of the growth. It was also found that the size of faceted crystals has the dependence on the holding time; the longer the annealing time is, the larger the crystals grow. Faceted crystals also grow larger during cooling from 1073 K.

On the other hand, while TiNiSn solely forms at the TiNi/Sn(L) interface, a Heusler compound $TiNi_2Sn$ layer forms together with TiNiSn at the $TiNi_3$/Sn(L) interface on the $TiNi_3$ side as shown in Fig. 2(e). The thickness of the $TiNi_2Sn$ layer estimated from the composition profiles is about 2 μm, i.e. much thinner than the 12 μm of the TiNiSn layer. Therefore, this $TiNi_2Sn$ layer is hardly distinguished in the present magnification of BEI, however, its presence is confirmed in the concentration profiles. In the case of the $Ti_2Ni(S)$/Sn(L) interface, a Ti_6Sn_5 layer and a two-phase layer consisting of TiNiSn and Ti_6Sn_5 have apparently formed on the

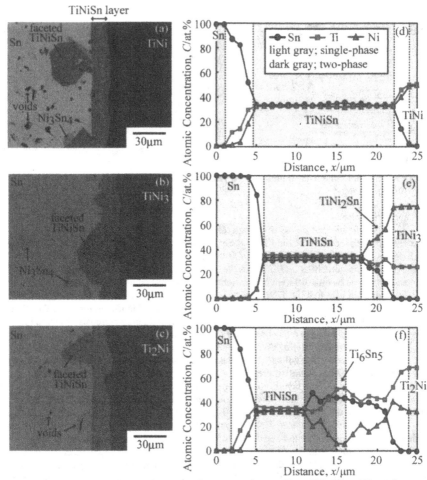

Figure 2. Back scattered electron images and concentration profiles of Ti, Ni and Sn at the interface; (a, d) TiNi/Sn(L), (b, e) TiNi₃/Sn(L) and (c, f) Ti₂Ni/Sn(L) of diffusion couples annealed at 1073 K for 1 h.

Ti_2Ni side as shown in Fig. 2(f). This two-phase layer is difficult to confirm by BEI or the concentration profiles since two-phase microstructure is smaller than spatial resolution of both BEI and EPMA, though its existence may be proven with the help of the diffusion path as will be explained in the next section.

Temperature dependence of the TiNiSn phase formation

We also investigated the temperature dependence of the TiNiSn phase formation at the TiNi/Sn(L) interface. It was found that after 1 h annealing a TiNiSn layer and faceted crystals formed in the temperature range from 873 K to 1173 K. At 1273 K the tendency of additional formation of TiNi$_2$Sn is observed. Furthermore, faceted TiNiSn single-crystals change their appearance to dendrite morphology at 1273 K. A BEI showing a typical microstructure at around the interface of a diffusion couple annealed at 1273 K for 1 h is represented in Fig. 4. This change in morphology also supports that the growth mechanism should be associated with the solidification during isothermal holding at annealing temperature. The growth rate of dendrites is higher than that of faceted crystals, though the detail remains unclarified in this work. Notice that annealing temperatures applied in this work are higher than the melting point of Sn and lower than that of TiNiSn phase. Similar morphology change due to the solidification during isothermal holding is reported for the formation of Ni$_3$Sn$_4$ phase at the Ni(solid)/Sn(liquid) interface [6].

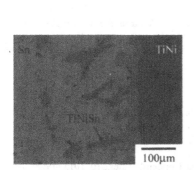

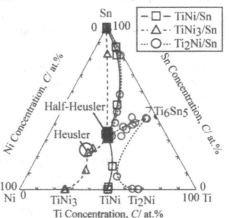

Figure 4. A back scattered electron image of the TiNi/Sn(L) diffusion couple annealed at 1273 K for 1 h.

Figure 5. Diffusion paths of TiNi/Sn(L), TiNi$_3$/Sn(L) and Ti$_2$Ni/Sn(L) annealed at 1073 K for 1 h.

Consideration of the diffusion path for the TiNiSn phase formation

The diffusion paths can be evaluated from diffusion couples by plotting chemical concentration profiles on a Ti-Ni-Sn ternary composition diagram. Three diffusion paths are depicted together in Fig. 5. The mass balance seems to be satisfied only in the diffusion path obtained from the Ti$_2$Ni/Sn(L) interface, since annealing for 1 h at 1073 K is not sufficient to observe the phase equilibria particularly in the other two diffusion couples, even if the diffusion rate in a liquid phase is relatively high. Though, diffusion paths determined in this work provide useful pieces of information for considering the formation of TiNiSn single-phase alloys. For instance, we have fabricated nearly single-phase TiNiSn alloys by sintering mixed powders of TiNi compound and Sn using hot-pressing.

183

CONCLUSIONS

Aiming at the development of a new fabrication process for Half-Heusler TiNiSn single-phase alloys used as thermoelectric materials, the formation of TiNiSn phase and corresponding diffusion paths were investigated using diffusion couples consisting of binary Ti-Ni intermetallic compounds and liquid Sn. The following conclusions are drawn in the present study.

1. The Half-Heusler TiNiSn single-phase layer forms at the TiNi/Sn(L) interface at annealing temperatures from 873 K to 1273 K. The layer consists of two regions; one is composed of fine grains on the TiNi side and the other is formed as coarse grains on the Sn side.

2. Faceted TiNiSn single-crystals grow on the TiNiSn layer towards the liquid Sn region by the solidification during isothermal holding at temperatures from 873 K to 1173 K. Most faceted crystals have the same crystallographic orientation, <001> is perpendicular to the interface.

3. A Heusler $TiNi_2Sn$ phase layer forms together with TiNiSn at the $TiNi_3$/Sn(L) interface on the $TiNi_3$ side, and a two-phase layer consisting of TiNiSn and Ti_6Sn_5 forms on the Ti_2Ni side of the Ti_2Ni/Sn(L) interface.

4. Diffusion paths for the formation of TiNiSn phase are depicted from the concentration profiles across the $TiNi_3$/Sn(L), TiNi/Sn(L) and Ti_2Ni/Sn(L) interfaces of diffusion couples annealed at 1073 K. These diffusion paths are useful for considering an ecological fabrication process of TiNiSn alloys

REFERENCES

1. S.-W. Kim, Y. Kimura and Y. Mishima, Intermetallics, 15, 349 (2007)
2. T.Katayama, S.-W. Kim, Y.Kimura and Y. Mishima, J. Electronic Mater., 32, 1160 (2003)
3. S. Bhattacharya, T. M. Tritt, Y. Xia, V. Ponnambalam, S. J. Poon and N.Thadhani, Appl. Phys. Lett., 81, 43 (1998)
4. J.Tobal, J.Pierre, S.Kaprzyk, R.V.Skolozdra, J. Phys. Condens. Mater., 10, 1013 (1998)
5. Yu. V. Stadnyk, R. V. Skolozdra, Inogranic Mater., 27, 1884 (1991)
6. D. Gur, M. Bamberger, Acta Mater., 46, 4917 (1998)

Mater. Res. Soc. Symp. Proc. Vol. 1128 © 2009 Materials Research Society   1128-U05-09

On the Effect of Strain Rate and Temperature on the Yield Strength Anomaly in L2$_1$ Fe$_2$AlMn

Markus W. Wittmann, Janelle M. Chang, Yifeng Liao and Ian Baker
Thayer School of Engineering, Dartmouth College, Hanover, NH 03755-8000, USA

ABSTRACT

The effects of strain rate and temperature on the yield strength of near-stoichiometric Fe$_2$AlMn single crystals were investigated. In the temperature range 600-800K the yield stress increased with increasing temperature, a response commonly referred to as a yield strength anomaly. No strain rate sensitivity was observed below 750K, but at higher temperatures the yield stress increased with increasing strain rate. Possible mechanisms to explaining the effects of temperature and strain rate are discussed.

INTRODUCTION

Interest in the microstructure and mechanical properties of Fe$_2$AlMn arises from previous investigations on Fe-26Mn-19Al which revealed that the alloy had some intrinsic ductility and displayed a yield strength anomaly (YSA) in the range from 600-800K [1]. It was found that alloys with a composition near stoichiometric Fe$_2$AlMn have the L2$_1$ or Heusler structure, shown in Figure 1. At room temperature the microstructure contains two sets of thermal anti-phase boundaries (APBs), *viz.*, a/4<111> APBs enclosing smaller a/2<100> APBs. These two sets of APBs are a result of the alloy first solidifying as a disordered b.c.c. phase, subsequently ordering to a B2 phase at ~ 1300K, and finally ordering to a L2$_1$ structure at ~895K [1]. Plastic deformation was accommodated by the motion of four anti-phase boundary (APB) coupled a/4<111> dislocations [1].

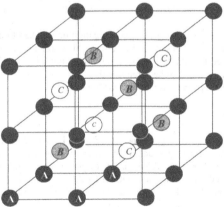

Figure 1: Atom positions for the L2$_1$ structure adopted by some A$_2$BC compounds. The structure consists of eight B2 unit cells with A atoms at the corner of each unit cell and the center atom in alternate unit cells alternating between B and C atoms.

Since the $L2_1$-B2 transformation temperature (891K) occurs just above the 600-800K temperature range of the YSA, and plastic deformation involves the motion of coupled partial dislocations, numerous proposed strengthening mechanisms can account for the YSA. These include increased dislocation friction arising from the order-disorder transformation [2-4], temperature and stress-induced reconfiguration of partial dislocations [5-7], or vacancies [8]. Due to the absence of serrated yielding or yield points in the stress strain curves, dynamic strain aging was discounted as a probable strengthening mechanism.

Transmission electron microscopy of polycrystalline Fe_2AlMn specimens strained at the peak strength temperature (800K) contained only <111> dislocations with no obvious evidence of pinning [1]. Thus, it was tentatively concluded that models which rely on stress and temperature aided dislocation transformations [5-7] were not the origin of the YSA Also, these models offer no physical mechanism to account for the results of quenching experiment in which the room temperature yield stress increased monotonically with increasing quench temperature [1]. Both the vacancy-hardening model [8] and the order-strengthening model [2] were considered to be more plausible strengthening mechanisms. They are both consistent with the observed dislocations structures and both models can account for the increasing strength observed with increasing quench temperature by increased vacancy concentrations for the vacancy-hardening model or increasing lattice disorder for the order-strengthening model. It was therefore decided that one of these mechanisms gives rise to the YSA [1].

Working out which of these mechanisms is operative should be possible by investigating the effects of strain rate on the magnitude and temperature of the peak yield stress. The vacancy-hardening model, which attributes the YSA to increasing, immobile vacancy concentrations, predicts that the magnitude of the peak yield strength will increase and move to higher temperatures with increasing strain rate [8]. This prediction was realized in compression tests of both single crystals and polycrystals of FeAl [9-11]. The order-strengthening model attributes the YSA to disorder-induced dissociation of superdislocations which must trail an APB when passing through ordered regions [2]. Strengthening is dependent on the degree of lattice ordering, and thus, the temperature of the peak yield strength should be relatively insensitive to changes in strain rate.

In this paper, we describe a brief investigation undertaken to clarify which strengthening mechanism most accurately describes the effects of strain rate and temperature on the yield stress. It was expected that the temperature and magnitude of the peak yield stress would either, be insensitive to strain rate if order strengthening were the cause of the YSA, or increase with increasing strain rate if vacancy-hardening was the cause.

EXPERIMENTAL

A cylindrical, master ingot (dia. ~ 2.5 cm, 1 ~10 cm) of nominally stoichiometric Fe_2AlMn was melted in an arc furnace under argon, and drop cast into a chilled copper mold at Oak Ridge National Laboratory. A single crystal was grown under argon from part of the cast ingot in an alumina crucible using a modified Bridgman technique. The resulting crystal was given a 120 h anneal at 673K in order to increase the $L2_1$ ordering and reduce the vacancy concentration [12]. The chemical composition of the resulting crystal was determined using a Cameca SX50 electron probe microanalyser (EPMA) to be 51.9 ± 0.2 at. % Fe, 23.0 ± 0.2 at. % Mn, and 25.1 ± 0.1 at. % Al.

Cuboidal compression specimens with dimensions of 2.5 x 2.5 x 6.5 mm high were cut from the single crystal using a high-speed, alumina cutting wheel, and polished to a 600 grit finish using silicon carbide paper. The compression axis was determined by Laue X-ray back reflection to be [$7\overline{2}6$]. Compression tests were performed at temperatures ranging from 298-1000K either in air on a furnace-equipped servo-hydraulic MTS 810 at strain rates of 5 x 10^{-4} s^{-1} and 1 s^{-1} or under vacuum on a furnace-equipped, screw-driven INSTRON model TTDML at a strain rate of 1.3 x 10^{-5} s^{-1}. The yield stress was defined as the 0.2% offset of the stress versus cross-head displacement curve.

In order to determine the L2$_1$ disorder temperature, calorimetric measurements were performed using a Perkin Elmer DSC7 differential scanning calorimeter (DSC) at a heating rate of 20K/min. The order-disorder transformation temperature was defined as the temperature at which the peak endothermic heat flow occurs.

DISCUSSION

The composition of the ingot used in this study, Fe-25Al-23Mn, was somewhat different from the alloy used in a previous study, which had a composition of Fe-26Al-19Mn [1]. However, DSC measurements, shown in Figure 2, demonstrated that the two alloys had very similar L2$_1$ to B2 transformation temperatures.

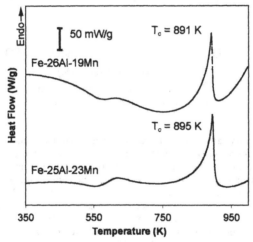

Figure 2: DSC heating curves showing the L2$_1$ to B2 transformation temperature (T$_c$) for two Fe$_2$AlMn alloys.

A plot comparing the critical resolved shear stress (c.r.s.s.) as a function of temperature for the two alloys (Figure 3) reveals that from 298-600K the alloy currently being investigated has a slightly lower yield strength, but in the temperature range of interest, 600-1000K, there is very little difference in yield strength. Based on the similar mechanical properties, and ordering temperatures of the two alloys, it is assumed that the dislocation configurations, and ultimately the mechanism responsible for the YSA is the same for these two alloys.

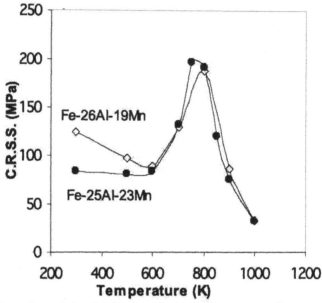

Figure 3: Comparison of the C.R.S.S. as a function of temperature for two compositions (indicated) of L2$_1$ structured Fe$_2$AlMn single crystals strained under compression at 5 x 10^{-4} s^{-1}.

Figure 4 shows a plot of yield stress as a function of temperature for three different strain rates. At temperatures below ~750K the yield stress is strain rate insensitive within the accuracy of the measurement technique. Above 750K there is a positive strain rate sensitivity of the yield stress. While there is some suggestion that the temperature of the peak yield stress moves to slightly higher temperatures at the strain rate of 1 s^{-1}, there is little evidence to suggest that the magnitude of the peak yield stress increases. These results are markedly different than those reported for the related compounds, B2-structured FeAl and DO$_3$-structured Fe$_3$Al, which show a clear increase in both the magnitude of the peak yield stress and the temperature at which it occurs with increasing strain rate [8, 10, 11, 13].

The relative insensitivity of the peak yield stress and peak temperature in Fe-25Al-23Mn to strain rate suggests that the YSA cannot be attributed to vacancy hardening. Fe$_2$AlMn could also be considered to be DO$_3$-ordered (Fe,Mn)$_3$Al in which Mn substitutes for Fe. The vacancy concentrations in (Fe)$_3$Al are much lower than FeAl [14], also making vacancy hardening a less likely mechanism. Order strengthening can adequately explain the insensitivity of the peak yield strength to strain rate, however it cannot account for mechanical properties above the disorder temperature. Most notably, it is difficult to reconcile the fact that at strain rates of 1 s^{-1} the yield stress at 900K, when the alloy is fully disordered, is 200 MPa larger than the yield stress of the fully ordered alloy at 500K. It must be concluded that at high strain rates the strengthening mechanism remains effective above the $L2_1$ to B2 disordering temperature, an observation at odds with order strengthening.

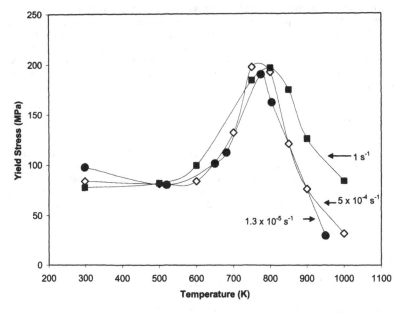

Figure 4: Plot of C.R.S.S. versus temperature for Fe_2AlMn single crystals strained under compression at the three different strain rates shown.

Recent investigations on $L2_1$-ordered Fe-Al-Ti alloys reveal this alloy also shows a yield anomaly and order strengthening does not appear to be the cause [15,16]: varying the alloy composition changes the $L2_1$/B2 transformation temperature, but this shows no correlation to changes in the temperature of the yield stress peak, with the transformation temperature being as much as 570K higher than the temperature of the yield stress peak [16]. These observations are similar to those for DO_3-ordered Fe_3Al in which variations in DO_3/B2 transformation temperature produced by alloying showed no correlation to the temperature of the yield strength peak [17]. Therefore, it appears likely that the occurrence of the $L2_1$/B2 transformation temperature near the peak yield strength in these Fe-Mn-Al alloys is coincidental, and not related to order-strengthening.

The difficulties in reconciling either the vacancy hardening model, or order-strengthening to the observed behavior suggests a re-evaluation of dislocation structures is warranted. Slip plane sectioning of crystals strained above the peak yield strength would be helpful to clarify the dislocation structures. Analysis of material strained at temperatures above the peak temperature

189

would clarify whether dislocations other than <111> become active, and dislocation pinning points could be more clearly identified in TEM samples sectioned along the slip plane.

CONCLUSIONS

An investigation into the effects of both temperature and strain rate on the yield strength of $L2_1$-ordered Fe_2AlMn was performed in an attempt to determine whether the vacancy-hardening model, or the order-strengthening model could accurately describe the yield strength anomaly. It was found that at temperatures below 750K the yield stress is largely strain-rate insensitive, while at higher temperatures it shows positive strain rate sensitivity. The magnitude of the yield strength peak was insensitive to strain rate. It appears that neither model can fully account for the observed yield stress behavior, suggesting that an alternative mechanism or mechanisms are operative.

ACKNOWLEDGMENTS

This research was supported by NIST grant 60NANB2D0120 and NSF grant DMR0314209. The authors would like to acknowledge Dr. Easo George of the Oak Ridge National Laboratory for supplying the cast material, and Prof. Paul R. Munroe of the University of New South Wales for compositional analysis of the alloys. The views and conclusions contained herein are those of the authors and should not be interpreted as necessarily representing official policies, either expressed or implied, of the National Science Foundation, the National Institute of Standards and Technologies, or the U.S. Government.

REFERENCES

1.   Wittmann M, Munroe PR, Baker I. Philos Mag A 2004; 84: 3169
2.   Stoloff NS, Davies RG. Prog Mat Sci 1966; 13: 3.
3.   Beauchamp P, Dirras G, Veyssière P. Philos Mag A 1992; 65: 477.
4.   Flinn PA. Trans AIME 1960; 218: 145.
5.   Morris DG. Philos Mag A 1995; 71: 1281.
6.   Yoshimi K, Hanada S. In: Darolia R, editor. Structural Intermetallics. Warrendale PA: The Minerals, Metals and Materials Society 1993 p. 475.
7.   Umakoshi Y, Yamaguchi M, Namba Y, Murakami K. Acta Metall 1976; 24: 89.
8.   George EP, Baker I. Philos Mag A 1998; 77: 737.
9.   Baker I, Yang Y. Mat Sci Eng 1997; 240: 109.
10.   Yang Y, Baker I, Gray GT, Cady C. Scripta Mater 1999; 40: 403.
11.   Li X, Baker I. Scripta Mater 1997; 36: 1387.
12.   Nagpal P, Baker I. Metallurgical Transactions 1990; 21A: 2281.
13.   Morris DG, Zhao P, Muñoz-Morris MA. Mat Sci Eng 2001; 297: 256.
14.   Morris DG, Muñoz-Morris MA. Intermetallics 2005; 13; 1269.
15.   Prakash U, Sauthoff G. Intermetallics 2001; 9: 107.
16.   Palm M, Sauthoff G. Intermetallics 2004; 12: 1345.
17.   Stein F, Schneider A, Frommeyer G. Intermetallics 2003; 11: 71.

Mater. Res. Soc. Symp. Proc. Vol. 1128 © 2009 Materials Research Society   1128-U05-11

The Effect of Annealing Temperature on Tensile Properties in a Fine-Grained Fe₃Al-Based Alloy Containing κ-Fe₃AlC Carbide Particles

Akira Takei[1], Satoru Kobayashi[2], Takayuki Takasugi[1,2]
[1]Department of Materials Science, Graduate School of Engineering, Osaka Prefecture University, 1-1Gakuen-cho Naka-ku, Sakai, Osaka 599-8531, JAPAN
[2]Osaka Center for Industrial Materials Research, Institute for Materials Research, Tohoku University, 1-1Gakuen-cho Naka-ku, Sakai, Osaka 599-8531, JAPAN

ABSTRACT

The effect of annealing temperature on the room temperature tensile properties was investigated in a fine-grained Fe₃Al based-alloy, and the correlation between microstructure, texture and tensile properties is discussed. Tensile elongation showed a peak as a function of annealing temperature. The highest elongation was obtained in the recrystallized samples annealed at 700 °C. The decrease in the elongation with increasing annealing temperature above 700 °C is attributed to the increase in the fraction of <100> oriented grains with respect to the tensile direction. A high sensitivity to environmental embrittlement was observed in the partially recrystallized samples annealed at 650 °C.

INTRODUCTION

Fe₃Al-based alloys have been studied so far because of their excellent oxidation and sulphidation resistance at high temperatures, low materials costs and light weight compared with conventional stainless steels [1]. One of the serious disadvantages of the alloys is, however, their poor room temperature strength and ductility, which are thought to be caused by environmental embrittlement.

It was reported that the room temperature strength and ductility were improved by thermomechanical processing (TMP) [2-5]. In the TMP, well recovered/pancake grain structure is introduced by warm rolling and annealing treatment at low temperatures since annealing at high temperatures causes recrystallization and reduces both tensile strength and ductility. It is, therefore, believed that well recovered/pancake grains are less sensitive to embrittlement than recrystallized grains in Fe₃Al alloys [2]. The relationship between the microstructure, texture and tensile properties are, however, not fully understood.

We have recently studied the texture and recrystallization of Fe₃Al-based alloys containning carbide precipitate particles aiming at grain refinement and thereby improving strength and toughness of the alloys [6-10], and proposed to insert a grain refinement process prior to the process to fabricate recovered/pancake grains. Fine-grained recovered structures with κ-Fe₃AlC particles were found to show more than 1200 MPa tensile strength and ~8% tensile ductility in air at room temperature [11]. In this paper, the effects of annealing temperatures on tensile properties were investigated in a fine-grained Fe₃Al based alloy and the correlation between microstructure, texture and tensile properties is discussed.

EXPERIMENTAL PROCEDURES

The nominal composition of the alloy investigated was Fe-27Al-1.2C-1.0Cr (at. %). The alloy was prepared by arc melting and casting into a block of 50 mm (in length) x 25 mm (in width) x 15 mm (in thickness). The cast block was cut to 10 mm in thickness and homogenized at 1200 °C followed by water quenching. The block was subsequently annealed at 1000 °C/1h to obtain an optimized κ-Fe$_3$AlC particle size distribution for grain refinement [10], then warm rolled and annealed to obtain fine-grained materials (see process 1 in Fig. 1). The fine-grained sample was subsequently warm rolled and annealed followed by air cooling (see process 2 in Fig. 1).

The tensile specimens were cut using an electrical discharge machine and mechanically polished with 1500 grade abrasive paper. Tensile tests were conducted at room temperature at a strain rate of 1.66 x 10^{-4} s^{-1} in air or in vacuum with the tensile direction parallel to the rolling direction (RD). At least two specimens were tested in each condition.

The microstructure was observed on the normal (ND)-RD section by optical microscopy and field emission scanning electron microscopy with an electron backscatter diffraction (EBSD) camera. EBSD measurements were conducted near the fractured position of tensile tested samples.

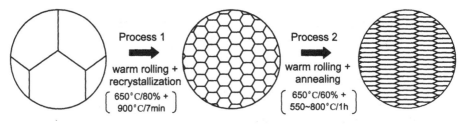

Fig. 1 Schematic diagram of thermomechanical process

RESULTS AND DISCUSSION

Microstructure

Figure 2 shows the recrystallized microstructure of the alloy obtained after process 1. Almost equiaxed grains were obtained. The average grain size measured by EBSD was 29 μm. Similar grain sizes were obtained in the center and the surface areas. It can be seen that rod-like κ-Fe$_3$AlC particles are aligned parallel to the rolling direction.

Figure 3 shows change in microstructure with annealing temperature during process 2. Deformed/ pancake grains are observed and recrystallized grains are rarely seen in the samples annealed at 600 °C (Fig.3 (a)). Recrystallized

Fig. 2 Optical micrograph of the alloy warm rolled and annealed via process 1. This micrograph was taken in the center area of the sample with respect to the thickness direction.

grains are seen in the deformed/pancake grains in the alloys annealed at 650°C (Fig.3 (b)), and they dominate in the samples annealed above 700 °C. It is noted that recrystallized grain sizes are not dependent on the annealing temperatures (Fig.3 (c), (d)), probably due to κ particles pinning grain boundaries.

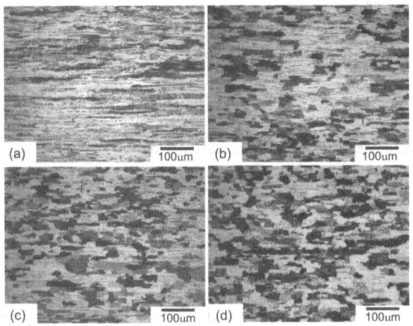

Fig. 3 Optical micrographs of the alloy annealed for 1h in the process 2: (a) 600 °C, (b) 650 °C, (c) 700 °C and (d) 800 °C.

Tensile properties in vacuum and air

Figure 4 shows the change in the room temperature tensile properties of the alloy with annealing temperature in the process 2. With increasing annealing temperature tensile elongation in vacuum increased and reached a peak at 700 °C, and then decreased above 700 °C. The trend in air was similar to that in vacuum, but the ductility in air was smaller than in vacuum, due to environmental embrittlement. It should be noted that a small drop in ductility is observed for the samples annealed at 650 °C in air. 0.2% proof stress decreased with increasing annealing temperature to 650 °C from 1350 MPa to 550 MPa, but it remained constant above 650 °C. No significant difference in the stress was seen in the tests in vacuum and air. Ultimate tensile strength (UTS) slightly decreased with increasing temperature to 650 °C and becomes constant above 650 °C. A small drop in UTS is visible at 650 °C.

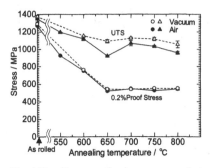 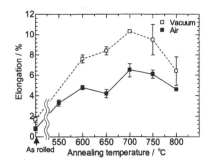

Fig. 4 Tensile properties in vacuum and air as a function of annealing temperature in the process 2: (a) 0.2% proof stress and Tensile strength, (b) Elongation.

The effects of annealing temperature on tensile properties

The tensile tests revealed that the room temperature tensile properties strongly depend on the annealing temperatures. We discuss the relationship between the tensile properties, microstructure and texture in the alloy.

With increasing annealing temperature tensile elongation increased to 700°C, while the 0.2% proof stress decreased to 650 °C. The changes in these properties with annealing are attributed to the decrease in the dislocation density in the microstructure. It should be noted that the best ductility was obtained at 700 °C annealing by which fully recrystallized grains were formed. This fact indicates that the recrystallized microstructure is not the cause of the embrittlement. The decrease in the ductility took place with increasing annealing temperature above 700 °C. Our texture analysis have revealed that the ductility decrease corresponds to the increase in the fraction of <100> oriented grains with respect to the RD, i.e. tensile direction. Figure 5 is an example of the texture analysis in the alloy annealed at 800 °C in the process 2. Grains can be classified into the three categories by crystallographic orientations with respect to the RD.

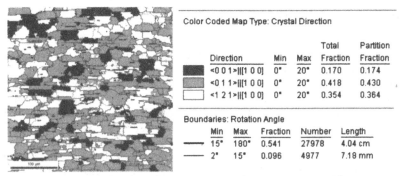

Fig. 5 EBSD Crystal Direction Map of the sample after annealing at 800 °C for 1h. Grains are classified by their crystal directions with respect to the RD.

194

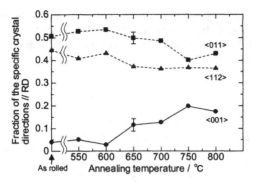

Fig. 6 The fraction of the specific crystal directions
parallel to the RD vs. annealing temperature.

Figure 6 shows the change in the fraction of the specific crystal direction with annealing temperature. It can be clearly seen that the fraction of <001> grains remains low below 600 °C, but it increases with annealing temperature above 650 °C. It is reported that (001) planes are cleavage planes in bcc base materials including Fe_3Al-based alloys [3,12]. It is, therefore, reasonable to consider that the decrease in the ductility is due to the formation of <001> oriented grains.

A high sensitivity to environmental embrittlement was observed at 650 °C annealing in which deformed grains and recrystallized grains are mixed. Taking into the account the higher 0.2% proof stress in the deformed/recovered samples than in recrystallized samples (Fig. 4 (a)), recrystallized grains start to deform plastically while recovered grains deform elastically, leading to the between the two types of grains. Hydrogen atoms might concentrate at the interface, causing high environmental embrittlement.

SUMMARY

The effect of annealing temperature on room temperature tensile properties was investigated in a fine-grained Fe_3Al based-alloy, and the correlation between microstructure, texture and tensile properties was discussed. The results obtained are summarized as follows:

(1) Tensile elongation showed a peak as a function of annealing temperature.

(2) The highest tensile elongation was obtained in the recrystallized samples annealed at 700 °C.

(3) The decrease in the elongation with annealing temperature above 700 °C corresponds to the increase in the fraction of <100> oriented grains with respect to the tensile direction.

(4) A high sensitivity to environmental embrittlement was observed in the partially recrystallized samples annealed at 650 °C.

ACKNOWLEDGMENTS

195

This research was supported in part by the Grant-in-aid for Scientific Research from the Ministry of Education, Culture, Sports and Technology, JAPAN. The help in the use of FESEM-EBSD system by Dr. Masahiko Demura at National Institute for Materials Science (NIMS), JAPAN is highly acknowledged.

REFERENCES

1. N.S. Stoloff, Mater. Sci. Eng. A 258, 1 (1998).
2. C.G. McKamey and D.H. Pierce, Scr. Metall. 28, 1173 (1993).
3. Y.D. Huang, W.Y. Yang, G.L. Chen, Z.Q. Sun, Intermetallics 9, 331 (2001).
4. D.G. Morris and M. Leboeuf, Acta mater. 42, 1817 (1994).
5. R.J. Lynch, K.A. Gee and L.A. Heldt, Scr. Metall. 30, 945 (1994).
6. S. Kobayashi, S. Zaefferer, Intermetallics 14, 1252 (2006).
7. S. Kobayashi, S. Zaefferer, D. Raabe, Mater. Sci. Forum 550, 345 (2007).
8. S. Kobayashi, S. Zaefferer, Mater. Sci. Forum 558-559, 235 (2007).
9. S. Kobayashi, S. Zaefferer, Proc of Mater. Res. Society Symp. proc. 980, II01 (2007).
10. S. Kobayashi, T.Takasugi, Intermetallics 15, 1659 (2007)9.
11. S. Kobayashi, A. Takei, T. Takasugi, Structural Aluminides for Elevated Temperature Applications, edited by Kim, Young-Won, (TMS Publishers, New Orleans, 2008), p. 383.
12. J. Chao, D.G. Morris, M.A.Munoz-Morris, J.L. Gonzalez-Carasco, Intermetallics 9, 299 (2001).

Mater. Res. Soc. Symp. Proc. Vol. 1128 © 2009 Materials Research Society   1128-U05-14

Microscale Fracture Toughness Testing of TiAl PST Crystals

Daisuke Miyaguchi[1], Masaaki Otsu[1], Kazuki Takashima[1] and Masao Takeyama[2]
[1]Department of Materials Science and Engineering, Kumamoto University, 2-39-1, Kurokami, Kumamoto, Japan
[2]Department of Metallurgy and Ceramics Science, Tokyo Institute of Technology, 2-12-1, Ookayama Meguro-ku, Tokyo, Japan

ABSTRACT

A microscale fracture testing technique has been applied to examine the fracture properties of lamellar in TiAl PST crystals. Micro-sized cantilever specimens with a size ≈ 10×20×50 μm^3 were prepared from Ti-48Al two-phase single crystals (PST) lamellar by focused ion beam (FIB) machining. Notches with a width of 0.5 μm and a depth of 5 μm were also introduced into the specimens by FIB. Two types of notch directions (interlamellar and translamellar) were selected when introducing the notches. Fracture tests were successfully completed using a mechanical testing machine for micro-sized specimens at room temperature. The fracture toughness (K_Q) values of the interlamellar type specimens were obtained in the range 1.5–3.6 MPam$^{1/2}$, while those of the translamellar specimens were 5.0–8.1 MPam$^{1/2}$. These fracture toughness values are lower than those having been previously reported in conventional TiAl PST samples. For macro-sized specimens, extrinsic toughening mechanisms, including shear ligament bridging, act in the crack wake, and the crack growth resistance increases rapidly with increasing length of crack wake for lamellar structured TiAl alloys. In contrast, the crack length in microsized specimens is only 2–3 μm. This indicates that extrinsic toughening mechanisms are not activated in micro-sized specimens. This also indicates that intrinsic fracture toughness can be evaluated using microscale fracture toughness testing.

INTRODUCTION

TiAl based alloys are promising for gas turbine engine components. This is because of their high specific strength and modulus coupled with their relatively good elevated temperature strength [1, 2]. Their mechanical properties, however, depend strongly upon microstructure, which varies widely with differing heat-treatments [3]. The microstructures of these alloys can be roughly divided into lamellar and duplex microstructures. The lamellar microstructure is expected to practical use, because it especially has good balance of elevated temperature properties, crack growth resistance and fracture toughness [4]. This increased fracture toughness, relative to duplex microstructure, is attributed to crack tip shielding by extrinsic toughening mechanisms, including shear ligament bridging. These extrinsic toughening mechanisms are dominated by microscopic lamellar structures including lamellar orientation [5, 6], colony size [7] and plate thickness [8]. It is therefore important to investigate the fracture properties of lamellar materials on the micro meter scale. In particular, the measurement of the interlamellar fracture toughness is required for designing of these alloys. However, it is extremely difficult to measure the interlamellar fracture toughness because the interlamellar spacing (the thickness of lamellar plate) is usually less than 1 μm.

We have developed a testing machine, which enables the mechanical testing of micro-sized materials [9]. Micro scale tests, for example tensile, bending, fracture toughness and fatigue tests, have been carried out for specimens with dimension of 10–50 μm [10-12]. The size of these specimens is smaller than one lamellar colony in the TiAl lamellar structure, making it possible to directly measure the mechanical properties of lamellar structure. We have applied this testing technique to investigate the microfracture behavior within one lamellar [13, 14]. However, it is difficult to control lamellar orientation in micro-sized specimen, and the fracture properties, particularly interlamellar fracture strength of lamellae, have not yet been clarified.

In this investigation, microscale fracture toughness tests of micro-sized specimens prepared from Ti-48Al PST crystals were performed, and fracture toughness values of lamellar structures were measured on the micrometer scale. The effect of size on the fracture toughness was also discussed.

EXPERIMENTAL PROCEDURE

The material used was a Ti-48 at%Al PST crystal rod. This alloy was prepared by unidirectional solidification technique. Thin slices with a thickness of 500 μm were cut from the rod by electro-discharge machining, and polished on both sides to make foils with a thickness of approximately 20 μm. Micro-sized cantilever beam type specimens were prepared by focused ion beam (FIB) machining. Figure 1 shows a scanning electron micro graph of a micro-sized specimen. The length (L), breadth (B) and width (W) of specimens were 50, 10 and ~20 μm' respectively. Furthermore, notches with a width of 0.5 μm and a depth of 5~10 μm were also introduced into the micro-sized specimens by FIB. The notch position was set to be 10 μm from the fixed end of the specimen. The loading position was located 30 μm from the notch. When preparing specimens, two types of specimens were prepared. The notch direction of one type specimen is parallel to lamellar as shown in Fig. 2 (a), and that of another type of specimen is perpendicular to lamellar structure as shown in Fig. 2 (b). The average lamellar spacing of this material was approximately 1.0 μm.

Fracture toughness tests were carried out using a mechanical testing machine for micro-sized specimens. The load resolution of this testing machine is 20 μN and the displacement resolution is 10 nm. The loading position can be adjusted by a precise X-Y stage at a translation resolution of 0.05 μm. Fracture surfaces after the fracture tests were observed by a scanning electron microscope (SEM).

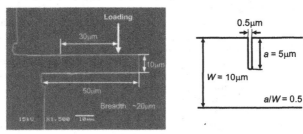

Figure 1. Scanning electron micrograph of micro-sized specimen and the notch geometry.

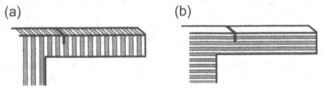

Figure 2. Schematic drawings of crack orientation in the lamellar structure.
(a) interlamellar and (b) translamellar type specimens.

RESULTS AND DISCUSSION

Figures 3 and 4 show load-displacement curves with fracture toughness tests for interlamellar and translamellar type specimens, respectively. For interlamellar type specimens, the relations of load and displacement were approximately linear, and the fracture occurred in a brittle manner. In contrast, ductile-like fracture occurred for translamellar type specimens, and some load drops are observed in the load-displacement curves.

Fracture toughness values were calculated using the following equations for stress intensity (K) for notched cantilever beam [15].

$$K = \frac{6PS}{W^2 B}\sqrt{\pi a}\,F(a/W),\quad (a/W < 0.6) \tag{1}$$

where

$$F(a/W) = 1.122 - 1.40 - 1.40(a/W) + 7.33(a/W)^2 - 13.08(a/W)^3 + 14.0(a/W)^4 \tag{2}$$

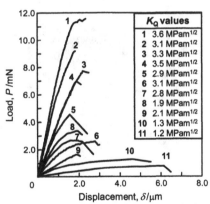

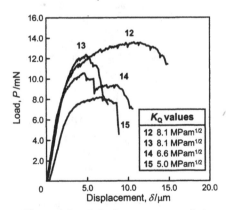

Figure 3. Load-displacement curves during fracture toughness testing of interlamellar type specimens.

Figure 4. Load-displacement curves during fracture toughness testing of translamellar type specimens.

In equations (1), a, P and S are total crack length, failure load and distance between the loading point and notch position, respectively. The failure load (P_Q) was taken as a maximum load of the load-displacement curve. The total crack length was measured from the fracture surface by scanning electron micrographs after testing. Calculated fracture toughness values for each specimen are also shown in Figs. 3 and 4. As some of these values do not satisfy small scale yielding conditions, fracture toughness values are indicated by K_Q. The fracture toughness values were in the range 1.2–3.6 MPam$^{1/2}$ for interlamellar type specimens, and were in the range 5.0–8.1 MPam$^{1/2}$ for translamellar type specimens. K_Q values of the translamellar type specimens were higher compared to those of interlamellar type specimens, and this trend is consistent with the results obtained for bulk TiAl specimens with lamellar structures. Although, these fracture toughness values are lower than those having been previously reported in conventional TiAl PST samples, especially, K_Q values of the translamellar specimens were remarkably lower compared with K_Q values of macro-sized specimens (~ 20 MPam$^{1/2}$) [16].

Figure 5 shows fracture surfaces after fracture toughness testing of interlamellar type specimens. The fracture occurs interlamellar mode, but two types of fracture surfaces are seen. One type is a flat surface as shown in Fig. 5 (a) and the other is a stepped surface as shown in Fig. 5 (b). K_Q values were in the range 1.5–3.6 MPam$^{1/2}$, but the K_Q values of specimens with flat fracture surface are approximately 1.8 MPam$^{1/2}$. In contrast, the K_Q values of specimens with stepped fracture surface are approximately 3.0 MPam$^{1/2}$. The existence of stepped surface suggests that the initiation of fracture may occur at several lamellae plates. On the other hand, flat fracture surface suggests that the fracture occurred between lamellar plates. Actually, a K_C value of cleavage fracture for (111) planes in γ-TiAl was calculated to be 1.47 MPam$^{1/2}$ [17] based on the cleavage energy calculation [18]. This K_C value is consistent with the K_Q values obtained in this investigation. This suggests that more accurate K_Q of the interlamellar mode for TiAl can be evaluated by micro-size fracture testing.

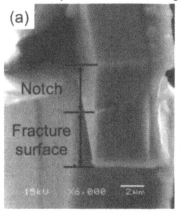
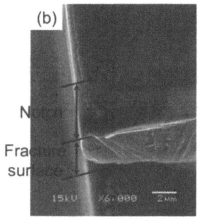

Figure 5. Scanning electron micrographs of fracture surface of interlamellar type specimens. (a) K_Q value: 2.1 MPam$^{1/2}$ and (b) K_Q values: 3.1 MPam$^{1/2}$.

Figure 6 shows a scanning electron micrograph of translamellar specimens after fracture toughness testing. The branching is observed during crack propagation. K_Q values were in the range 5.0–8.1 MPam$^{1/2}$. These fracture toughness values are much lower than those having been previously reported in macro-sized specimens (~ 20 MPam$^{1/2}$) [16]. For macro-sized specimens, extrinsic toughening mechanisms, including shear ligament bridging, act in the crack wake, and crack growth resistance increases rapidly with increasing length of crack wake for lamellar structured TiAl alloys [19]. In contrast, the crack length in micro-sized specimens is only 2–3 μm. This indicates that the specimen fractured immediately when the crack starts to grow in micro-sized specimens. This also shows that extrinsic toughening mechanisms are not activated in micro-sized specimens. This is one reason that K_Q values are lower compared to those of macro-sized specimens. This also suggests that intrinsic fracture toughness can be evaluated using micro scale fracture toughness testing. The results obtained in this investigation give important and fundamental information for designing TiAl based alloys with high fracture toughness.

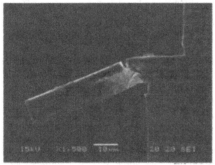

Figure 6. Scanning electron micrograph of translamellar specimens after fracture toughness testing.

CONCLUSIONS

Fracture tests have been carried out on micro-sized specimens prepared from Ti-48Al PST crystals. Fracture tests were successfully completed using a mechanical testing machine for micro-sized specimens at room temperature. The fracture toughness (K_Q) values of the interlamellar type specimens were obtained in the range 1.5–3.6 MPam$^{1/2}$, while those of the translamellar specimens were 5.0–8.1 MPam$^{1/2}$. These fracture toughness values are lower than those having been previously reported in conventional TiAl PST samples. For macro-sized specimens, extrinsic toughening mechanisms, including shear ligament bridging, act in the crack wake, and the crack growth resistance increases rapidly with increasing length of crack wake for lamellar structured TiAl alloys. In contrast, the crack length in microsized specimens is only 2–3 μm. This indicates that extrinsic toughening mechanisms are not activated in micro-sized specimens. This also indicates that intrinsic fracture toughness can be evaluated using microscale fracture toughness testing.

ACKNOWLEDGMENTS

This work was supported by a "Grants-in-Aid for Scientific Research (B)" from Japan Society for Promotion of Science (JSPS).

REFERENCES

1. R. E. Schafrik, in *Proceedings of third International Symposium on Structural Intermetallics (ISSI 3)*, edited by K. J. Hemker, D. M. Dimiduk, et al. (TMS, Warrendale, PA, 2002), pp.13-17.
2. R. Pather, A. Wisbey, A. Partridge, T. Halford, D. N. Horspool, P. Bowen and H. Kestler, in *Proceedings of third International Symposium on Structural Intermetallics (ISSI 3)*, edited by K. J. Hemker, D. M. Dimiduk, et al. (TMS, Warrendale, PA, 2002), pp.207-215.
3. Y. Kim, *Acta Metall.*, **40**, 1121 (1992).
4. R. O. Ritchie, *In. J. Fracture*, **100**, 55 (1999).
5. T. Nakano, T. Kawanaka, H. Y. Yasuda and Y. Umakoshi, *Mater. Sci. Eng., A*, **194**, 43 (1995).
6. S. Yokoshima and M. Yamaguchi, *Acta Metall. Mater.*, **44**, 873 (1996).
7. K. S. Chan, *Metall. Trans. A*, **24A**, 569 (1993).
8. K. S. Chan and Y. W. Kim, *Acta Metall.*, **43**, 439 (1995).
9. K. Takahima, Y. Higo, S. Sugiura and M. Shimojo, *Mat. Trans.*, **42**, 68 (2001).
10. K. Takashima and Y. Higo, *Fatigue Fract. Engng. Mater Strct.*, **28**, 10 (2005).
11. T. Tuchiya, M. Hirata, N. Chiba, R. Udo, Y. Yoshitomo, T. Ando, K. Sato, K. Takashima, Y. Higo, Y. Saotome, H. Ogawa and K. Ozaki, *J. Microelectromech. Syst.*, **14**, 1178 (2005).
12. G. P. Zhang, K. Takashima and Y. Higo, *Mater. Sci. Eng., A*, **426**, 95 (2006).
13. T. P. Halford, K. Takashima, Y. Higo and P. Bowen, *Fatigue Fract. Engng. Mater. Struct.*, **28**, 695 (2005).
14. K. Takashima, T. P. Halford, D. Rudinal and Y. Higo, M. Takeyama, in *Integrative and Interdisciplinary Aspects of Intermetallics*, edited by M. J. Mills, H. Inui, H. Clemens, C-L. Fu, (Mater. Res. Soc. Proc. 842, Pittsburgh, PA, 2005), pp. S5.44.1-6.
15. H. Okamura, *Introduction to Linear Fracture Mechanics*, (Baifukan, Tokyo, 1976) pp.218 (in Japanese).
16. T.P.Halford, *Fatigue and Fracture of a High Strength, Fully Lamellar γ-TiAl based Alloy*, PhD Thesis, The University of Birmingham, (2003).
17. K.S. Chan, P. Wang, N. Bhate and K. S. Kumar, *Acta Metall. Mater.*, **52**, 4601 (2004).
18. M. H. Yoo and K. Yoshimi, *Intermetallics*, **8**, 1215 (2000).
19. K. T. V. Rao, Y. W. Kim, C. L. Muhlstein and R. O. Ritchie, *Mat. Sci. Eng., A*, **192/193**, 474 (1995).

Mater. Res. Soc. Symp. Proc. Vol. 1128 © 2009 Materials Research Society 1128-U05-15

Microstructure and Compression Behavior of Ti₃Al Based Ti-Al-V Ternary Alloys

Tohru Takahashi, Ayumu Kiyohara, Daisuke Masujima, and Jun Nagakita
Department of Mechanical Systems Engineering, Tokyo University of Agriculture and
Technology, Naka-cho 2-24-16, Koganei, Tokyo 184-8588, JAPAN

ABSTRACT

The ordered alloy phase of Ti_3Al shows a rather wide solid solution range in aluminum and also in vanadium. Several Ti-Al-V ternary alloys have been prepared to investigate the alloy composition effect upon microstructure, crystallography, and mechanical characteristics. The materials containing 75, 70, and 65 at.% titanium, and 0 or 5 at.% vanadium were prepared by arc melting. Metallographic observation has revealed that the binary Ti-Al alloys contained somewhat coarse grains with about 100 μm grain diameter. In contrast to this, ternary alloys containing 5 at.% vanadium showed smaller grained microstructures with grain diameters around 15 μm. The grain size could not be adjusted to a unified value in the present study. X-ray diffraction study and microanalysis showed that the alloys contained single phase α_2. Not every possible diffraction peak of the DO_{19} ordered structure has been observed in the XRD patterns. The lattice parameters, a and c, were observed to decrease as the aluminum content increased and also when vanadium was added. Compression tests have been performed at various temperature ranging from an ambient temperature up to 1300K on rectangular parallelepiped specimens with 2mm×2mm×3mm dimensions. Alloys containing more aluminum showed higher strength, and vanadium addition enhanced the strength of the alloys. In some alloys deformability and strength are both enhanced by vanadium addition in some alloys. Temperature dependence of strength showed a little variation upon chemical compositions.

INTRODUCTION

The intermetallic Ti_3Al ordered phase is a prospective candidate material for application in light-weight and heat-resistant structures [1,2]. Concerning the α_2 phase whose stoichiometric composition is 75 at.% titanium and 25 at.% aluminum, the DO_{19} type ordered structure is known to be stable for a wide range of aluminum content from 22 to 35 at.% [3]. One of the present authors previously reported the dependence of strength on aluminum content of the α_2 ordered phase for titanium-aluminum binary compositions [4]. In the Ti-Al-V ternary alloy system the ordered phase of Ti_3Al has a relatively wide solid solution range in aluminum and also in vanadium [5]. In the present study, materials containing 75, 70, 65 or 60 at.% titanium, and 0 or 5 at.% vanadium were prepared by arc melting in order to investigate the alloy composition effect upon microstructure, crystallography, and mechanical characteristics within the solid solution range of the α_2 ordered phase. The materials were hot deformed and vacuum annealed to obtain chemical homogeneity in the recrystallized microstructure with equiaxed grains.

EXPERIMENT

Materials and materials preparation

Aluminum (99.999% purity), titanium (99.99% purity), and vanadium (99.7% purity) metal pellets were blended and arc-melted under an argon atmosphere into small ingots of about 10 grams weight. The chemical compositions of the alloys studied in this paper were divided into two categories; (a) three titanium-aluminum binary alloys ($Ti_{75}Al_{25}$, $Ti_{70}Al_{30}$, and $Ti_{65}Al_{35}$) and (b) three titanium-aluminum-vanadium ternary alloys containing 5 at.% vanadium ($Ti_{70}Al_{25}V_5$, $Ti_{65}Al_{30}V_5$, and $Ti_{60}Al_{35}V_5$). In the following, the above alloys are denominated as A25, A30, and A35 for the binary compositions and A25V5, A30V5, and A35V5 for the ternary alloys, respectively. All compositions are stated in at.%. Small pieces of the ingot were compressed at 1300K by about 50% reduction of height in air and subsequently vacuum annealed at 1400K for 10ks to obtain a recrystallized microstructure with equiaxed grains. After annealing at 1400K the binary alloys were single phase polycrystals of α_2. In contrast to the binary alloys, the ternary alloys with 5 at.% vanadium addition contained small amount of β phase grains segregated at the α_2 grain boundaries. The ternary alloys were additionally vacuum annealed at 1150K for 100ks to eliminate the remnant β phase formed during the prior annealing at 1400K. Small rectangular parallelepiped pieces with dimensions of about 2mm×2mm×3mm were spark cut and finished by emery paper polishing.

Microstructural characterization

Microstructures of the materials were observed under polarized light by an optical microscope. Electron probe micro analyzer were also utilized to reveal composition images of the samples to make sure that the materials were well homogenized into α_2 single phase. Crystallographic characterizations were carried out by X-ray diffraction using copper $K\alpha_1$ characteristic radiation; $K\beta$ radiation was filtered by a nickel foil and the $K\alpha_2$ contribution was stripped off by the analyzing software of RIGAKU RAD-IIC diffractometer.

Compression tests

Compression tests were carried out in air. The amount of deformation was measured by LVDT sensors attached to the compression jigs. Based on the recorded contraction in height of the specimen, the true stress vs. true strain curves were calculated. The test temperature ranged from an ambient temperature to 1300K for the binary alloys that were recrystallized and homogenized at 1400K. For the ternary alloys, that were homogenized at 1150K to eliminate the β phase, the upper limit of the test temperature was set at 1100K to prevent the recurrence of β phase. The test temperature was monitored with an alumel-chromel thermocouple directly welded to the specimen and was kept constant.

RESULTS and DISCUSSION

Microstructures

Figure 1 shows the microstructures in the materials that were hot deformed and vacuum annealed. The microstructures in the titanium-aluminum binary alloys as shown in (a), (b), and (c) were coarse grained and equiaxed microstructure, and their average grain diameters were measured to be about 100 μm. The titanium-aluminum-vanadium ternary alloys in (d), (e), and (f) contained finer grains of about 15 μm in diameter. The ternary alloys showed quite uniform grain shape and size in each material. The aluminum content did not show large effect on the microstructure in both binary and ternary alloys. On the other hand, vanadium addition caused a grain refining effect irrespectively on the aluminum content of the materials. In the present study, the average grain diameters could not be totally unified. Grain size could possibly be further controlled by applying different conditions for hot deformation and subsequent annealing.

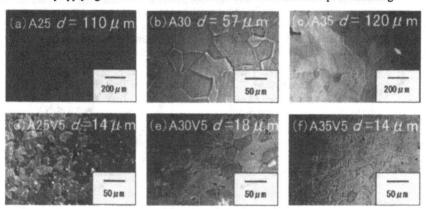

Figure 1. Microstructures of the tested materials; (a) A25, (b) A30, (c) A35, (d) A25V5, (e) A30V5, and (f) A25V5. The average grain diameter, d, is also shown on each picture.

Crystallography

Figure 2 shows the diffraction pattern as observed for the A25V5 ternary alloy.

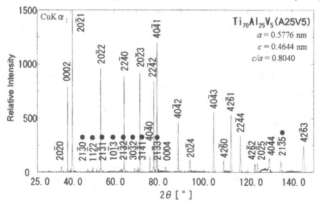

Figure 2. X-ray diffraction pattern of A25V5 ternary alloy. The Kα₁ characteristic radiation from a copper target was used. The determined values for a and c are indicated.

The numbers attached to the peaks show the indices of their reflecting planes. The prominent peaks are from basic reflections from the hcp structure, and relatively weak superlattice reflections from the DO_{19} ordered structure are marked by solid circles. Not all possible superlattice reflections were observed. Similar results were obtained for the other materials.

Lattice parameters evaluated from the diffraction patterns are summarized in Figure 3. The lattice parameters, a and c, both decreased almost linearly with increasing aluminum content. A vanadium addition of 5 at.% slightly decreased both a and c lattice parameters. The c/a axial ratio of the binary alloys were calculated to be 0.8033, 0.8021, and 0.8019, in A25, A30, and A35 alloys, respectively. The vanadium containing ternary alloys, A25V5, A30V5, and A35V5, show 0.8040, 0.8023, and 0.8025, respectively. Effect of vanadium addition is very small, but it seems that the axial ratio is little increased when compared with the ratios in the binary alloys.

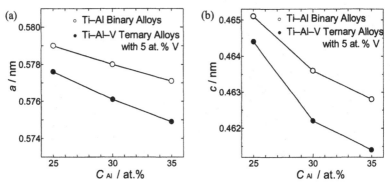

Figure 3. Lattice parameters, a in (a) and c in (b), evaluated from the XRD patterns of the alloys.

Compression behavior

Figures 4 (a) and (b) show the observed stress vs. strain curves of the 6 alloys at a temperature of 293K and 1100K, respectively.

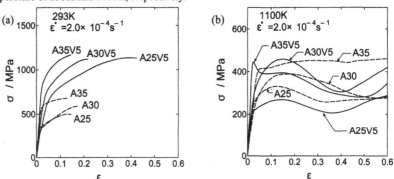

Figure 4. Stress vs. strain curves as observed in the studied alloys at (a) 293K and (b) 1100K.

The highest test temperature was set to 1300K for binary alloys (A25, A30, and A35) and to 1100K for ternary alloys with vanadium addition (A25V5, A30V5, and A35V5). For the case of ternary alloys containing 5 at.% vanadium, test temperature of 1200K and higher would induce recurrence of β phase, resulting in α2 + β dual phase constitutions. Thus the test temperature for the ternary alloys was limited to 1100K where the solid solution effect of vanadium upon deformation behavior of α2 single phase polycrystals could simply be derived. It is quite apparent that vanadium addition enhanced the room temperature strength. Such a strengthening effect of vanadium in Ti3Al ordered alloy could be explained by solid solution strengthening. In the present study, however, the strengthening effect of finer grains in the ternary alloys could not be assessed critically. The ternary alloy A25V5 showed attractive combination of strength and deformability at room temperature.

Figure 5(a) shows the stress vs. strain curves of A25V5 ternary alloy as observed at various temperatures ranging from room temperature up to 1100K. Vanadium addition is most effective in enhancing the both room temperature strength and deformability. In other ternary alloys with higher aluminum content, the strengthening effect is similar, but the enhancement in deformability is smaller as compared with that in Al25V5 alloy.

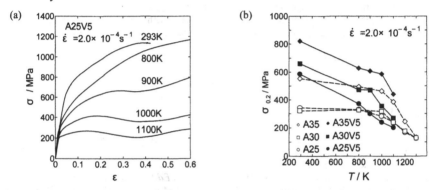

Figure 5. Temperature dependence of stress vs. strain curves in Al25V5 alloy in (a), and temperature dependence of 0.2% proof stress in the binary and ternary alloys in (b).

Figure 5(b) shows the temperature dependence of 0.2% proof stress of the alloys. In the binary alloys, the 0.2% proof stress shows a flat behavior at temperatures up to 1000K, and the strength of A25 and A 30 alloys show only a little difference in this temperature range. In contrast to this, the ternary alloys containing finer grains show a remarkable strengthening effect by vanadium addition at the lower temperature range. Among the ternary alloys containing 5 at.% vanadium, the strength is enhanced with increasing aluminum content. The temperature dependence of strength in these ternary alloys can be characterized by a somewhat monotonous decrement with increasing temperature. The A25V5 ternary alloy became weaker than the binary alloy at temperatures higher than 900K. The other two ternary alloys showed a strengthening effect of vanadium irrespectively on temperature. At still higher temperatures above 1100K, the phase equilibrium in the ternary alloys would shift from α2 single phase to α2+β dual phase, and the materials will become softer due to the high temperature softness of the β phase.

207

CONCLUSIONS

Binary titanium-aluminum alloys (A25, A30, and A35) and ternary titanium-aluminum-vanadium alloys (A25V5, A30V5, and A35V5) containing α_2 single phase equiaxed polycrystals were prepared to investigate microstructure and compression behavior. Main conclusions of the present study can be summarized as follows:

(1) Coarse grained microstructures containing equiaxed grains were obtained in the binary alloys.

(2) 5 at.% vanadium was homogeneously dissolved into the α_2 phase for all ternary alloys.

(3) Ternary alloys containing 5 at.% vanadium showed up to ten times smaller grain sizes as binary alloys containing the same amount of aluminum.

(4) Lattice parameters, a and c, both decreased with increasing aluminum content.

(5) Lattice parameters, a and c, both decreased slightly by adding vanadium.

(6) Vanadium additions increased the strength at room temperature, and the alloys containing more aluminum showed higher strength and poorer deformability.

(7) At 1100K a vanadium addition is not effective in enhancing the strength of the A25V5 alloy, however, in the A30V5 and A35V5 alloys a vanadium addition was still effective in increasing the strength, especially in the smaller strain region.

(8) The strength of ternary alloys decreased monotonically with increasing temperature.

(9) The effect of grain size on the strength of the materials could not be separated.

ACKNOWLEDGMENTS

The authors would like to thank Mr. R. Miyashita for preparing the figures in this paper.

REFERENCES

1. H. A. Lipsitt in *High-Temperature Ordered Intermetallic Alloys*, edited by C.C. Koch, C.T. Liu, and N.S. Stoloff, (Mater. Res. Soc. Symp. Proc. **39**, Pittsburgh, PA, 1985) pp.351-364.
2. M.Yamaguchi and Y.Umakoshi, Progress in Materials Science, **34**,1(1990).
3. J.L. Murray, in *Binary Alloy Phase Diagrams, Second Edition*, edited by T.B. Massalski, H. Okamoto, P.R. Subramanian and L. Kacprzak (ASM International, Materials Park,1990) vol. **1**, p.225.
4. T. Takahashi, Y. Sakaino, and S. Song in *Integrative and Interdisciplinary Aspects of Intermetallics*, edited by M. J. Mills, H. Inui, H. Clemens, and C.L. Fu, (Mater. Res. Soc. Symp. Proc. **842**, Pittsburgh, PA, 2005) pp.169-174.
5. F. H. Hayes, in *Ternary Alloys - A Comprehensive Compendium of Evaluated Constitutional Data and Phase Diagrams*, edited by G. Petzow and G. Effenberg (VCH Publishers, New York, NY, 1993) vol.**8**, p.426.

Mater. Res. Soc. Symp. Proc. Vol. 1128 © 2009 Materials Research Society   1128-U05-22

Application of the Fluorine Effect to TiAl-Components

A. Donchev, R. Pflumm, M. Schütze
DECHEMA e.V. Karl-Winnacker-Institut
Theodor-Heuss-Allee 25
D-60486 Frankfurt/Germany

ABSTRACT

The oxidation resistance of TiAl-alloys can be improved by several orders of magnitude by fluorine doping of the surface zone of the material. The oxidation mechanism changes from the formation of a thick mixed oxide scale to a protective alumina layer. This fluorine treatment influences only the surface region of the components so that the bulk properties are not affected. Recent results achieved with TiAl-components showed the potential of a fluorine treatment for the use of TiAl in several high temperature applications. Turbine blades for aero engines made of TiAl were treated with fluorine by different methods and their performance during high temperature oxidation tests in air is shown. Further on by selective local fluorination a structured oxide scale develops on TiAl above 800°C. A simple high temperature activation cause the formation of areas covered by a thin alumina layer alternating with a thick mixed oxide scale where no fluorine was applied before oxidation. The aim is to reproduce a shark-skin pattern (tiny parallel ridges) on the surface in order to minimize the aero dynamic resistance of turbine blades rotating in a gas flow. Different methods used for this attempt and the corresponding results are also presented.

INTRODUCTION

Among the known light materials, the intermetallic titanium aluminides have suitable properties for high temperatures applications [1]. The specific weight of the intermetallic titanium aluminides is only half of that of Ni-based super alloys or high temperature steels and their thermal expansion coefficient is comparable with the one of ferritic steel. TiAl-alloys qualify for applications like turbocharger rotors for the automobile industry, turbine blades in aero engines or rotating components in land based gas turbines. A first application of the halogen effect on TiAl-components led to improved oxidation behavior. Direct consequences of this are an extension of the operating temperature range from 700 up to 900°C or even 1050°C. The halogen effect arises by alloying very small amounts of halogens at the materials surface and causes a complete change in the oxidation mechanism [2]. At temperatures above 700°C, in the absence of halogens a porous mixed oxide will develop rapidly and once achieving a critical thickness it will spall leading to material losses [3]. The right amount of halogen on the surface of titanium aluminides leads to a preferential oxidation of aluminum, which means the development of a slow growing Al-oxide barrier on the surface preventing the substrate material from further oxidation [4]. The second application of the halogen effect on TiAl refers to the development of functional materials. By this application a controlled change in the surface topography during the oxidation process, which means temperatures between 800 and 1000°C, should be achieved. Halogens applied selectively to TiAl led to oxides with different grow rates along the surface. The aim is to generate a riblet structure similar to the one of a shark skin. If the flow condition is turbulent, such aligned riblets hamper the cross flow in the boundary layer and lead to a drag reduction [5]. The advantages of a riblet structure have been usedd until now onlx at low temperatures [6]. Such structures could be used e.g. in the aero engines and contribute to the development of a new generation of environmental friendly engines.

EXPERIMENTAL

Coupon specimens and turbine blades of different TiAl-alloys (Table I) were investigated.

Table I: Chemical compositions of the examined alloys (at. %)

Alloy	Ti-content (at.%)	Al-content (at.%)	Other elements (at.%)
γ-TiAl	50	50	-
γ-MET	49.5	46.5	4 (B, Cr, Nb, Ta)
TNB	48.5	41.5	10 Nb
TNB V5	49.3	45	5.7(B, C, Nb)

The coupons were cut from bars or sheets, ground to 500, 1200 or 4000 grit respectively with SiC-paper, rinsed with distilled water, cleaned ultrasonically in ethanol and dried in air. Such specimens were oxidized to investigate the oxidation behavior of the untreated material. Fluorine treatment was performed by PI³ (plasma immersion ion implantation), BLI² (beam line ion implantation), gas phase and liquid phase treatment.

For the generation of a riblet topography of the samples just by high temperature oxidation, the halogens have been applied selectively on only one side of coupon sample by using the following methods: BLI² (ion dosage 10^{16} to 2×10^{17} F^-/cm^2, ion energy 30 keV) and liquid phase application of diluted HF (0.1 wt.%) using the paintbrush technique. In order to perform a selective halogenation of the sample by ion implantation we used masks. The masks are laser cut in an Al-foil with a thickness of ca. 100 μm and have been fixed with tungsten clamps on the sample prior to implantation. The spatially resolutions of the implantation geometry are: 600 μm wide halogen-rich and 200 μm wide halogen-free stripes and alternatively 300 μm wide implanted and 100 μm wide not implanted stripes. In the case of the liquid phase application, the surface has been treated selectively with HF by using the paintbrush technique. After the completion of the selective halogenation, the samples were oxidised at 900°C in dry or wet air for different times, according to their specific oxidation behaviour.

Post-experimental investigations of the oxide scale morphology, topography and composition included metallography, LM (light microscopy), SEM (scanning electron microscopy), EDX (energy dispersive x-ray), EPMA (electron probe micro analysis) and WLI (white light interferometry).

RESULTS AND DISCUSSION

The oxidation behavior of TiAl-alloys differs with composition [7]. Isothermal investigations are not so meaningful because they usually do not show spallation. Due to e.g. mismatch of the thermal expansion coefficients of the substrate and oxide the oxide scales can spall during cooling if a critical stress level is reached [8]. In figure 1 the mass change behavior of untreated and F-treated γ-MET in a thermocyclic 25h-cycle test (24h on temperature, one hour cooling and weighing) and metallographic cross sections after 50 days of testing are shown. The mixed oxide scale on the untreated sample is about 20 μm thick while the Al_2O_3 layer on the fluorine treated sample is less than 5 μm thin.

210

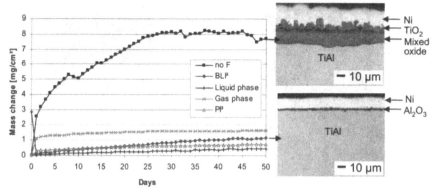

Figure 1: Mass change behavior of the alloy γ-MET during a 25h-cycle test at 900°C in laboratory air and metallogravic images of the untreated sample and the BLI²-samples after 50 days

    The high temperature corrosion of untreated TiAl in oxidizing atmospheres leads to the formation of a mixed oxide scale (TiO₂/Al₂O₃) which is fast growing and non protective. During cooling oxide flakes spalled from the untreated sample which caused the mass losses in t' e curve. The treatment with halogens changes the oxidation mechanism to exclusive alum. na formation but only the fluorine effect works under thermocyclic conditions [9]. Long term oxidation tests up to one year on exposure proved that only a total coverage of the sample with fluorine is able to withstand the oxidation [10]. BLI² does not provide a total coverage because it is limited to flat samples. The PI³-technique on the other side works also for complex components, therefore turbine blades were implanted with fluorine after optimization of the process. In figure 2 the mass change behavior of two turbine blades one untreated and the second with F-PI³ are presented and photos of the samples after testing are shown (25h-cycle test/900°C/laboratory air).

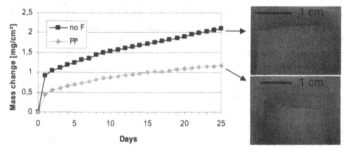

Figure 2: Mass change behavior of two turbine blades during a 25h-cycle test at 900°C in laboratory air and photos after the test

    The optical appearance of the two blades after oxidation is similar. SEM-investigations of the surface reveal the differences. The morphology and the crystal structure are different (fig.3a, c). EDX-analysis show a mixed oxide on the untreated sample while the implanted sample is covered by a thin alumina layer and the substrate was also detected (fig. 3b, d).

211

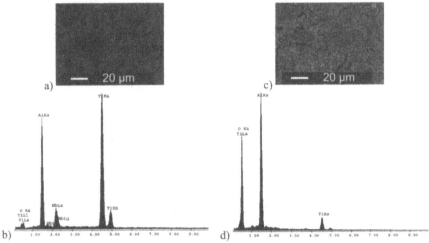

Figure 3: SEM-micrographs and corresponding EDX-spectra of the turbine blades after 25 days of thermocyclic oxidation at 900°C in laboratory air (a + b = untreated, c + d = F-PI³)

Regarding to the self growing structured oxide scale with a riblet pattern the results can be divided in two categories:

Liquid phase application: A ground TiAl (50%Al, 500 grit SiC-paper) has been selectively halogenated with diluted HF (0.1 wt.%) using a paint brush. After drying, the halogenrich domains could be optically distinguished from the halogen-free ones due to their bluebrown colour. EDX, measurements show F-concentration between 30 and 40 at% on the halogenated domains. Further on, this sample has been oxidised at 900°C in air for three consecutive cycles: 15 h, 35 h and 50 h. The morphology of the oxide scale corresponds to a selective halogen effect: yellow stripes of Ti-oxide crystals developed on the halogen-free areas and dark grey Al-oxide on the halogen-rich ones (Figure 4). On the light grey domains observed after 100 h cyclic oxidation the oxide had spalled, which can be interpreted as edge effects. Although the oxide layer topography reflects the halogenation pattern, a lateral expansion of Ti-oxide over Al-oxide has been measured. The rate varies along the surface and reaches values from 1.5 to 3.5 µm/h after 100 h cumulated oxidation time.

Figure 4: TiAl (50%Al) sample (surface 2x3 cm²) halogenated selectively with diluted HF solution. a) before oxidation, b) after 15 h oxidation and c) after 100 (15+35+50) h oxidation in air at 900°C. Yellow: Ti-Oxide. Dark grey: Al-oxide. Light grey: spalled domains.

The stoichiometric TiAl-alloy has the fastest oxidation kinetics compared to the other investigated alloys. This means that the size difference between the Ti-oxide crystals and the Al-oxide ones is remarkable. WLI (white light interferometry) measurements showed that the arithmetic average of the roughness profile along a Ti-oxide domain is about R_a=4.4 µm (0.9 µm standard deviation). Along the protective Al-oxide a R_a=0.3 µm (0.1 µm standard deviation) has been measured. The height difference at the transition from Al-oxide to Ti-oxide reaches 20 μm. Previous experiments on samples ground to 1200 grit showed a discontinuously growth of Ti-oxide on the halogen-free domains. This means that a rougher surface has a better structuring potential. This is beneficial for a later application on a larger scale. The fast oxidation kinetic of this alloy could be a disadvantage for long time oxidation experiments. Technical TiAl-alloys with a slower oxidation kinetic are under investigation.

Ion implantation: For TNB-V5 1200 grit samples were covered by Al-masks and implanted with the given implantation geometry and implantation energy of 30 keV; a dose of 10^{16} F$^-$/cm^2 represents the minimum amount of halogen needed to observe a selective halogen effect over 100 h. Cyclic oxidation tests performed at 900°C in dry air showed that for this dose the lateral extent of the Ti-oxide crosses the boundary of the not implanted domains as you can clearly see in Figure 5.

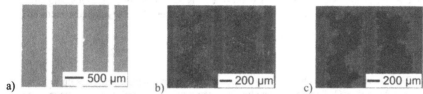

a)   — 500 μm   b)   — 200 μm   c)   — 200 μm

Figure 5: TNB-V5 sample selectively implanted with a dose of 10^{16} F$^-$/cm^2. a) LM-micrograph of the sample surface prior to oxidation (grey means halogen rich) b) and c) LM-micrographs with a higher magnification which show the same domain of the sample surface after 24 h (b) and 100 (24+76) h (c). Light grey: Ti-oxide

The amount of this lateral extent varies along the surface and the most of it took place in the first 24 h of oxidation. The maximal calculated value for this lateral growth rate was 16 μm/h and the smallest 3.9 μm/h. After another oxidation cycle of 76 h under the same conditions, this lateral rate converged to an average value of 0.9 μm/h (0.5 μm/h standard deviation). Three different points on the surface showed in the figures 5b and 5c have been taken into consideration for these estimations: one with the highest, one with the smallest and one with an asymmetrical lateral growth rate of the boundary between Ti- and Al-oxide. For a 20 times higher implantation dose the topography of the oxide scale changed completely. The Ti-oxide developed on the halogen-free domains but not at once. By cyclic oxidation under the same conditions as before one could observe an evolution pattern of the oxide morphology grown along the not implanted domains from a strong convoluted Al-oxide, to a strong convoluted Al-oxide with Ti-oxide crystals in between to a complete Ti-oxide scale (Figure 6). This transition was not completed along the not implanted domains even after a cumulated oxidation time of 199 h.

a)   — 10 μm   b)   — 5 μm   c)   — 10 μm

Figure 6: Evolution of the oxide scale morphology on a TNB-V5 sample selectively implanted with a dose of 2×10^{17} F$^-$/cm^2 during cyclic oxidation. a) Bundary between a halogen-free domain (strong convoluted Al-oxide) and a halogen-rich domain (less convoluted Al-oxide). b) Convoluted Al-oxide with Ti-oxide crystals between the convolutions. c) Boundary between implanted domain (less convoluted Al-oxide) and not implanted domain (Ti-oxide crystals).

An interesting result is that the spatial resolution of the implantation geometry which leads to a selective oxidation of the surface increases with the implanted F-dose. For 2×10^{17} implanted F-ions per unit area this has been 110 μm for the halogen-free domains and 290 μm for the halogen-rich ones. A close examination of this sample after each oxidation cycle showed a lateral extent of Ti-oxide over the Al-oxide with an estimated rate lower than 0.5 μm/h. The uniformity of the fluorine distribution along implanted domains has been proven by EDX. No shadow effect due to the mask thickness could be detected with this method. Further investigations about a potential correlation between the alloy microstructure and the lateral extent of Ti-oxide will follow. Nevertheless the aerodynamic improvement of TiAl-components demands a refinement of the structures and also a precisely defined riblet aspect ratio. The refinement must consider the influence of the halogen lateral diffusion under the halogen-free domains, which influences not only the lateral growing of Ti-oxide over the protective Al-oxide but also the riblet aspect ratio. This is the subject of current investigations. Further research about the thermal and mechanical properties of the developed structures is also planned.

CONCLUSIONS

The fluorine effect protects TiAl-components under oxidizing conditions up to temperatures of at least 900°C. The protective alumina layer works as a barrier against the environment. The fluorine treatment takes place after manufacturing of the components. In the present paper the fluorine effect has been applied to air foils and proved to provide a significant increase in oxidation resistance. The different oxidation mechanism of pure TiAl and TiAl+F can also be used for structuring the surface. Further development of the structures for surface aerodynamic improvements of such components is the subject of the current work.

ACKNOWLEDGEMENTS

The authors would like to thank the German Research Society (DFG) and the German Ministry of Economics and Technology (BMWi) for funding the projects. Additionally the cooperation with the research center Dresden is gratefully acknowledged. The materials were given by GfE, Plansee and Rolls-Royce which also appreciative mentioned.

References
1. F. Appel and M. Oehring, in *Titanium and Titanium Alloys*, Edited by C. Leyens and M. Peters (Wiley-VCH GmbH & Co. KGaA, 2003), Chap. 4, p. 89.
2. M. Hald and M. Schütze, Material Science Forum **251-254,** 179 (1997).
3. C. Leyens, in *Titanium and Titanium Alloys*, Edited by C. Leyens and M. Peters 2003), Chap. 6, p. 187.
4. A. Donchev, B. Gleeson, and M. Schütze, Intermetallics **11,** 387 (2003).
5. D. W. Bechert, M. Bruse, and W. Hage, Experiments in Fluids **28,** 403 (2000).
6. D. W. Bechert, M. Bruse, W. Hage, and R. Meyer, Naturwissenschaften **87,** 157 (2000).
7. Y. Shida and H. Anada, Oxidation of Metals **45,** (1996).
8. D. L. Douglass, P. Kofstad, A. Rahmel, and G. C. Wood, Oxidation of Metals **45,** 529 (1996).
9. A. Donchev and M. Schütze, Material Science Forum **461-464,** 447 (2004).
10. A. Donchev, E. Richter, M. Schütze, and R. Yankov, Journal of Alloys and Compounds **452,** 7 (2008).

Mater. Res. Soc. Symp. Proc. Vol. 1128 © 2009 Materials Research Society   1128-U05-25

Processing, Microstructure, and Thermal Expansion Measurements of High Temperature Ru-Al-Cr B2 Alloys

Y. Hashimoto[1], N.L. Okamoto[1], M. Acosta[2], D.R. Johnson[2], and H. Inui[1]

[1]Department of Materials Science and Engineering, Kyoto University,

Sakyo-ku, Kyoto 606-8501, Japan

[2]School of Materials Engineering, Purdue University, West Lafayette IN 47907

ABSTRACT

The B2 intermetallic compound RuAl has a melting temperature above 2000 °C and is a candidate for high temperature structural applications. A large extension of the B2 phase field is found in the Ru-Al-Cr system as was documented by the characterization of arc-melted and heat treated alloys. Two compositions consisting of Ru-35Al-19Cr and Ru-20Al-38Cr (at. %) were directionally solidified in an optical floating zone furnace. Depending upon the processing conditions, single phase, polycrystalline, B2 microstructures could be produced.   The coefficient of thermal expansion (CTE) was measured from room temperature to 1250 °C for the Ru-20Al-38Cr alloy, and an average value of 11×10^{-6} K^{-1} was found. Additionally, the thermal conductivity was measured as 27 W/mK at room temperature for the Ru-20Al-38Cr B2 alloy and as 89 W/mK for binary RuAl.

INTRODUCTION

Ruthenium aluminide with a CsCl (B2) crystal structure has been identified as a candidate material for high temperature structural applications [1-4]. From a recent investigation of Ru-Cr-Al alloys [5, 6], extensive solubility of Cr within the B2 phase was observed. The addition of Cr to RuAl improves the high temperature oxidation resistance, but it has been reported to decrease the toughness [7].   From four-point bend tests of notched samples of two-phase (HCP+B2) Ru-Al-Cr alloys, a fracture toughness of less than 13 MPa√m was estimated for the Cr-containing B2 phase [5]. In this investigation, a range of Ru-Al-Cr alloys are investigated to better describe the phase relationships at 1100 and 1500 °C, to investigate if single crystals of RuAl(Cr) could be grown, and to measure selected physical properties of the B2 phase. For example, attempts to grow single crystals of binary RuAl have been hampered by the rapid evaporation of Al from the melt resulting in a loss of stoichiometry and the formation of an intergranular eutectic film [5]. However, decreasing the Al content by the substitution of Cr may improve the solidification processing of these, Ru-Al-Cr, B2 alloys. Although single crystals were not grown in this study, the solidification behavior was detailed, and the coefficient of thermal expansion (CTE) to temperatures of 1250 °C and the room temperature thermal conductivity were measured.

EXPERIMENTAL PROCEDURE

Approximately 10 g buttons of Ru-Al-Cr alloys with compositions A to D and H to K (as marked in Figure 1) were produced by arc-melting in flowing Ar-5% H_2. Compositions E, F and G were prepared by induction melting and directionally solidified as described previously [5, 6]. The arc-melted and induction melted materials were sectioned and heat treated at 1100 °C for 200 h (compositions A to G) or 100 h (compositions H to K) and at 1500 °C for 6 in vacuum. The samples heat treated at 1100 °C were encapsulated in evacuated quartz tubes while a vacuum annealing furnace was used for the heat treatments at 1500°C.

X-ray diffraction was used for phase identification of the different alloys. Diffraction runs were performed on polycrystalline samples with a scan range of 20 to 120 degrees at a scan rate of 6°/minute under Cu K_α radiation in a D500 SIEMENS Diffraktometer. Phase compositions were measured by energy dispersive spectroscopy (EDS) in a T-300 JEOL SEM using elemental standards of Ru, Al, and Cr. EDS spectra were taken from each standard at a working distance of 39 mm, a takeoff angle of 30° and an accelerating voltage of 25 kV with a beam current of 4 mA and used to analyze phase compositions of the as-cast and heat treated samples [6].

Compositions of Ru-35Al-19Cr and Ru-20Al-38Cr (at. %) were chosen for crystal growth in an optical floating zone furnace at growth rates between 2.5 and 5 mm/h. Two furnaces were used. One employed halogen lamps and dual elliptical mirrors. The other employed a xenon lamp source and a single elliptical mirror. The latter setup has the advantage of maintaining a more sharply focused hot zone, which is beneficial for alloys having a wide melting range as in this case. Rectangular parallelepiped specimens for measurements of CTE and thermal conductivity were cut from the grown crystals by spark-machining. Measurements of CTE were carried out with a push-rod type differential dilatometer (Rigaku TMA8310) in the temperature range from room temperature to 1250 °C at a heating rate of 5 K/min under an Ar gas flow while those of thermal conductivity by the static method in vacuum at room temperature using 304 stainless steel cubes (41 W/mK) as reference samples.

RESULTS AND DISCUSSION

Representative microstructures of the heat treat arc-melted buttons are shown in Figure 1. After heat treating, evidence of the prior dendritic microstructure is still visible for the Ru-17Al-36Cr (at. %) alloy as shown in Figs. 1(a) & (b). For the 1500 °C anneal, the microstructure consists of the primary B2 phase and the secondary HCP solid solution, Fig. 1(a). After heat treating at 1100 °C, B2 precipitates are observed in the HCP phase, Fig. 1(b). From EDS analysis, the composition of the B2 phase was found to be Ru-19Al-35.2Cr (at. %) for the sample heat treated at 1500 °C. Subsequently, an alloy with a composition of Ru-20Al-38Cr (composition C in Fig. 1) was found to be single phase, consisting of only the B2 phase after heat treating at 1500 and 1100 °C. Following a similar analysis, the partial 1100 and 1500 °C isothermal sections were constructed [6] as shown in Figs. 1(c)&(d).

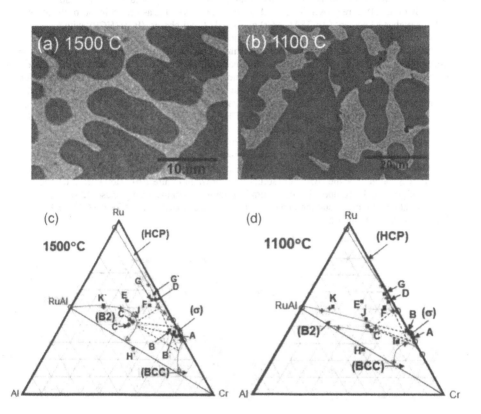

Figure 1: Microstructures (SEM backscattered electron images) from an arc-melted Ru-17Al-36Cr (at.%) alloy heat treated at (a) 1500 °C, (b) 1100 °C, and (c)&(d) the corresponding partial isothermal sections. The compositions of the alloys investigated are labeled by the letters A-K.

As seen in Figure 1, a large solubility range exists for the B2 phase in the ternary Ru-Al-Cr system. Two alloys with compositions within the B2 phase field (Ru-35Al-19Cr and Ru-20Al-38Cr (at. %)) were chosen for crystal growth. However, the as-grown ingot for the alloy with the higher Al content (Ru-35Al-19Cr) consisted of two-phases, (B2)+(HCP), with a microstructure similar to those shown Fig. 1. The development of the two-phase microstructure was attributed to the loss of Al during zone melting as is commonly observed in the solidification processing of RuAl alloys [2, 5]. On the other hand, a single phase B2 microstructure could be produced from the Ru-20Al-38Cr alloy upon zone melting, but the wide freezing range found for this composition prevented the growth of a single crystal. Evidence for this difficulty in crystal growth is shown in Figure 2. The arc-melted microstructure of the Ru-20Al-38Cr alloy consists of primary B2 dendrites and an interdendritic HCP phase, Fig. 2(a). Upon heat treating, a single phase B2 microstructure is obtained, such as the seed material shown in Fig. 2(b). Unfortunately, a large difference in melting temperature exits between the B2 and HCP phases. As a result, the initial molten zone that forms between the seed and the precursor ingot is richer in Ru and Cr when compared to the overall alloy composition. Consequently, the first phase to solidify was found to be the HCP solid solution as shown in Figure 2(b). However, after about 10 mm of growth, the composition of the liquid zone was such that single phase B2 was again obtained, although polycrystalline in nature. This material was then used for the CTE and thermal conductivity measurements.

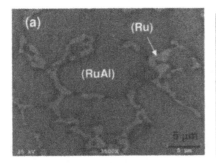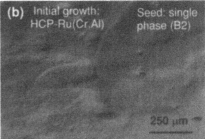

Figure 2: Microstructures from a Ru-20Al-38Cr alloy showing (a) the typical arc-melted microstructure (SEM backscattered electron image) and (b) the change in microstructure (optical light microcopy) between the seed and the initial solid to solidify during zone melting at a growth rate of 5 mm/h.

Figure 3 shows the relative elongation of the Ru-20Al-38Cr alloy plotted as a function of temperature. The relative elongation above 700 °C increases more rapidly than that below 700 °C. The CTE values averaged over some temperature ranges from room temperature to selected temperatures are tabulated in Figure 3. The CTE value averaged over the temperature range from room temperature to 1250 °C is 11.8×10^{-6} K^{-1}, which is larger than that for binary RuAl in the same temperature range as reported by Tryon et al. [8] but comparable to that for CoAl and NiAl [9].

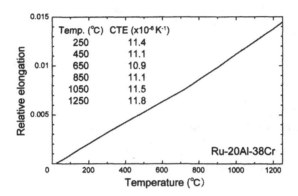

Figure 3. Relative elongation of the Ru-20Al-38Cr alloy consisting of single phase B2 plotted as a function of temperature.

We have measured the values of thermal conductivity of the ternary Ru-20Al-38Cr and binary RuAl alloys at room temperature. The value of thermal conductivity for the Ru-20Al-38Cr alloy is around 27 W/mK, which is comparable to that for Ni_3Al with the $L1_2$ structure [10] but considerably smaller than that of the binary RuAl alloy (89 W/mK). Anderson and Lang [11] have reported that off-stoichiometry in binary RuAl alloys introduces a large reduction in the thermal conductivity: an Al-rich RuAl alloy ($Ru_{47}Al_{53}$) exhibits almost half of the value of thermal conductivity for the stoichiometric RuAl alloy ($Ru_{50}Al_{50}$). The considerably small value of thermal conductivity for the Ru-20Al-38Cr alloy is considered to come from the large degree of alloying with Cr.

CONCLUSIONS

The phase relationship, solidification behavior, thermal expansion, and thermal conductivity of B2, Ru-Al-Cr, alloys were investigated. The main findings are:
(1) Partial isothermal sections at 1500 and 1100 °C were constructed and an extensive compositional range for the B2 phase was identified.
(2) A single phase, B2, microstructure can be obtained for a Ru-20Al-38 Cr alloy although the wide freezing rage for this composition makes crystal growth via zone melting difficult.
(3) The coefficient of thermal expansion of the B2 phase with a composition of Ru-20Al-38Cr was measured and an average value of 11.8×10^{-6} K^{-1} was found over the temperature range from room temperature to 1250 °C.
(4) The room temperature thermal conductivity for the B2 phase with a composition of Ru-20Al-38Cr was found to be 27 W/mK while that of binary RuAl was found to be 89 W/mK.

ACKNOWLEDGMENTS

We wish to acknowledge Dr. T. Hirano and Mr. Bannai of NIMS for the use of the xenon lamp optical floating zone furnace and for helpful discussions. DRJ and MA wish to acknowledge support from National Science Foundation (DMR-0076219) and from the Technical Assistance Program (TAP) at Purdue University.

REFERENCES

[1] R.L. Fleischer, R. J. Zabala, *Met Trans. A* **21A**, 2709 (1990).
[2] I.M. Wolff, G. Sauthoff , *Acta Mater.* **45**, 2949 (1997).
[3] D.C. Lu, T. M. Pollock, *Acta Mater.* **47**, 1035 (1999).
[4] F. Mücklich and N. Ilić, *Intermetallics* **13**, 5 (2005).
[5] T.D. Reynolds and D.R. Johnson, *Mater. Res. Symp. Proc.* **842**, S6.2.1 (2005).
[6] M. Acosta, MS thesis, 2008, Materials Engineering, Purdue University.
[7] R.L. Fleischer and D.W. McKee *Met. Trans. A* **24A**, 759 (1993).
[8] B. Tryon, T.M. Pollock, M.F.X. Gigliotti and K. Hemker, *Scripta Mater.* **50**, 845 (2004).
[9] R.W. Clark and J.D. Whittenberger, *Thermal expansion*, 8. (Plenum Press, 1981) p. 189.
[10] S. Hanai, Y. Terada, K. Ohkubo, T. Mohri, and T. Suzuki, *Intermetallics* **4**, S41 (1996).
[11] S.A. Anderson and C.I. Lang, *Scripta Mater.* **38**, 493 (1998).

Mater. Res. Soc. Symp. Proc. Vol. 1128 © 2009 Materials Research Society   1128-U05-26

Microstructures and Mechanical Properties of Co₃(Al,W) With the L1₂ Structure

Takashi Oohashi, Norihiko L. Okamoto, Kyosuke Kishida, Katsushi Tanaka and Haruyuki Inui
Department of Materials Science and Engineering, Kyoto University, Sakyo-ku, Kyoto 606-8501,
Japan

ABSTRACT

Microstructures and deformation behavior of γ/γ' Co-Al-W single-crystals and γ'-Co$_3$(Al,W) polycrystals have been investigated. Single crystals with the γ/γ' two-phase microstructure with a large volume fraction of the γ' phase are obtained by the Bridgman method. These single crystals exhibit anomalies in yield stress in compression at temperatures above 700 °C. Polycrystalline Co$_3$(Al,W) alloys can be rolled in ambient atmosphere such that the nominal thickness was reduced by 40%. The values of tensile elongation obtained for the polycrystalline Co$_3$(Al,W) alloys in air is slightly smaller than that obtained in vacuum, indicating that the Co$_3$(Al,W) alloys are not severely susceptible to the environmental embrittlement in air.

INTRODUCTION

The most widely used high-temperature structural materials in aircraft engines and power generation systems are Ni-base superalloys consisting of the solid-solution based on Ni with a face-centered cubic structure (γ phase) and the stable L1₂-ordered intermetallic compound based on Ni$_3$Al (γ' phase) [1-3]. The excellent high-temperature mechanical properties of these Ni-base superalloys have been believed to be closely related to the γ/γ' two-phase microstructure in which the cuboidal γ' precipitates are aligned coherently in the γ matrix [1-3]. Since the ternary intermetallic compound Co$_3$(Al,W) with the L1₂ structure was discovered [4], two-phase Co-base alloys composed of γ-Co solid-solution and γ'-Co$_3$(Al,W) phases have been investigated as promising high-temperature materials [5,6]. Some Co-base γ/γ' two-phase alloys have been reported to exhibit high-temperature strength greater than those of conventional Ni-base superalloys [5]. However, almost nothing is known about the pristine physical properties of γ'-Co$_3$(Al,W), supposedly due to the difficulties in obtaining single-phase Co$_3$(Al,W). In the present study, we examine microstructures and compression deformation behavior of single-crystals with a large volume fraction of the γ' phase as well as tensile deformation behavior of polycrystalline Co$_3$(Al,W) alloys in different test environments in order to clarify environmental effects on the tensile ductility.

EXPERIMENTAL PROCEDURES

Polycrystalline specimens with compositions of Co-9Al-9W (A), Co-10.5Al-11.5W (B), Co-10.5Al-11.75W (C), Co-10.5Al-12W (D), Co-11Al-11W (E), Co-11Al-12W (F) and Co-12Al-11W (G) (at%: as marked in figure 1) were prepared by arc-melting elemental Co, Al and W under an Ar gas flow (figure 1). Directional solidification was made for the alloys A-D by the Bridgman method in vacuum at a growth rate of 4.2 mm/h. The alloys D-G were annealed in Ar gas at 900 °C for 168 h. Microstructures of the as-grown and annealed samples were examined by scanning electron microscopy (SEM). Compression tests were conducted for the single

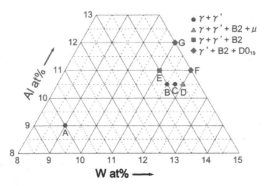

Figure 1. Compositions of the alloys investigated in the present study.

crystals on an Instron-type testing machine in vacuum in the temperature range from room temperature to 1050 °C at a strain rate of 1×10^{-4} s^{-1}. The [001] direction was chosen for the compression axis. As-grown polycrystalline specimens (E and F) with a thickness of 2.5 mm were rolled at room temperature to reduce the thickness by ~40%. Tensile specimens having a gauge section of $5 \times 2 \times 1.8$ mm^3 were electro-discharge machined from the as-annealed polycrystalline specimens, and ~0.1 mm was then removed from the surfaces by grinding followed by electro-chemical etching. Tensile tests were conducted in ambient atmosphere (air), vacuum (~10^{-3} Pa), and dry hydrogen gas (~0.1 MPa) at room temperature at a strain rate of 1×10^{-4} s^{-1}.

RESULTS AND DISCUSSION

Microstructures and compression deformation behavior of single crystal alloys

Large single crystals with a diameter of 20 mm were successfully obtained for the alloys A-D. Figures 2(a)-(d) show the SEM back-scattered electron images (BEIs) of the (001) suface of the grown crystals. As shown in figure 2(h), μ (Co_7W_6) and B2 (CoAl) phases are precipitated in the single crystal D. The volume fraction of the γ' phase in the single crystals A-D was estimated to be approximately 70, 94, 98 and 96%, respectively. Although the γ' volume fraction increases with the increase in the W content (A→C), it decreases again for the alloy D, mostly due to the precipitation of the μ and B2 phases. According to the ternary phase diagram reported by Sato et al. [4], the γ phase is equilibrated with the μ and B2 phases at high temperatures above 1000 °C, which makes it very difficult to obtain γ' single-phase single crystals when the temperature gradient during the crystal growth is small. A crystal growth method with a large temperature gradient other than the Bridgman method may be more suitable for the growth of $Co_3(Al,W)$ single crystals. The inset of figure 2(c) indicates the microstructure of the annealed sample of C, showing the cuboidal feature of the γ' precipitates with an average edge length of approximately 300 nm.

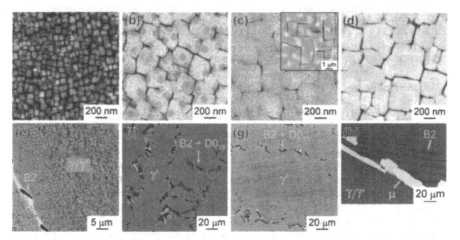

Figure 2. (a)-(c) SEM BEIs of the grown single crystals with the nominal compositions A-C. (d) and (h) SEM BEI of the annealed single crystal with the composition C. (e)-(g) SEM BEIs of as-annealed samples with the compositions E, F and G.

Compression tests were conducted for the as-grown single crystals with the nominal compositions A-C and the annealed single crystal with the composition C. The temperature dependence of the 0.2% yield stress obtained for the four crystals is shown in figure 3. The yield stress for the crystal A with the γ' volume fraction of 70% decreases monotonically and is much larger than those for the crystals at lower temperature below 800 °C. On the other hand, the yield stress for the crystals B and C decreases with the increase in temperature until the temperature

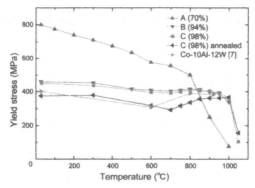

Figure 3. Temperature dependence of the yield stress for the single crystals with the nominal compositions A-C.

reaches 700 °C, and increases again to a small extent, followed by a rapid decrease above 1000 °C. The anomaly in yield stress for the annealed crystal C is similar to that for polycrystalline alloys Co-10Al-12W reported by Miura et al. [7]. The degree of the anomaly for the annealed crystal C is larger than that for the non-annealed counterpart, implying that the size of most cuboidal γ' precipitates in the annealed single crystal seems to be comparable to the heights of superkinks on APB(anti-phase boundary)-coupled <110> dislocations, although the dissociation scheme for <110> dislocations has not yet clarified.

Microstructures and tensile deformation behavior of polycrystalline alloys

SEM BEIs of as-annealed samples with the compositions E, F and G are shown in figures 2(e)-(g). A three-phase microstructure consisting of $\gamma + \gamma'$ + B2 phases is observed for the alloy with the composition E, whereas a three-phase microstructure consisting of γ' + B2 + D0$_{19}$ (Co$_3$W) phases is observed for the alloys with the compositions F and G. Observations of the alloys F and G by transmission electron microscopy confirm that the matrix region consists of only the γ' phase. This indicates that the boundary between the γ' + B2 two-phase region and the γ' + B2 + D0$_{19}$ three-phase region must be located between the compositions E and G. The volume fraction of the γ' phase was estimated to be around 90% and 95% for the alloys F and G. The polycrystalline specimen G of a strip shape was rolled at room temperature. Figures 4(a) and (b) show pictures of the specimen G respectively before and after cold-rolling. Though some small cracks are observed at the rim of the rolled specimen (indicated by arrows), it is possible to roll polycrystalline Co$_3$(Al,W) alloys in ambient atmosphere, inferring that the Co$_3$(Al,W) alloys are not susceptible to the environmental embrittlement in air, which is contrary to the case in polycrystalline Ni$_3$Al alloys with the L1$_2$ structure [8].

Tensile tests were conducted on the annealed sample G in order to further check the environmental susceptibility of the Co$_3$(Al,W) alloys. The stress-strain curves obtained for the tensile tests in ambient atmosphere (air), vacuum, and dry hydrogen gas are shown in figure 5. The yield stress and work hardening rate are almost insensitive to the testing environment. However, the values of tensile elongation observed in vacuum and air are larger than 20% whereas that observed in dry hydrogen gas is as small as 11%. The value of tensile elongation observed in air is slightly smaller than that observed in vacuum. This indicates that the Co$_3$(Al,W) alloys are not severely susceptible to the environmental embrittlement in air but may

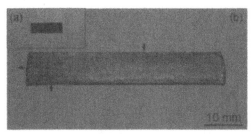

Figure 4. Photos of the specimen with the nominal compositions G (a) before and (a) after rolling at room temperature.

be susceptible to that in hydrogen gas. The reasons for the different ductility in air and hydrogen gas are now under investigation.

CONCLUSIONS

Microstructures and deformation behavior of γ/γ' Co-Al-W single-crystals and γ'-Co$_3$(Al,W) polycrystals were investigated. The main findings include:
(1) γ/γ' single crystals with a large volume fraction of the γ' phase (~98%) were obtained by the Bridgman method. These single crystals exhibit anomalies in yield stress in compression at temperatures above 700 °C.
(2) Polycrystalline Co$_3$(Al,W) alloys with the nominal composition of Co-12Al-11W can be rolled in ambient atmosphere such that the nominal thickness was reduced by 40%.
(3) Tensile tests were conducted on Co$_3$(Al,W) polycrystals in three different atmospheres. The values of tensile elongation obtained in air and vacuum are larger than 20% whereas that obtained in dry hydrogen gas is as small as 11%, indicating that the Co$_3$(Al,W) alloys are not severely susceptible to the environmental embrittlement in air but may be susceptible to that in hydrogen gas.

ACKNOWLEDGMENTS

This work was supported by Grant-in-Aid for Scientific Research (A) (No. 19656179) from the Ministry of Education, Culture, Sports, Science and Technology (MEXT), Japan and in part by the Global COE (Center of Excellence) Program of International Center for Integrated

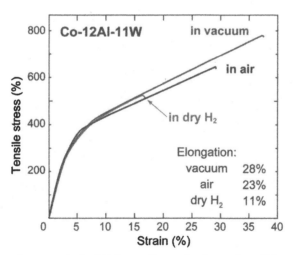

Figure 5. Stress-strain curves obtained in the tensile tests for the specimen G in ambient atmosphere (air), vacuum, and dry hydrogen gas.

225

Research and Advanced Education in Materials Science from the MEXT, Japan. The authors would like to thank Mr. Koyanagi of Osaka Titanium Technologies Co., Ltd. for supplying titanium sponge.

REFERENCES

1. D.P. Pope and S.S. Ezz, *Intl. Metals Rev.*, **29**, 136 (1984).
2. T.M. Pollock and R.D. Field, in F.R.N. Nabaroo and M.S. Duesbery (Eds.), *Dislocations in Solids*, Vol. 11, (Elsevier, Amsterdam, 2002), p.546.
3. F.R.N. Nabarro and M.S. Duesbery (Eds.), *Dislocations in Solids*, Vol. 10, (Elsevier, Amsterdam, 1997).
4. J. Sato, T. Omori, K. Oikawa, I. Ohnuma, R. Kainuma and K. Ishida, *Science*, **312**, 90 (2006).
5. A. Suzuki, G.C. DeNolf and T.M. Pollock, *Scripta Mater.*, **56**, 385 (2007).
6. A. Suzuki and T.M. Pollock, *Acta Mater.*, **56**, 1288 (2008).
7. S. Miura, K. Ohkubo and T. Mohri, *Mater. Trans. JIM*, **48**, 2403 (2007).
8. K. Aoki and O. Izumi, *Nippon Kinzoku Gakkaishi*, **41**, 170 (1977).

Mater. Res. Soc. Symp. Proc. Vol. 1128 © 2009 Materials Research Society   1128-U05-27

Influence of Elastic Properties on the Morphology of the γ' Precipitates in Ni-Base Superalloys

Wataro Hashimoto[1], Katsushi Tanaka[1], Kyosuke Kishida[1], Norihiko L. Okamoto[1], and Haruyuki Inui[1]

[1] Department of Materials Science and Engineering, Kyoto University Sakyo-ku, Kyoto 606-8501, Japan

ABSTRACT

Influence of elastic properties on the morphology of the γ' precipitates in Ni-base superalloys have been investigated by a theoretical and 3-dimantional elastic energy calculation. Large elastic anisotropy makes the shape of the precipitates more angular. The difference in the elastic properties between the matrix and the precipitates also has a significantly effect on the shape of the precipitates.

INTRODUCTION

Ni-base single-crystal superalloys are widely used for turbine blades of aircraft engines and gas turbine generators, because the alloys exhibit superior performance at high temperatures. The excellent mechanical properties at high temperatures are significantly owing to the γ' precipitates distributed in the γ matrix. Since γ/γ' interfaces efficiently block the motion of creep dislocations, the morphology of the γ' precipitates is an important factor for strengthening superalloys. When the alloys are crept under a low tensile stress at a high temperature, a directional coarsening of the γ' precipitates occurs and forms the lamellar structure, so-called "raft-structure", consisting of γ and γ' plates alternately stacked along the tensile stress direction. It is known that the raft structure efficiently suppresses creep deformation because creep dislocations are hard to penetrate into the γ' phase. The raft structure gradually collapses by further creep deformation and the collapse governs the lifetime of superalloys. Our elastic analysis of the raft structure previously performed [1,2] implies that the well aligned raft structure (γ/γ' interfaces are flat and lie on the (001) crystallographic plane) is expected to have a long lifetime under the creep condition. The initial shape and arrangement of γ' precipitates before creep deformation significantly affect to the morphology of the raft structure, and are closely related to elastic properties of both γ and γ' phases. Some assessments have been reported [3-5] but most have made qualitative analysis with 2-dimensional models or a small volume fraction of γ' precipitates.

The purpose of this study is to understand relationships between the elastic properties of the γ and γ' phases and the morphologies of the γ' precipitates distributed in the γ phase. We calculated the elastic energies for various shape and volume fraction of precipitates, and for various combinations of elastic moduli of γ and γ' phases. This calculation reveals the influence of elastic anisotropy and elastic misfit of γ and γ' phases on the shape of the γ' precipitates.

METHOD OF CALCULATION

Ni-base single-crystal superalloys have three independent elastic stiffness constants, C_{11}, C_{12} and C_{44}, because they have a cubic structure. In this study, we use a different set of three independent moduli, that is, bulk modulus K, Poisson's ratio for <100> extension v, and elastic anisotropy factor A, which are expressed as,

$$K = (C_{11} + 2C_{12})/3,$$ (1)

$$v = C_{12}/(C_{11} + C_{12}) \text{ and}$$ (2)

$$A = 2C_{44}/(C_{11} - C_{12}).$$ (3)

When there is no external applied stress, elastic energy is generated only by an internal elastic strain field caused by the lattice misfit. We have calculated the elastic strain and energy by Mura's [6] and Hu and Chen's [3] method with 128×128×128 grids. In this calculation, we use the eigen strains derived from the lattice misfit δ which are expressed as

$$\varepsilon_{ij}^*(r) = \begin{cases} V_f \delta & (r \text{ points the } \gamma \text{ phase}) \\ -(1 - V_f)\delta & (r \text{ points the } \gamma' \text{ phase}) \end{cases}$$ (4)

where V_f is volume fraction of the γ' phase. In elastic homogeneous case, that is all the elastic constants of the γ and γ' phases are identical, total strain (the sum of elastic and eigen strain) is given by

$$\varepsilon_{kl}(r) = \int C_{mnpq} \tilde{\varepsilon}_{pq}^*(g) g_n \frac{g_n N_{km}(g) + g_k N_{lm}(g)}{2D(g)} e^{ig\cdot r} dg$$ (5)

where g is wave vector, $\tilde{\varepsilon}_{pq}^*(g)$ is the Fourier transformed eigen strain, N_{ij} is the cofactor matrix of the matrix of $K_{pm} = C_{pqmn}g_q g_n$, and D is the determinant of the matrix K_{pm}. In elastic inhomogeneous case, we use the non-linear partial differential equation of elastic strain derived from the equilibrium equation of stress. We have solved this equation with perturbative approach using Equation (5). Hu and Chen have proposed the detail procedure for solving this differential equation [3].

In this calculation, we ignored the γ/γ' interfacial energy and assume the precipitates regularly aligned along the <100> directions. We express the shape of the precipitates with the "supersphere" expression proposed by Onaka et al [4,7] as

$$\left(x^2\right)^{p/2} + \left(y^2\right)^{p/2} + \left(z^2\right)^{p/2} = \left(a^2\right)^{p/2}$$ (6)

where a is radius of the precipitates, and p is shape exponent factor. As shown in Figure 1, Equation 6 can express the shape between cube, sphere, and regular-octahedron continuously. We use the shape factor η instead of p:

$$\eta = \sqrt{2} \cdot 2^{(-1/p)}$$ (7)

The elastic energy is normalized by bulk modulus K and eigen strain (lattice misfit) ε^* because

p	1.0	1.6	2.0	2.5	4.9	10.0	∞
η	0.707	0.917	1	1.07	1.23	1.32	$\sqrt{2}$
shape	octa-hedron		sphere				cube

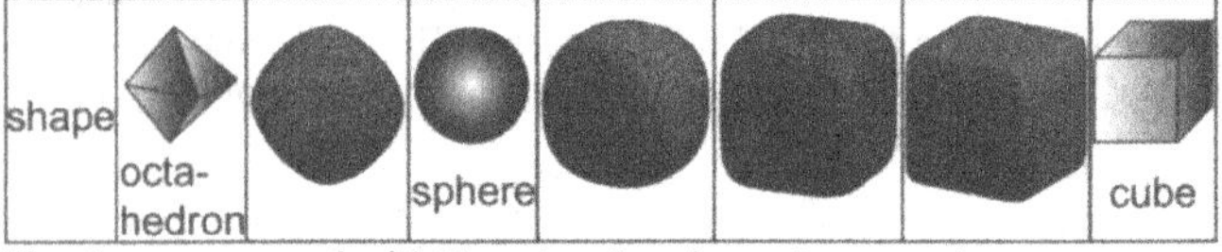

Figure 1. The shape expressed with Equation 6 under various p and η.

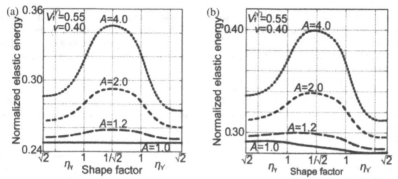

Figure 2. The variation of the normalized elastic energy as a function of the shape factor η. (a) elastic homogeneous case and (b) bulk modulus of the γ' phase is 5% larger than that of the γ phase. The volume fraction of the γ' phase is 55% and Poisson's ratio of both phases is 0.40, respectively.

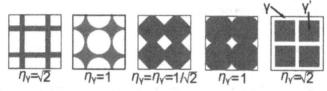

Figure 3. The variation of the morphologies of the γ and γ' phases with the shape factor η.

the elastic energy is proportionate to the bulk modulus and to the square of the magnitude of eigen strain.

RESULTS AND DISCUSSION

Case 1: Elastically Homogeneous System

Figure 2(a) shows the variation of normalized elastic energy with the shape factor η under the condition where both γ and γ' phases have the same elastic constants. The relationship between shape factor and realistic microstructure assumed is illustrated in Figure 3. The left end of the Figure 2(a) corresponds to the cubic precipitates of the γ phase ($\eta_\gamma = \sqrt{2}$). With decreasing the value of η_γ, corners of precipitates of the γ phase dented with keeping the constant volume fraction of both phases. The shape of the γ precipitates becomes spherical ($\eta_\gamma = 1$), and the γ precipitates begin to be combined with the neighbor γ precipitates. The shape becomes octahedral at $\eta_\gamma = \eta_{\gamma'} = 1/\sqrt{2}$, where the distribution of γ and γ' phases is symmetrical. Then the shape parameter is switched from η_γ to $\eta_{\gamma'}$. With increasing the value of η_γ, the shape of the γ' precipitates changes from octahedral, sphere and finally cubic.

When elastic anisotropy factor, A, equal to the unity (isotropic elasticity), elastic energy stays constant at any shape. When $A>1.0$ (anisotropic elasticity), the cubic γ' precipitates that

229

correspond to the case of $\eta_{\gamma'} = \sqrt{2}$ has the smallest elastic energy indicating that the morphology consisting of cubic γ' precipitates is the stable one. This result is well explained by the orientation dependence of the Young modulus in the crystal. Young's modulus along the <100> and the <111> directions, E_{100} and E_{111}, for cubic symmetry are expressed as the following equations,

$$E_{100} = 3(1 - 2\nu)K \quad \text{and} \tag{8}$$

$$E_{111} = 9(1 - 2\nu)AK /(2(1 + \nu) + (1 - 2\nu)A). \tag{9}$$

The relationship of $E_{100}= E_{111}$ holds when A=1.0, and the relationship of $E_{100}< E_{111}$ holds when A>1.0. When γ/γ' interfaces are inclined from {100} the elastic strain component along the <111> direction increase resulting to the relatively large elastic energy. Let's recall the fact that the elastic constants of both phases are set to the same value, and it is easy to understand that the asymmetry of the elastic energy with respect to the shape factors of γ and γ' phases comes from the volume fractions. The phase having larger volume fraction is preferred to form precipitates in the other phase having smaller volume fraction as illustrated in Figure 4(a). This is the opposite sense to the consideration of the interfacial energy. The asymmetry of the elastic energy is the stronger for the larger elastic anisotropy factor.

Case 2: Elastically Inhomogeneous System

Figure 2(b) shows the relationship between normalized elastic energy and the shape factor η as in the case of Figure 2(a) but the bulk modulus of the γ' phase is 5% larger than that of the γ phase. Similar to the elastic homogeneous case, cubic γ' precipitate has the smallest elastic energy and is the most stable microstructure. In this case, cubic γ' precipitate is the stable microstructure even when there is no elastic anisotropy, A=1.0. This energy difference between cubic γ and γ' precipitates biases the elastic energy asymmetry caused by volume fractions of the phases that mentioned in the above section.

Figure 4(b) shows the map of the stable microstructure plotted as functions of anisotropy factor and the volume fraction of the γ' phase. Due to the energy difference caused by the difference in the bulk modulus, the area of cubic γ precipitates distributed in the γ' matrix being stable shrinks to the left upper side.

Effect of Interfacial Energy

The calculations mentioned above have ignored an interfacial energy. In order to obtain more realistic description, an interfacial energy is taken into account in the followings. The total energy per unit volume, E_{tot}, express the sum of the elastic energy, E_{el}, and the interfacial energy "per unit volume", E_{inter} where the interfacial energy per unit area is converted to the energy per unit volume by introducing a factor representing area containing in a unit volume. Figure 5 shows the variation of these energies as a function of the shape factor. The shape is closed to the sphere, if the interface energy, E_{inter}, increases or the elastic energy E_{el} decreases. The energy change with the shape change from sphere to cubic, ΔE_{el} in Figure 5, significantly depends on the elastic anisotropy factor as indicated in Figure 2. The fact that the value of ΔE_{el} is larger for larger elastic anisotropy factor leads the shape of precipitates become more angular for a constant interfacial energy, E_{inter}. The more angular shape of γ' precipitates allow the larger volume fractions of the γ' phase without connecting the precipitates, resulting in the narrower γ

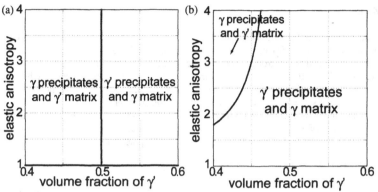

Figure 4. The stable morphologies under various elastic anisotropy and the volume fraction of the γ' phase. (a) homogeneous elastic property and(b) the bulk modulus of the γ' phase is 5% larger than that of the γ phase. Poisson's ratio of each phase is set to 0.40.

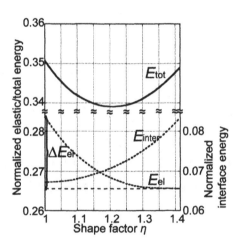

Figure 5. The shape dependence of the elastic energy E_{el}, the interface energy per unit volume E_{inter}, and the total energy E_{tot} between the spherical shape and the cubic shape. ΔE_{el} is the energy change with the shape change from sphere to cubic. All energies are normalized with the parameters that $K=200$GPa, $\delta=0.2\%$. This is in the case that $A=2.0$, $\nu=0.40$, volume fraction of γ' is 0.50, the interface energy per unit area is 10mJ/m^2, and the particle size of γ' is 0.45μm in elastically homogeneous system.

channel where creep dislocations passing through [5]. Such microstructure is preferred from a view point of creep strength, though the control of the elastic anisotropy is not easy.

The elastic anisotropy of the Ni-base superalloy, TMS-26, is reported as 3.6 at 1373K [8]. On the other hand, the value of a newly developed Co/Co$_3$(Al, W) alloy that is one of the candidate to a new class of superalloy [9] is 3.8 at 1173K. This indicates that the Co/Co$_3$(Al, W) alloy is expected to form more angular precipitates than Ni-base superalloys, which is a preferred properties for a high temperature use.

SUMMARY

Elastic energies for various shape of precipitate and for the elastic anisotropy factor have been calculated in a three-dimensional model. The results indicate as follows.

(1) Large elastic anisotropy makes the shape of the γ' precipitates angular.

(2) The elastically stable microstructure consists of precipitates of the phase having larger volume fraction distributed in the phase having smaller volume fraction. This is the opposite sense to the consideration of the interfacial energy.

(3) The difference of bulk modulus between the two phase makes the microstructure that consists of elastically harder precipitates distributed in elastically soft γ matrix stable even for smaller volume fractions of the harder phase than 50%.

ACKNOWLEDGMENTS

This work was supported by Grant-in-Aid for Scientific Research (A) (19656179) from the Ministry of Education, Culture, Sports, Science and Technology (MEXT), Japan and in part by the Grobal COE (Center of Excellence) Program of International Center for Integrated Research and Advanced Education in Materials Science from the MEXT, Japan.

REFERENCES

[1] K. Tanaka, T. Ichitsubo, K. Kishida, H. Inui, E. Matsubara, Acta Mater., **56**, 3786 (2008).

[2] T. Inoue, K. Tanaka, H. Adachi, K. Kishida, N. L. Okamoto, H. Inui, T. Yokokawa, H. Harada, to be published.

[3] S. Y. Hu, L. Q. Chen, Acta Mater., **49**, 1879 (2001).

[4] S. Onaka, N. Kobayashi, T. Fujii, M. Kato, Mater. Sci. Eng., **A347**, 42 (2003).

[5] T.M. Pollock and A.S. Argon, Acta Mater., **40**, 1 (1992).

[6] T. Mura, "*Micromechanics of defects in solids*", Martinus Nijhoff, The Hague, (1987).

[7] S. Onaka, T. Fujii, M. Kato, Mech. Mater., **37**, 179 (2005).

[8] K. Tanaka, T. Kajikawa, T. Ichitsubo, M. Osawa, T. Yokokawa, H. Harada, Mater. Sci. Forum, **475-479**, 619 (2005).

[9] J. Sato, T. Omori, I. Ohnuma, R. Kainuma, K. Ishida, Science, **312**, 90 (2006).

Mater. Res. Soc. Symp. Proc. Vol. 1128 © 2009 Materials Research Society   1128-U05-28

New Pt-Based Superalloy System Designed from First Principles

Vsevolod I. Razumovskiy[1,2], Eyvaz I. Isaev[2,3,4], Andrei V. Ruban[1] and Pavel A. Korzhavyi[1]
[1] Applied Materials Physics, Department of Materials Science and Engineering, Royal Institute of Technology, SE-100 44 Stockholm, Sweden.
[2] Department of Theoretical Physics, Moscow Institute of Steel and Alloys, 119049 Moscow, Russia.
[3] Condensed Matter Theory Group, Department of Physics, Uppsala University, Box 530, SE-751 21 Uppsala, Sweden.
[4] Department of Physics, Chemistry and Biology (IFM), Linköping University, SE-581 83 Linköping, Sweden.

ABSTRACT

Pt-Sc alloys with the γ-γ' microstructure are proposed as a basis for a new generation of Pt-based superalloys for ultrahigh-temperature applications. This alloy system was identified on the basis of first-principles calculations. Here we discuss the prospects of the Pt-Sc alloy system on the basis of calculated elastic properties, phonon spectra, and defect formation energies.

INTRODUCTION

Platinum-based high-temperature alloys have been successfully used in glass-making industry and for aerospace applications [1-3]. Modern platinum-based alloys have a homogeneous microstructure, quite similar to that of the first generation Ni-based superalloys, and are strengthened by means of solid-solution hardening. However, today's Ni-based superalloys are precipitation-hardened: their two-phase microstructure consists of an fcc γ-matrix that is strengthened by $L1_2$-ordered precipitates of the γ'-phase (Ni_3Al-based solid solution). The famous γ-γ' microstructure of Ni-based superalloys gives them the unique combination of high strength, creep resistance, and resistance against fatigue and corrosion, enabling a long service lifetime of the alloys at high temperatures and stresses. It is, therefore, clear that precipitation hardening offers a greater potential for increasing the high-temperature strength of structural alloys, as compared to what is achievable using solid solution hardening.

A recent combinatorial study, based on first-principles total-energy calculations and employing a "genetic" selection algorithm, has identified Pt_3Sc as one of the strongest intermetallic compounds. The compound has a large negative enthalpy of formation [4]. Its $L1_2$-type crystal structure is stable at low as well as at high temperatures; the structure remains ordered up to the melting point (1850 °C) [5]. According to the phase diagram [5], a two-phase region exists between a solid solution of Sc in fcc Pt and the ordered Pt_3Sc phase. An unconstrained lattice misfit between fcc Pt and $L1_2$ Pt_3Sc is estimated to be less than 1% [6]. Therefore, one may suggest that the intermetallic compound Pt_3Sc should be highly suitable for the role of a strengthening γ'-phase in a Pt-based superalloy intended for ultrahigh-temperature applications. In this work, we analyze and discuss the prospects of using the Pt-Sc binary system as a new base for designing Pt-based superalloys. As input data for our analysis we use the calculated elastic properties of disordered and ordered Pt–Sc alloys, the phonon spectrum of Pt_3Sc calculated using linear-response theory, and the formation energies of intrinsic point defects in off-stoichiometric Pt_3Sc obtained from supercell calculations.

METHODS OF CALCULATIONS

The electronic-structure calculations of the elastic properties of Pt–Sc alloys were based on density functional theory (DFT) [7] and were performed by means of the exact muffin-tin orbitals (EMTO) method [8,9]. Substitutional disorder in random fcc Pt–Sc solid solutions was treated within the coherent potential approximation (CPA) [10]; the treatment included screening corrections to the electrostatic potential and energy [11]. The screening parameters were evaluated using supercell calculations, which were performed within the locally self-consistent Green's function (LSGF) technique [12].

The self-consistent DFT calculations of the electron density and total energy were performed within the local density approximation (LDA) [13] for the exchange-correlation potential. During the self-consistency procedure, the integrations over the Brillouin zone were performed using a 37×37×37 grid of special k-points determined according to the Monkhorst–Pack scheme [14]. A finer 41×41×41 k-point grid was used in the final calculations of the electron density. The core states were recalculated at each self-consistency loop. Local lattice relaxations in random Pt-Sc alloys were neglected. Further details about the present elastic property calculations can be found in Ref. [15].

The phonon spectrum of Pt_3Sc was calculated using ultrasoft pseudopotentials [16], the harmonic approximation for the force constants, and the linear response method [17] as implemented in the Quantum Espresso code [18]. In these calculations, the exchange-correlation effects were treated within the generalized gradient approximation [19]. The integrations over the Brilluoin zone were performed by means of the Monkhorst-Pack special point technique [14], using the 16×16×16 k-point mesh and averaged on a grid of 6×6×6 q-points.

LSGF calculations were also used in order to calculate the formation energies of intrinsic point defects in Pt_3Sc. The calculations employed a supercell approach (single defect in a 256-site supercell). The Wagner-Schottky model is then used in order to estimate the equilibrium concentrations of point defects in $Pt_{3-\delta}Sc_{1+\delta}$ as a function of temperature and non-stoichiometry δ. The details of our LSGF calculations have been published elsewhere [20].

RESULTS AND DISCUSSION

Elastic constants

Table 1
Calculated (LDA) bulk modulus B (GPa), Young's modulus E (GPa), shear modulus G (GPa), ratio G/B, Poisson's ratio ν and anisotropy constant $A_G = 2C_{44}/(C_{11}-C_{12})$ of the γ and γ' phases in binary alloy systems: Pt–Sc (present work), Ni–Al, and Ir–Nb

	B	E	G	G/B	ν	A_G
Pt	288.4	167.1	59.5	0.21	0.40	1.49
$Pt_{87.5}Sc_{12.5}$	256.1	172.5	62.2	0.24	0.39	1.37
Pt_3Sc	246.2	203.6	74.7	0.30	0.36	0.93
Ni, Ref. [22]	184.0	221.0	85.0	0.46	0.30	2.48
Ni_3Al, Ref. [23]	173.9	203.1	77.8	0.45	0.30	3.49
Ir, Ref. [24]	381.3	590.0	237.5	0.62	0.24	1.41
Ir_3Nb, Ref. [24]	379.3	630.6	257.8	0.68	0.22	1.60

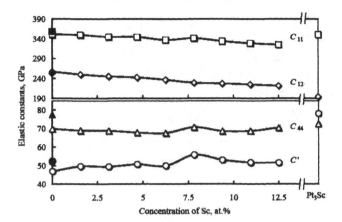

Figure 1: Calculated (LDA) elastic constants of pure Pt, of substitutionally disordered fcc Pt–Sc alloys, and of L1₂-ordered compound Pt₃Sc. Open squares correspond to the C_{11}, open diamonds correspond to C_{12}, open triangles correspond to C_{44} and open circles correspond to C'. The filled symbols show the respective experimental data for pure Pt [21].

In order to eliminate the errors of DFT approximations in predicting the molar volumes of solids, the present elastic-constant calculations have been performed at the experimental lattice parameters. Such experimental data exist for pure Pt and Pt₃Sc, but are absent for the disordered Pt–Sc alloys. In our calculations, the lattice parameter of random Pt-Sc alloys is found to exhibit linear concentration dependence. Therefore, we approximated the lattice parameter of random fcc Pt–Sc alloys using the experimental room-temperature value for pure Pt and the calculated linear slope, supposing that the slope in real Pt-Sc alloys will be the same. The calculated elastic constants of Pt-Sc alloys are presented in Figure 1.

In Table 1 the calculated elastic properties of γ and γ' phases of the present alloy system are summarized and compared with the corresponding properties of two other binary alloy systems, Ni–Al (base system for superalloys) and Ir–Nb (base system for ultrahigh-temperature alloys). The data for the Ni–Al alloy system are experimental. The elastic properties of the Pt–Sc phases are similar to or higher than those of the corresponding Ni–Al phases, but lower than those of the Ir–Nb phases. Taking also into account the high melting temperature (1850 °C) of γ–γ' Pt–Sc alloys [5], one may expect that the high-temperature stability and strength of Pt-Sc alloys will be intermediate between those of Ni–Al and Ir–Nb alloys. The ductility of Pt–Sc alloys is a very important advantage over the Ir–Nb alloys which have a high G/B ratio and therefore are predicted to be intrinsically brittle.

Phonon spectrum

The calculated phonon dispersion relations and phonon density-of-states (DOS) are presented in Figure 2. The phonon spectrum of Pt₃Sc is very similar to that of L1₂ Ni₃Al [25].

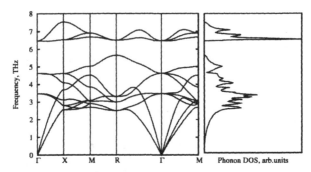

Figure 2. Phonon dispersion relations (on the left) and phonon DOS (on the right) for Pt$_3$Sc.

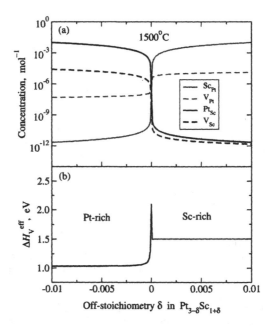

Figure 3. (a) Equilibrium concentrations of point defects in Pt$_{3-\delta}$Sc$_{1+\delta}$ at 1500°C (1773 K) estimated within the Wagner-Schottky model. Solid/dashed lines indicate concentrations of antisite defects/vacancies. (b) The effective vacancy formation energy $\Delta H_V^{eff} = -d\ln(x_V)/dT^{-1}$ in Pt$_{3-\delta}$Sc$_{1+\delta}$, as a function of deviation from stoichiometry δ. Here x_V is the total vacancy concentration (mole fraction). Note that the dominant kind of vacancies is predicted to be different for different compositional regions (Pt-rich and Sc-rich) of Pt$_3$Sc.

Let us first note that acoustic modes are not predicted to exhibit any dynamic instabilities. There is a slight mode softening near the high-symmetry point R. A similar behavior was reported for Ni_3Al, for which a large discrepancy (of about 0.5 THz) between the calculated and the experimental frequencies near point R was ascribed to the existence of flat bands near point R in the electronic spectrum. Regarding the phonon dispersion relations one also notes almost dispersionless acoustic modes between points X and M, as well as between Γ and X points for some optical modes which are mainly due to vibration of light Sc atom. The sharp DOS peak at around 6.7 THz is due to optical vibrations of the relatively light Sc atoms surrounded by heavy Pt atoms.

Point defects

Antisite defects and vacancies in compounds make important contributions to their properties. Since individual defects do not conserve the composition, their formation energies need to be expressed with respect to some standard states, which in this work are chosen to be pure fcc-Pt and stoichiometric $L1_2$-Pt_3Sc. The calculated defect-formation energies are as follows: antisite Sc (Sc_{Pt}, 4.1 eV), antisite Pt (Pt_{Sc}, 0.4 eV), Pt vacancy (V_{Pt}, 2.5 eV), and Sc vacancy (V_{Sc}, 1.3 eV). Fig. 3(a) shows the results of Wagner-Shottky modeling (ideal lattice gas) for point defects in $Pt_{3-\delta}Sc_{+\delta}$ compound at 1500°C. Constitutional defects (the defects that mainly account for the deviation from stoichiometry δ) are antisites, Pt_{Sc} ($\delta<0$) and Sc_{Pt} ($\delta>0$). Vacancies at all compositions are predicted to appear as primary thermal defects.

As follows from Fig. 3(b), the effective vacancy formation energy in Pt-rich $Pt_{3-\delta}Sc_{1+\delta}$ ($\delta<0$) is as low as 1.0 eV, whereas at Sc-rich composition ($\delta>0$) it is about 1.5 eV. Note that the vacancies that form in Pt-rich and Sc-rich Pt_3Sc preferentially reside on different sublattices (the deficient element sublattice is expected to contain more vacancies).

CONCLUSIONS

We have calculated the elastic properties of pure Pt, disordered Pt-Sc alloys and ordered intermetallic compound Pt_3Sc using EMTO method. For Pt_3Sc we also calculated the phonon spectrum and defect formation energies. Our calculations predict vacancies to be abundant point defects in Pt_3Sc at high temperatures, especially for Pt-rich compositions. The combination of elastic properties calculated here for Pt–Sc alloys with the γ– γ' microstructure suggests that these alloys should combine a good strength with an exceptional ductility. Moreover, this system has a very high melting temperature (Pt_3Sc is stable up to 1850 °C) which means that these alloys could be used as a material for ultrahigh temperature applications where the ductility offered by Pt–Sc alloys is an advantage in comparison with the brittleness of Ir-based alloys [24].

ACKNOWLEDGMENTS

Authors would like to thank I.M. Razumovskiy for useful discussions and fundamental ideas. This study has been supported by the Royal Swedish Academy of Sciences (KVA), through the grant for cooperation between Sweden and the former Soviet Union, and the Russian Foundation for Basic Researches (grant #07-02-01266). P.A.K. and A.V.R. acknowledge financial support by the Swedish Foundation for Strategic Research (SSF, program INALLOY). E.I.I. acknowledges the Netherlands Organization for Scientific Researches (NWO, grant

#047.016.005). Computer resources for this study have been provided by the Swedish National Allocation Committee (SNAC) at the Center for Parallel Computing (PDC), Stockholm, and the Joint Supercomputer Center of RAS (JSCC), Moscow.

REFERENCES

1. D. Lupton, Adv. Mater. **5**, 29 (1990).
2. M.V. Whalen, Platinum Met. Rev. **32(1)**, 2 (1988).
3. E.I. Rytvin, *High-temperature strength of platinum-based alloys* (Metallurgy Press, Moscow, 1987) p. 5.
4. G.H. Johannesson, T. Bligaard, A.V. Ruban, H.L. Skriver, K.W. Jacobsen, J.K. Norskov, Phys. Rev. Lett. **88**, 255506 (2002).
5. T.B. Massalski, *Binary alloy phase diagrams*. 2nd ed., vol. **1** (Materials Park, ASTM International, Ohio, 1990) p. 195.
6. *Pearson's handbook of crystallographic data for intermetallic phase*, 2nd ed., edited by P. Villars and L.D. Calvert (Materials Park, ASTM International, Ohio, 1991).
7. P. Hohenberg, W. Kohn, Phys. Rev. B **136**, 864 (1964).
8. L. Vitos, H.L. Skriver, B. Johansson, J. Kollar, Comput. Mater. Sci. **18**, 24 (2000).
9. L. Vitos, Phys. Rev. B **64**, 014107 (2001).
10. B.L. Gyorffy, Phys. Rev. B **5**, 2382 (1972).
11. A.V. Ruban, H.L. Skriver, Phys. Rev. B **66**, 024201 2002; A.V. Ruban, S.I. Simak, P.A. Korzhavyi, H.L. Skriver ibid., **66**, 024202 (2002).
12. I.A. Abrikosov, A.M.N. Niklasson, S.I. Simak, B. Johansson, A.V. Ruban, H.L. Skriver, Phys. Rev. Lett. **76**, 4203(1996); I.A. Abrikosov, S.I. Simak, B. Johansson, A.V. Ruban, H.L. Skriver, Phys. Rev. B **56**, 9319(1997).
13. J.P. Perdew, Y. Wang, Phys. Rev. B **45**, 13244 (1992).
14. H.J. Monkhorst, J.D. Pack, Phys. Rev. B **13**, 5188 (1972).
15. V.I. Razumovskiy, E.I. Isaev, A.V. Ruban and P.A. Korzhaviy, Intermetallics **16**, 982 (2008).
16. D. Vanderbilt, Phys. Rev. B **41**, R7892 (1990).
17. S. Baroni, S. De Gironcoli, A. Dal Corso, and P. Giannozzii, Rev. Mod. Phys. **73**, 515 (2001).
18. Quantum Espresso is a community project for high-quality quantum-simulation software based on density-functional theory and coordinated by Paolo Giannozzi. See http://www.quantumespresso.org and http://www.pwscf.org
19. J.P. Perdew, K. Bruke, M. Ernzerhof, Phys. Rev. Lett. **77**, 3865 (1996).
20. P.A. Korzhavyi, A.V. Ruban, A.Y. Lozovoi, Yu.Kh. Vekilov, I.A. Abrikosov and B. Johansson, Phys. Rev. B **61**, 6003 (2000).
21. S.M. Collard and R.B. McLellan, Acta Metall. Mater. **40**, 699 (1992).
22. E.S. Fisher, Scripta Metall. **20**, 279 (1986).
23. S.V. Prikhodko, H. Yang, A.J. Ardell, J.D. Carnes, D.G. Isaak, Metall. Mater. Trans. A **30**, 2403 (1999).
24. K. Chen, L.R. Zhao, J.S. Tse, J. Appl. Phys. **93**, 2414 (2003).
25. E.I. Isaev A.I. Lichtenstein, Yu.Kh. Vekilov, E.A. Smirnova, I.A. Abrikosov, S.I. Simak, R. Ahuja and B. Johansson Sol. St. Commun. **129**, 809 (2004).

Mater. Res. Soc. Symp. Proc. Vol. 1128 © 2009 Materials Research Society   1128-U05-30

Thermodynamic Re-Assessment of the Co-Nb System

C. He, F. Stein, M. Palm and D. Raabe

Max-Planck-Institut für Eisenforschung GmbH, Postfach 140444, D-40074 Düsseldorf, Germany

ABSTRACT

A new thermodynamic assessment of the Co-Nb system is presented. All experimental phase diagram data available from the literature have been critically reviewed and assessed using thermodynamic models for the Gibbs energies of the individual phases (Thermo-Calc). Compared to previous assessments more elaborate models for the description of the C14 and C36 Laves phases and for the μ phase were employed. Thereby a calculated phase diagram is obtained which satisfactorily agrees with the experimental data.

INTRODUCTION

The Laves phases AB_2 form the largest group of intermetallic compounds. Three different crystallographic polymorphs exist, namely cubic $MgCu_2$-type (C15) and hexagonal $MgZn_2$-type (C14) and $MgNi_2$-type (C36). Since the last decade, there is renewed interest in Laves phases as they have become candidates for several functional applications such as hydrogen storage as well as structural materials [1, 2]. Though, development of such materials is considerably hindered by the fact that the stability of the Laves phases in certain systems, the occurrence of the various polymorphs in dependence of composition and temperature, and the transformation from one polymorph into another are up to now not well understood [3,4].

The Co-Nb system is an excellent candidate for fundamental studies of the stability of Laves phases because all three polymorphs occur within this system. Moreover, Co and Nb are also two important alloying elements for high-temperature superalloys, which make this system of considerable interest. Therefore, phase equilibria in the Co–Nb system have been studied experimentally quite frequently before [5-13], but reported data show noticeable discrepancies. In order to settle these uncertainties and specifically the stability of the different Laves phase polymorphs, the system has been recently reinvestigated [14].

Based on the earlier literature data, several thermodynamic assessments of the Co-Nb system have been performed [15-18] and the latest and most comprehensive work was carried out by Hari Kumar et al. [18]. They considered the five binary compounds C14, C15, C36, Co_7Nb_2, and μ phase in their calculation. However, the homogeneity ranges of the C14 and C36 Laves phases were not considered and both phases were simply modeled as stoichiometric compounds ($Co_{16}Nb_9$ and Co_3Nb, respectively) due to the lack of respective data at that time.

The purpose of the present work is to obtain a self-consistent set of thermodynamic parameters of the Co-Nb system by considering the latest experimental work with specific emphasis on modeling the homogeneity ranges for the Laves phases.

EXPERIMENTAL DATA

The literature data up to 1997 were critically reviewed by Hari Kumar et al. [18]. At that time the existence of the μ phase and C15 were accepted, but the existence of C36, C14 and Co_7Nb_2 was debatable. Recently, Stein et al. [14] redetermined the Co-Nb phase diagram in the whole composition range by differential thermal analysis (DTA), metallography, X-ray diffraction (XRD), and electron probe microanalysis (EPMA). The existence of the μ phase, the three Laves phase polymorphs, and of Co_7Nb_2 was confirmed and the homogeneity ranges of all phases were carefully determined. According to this the solubility limit of Nb in (Co) is about 5.5 at.%, consistent with the data of Pargeter & Hume-Rothery [11], Co_7Nb_2, C36, and C14 have narrow homogeneity ranges, similar to those reported earlier [12], while the homogeneity ranges of C15 (26 to 35.3 at.% Nb), the μ phase (46.5 to 56.1 at.% Nb) and of (Nb) (maximum solubility of about 5.4 at.% Co) are larger than according to previous reports [11,13].

The present assessment is mainly based on the new experimental data of [14]. In addition, experimental data for the congruent melting point of the μ phase, the invariant reaction temperature of L ↔ μ + (Nb), and the liquidus in the Nb-rich region from [11] were used in the present optimization and given a high weight. According to the review by Hari Kumar et al. [18] only a very limited number of thermodynamic data exists. As no new thermodynamic data have become available since then, no thermodynamic data were used in the present assessment.

THERMODYNAMIC MODELS

There are eight stable phases in the Co-Nb system, namely liquid, (Co), (Nb), C14, C15, C36, μ phase, and Co_7Nb_2. The models employed to describe these phases are presented in Table I. For comparison the models used by Hari Kumar et al. [18] are listed in Table I as well. The differences between the models used in the present work and those used in [18] are that 2-sublattice models instead of stoichiometric compounds model are used for C36 and C14, and Co, Nb coexisting in three sublattices of 4-sublattice model is used instead of two sublattices of 4-sublattice model for the description of the μ phase.

Table I. The models used for the various phases of the Co-Nb system in the present work compared to those used in [18].

Symbols	Phase name	Present model	Model used in [18]
Liq	liquid	(Co,Nb)	(Co,Nb)
Bcc	bcc (Nb)	$(Co,Nb)_1 Va_3$	$(Co,Nb)_1 Va_3$
Fcc	fcc (Co)	(Co,Nb)Va	(Co,Nb)Va
Hcp	hcp (Co)	$(Co,Nb)Va_{0.5}$	$(Co,Nb)Va_{0.5}$
C15	Laves C15	$(Co,Nb)_2(Co,Nb)$	$(Co,Nb)_2(Co,Nb)$
C36	Laves C36	$Co_2(Co,Nb)$	Co_3Nb
C14	Laves C14	$(Co,Nb)_2Nb$	$Co_{16}Nb_9$
μ	CoNb μ	$(Co,Nb)_1(Co,Nb)_2Nb_4(Co,Nb)_6$	$(Co,Nb)_1(Co,Nb)_2Nb_4Co_6$
Co_7Nb_2	Co_7Nb_2	Co_7Nb_2	Co_7Nb_2

RESULTS AND DISCUSSION

The evaluation of the model parameters is attained by recurrent runs of the PARROT module [19] in Thermo-Calc [20], which works by minimizing the square sum of the differences between experimental values and computed ones. In the assessment procedure, each piece of experimental information is given a certain weight. The weights were varied systematically during the assessment until most of the experimental data were accounted for within the estimated uncertainty limits. A complete and self-consistent thermodynamic description of the Co-Nb system is obtained with the present set of model. Both, calculated and experimental temperatures and the compositions of the phases at the invariant equilibria are listed in Table II.

The calculated phase diagram is shown in Fig. 1. In Fig. 2 the calculated phase diagram is compared with the experimental data by Stein et al. [14]. An excellent agreement with the data was obtained within the experimental uncertainties. Fig. 3a is the enlargement of the calculated Co-rich part of the phase diagram. All the experimental values used in the optimization are well reproduced. The enlargement of the calculated Nb-rich part of the phase diagram including the experimental data is shown in Fig. 3b. It shows that also the experimental data of Pargeter & Hume-Rothery [11] which were used in this assessment are reasonably described.

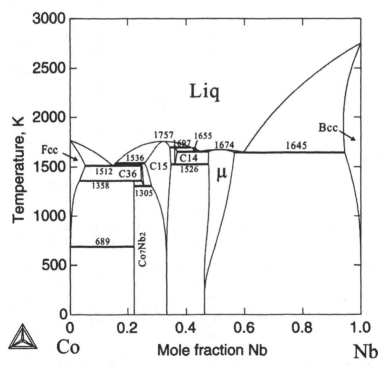

Figure 1. The calculated Co-Nb phase diagram according the present assessment.

241

Table II. Comparison of the calculated and experimentally determined temperatures and compositions of the phases at the various invariant reactions. x_1, x_2, and x_3 refer to the mole fraction of Nb in the three phases in the order they appear in the invariant reactions.

Invariant reaction	Type	T (K)	x_{Nb}	x_{Nb}	x_{Nb}	Ref.
L ↔ μ	Congruent	1674		0.514		This work
		1675		0.515		[11]
		1703		0.510		[12]
		1748		0.520		[9]
L ↔ C15	Congruent	1757		0.323		This work
		1757		0.333		[14]
		1753		0.32		[11]
		1793		0.34		[12]
		1813				[9]
		1823		0.286		[5]
L ↔ fcc (Co) + C36	Eutectic	1512	0.15	0.053	0.246	This work
		1512	0.14	0.055	0.245	[14]
		1510	0.139	0.055		[11]
		1483	0.150			[12]
L + C15 ↔ C36	Peritectic	1536	0.158	0.258	0.248	This work
		1537	0.145	0.26	0.25	[14]
		1520	0.145	0.273		[11]
		1513	0.165			[12]
L + C15 ↔ C14	Peritectic	1697	0.416	0.345	0.362	This work
		1697	0.42	0.345	0.365	[14]
		1693	0.470			[12]
		1713				[9]
L ↔ μ + C14	Eutectic	1655	0.448	0.476	0.369	This work
		1652	0.45	0.47	0.37	[14]
		1651				[11]
		1643	0.47			[12]
		1653				[9]
L ↔ bcc (Nb) + μ	Eutectic	1645	0.60	0.946	0.565	This work
		----		0.947	0.56	[14]
		1647	0.610	0.969	0.527	[11]
		1643	0.650			[12]
		1653				[9]
fcc (Co) + C36 ↔ Co$_7$Nb$_2$	Peritectic	1358	0.032	0.248	0.222	This work
		1359	0.037	0.247	0.222	[14]
		1323				[12]
C36 ↔ Co$_7$Nb$_2$ + C15	Eutectoid	1305	0.255	0.22	0.28	This work
		1313 ± 40	0.25	0.22	0.26	[14]
		1273				[12]
C14 ↔ C15 + μ	Eutectoid	1526	0.362	0.349	0.477	This work
		1523 ± 20	0.365	0.355	0.465	[14]
		1473				[12]
		1498				[9]
fcc (Co) ↔ hcp (Co) + Co$_7$Nb$_2$	Eutectoid	689	6.7E-4	5.2E-5	0.22	This work

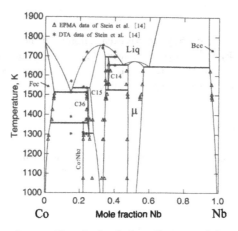

Figure 2. Comparison between the calculated phase diagram and the experiments by [14].

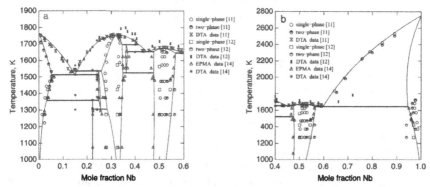

Figure 3. Comparison of the calculated Co-Nb phase diagram with experimental data in the Co-rich region (a), and in the Nb-rich region (b).

Compared to the assessment by Hari Kumar et al. [18] the present assessment is improved with respect to the three following points: (1) More recent descriptions of the crystallographic structures of the C14 and C36 Laves phase were used [21, 22] and their homogeneity ranges were taken into account. Therefore, the models for C14 and C36 were improved by using a 2-sublattice model. Thereby, homogeneity ranges for both phases can be modeled, which not only gives a correct description for the binary phases, but which is necessary to describe the substitution by third elements in higher-order systems. (2) The model used in [18] for the μ phase can only accommodate compositions in the range 30.8 - 53.8 at.% Nb. In order to describe the new experimental data for the composition range of 46.5 – 56.1 at.% Nb for the μ phase correctly, Co, Nb coexisting in three sublattices of 4-sublattice model was used in the current assessment. (3) More recent data for the phase equilibria were used in this assessment, and these data are well described by the present modeling.

243

SUMMARY

A re-assessment of the Co-Nb system has been performed on the basis of new experimental data. The crystallography and homogeneity ranges of the C14 and C36 Laves phases and the μ phase were taken into account by modeling with 2- and 4-sublattice models. A self-consistent and more reasonable description of the Co-Nb system is thereby obtained.

ACKNOWLEDGEMENTS

This research work is supported by the Max-Planck Society within the framework of the inter-institutional research initiative "The Nature of Laves Phases".

REFERENCES

1. G. Sauthoff, *Intermetallics*. Weinheim, Germany: VCH; (1995).
2. J.D. Livingston, *Phys. Status Solidi* **131(a)**, 415 (1992).
3. F. Stein, M. Palm and G. Sauthoff, *Intermetallics* **12**, 713 (2004).
4. F. Stein, M. Palm and G. Sauthoff, *Intermetallics*, **13**, 1056 (2005).
5. W. Köster and W. Mulfinger *Z. Metallkd.* **30**, 348 (1938).
6. H.J. Wallbaum, *Arch. Eisenhuettenwes.* **14**, 521 (1941).
7. H.J. Wallbaum, *Z. Kristallogr.* **103**, 391 (1941).
8. S. Saito and P.A. Beck, *Trans. Metall. Soc. AIME* **218**, 670 (1960).
9. A.K. Shurin and G.P. Dmitrieva, *Vopr. Fiz. Met. Metalloved.* **18**, 175 (1964).
10. A. Raman, *Trans. Metall. Soc. AIME* **236**, 561 (1966).
11. J.K. Pargeter and W. Hume-Rothery, *J. Less-Common Met.* **12**, 366 (1967).
12. S.K. Bataleva, V.V. Kuprina, V.Y. Markiv, V.V. Burnashova, G.N. Ronami, S.M. Kuznetsova, *Vest. Mosk. Univ., Khim.* **11**, 432 (1970).
13. W. Sprengel, M. Denkinger and H. Mehrer, *Intermetallics* **2**, 127 (1994).
14. F. Stein, D. Jiang, M. Palm, G. Sauthoff, D. Grüner and G. Kreiner, *Intermetallics* **16**, 785 (2008).
15. L. Kaufman and H. Nesor, *Metall. Trans.* **A6**, 2115 (1975).
16. L. Kaufman and H. Nesor, *CALPHAD* **2**, 81 (1978).
17. R. Bormann and R. Busch, *J. Non Cryst. Solids* **117/118**, 539 (1990)
18. K.C. Hari Kumar, I. Ansara, P. Wollants and L. Delaey. *J Alloys Compd.* **267**, 105 (1998).
19. B. Jansson, *Trita Mac-0234* (Royal Institute of Technology, Stockholm, Sweden, (1984).
20. B. Sundman, B. Jansson and J.-O. Andersson, *CALPHAD* **9**, 153 (1985).
21. I. Ansara, T.G. Chart, A.F. Guillermet, F.H. Hayes, U.R. Kattner, D.G. Pettifor, N. Saunders and K. Zeng, *CALPHAD* **21**, 171 (1997).
22. D. Grüner, F. Stein, M. Palm, J. Konrad, A. Ormeci, W. Schnelle, y. Grin and G. Kreiner, *Z. Kristallogr.* **221**, 319 (2006).

Mater. Res. Soc. Symp. Proc. Vol. 1128 © 2009 Materials Research Society 1128-U05-31

Alloying Effect on Mechanical Properties and Oxidation Resistance of Cold-Rolled Ni$_3$(Si,Ti) Foils

Yasunori Fujimoto, Yasuyuki Kaneno and Takayuki Takasugi

Department of Materials Science, Osaka Prefecture University, 1-1 Gakuen-cho, Naka-ku, Sakai, Osaka 599-8531, Japan

ABSTRACT

Four kinds of L1$_2$-type Ni$_3$(Si,Ti) intermetallics alloyed with a quaternary element X (X: Al, Cr, Co and Mo) were warm rolled accompanied by intermediate annealing and then cold rolled to thin foils. The effects of the alloying elements on microstructure, tensile properties and oxidation resistance of the cold-rolled Ni$_3$(Si,Ti) foils were investigated. The Al-alloyed Ni$_3$(Si,Ti) showed an L1$_2$ single-phase microstructure, while the Cr-, Co- and Mo-alloyed Ni$_3$(Si,Ti) exhibited a two-phase microstructure consisting of L1$_2$ and fcc Ni solid solution. Room-temperature strength of the Ni$_3$(Si,Ti) foils was slightly enhanced by the addition of the quaternary elements, whereas high-temperature strength was significantly enhanced especially by the addition of Mo and Co. High-temperature tensile elongation was remarkably improved by the additions of all the elements investigated. On the other hand, the oxidation resistance was improved by the addition of Al and Cr.

INTRODUCTION

Many ordered intermetallic alloys and compounds show a poor ductility due to their complex crystal structures and lack of sufficient slip systems. However, some ordered alloys with L1$_2$ structure such as Ni$_3$Al micro-alloyed with boron [1], Ni$_3$Si alloyed with Ti (i.e., Ni$_3$(Si,Ti)) [2] and Co$_3$Ti [3] show a fairly good ductility even at low temperature, and thereby heavy cold working is applicable to these intermetallic alloys. It has been reported that Ni$_3$Al foils were fabricated by heavy cold rolling of directionally solidified single crystals [4] and also by heavy cold rolling at liquid nitrogen temperature and subsequent annealing of coarse grained cast materials [5]. Recently, the present authors have found that Ni$_3$(Si,Ti) foils as well as Co$_3$Ti and Ni$_3$Al foils could be fabricated from arc-melted polycrystalline materials by thermomechanical processing and subsequent heavy cold rolling [6]. The obtained Ni$_3$(Si,Ti) foils were found to exhibit good mechanical properties: low- and intermediate-temperature tensile strength is extremely superior to that of conventional alloys such as nickel-based alloys and stainless steels. However, further improvement is desired for the high-temperature tensile properties and the oxidation resistance because these properties are important for high-temperature structural materials. Earlier studies have shown that some interstitial and substitutional elements were useful for the improvement of the mechanical properties of the warm-rolled Ni$_3$(Si,Ti) sheets [7-9]. However, the effect of alloying elements on the mechanical and chemical properties of the cold-rolled Ni$_3$(Si,Ti,) foils has been little studied [10].

In this study, microstructures, tensile properties and oxidization resistance of the cold-rolled Ni$_3$(Si,Ti) foils alloyed with various elements were investigated.

EXPERIMENTAL PROCEDURES

The composition of the base alloy (i.e., the unalloyed $Ni_3(Si,Ti)$) was 79.5 at.% Ni, 11 at.% Si, 9.5 at.% Ti and 50 wt.ppm % boron. Taking into account the substitution behavior of alloying elements in Ni_3Si [11], four kinds of quaternary $Ni_3(Si,Ti)$ alloys, i.e., $Ni_{79.5}Si_{11}Ti_{7.5}Al_2$, $Ni_{78.5}Si_{11}Ti_{8.5}Cr_2$, $Ni_{77.5}Si_{11}Ti_{9.5}Co_2$, and $Ni_{79.5}Si_{11}Ti_{7.5}Mo_2$ each with 50 wt.ppm % boron were arc-melted in an argon atmosphere and cast into a water-cooled copper hearth. The button ingots were homogenized at 1323 K for 48 h in a vacuum. The homogenized buttons were cut into plates with a thickness of about 10 mm, and were then warm rolled to sheets with a thickness of 2 mm by thermomechanical processing (i.e., repeated warm rolling at 573 K and annealing at 1273 K for 5 h in vacuum). These sheets were finally annealed at 1273 K for 1 h to prepare starting materials for cold rolling. Cold rolling was conducted to 90 % reduction in thickness without intermediate annealing. The thickness of the obtained cold-rolled foils was approximately 200 μm. Microstructural observation was carried out by scanning electron microscopy (SEM) and X-ray diffraction (XRD). Room-temperature tensile testing was done in air using the cold-rolled and fully recrystallized foils. High-temperature tensile tests were also conducted from room temperature to 973 K in vacuum using fully recrystallized foils. A strain rate of 8.4×10^{-5} s^{-1} was used in both room-temperature and high-temperature tensile tests. The fracture surfaces of the tensile-tested specimens were observed by SEM. The oxidation resistance of the cold-rolled foils was evaluated using thermogravimetric-differential thermal analysis (TG-DTA). Also, an isothermal oxidation test was performed in air at 1173 K for 48 h and the observed mass gains were evaluated.

RESULTS AND DISCUSSION

Microstructure

All alloyed $Ni_3(Si,Ti)$ ingots were successfully cold-rolled to thin foils with a thickness of 200 μm, indicating that the alloying elements used in this study did not deteriorate the cold-rolling workability of the $Ni_3(Si,Ti)$ alloy. Figure 1 shows representative SEM microstructures of alloyed $Ni_3(Si,Ti)$ foils annealed at 1173 K for 1 h. The Al-alloyed $Ni_3(Si,Ti)$ as well as the unalloyed $Ni_3(Si,Ti)$ showed an $L1_2$ single phase microstructure while the Cr-, Co-, and Mo-alloyed $Ni_3(Si,Ti)$ showed a two-phase microstructure consisting of $L1_2$ and fcc nickel solid solution. For the latter alloys, the fcc Ni solid solution exsits within the $L1_2$ grains. The XRD

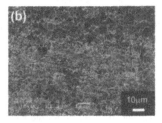

Figure 1. SEM microstructures of (a) Al-alloyed and (b) Mo-alloyed $Ni_3(Si,Ti)$ foils annealed at 1173 K for 1 h.

analysis showed that the observed reflection peaks can be allocated to $L1_2$ and/or fcc phases, and no other peaks were obtained in all the alloys investigated. Also, the lattice parameters of the $L1_2$ and fcc Ni solid solution were almost identical to each other for the Cr-, Co- and Mo-alloyed $Ni_3(Si,Ti)$. Grain sizes were smaller in the Cr-, Co- and Mo-alloyed $Ni_3(Si,Ti)$ with an $L1_2$/fcc two-phase microstructure than in the unalloyed and Al- alloyed $Ni_3(Si,Ti)$ with an $L1_2$ single-phase microstructure (Fig. 1a). Among the alloys with a two-phase microstructure, the Mo-alloyed $Ni_3(Si,Ti)$ foil which has the largest volume fraction of fcc Ni solid solution phase showed a very fine microstructure (Fig. 1b), suggesting that grain growth during annealing was suppressed by the Ni solid solution.

Tensile properties at room temperature

Figure 2 shows the nominal stress versus nominal strain curves for cold-rolled and fully recrystallized foils which were tensile deformed in air at room temperature. The cold-rolled foils including the unalloyed $Ni_3(Si,Ti)$ foil showed a high fracture strength (over 2 GPa), although no plastic elongation was observed. The fracture strength of the alloyed foils was slightly higher than for the unalloyed foil. On the other hand, the alloying elements mostly enhanced the tensile strength of the fully recrystallized foils. In particular, the yield strength of the Mo-alloyed $Ni_3(Si,Ti)$ foil annealed 1173 K was much higher than that of the other alloyed foils including the unalloyed $Ni_3(Si,Ti)$ foil. The enhanced yield strength is attributed to the fine-grained two-phase microstructure in addition to solid solution hardening. Here, it is noted that the Ni solution with a *disordered* fcc structure is not harmful to the strength of $Ni_3(Si,Ti)$ foils with an *ordered* $L1_2$ structure.

Figure 3 shows SEM fractography of the Al-alloyed foil with an $L1_2$ single-phase microstructure and the Cr-alloyed foil with a two-phase microstructure which were tensile-deformed at room temperature. Dimple patterns prevailed on the fracture surfaces of both the alloyed foils, although the depth of the dimples is quite shallow. Regardless of the prevailing microstructures or alloying elements used, the fracture surfaces at room temperature were not so much different.

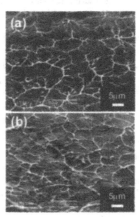

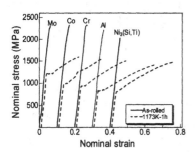

Figure 2. Nominal stress versus nominal strain curves for cold-rolled and fully-recrystallized foils which were tensile deformed at room temperature in air.

Figure 3. SEM-fractography of cold-rolled (a) Al-alloyed and (b) Cr-alloyed $Ni_3(Si,Ti)$ foils tensile deformed at room temperature.

Tensile properties at high temperatures

Figure 4 shows changes in tensile strength, yield stress (0.2 % proof stress) and tensile elongation with temperature for the alloyed and unalloyed $Ni_3(Si,Ti)$ foils that were annealed at 1173 K for 1 h and then deformed at various temperatures. The tensile strength of all alloys monotonously decreased with increasing testing temperature. On the other hand, the yield stress of all the alloys increased with increasing temperature, reached maximum around at 773 K and then decreased with further increasing temperature. It was shown that the $Ni_3(Si,Ti)$ alloys display the positive temperature dependence of strength caused by Kear-Wilsdorf (K-W) locking [2]. The alloying elements used in this study more or less enhanced both tensile strength and yield stress in a wide range of testing temperatures. Particularly, Co and Mo were effective in the enhancement of the high-temperature strength of the $Ni_3(Si,Ti)$ foils. It is noted that the yield stress of these two alloys is beyond 1 GPa even at 873 K. For the Al-alloyed $Ni_3(Si,Ti)$ foil with an $L1_2$ single-phase microstructure, the enhanced strength is attributed to solid solution hardening. For the other alloys (i.e. Cr-, Co- and Mo-alloyed $Ni_3(Si,Ti)$ foils with a two-phase microstructure), the enhanced strength is attributed to the two-phase microstructure of $L1_2$ and fcc Ni solid solution and the resultant fine-grained microstructure, in addition to solid solution hardening. On the contrary, the tensile elongation of the alloyed foils showed a quite different temperature dependence from that of the unalloyed foil. While the tensile elongation of the unalloyed foil steadily decreased from room temperature with increasing temperature and dropped to 0 % at 973 K, that of the alloyed foils remained at high values beyond 30 % up to 773 K and then showed a minimum, followed by rapid increase at the highest temperature tested (973 K).

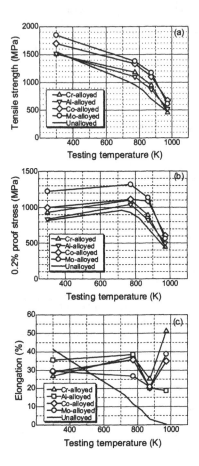

Figure 4. Changes in tensile strength, yield stress (0.2 % proof stress) and tensile elongation with temperature for the alloyed and unalloyed $Ni_3(Si,Ti)$ foils that were annealed at 1173 K for 1 h and then tested at various temperatures.

248

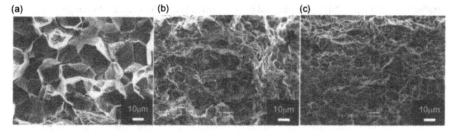

(a)   (b)   (c)

Figure 5. SEM fractography of (a) the unalloyed, (b) the Al-alloyed and (c) the Mo-alloyed Ni₃(Si,Ti) foils that were fully annealed at 1173 K for 1 h and then tensile deformed at 873 K.

Figure 5 shows SEM fractography of the unalloyed and alloyed Ni₃(Si,Ti) foils that were annealed at 1173 K for 1 h and then tensile tested at 873 K. For the unalloyed Ni₃(Si,Ti) foil, intergranular fracture was dominating at 873 K (Fig. 5a) although ductile dimple patterns were present at room temperature (not shown here). Thus, the temperature dependence of the fracture mode observed for the unalloyed Ni₃(Si,Ti) foil corresponds well to that of the observed tensile elongation (Fig. 4c). For the Al-alloyed foil with the same single-phase microstructure as the unalloyed foil, brittle intergranular fracture at high temperature was apparently suppressed, as shown in Fig. 5b. However, the reason for the suppression of the propensity to intergranular fracture by the addition of Al is unknown at the moment. It is suggested that grain boundary cohesion is improved by the addition of Al. For the Cr-, Co- and Mo-alloyed foils with a two-phase microstructure, the transgranular fracture with dimple pattern dominated at all temperatures (Fig. 5c). In this case, the observed high tensile ductility at high temperature may be attributed to suppression of the propensity to intergranular fracture due to the presence of the Ni solid solution.

Oxidation properties

Figure 6 shows the mass gain in air as a function of exposure time at 1173 K for the cold-rolled foils. The mass gain of the alloyed foils was generally reduced compared to that of the unalloyed foil. Particularly, the addition of Al and Cr were more effective in improving the oxidation resistance than that of Co and Mo. It is presumed that protective layers such as Al₂O₃ and Cr₂O₃ may be formed in the Al- and Cr-alloyed foils, leading to the improved oxidation resistance. Also, there is a possibility that the oxidation resistance may be influenced through the change of the Ti content. It has been reported that the

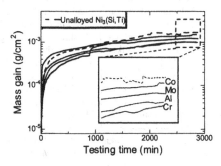

Figure 6. Mass gain in air as a function of exposure time at 1173 K for the cold-rolled foils.

249

oxidation resistance of Ni_3Si was reduced by Ti addition [12]. The Ti content is larger in the unalloyed and Co-alloyed $Ni_3(Si,Ti)$ (9.5 at. % Ti) than in the other alloys (7.5 or 8.5 at. % Ti), consequently resulting in poor oxidation resistance in the former alloys.

CONCLUSIONS

The effects of alloying elements (Al, Cr, Co and Mo) on mechanical properties and oxidation resistance of the $Ni_3(Si,Ti)$ foil were investigated. The following results were obtained:

1. All alloys used in this study were successfully cold-rolled to thin foils with a thickness of 200 μm, revealing that these alloying elements did not spoil the cold-rolling workability of the $Ni_3(Si,Ti)$ alloy.
2. The Al-alloyed $Ni_3(Si,Ti)$ consisted of an ordered $L1_2$ single phase similar to the unalloyed $Ni_3(Si,Ti)$, while the other alloys (i.e. Cr-, Co- and Mo-alloyed $Ni_3(Si,Ti)$) contained fcc Ni solid solution within the $L1_2$ grains. High-temperature tensile strength was enhanced especially by the addition of Co and Mo, while high-temperature elongation was remarkably increased by all the alloying elements via the transition from brittle intergranular fracture mode to ductile transgranular fracture mode.
3. The oxidation resistance was improved particularly by the addition of Al and Cr.

ACKNOWLEDGEMENTS

This work was supported in part by Grant-in-aid for Scientific Research for the Ministry of Education, Culture, Sports and Technology and by the Amada Foundation for Metal Work Technology.

REFERENCES

1. K. Aoki and O. Izumi, *J. Jpn. Inst. Met.* **43,** 1190 (1979).
2. T. Takasugi, M. Nakajima and O. Izumi, *Acta metal. mater.* **38,** 747 (1990).
3. T. Takasugi and O. Izumi, *Acta metal. mater.* **33,** 39 (1985).
4. T. Hirano, M. Demura, K. Kishida, H.U. Hong, and Y. Suga, in *Structural Intermetallics*, edited by K.J. Hemker, D.M. Dimiduk, H. Clemens, R. Daroria, H.Inui, J. M. Larsen, V.K. Sikka, M.Thomas, and J.D. Whittenberger, (TMS, Warrenale,PA, 2001), 765-774.
5. Z. Bojar, P. Jozwik and J. Bystrzycki, *Scr. Mater.* **55,** 399 (2006).
6. Y. Kaneno, T. Myoki and T. Takasugi, *Int. J. Mater. Res. (Z. Metallkd.)*, **99,** 1229 (2008).
7. T. Takasugi and M. Yoshida, *J. Mater. Sci.* **26,** 3032 (1991).
8. T. Takasugi and M. Yoshida, *J. Mater. Sci.* **26,** 3517 (1991).
9. T. Takasugi, *Intermetallics* **8,** 575 (2000).
10. Y. Kaneno and T. Takasugi, *Mater. Sci. Forum* **561-565,** 411 (2007).
11. S. Ochiai, Y. Oya and T. Suzuki, *Acta metal.* **32,** 289 (1984).
12. T. Takasugi, H. Kawai and Y. Kaneno, *Mater. Sci. Eng.* **A329-331,** 446 (2002).

Mater. Res. Soc. Symp. Proc. Vol. 1128 © 2009 Materials Research Society   1128-U05-32

Plasma-Assisted Surface Modification of Two-Phase Intermetallic Alloy Composed of Ni₃X Type Structures

Y. Kaneno[1], N. Matsumoto[1], N. Tsuji[2], S. Tanaka[3] and T. Takasugi[1]

[1]Department of Materials Science, Osaka Prefecture University, 1-1 Gakuen-cho, Naka-ku, Sakai, Osaka, 599-8531, Japan

[2]Tanaka Ltd., 1-10-6 Tedukayama-naka, Sumiyoshiku, Osaka 558-0053, Japan

[3]SDC Inc., 4-132-1 Kannabe-cho, Sakai-ku, Sakai, 590-0984, Japan

ABSTRACT

A two-phase intermetallic alloy composed of Ni_3Al ($L1_2$) and Ni_3V ($D0_{22}$) was plasma-nitrided or -carburized in dependence of temperature and time. It was found that the hardness of the surface layer of the present intermetallic alloy was enhanced by both plasma-nitriding (PN) and -carburizing (PC), and primarily depended on treating temperature; the maximum surface hardness of the alloy was shown by nitriding at around 850 K and by carburizing at 1025 K. In addition, the hardened layers due to PN and PC effectively kept their hardness up to a high temperature. The XRD analysis revealed that vanadium nitride (VN) and vanadium carbide (VC) were formed in the surface of the nitrided and carburized alloy, respectively, suggesting that the enhanced surface hardness was attributed to the dispersion hardening due to the nitrides and carbides.

INTRODUCTION

Recently, the present authors have succeeded in developing a new type of two-phase intermetallic alloys composed of Ni_3Al ($L1_2$) and Ni_3V ($D0_{22}$) phases [1, 2]. The microstructure is composed of Ni_3Al and Ni solid solution (A1) phases at high temperatures, and the Ni solid solution phase is decomposed into $Ni_3Al + Ni_3V$ by a eutectoid reaction at a low temperature. The two-phase microstructure is thus comprised of only intermetallic (ordered) phases with high structural coherency, leading to a high microstructural stability at high temperatures. The two-phase intermetallic alloys have been found to exhibit excellent mechanical and creep properties especially at high temperatures due to their stable microstructures [3], and thereby are expected to be used as high-temperature structural materials such as jet engine and gas turbine parts. However, wear resistance property of the present intermetallic alloys is insufficient for application of conventional heat-resistant elements because hardness is not so high. Therefore, surface modification to improve tribological property of the present intermetallic alloys is desired. Some studies on nitriding and carburizing of nickel based alloys including superalloys have been reported [4-6]. However, whether or not nitriding and carburizing are useful in surface modification of the intermetallic alloys including the present multi-phase intermetallic alloys is unknown. In this study, the capability of the surface hardening by means of plasma-nitriding and -carburizing has been investigated in dependence of temperature and time. The alloy contained Nb, Co and Cr to enhance hardness or to improve oxidation resistance.

EXPERIMENT

The nominal composition of the material used in this study was $Ni_{70.5}Al_{10}V_{10.5}Nb_3Co_3Cr_3$ doped with 500 wt.ppm boron (expressed by at.% except for boron). Ingot with a weight of approximately 30 kg was produced by vacuum induction melting method (VIM). The ingot was heat-treated at 1273 K for 10 h to relieve the internal strain introduced through solidification and to attain the two-phase microstructure. Test pieces with a typical size of $10 \times 10 \times 1$ mm^3 were cut from the heat-treated ingot by an electric discharge machine (EDM). They were abraded with a SiC paper and buff-polished with fine alumina powers, and then plasma-nitrided or -carburized.

Plasma-nitriding (PN) and plasma-carburizing (PC) were conducted on a universal apparatus using a pulsed D.C. glow discharge. The test piece was pre-cleaned by sputtering in argon plasma for 1.8 ks prior to nitriding and carburizing process. The plasma-nitriding and plasma-carburizing were performed at various temperatures and for various periods in a ~200 Pa gas mixture of 50% N_2–50% H_2 and in a ~270 Pa gas mixture of 10% C_3H_8–10% Ar–80% H_2, respectively.

The room-temperature hardness profiles of the plasma-nitrided and -carburized samples were examined in the cross section by means of a Vickers micro-hardness tester with a load of 10 or 25 g. Also, high-temperature hardness was measured by Vickers hardness tester with a load of 1 kg in the samples nitrided at 848 K for 48 h and carburized at 1023 K for 48 h, respectively. The surface layer in the cross sections was observed by a scanning electron microscope (SEM). Also, the composition profiles in the surface layer were measured by an electron microprobe analyzer (EPMA). The phase identification of the surface layer was conducted by X-ray diffraction (XRD) using CuKα radiation.

RESULTS

Hardness profile

Figure 1 shows a back-scattered (BS)-SEM image of the sample heat-treated at 1273 K for 10 h. The two-phase microstructure which was comprised of rectangle-shaped $L1_2$ precipitates with a light-grey contrast and the ($L1_2$ + $D0_{22}$) *channel* with dark contrast was confirmed in the samples before PN and PC process. Average hardness of the heat-treated sample was 460 HV.

Surface hardening behavior of the present intermetallic alloy by PN and PC was first surveyed as a function of treating temperature. **Figure 2** shows the change in hardness with treating temperature for the samples nitrided and carburized at a constant time of 2h. The surface hardness of the nitrided and carburized samples showed a peak around at 850 K and at 1025 K, respectively. Thus, the optimum treating temperature is different between the PN and PC treatment.

Next, the hardness profile of the samples nitrided at 848 K and carburized at 1023 K was investigated as a function of treating time. **Figure 3** shows hardness profile as a function of distance from the surface for the samples plasma-nitrided at 848 K for 2 h, 48 h and 100 h. The surface hardness and thickness of the hardened layer increase with increasing treating time until 48 h, but they little changed for times more than 48 h. This result suggests that treating time of more than 48 h is not so effective in surface hardening for the present intermetallic alloy.

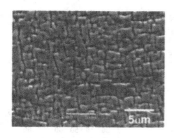

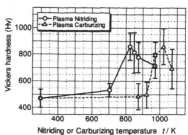
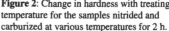

Figure 1: BS-SEM image of the sample heat-treated at 1273 K for 10 h.

Figure 2: Change in hardness with treating temperature for the samples nitrided and carburized at various temperatures for 2 h.

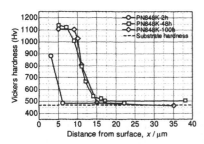

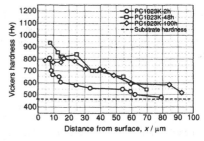

Figure 3: Variation of hardness with distance from the surface for the samples nitrided at 848 K for 2 h, 48 h and 100 h. Dashed line shows the hardness of the substrate i.e., the untreated

Figure 4: Variation of hardness with distance from the surface for the samples carburized at 1023 K for 2 h, 48 h and 100 h. A dashed line shows the hardness of the substrate i.e., the untreated sample.

Figure 4 shows hardness profile for the samples plasma-carburized at 1223 K for 2 h, 48 h and 100 h. Similar to the PN treatment, the prolonged treatment (i.e., 48 h- and 100 h-PC) led to a deep hardened layer. Also, like the PN treatment, the treating time of more than 48 h is unnecessary in the case of the PC treatment. When comparing the PN and PC treatment, the surface hardness is higher in the PN samples than in the PC samples while the hardened layer is deeper in the PC samples than in the PN samples.

Nitrogen and carbon concentration profile

Figures 5 and **6** show variation of nitrogen and carbon concentration as a function of distance from the surface. Figure 5 clearly shows that nitrogen diffuses towards the inward matrix with increasing treating time. These nitrogen concentration profiles basically correspond to the hardness profiles shown in Fig. 3. In the carburized samples, the carbon concentration rapidly dropped near the surface and then gradually decreased with increasing distance from the surface. Similar to the nitrided samples, these carbon concentration profiles also correspond well to the hardness profiles (Fig. 4). Comparing Fig. 5 with Fig. 6, carbon obviously diffuses much

253

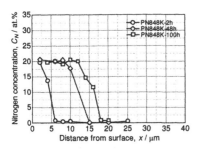
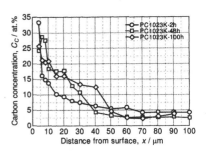

Figure 5: Change in nitrogen concentration with distance from the surface of the samples nitrided at 848 K for 2 h, 48 h and 100 h.

Figure 6: Change in carbon concentration with distance from the surface of the samples carburized at 848 K for 2 h, 48 h and 100 h.

more toward the inward region than nitrogen. This is primarily due to the fact that the treating temperature of the carburizing is higher than that of the nitriding. Also, the difference in the solubility limit in the constituent phases between nitrogen and carbon atoms may result in the difference in the observed concentration profiles.

XRD profile

Figure 7 shows the XRD profiles of the untreated, 848 K-48 h nitrided and 1023 K-48 h carburized samples. For the untreated sample, the principal constituents of the present intermetallic alloy, i.e., Ni_3Al and Ni_3V were identified. Ni_3Nb (DO_a) was also detected because the present intermetallic alloy contained small amount of Nb. For the nitrided and carburized samples, vanadium nitride (VN) and vanadium carbide (VC) with a cubic-type structure were identified respectively, in addition to Ni_3Al and Ni_3V. However, nitrides and carbides combined with other constituent elements (i.e., Al, Cr and Co) were not detected in the present samples.

Stability of nitrided and carburized layers

Figure 8 shows the changes in hardness with testing temperature for the untreated, nitrided and carburized samples. For comparison, the data for the stainless steel of type 440C (Fe-18Cr-

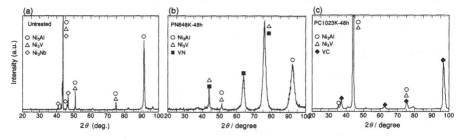

Figure 7: X-ray diffraction (XRD) profiles of the samples (a) untreated, (b) plasma-nitrided at 848 K for 48 h and (c) plasma-carburized at 1023 K for 48 h.

254

1.0C-1.0Mo, wt.%) was also included in this figure. For the untreated sample, the hardness of the present intermetallic alloys with two phase microstructure little decreased with increasing testing temperature. For the nitrided and carburized samples, although a similar change of hardness with testing temperature was observed, the hardness value of the PN and PC samples was apparently high compared with the untreated samples at all temperature range investigated. This result indicates that the hardened layers due to PN and PC were still effective even at high temperature. Also, it is noted that the hardness at high temperature is higher in not only the nitrided and carburized but also the untreated samples than in the Type 440C stainless steel which is the hardest alloy among stainless steels.

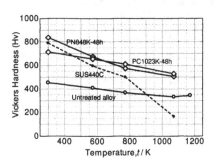

Figure 8: Changes in hardness with testing temperature for the samples untreated, plasma-nitrided at 848 K for 48 h and plasma-carburized at 1023 K for 48 h. For comparison, hardness data for stainless steel type 440C is also included.

DISCUSSION

It was confirmed from the present study that the surface hardness of the two-phase intermetallic alloys composed of Ni_3Al and Ni_3V was enhanced by both PN and PC treatments. It has been reported that the VN precipitates were formed in the nitrided Ni-V (disordered) alloys [4, 7] and also that surface hardness was significantly increased by the VN precipitates [4]. Therefore, the observed surface hardening of the present intermetallic alloys by nitriding or carburizing is suggested to be the dispersion hardening due to the VN or VC precipitates although the role of solid solution hardening due to nitrogen or carbon atoms in the constituent phases cannot be excluded.

Regarding the formation of the nitrides (VN) or the carbides (VC), the internal nitriding or carburizing mechanism is suggested; nitrogen or carbon diffuses towards the inside from the surface and combines with the constituent element V of Ni_3V, consequently resulting in the VN or VC precipitates. Although the present intermetallic alloy contains Al content almost identical with V content, no aluminum nitrides and carbides such as AlN and Al_4C_3 were formed after PN and PC. **Table 1** gives enthalpy of formation of the compounds related to the present intermetallic alloy. In this table, the data for the nitrides and carbides are referred from [8] while the data for Ni_3Al and Ni_3V are calculated according to [9]. When comparing formation

Table 1 Enthalpy of formation of compounds related to the present intermetallic alloys. Data for VN and VC are referred from [8]. Data for Ni_3Al and Ni_3V are calculated according to [9].

Compound	- H_{298}(kJ/g·atom)
AlN	159.3 [8]
VN	108.7 [8]
VC	50.5 [8]
Ni_3Al	31.6
Al_4C_3	30.8 [8]
Ni_3V	17.7

enthalpy of Ni_3Al with that of Ni_3V, the former value is negatively larger than the latter value, indicating that Ni_3Al is stable compared with Ni_3V. In other words, Ni_3V is decomposed more easily than Ni_3Al. This estimation is supported by the fact that Ni_3V is a Kurnakov-type compound with a transformation temperature of 1318 K (i.e., its binding energy is not so high) whereas Ni_3Al is a Berthollide-type compound with a melting temperature of 1668 K (i.e., its binding energy is high). Moreover, when comparing formation enthalpy of VN and VC with that of Ni_3V (and Ni_3Al), VN and VC are stable compared with Ni_3V. Therefore, it is suggested that selective nitrization and carburization take place only with V. Furthermore, it is considered that since VN and VC are thermally stable because of their high binding energy, the dispersion hardening due to these nitride and carbide is effective even at high temperature.

CONCLUSIONS

1. Both plasma-nitriding (PN) and -carburizing (PC) enhanced the surface hardness of the multi-phase intermetallic alloy composed of Ni_3X type structures.
2. The maximum surface hardness of the multi-phase intermetallic alloy was attained by nitriding at around 850 K and by carburizing at around 1025 K, respectively.
3. The hardened layers due to PN and PC remained even at high temperature.
4. The surface hardening of the multi-phase intermetallic alloy by PN and PC was primarily attributed to the dispersion hardening due to the formation of vanadium nitride (VN) and carbide (VC), respectively.

ACKNOWLEDGMENTS

This work was supported in part by Osaka Prefecture Government and by Grant-in-aid for Scientific Research for the Ministry of Education, Culture, Sports and Technology.

REFERENCES

[1] Y. Nunomura, Y. Kaneno, H. Tsuda and T. Takasugi, *Intermetallics* **12** (2004) 389-399.
[2] Y. Nunomura, Y. Kaneno, H. Tsuda and T. Takasugi, *Acta Mater.* **54** (2006) 851-860.
[3] S. Shibuya, Y. Kaneno, M. Yoshida and T. Takasugi, *Acta Mater.* **54** (2006) 861-870.
[4] T. Makishi and K. Nakata, *Met. Mat. Trans.* **35A** (2004) 227-238.
[5] V. Singh and E.I. Meletis, *Surf. Coat. Technol.* **201** (2006) 1093-1101.
[6] S. Baba, H. Kitagawa, A.S. Saleh, K. Hasezaki, T. Kato, K. Arakawa and Y. Noda, *Vacuum* **78** (2005) 27-32.
[7] A.T. Allen and D.L. Douglass, *Oxid. Met.* **51** (1999) 1-22.
[8] *Smithells Metals Reference Book* (8th Ed.), Ed. by W.F. Gale and T.C. Totemeier, Elsevier Butterworths-Heinemann, Oxford, (2004).
[9] A.R. Miedema, *J. Less-Common Met.* **46** (1976) 67-83.

Mater. Res. Soc. Symp. Proc. Vol. 1128 © 2009 Materials Research Society   1128-U05-33

Spontaneous Catalytic Activation of Ni3(Si,Ti) Intermetallic Foils in Methanol Decomposition

Y. Kaneno[1], H. Tsuda[1], Y. Xu[2], M. Demura[2], T. Hirano[2], H. Iwai[2], T. Takasugi[1]

[1]Department of Materials Science, Graduate School of Engineering, Osaka Prefecture University, 1-1 Gakuen-cho, Naka-ku, Sakai, Osaka 599-8531, Japan
[2]National Institute for Materials Science, 1-2-1 Sengen, Tsukuba, Ibaraki 305-0047, Japan

ABSTRACT

Methanol decomposition tests were carried out over cold-rolled Ni3(Si,Ti) foils in a temperature range 513-793 K. The catalytic ability of the cold-rolled Ni3(Si,Ti) foils for the methanol decomposition reaction ($CH_3OH \rightarrow 2H_2 + CO$) was significant at temperatures above 713 K. Also, the catalytic activity (i.e., methanol conversion) at 793 K spontaneously increased with increasing reaction time. Surface analysis revealed that fine nickel particles supported on carbon fibers were formed on the surfaces of the foils during the reaction. The observed catalytic ability of the Ni3(Si,Ti) foils for the methanol decomposition was attributed to the surface nanostructure that was spontaneously formed during the reaction.

INTRODUCTION

Hydrogen is attracting a growing attention as a clean energy source in fuel cell technologies. One of the critical issues about hydrogen is to develop an efficient and low-cost production. At present, hydrogen is mostly produced from hydrocarbon and alcohol via reforming and decomposition over heterogeneous catalysts. Therefore, development of high-performance catalysts is desired.

Recently, some of the present authors found a high catalytic activity of the Ni3Al intermetallic compound for methanol decomposition into H_2 and CO [1-3]. Ni3Al with the L1$_2$ ordered structure is well known as a high-temperature structural material because of its excellent high-temperature strength and oxidization/corrosion resistance. The catalytic activity in Ni3Al was first observed in alkali-leached powder samples and then in flat cold-rolled foils without any pretreatment [1,2]. The Ni3Al foils were found to be spontaneously activated during the reaction. The spontaneous activation can be attributed to the formation of active fine Ni particles which are induced possibly by the selective oxidation of Al in Ni3Al.

This finding suggests a high potential of other Ni-base intermetallic compounds as a catalyst. One of the promising compounds is Ni3(Si,Ti) which was developed by Ti addition to L1$_2$ Ni3Si [4]. Ni3(Si,Ti) is also known as a high-temperature structural material because of its excellent mechanical properties and corrosion/oxidation resistance in a wide temperature range. We recently have succeeded in fabricating thin foils of Ni3(Si,Ti) by cold rolling of the polycrystalline ingots [5]. Ni3(Si,Ti) and Ni3Al have some common features. The primary constituent element is catalytically active Ni. The secondary and ternary elements have a strong affinity to oxygen, and thus Si, Ti and Al oxides can be formed even in the low oxygen partial pressure. With these common features, Ni3(Si,Ti) is expected to exhibit a high catalytic activity similar to Ni_3Al. In this study, the catalytic properties of the $Ni_3(Si,Ti)$ foils for methanol decomposition were investigated.

EXPERIMENTAL

Foil samples were fabricated by a metallurgical process. A detailed procedure for the foil preparation is described in another paper [5]. An arc-melted button ingot with nominal composition of $Ni_{79.5}Si_{11}Ti_{9.5}$ doped with 50 wt.ppm boron was thermomechanically treated and then cold rolled to 0.2 mm in thickness with over 90 % reduction in thickness. After annealing at 1273 K for 1h, they were finished by cold rolling to 40 μm thickness with 80 % reduction.

The cold-rolled foils with a geometrical surface area of 7.6×10^{-4} m^2 were used for catalytic tests. Catalytic tests were carried out in a conventional fixed-bed flow reactor composed of a quartz tube chamber and an electric furnace. The procedure is almost the same as that conducted for Ni_3Al foils [1,2]. Before measurement, the samples were reduced at 773 K for 1h in flowing hydrogen mixed with nitrogen. After the hydrogen was flushed out with nitrogen, methanol was introduced to the quartz tube chamber with nitrogen carrier gas. Methanol was fully evaporated in a thermal evaporator at 423 K before introduction to the quartz tube. The catalytic properties of the foils were evaluated by measuring the outlet composition of the gaseous products with a gas chromatograph and the total flow rate of outlet gases with a soap bubble meter. Since the samples were in a foil form, the production rate of each gaseous product was presented by the unit of $mol \cdot h^{-1} \cdot m^{-2}$. Two types of reaction tests, isochronal and isothermal tests, were performed. For the isochronal test, methanol decomposition was performed with increasing reaction temperature stepwise from 513 K to 793 K at an interval of 40 K. The outlet composition and flow rate of the gaseous products were measured after holding at each temperature for 1.2 ks (20 min). Isothermal tests were conducted at 793 K.

After the isothermal tests, the sample surfaces were observed or characterized by scanning electron microscopy (SEM) with a field emission gun, energy dispersive X-ray spectroscopy (EDS) equipped to SEM, X-ray diffraction (XRD) using $CuK\alpha$ radiation, transmission electron microscopy (TEM) and X-ray photoelectron spectroscopy (XPS).

RESULTS

Isochronal test

Figure 1 shows methanol conversion and production rates of gaseous products as a function of reaction temperature. The methanol conversion was low below 713 K but significantly increased with increasing temperature above 713 K, indicating that that the $Ni_3(Si,Ti)$ foils were catalytically active above this temperature. Main products of the catalytic reaction were H_2 and CO, and very small amounts of CO_2 and CH_4. The production rate of H_2 was almost twice as large as that of CO, indicating the occurrence of the following methanol decomposition:

$$CH_3OH \rightarrow 2H_2 + CO \qquad (1)$$

In addition, reactions (2)–(4) were considered to occur.

$$2CO(g) \rightarrow C(s) + CO_2(g) \qquad (2)$$
$$CO_2(g) + H_2(g) \rightarrow CO(g) + H_2O(g) \qquad (3)$$
$$CO(g) + 3H_2(g) \rightarrow CH_4(g) + H_2O(g) \qquad (4)$$

Isothermal test

Figure 2 shows the methanol conversion and production rates of the gaseous products as a function of reaction time at 793 K. It is noted that the methanol conversion was low at the beginning of the reaction and increased with increasing reaction time, indicating that the cold-rolled $Ni_3(Si,Ti)$ foils were spontaneously activated during the reaction. The methanol conversion increased to about 10 % and remained constant for the first 7 h. After the incubation period, the methanol conversion rapidly increased with increasing reaction time and reached a relatively high value of 70% after 30 h, showing a local maximum value. Then, it slightly decreased in the following 10 h and again increased after 50 h. After 30 h, the internal pressure in the reaction chamber increased (not shown here), indicating the plugging of the chamber. Thus, the increase in the conversion after 50 h is attributed to the decrease in the methanol flow rate as a result of the plugging due to carbon deposition. Also, Fig.2 shows that the production rates of the main products, H_2 and CO, increased with increasing reaction time after the incubation period, then reached a maximum and then started to decrease, corresponding to the change of the methanol conversion. However, in contrast to the conversion, they monotonously decreased

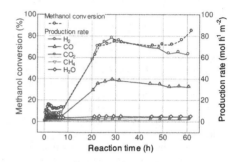

Figure 2. Methanol conversion and production rates of gaseous products in methanol decomposition reaction at 793 K over $Ni_3(Si,Ti)$ foils as a function of reaction time.

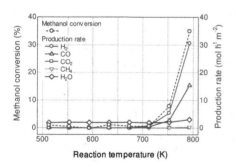

Figure 1. Methanol conversion and production rates of gaseous products over $Ni_3(Si,Ti)$ foils as a function of reaction temperature.

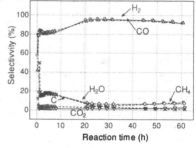

Figure 3. Selectivities to H_2, CO, CH_4, H_2O, CH_4, CO_2 and C in methanol decomposition over $Ni_3(Si,Ti)$ foils at 793 K as a function of reaction time.

259

Figure 4. Surface secondary-electron (SE) SEM images of Ni₃(Si,Ti) foil after reaction at 793 K for 2.5 h.

with increasing reaction time after 50 h. In comparison to the main products, the production rates of the minor products, CO_2, CH_4 and H_2O, remained very low during the entire reaction. **Figure 3** shows the selectivities of the main and minor products as a function of reaction time. The values were calculated according to Refs. [2,6]. Although the selectivities of the main products were low at the beginning of the reaction, they sharply increased to over 90 % with increasing reaction time. In contrast, the selectivities of the minor products were very low, below 10 %,

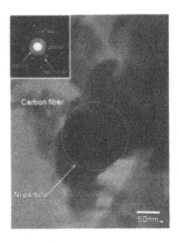

Figure 5. TEM bright field image and selected area diffraction pattern (SADP) taken from the region containing a single particle and the surrounding fibers. Observation was-conducted using the powders obtained from the surface of the Ni₃(Si,Ti) foil after reaction at 793 K for 2.5 h.

except at the beginning of the reaction. Thus, the results confirm that methanol decomposition expressed by Eq. (1) dominates as the main reaction during the entire reaction.

Surface characterization after isothermal test

Figure 4 shows the FE-SEM images of the foil surface at 793 K after 2.5 h (i.e., in the incubation period). After a reaction time of 2.5 h, small particles accompanied with fibrous products were observed all over the surface of the foil. These surface products were analyzed by TEM. **Figure 5** shows the bright-field (BF) image and selected area diffraction pattern (SADP) taken from the region containing a single particle and the surrounding fibers. From the SADP, the particle and fibers were identified as a single crystal of nickel with fcc structure and polycrystalline graphite with hexagonal structure, respectively. After 22 h, at which methanol

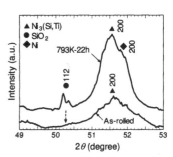

Figure 6. X-ray diffraction profiles for the Ni₃(Si,Ti) foils before (as cold-rolled) and after reaction at 793 K for 22 h.

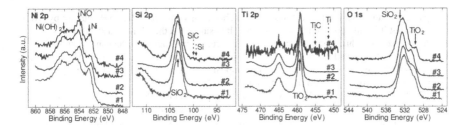

Figure 7. XPS spectra of each element obtained over Ni₃(Si,Ti) foils at different states of reaction at 793K. #1: after reduction treatment, #2: 1 minute reaction, #3: 3 minutes reaction and #4: 20 minutes reaction.

conversion showed the local maximum value, Ni particles as well as carbon fibers significantly grew. **Figure 6** shows the XRD profiles for the foils reacted after 22 h. In this stage, Ni and SiO_2 became detectable in the XRD pattern. The results indicate that Ni particles built up in this stage, and also that Si in the foil was oxidized during the reaction, although SiO_2 was not detectable in the foil before the reaction. **Figure 7** shows the XPS spectra after the reduction treatment, after 1 minute, 3 minutes and 20 minutes of reaction. After the reduction treatment, the spectra showed the presence of NiO, $Ni(OH)_2$, SiO_2 and TiO_2 on the foils surface. These oxides and the hydroxide were considered to be natively formed during the foil fabrication process and remained unreduced after the hydrogen reduction treatment. Metallic Ni was observed but both, elemental Si and metallic Ti, were not observed. Ni is formed from NiO and $Ni(OH)_2$ during the reduction treatment, nemely, NiO and $Ni(OH)_2$ are reduced to form metallic Ni. The coexistence of Ni, NiO and $Ni(OH)_2$ suggests that NiO and $Ni(OH)_2$ were partially reduced. In comparison to NiO and $Ni(OH)_2$, SiO_2 and TiO_2 are very stable, and thus they remained unchanged after the reduction treatment. These surface substances showed very little change in the early stage of the reaction. Any carbides such as SiC and TiC were not detected in all the samples, though Si and Ti are carbide forming elements.

DISCUSSION

It was found that the cold-rolled Ni₃(Si,Ti) foils were spontaneously activated for methanol decomposition as well as the Ni₃Al foils [1,2]. The surface characterization revealed that the spontaneous activation is closely related to the evolution of NiO, $Ni(OH)_2$, SiO_2, TiO_2, and resultant fine Ni particles on the foil surface. These oxides and hydroxides compounds were present before the reaction. It is suggested that as the reaction time increased, NiO and $Ni(OH)_2$ were reduced and Ni particles were produced instead. This understanding is reasonable because the formation energy of NiO and $Ni(OH)_2$ is not that largely negative, and also the amount of reducing gases, H_2 and CO, increased with reaction time (Fig.3). In other words, the reaction environment became more reductive for NiO and $Ni(OH)_2$ with the progress of the reaction. In contrast, SiO_2 was newly produced during the reaction. This is also reasonable considering the very large negative formation energy of SiO_2. Though TiO_2 has a large negative formation

261

energy, it was not observed in the XRD pattern. That TiO_2 was detected by XPS but not by XRD confirms that it is only present in trace amounts. It is suggested that Si in $Ni_3(Si,Ti)$ was selectively oxidized during the reaction, and therefore, the formation mechanism of Ni particles is similar to that observed in case of Ni_3Al foils in which Al is selectively oxidized [1,2]. Thus, the fine Ni particles induced the methanol decomposition expressed by Eq. (1), resulting in the production of hydrogen and carbon monoxide.

The long incubation period observed for activation at 793 K in the $Ni_3(Si,Ti)$ foils, which was not observed in the Ni_3Al foils, may be associated with the presence of NiO, $Ni(OH)_2$, SiO_2 and TiO_2 on the foil surface before reaction. The incubation period is considered to be the time necessary for the reduction of NiO and $Ni(OH)_2$. Thus, the SiO_2 and TiO_2 preexisted on the surface may act as a barrier for the selective oxidation of Si in $Ni_3(Si,Ti)$, and resulted in the delay of activation.

CONCLUSIONS

(1) The cold-rolled $Ni_3(Si,Ti)$ foils showed catalytic activity for methanol decomposition into H_2 and CO in a temperature range 713 K – 793 K.
(2) The catalytic activity increased with increasing reaction time via incubation period, indicating that the cold-rolled $Ni_3(Si,Ti)$ foils were spontaneously activated during the reaction.
(3) Fine Ni particles dispersed on carbon fibers were spontaneously formed on the surface of the foil through reaction. It was suggested that the observed spontaneous catalytic activation of the $Ni_3(Si,Ti)$ foil was attributed to the selective oxidation of Si (and/or Ti) on the foil surface and the concomitant formation of fine Ni particles dispersed on carbon fibers.

ACKNOWLEDGMENTS

This work was supported in part by Grant-in-aid for Scientific Research for the Ministry of Education, Culture, Sports and Technology and by the Amada Foundation for Metal Work Technology.

REFERENCES

[1]   D.H. Chun, Y. Xu, M. Demura, K. Kishida, M.H. Oh, T. Hirano, D.M. Wee, *Catalysis Letters* **106,** 71 (2006).
[2]   D.H. Chun, Y. Xu, M. Demura, K. Kishida, D.M. Wee, T. Hirano, *J. Catalysis* **243,** 99 (2006).
[3]   Y. Xu, S. Kameoka, K. Kishida, M. Demura, A. P. Tsai, T. Hirano, *Intermetallics* **13,** 151 (2005).
[4]   T. Takasugi, M. Nagashima, O. Izumi, *Acta Metal. Mater.* **38,** 747 (1990).
[5]   Y. Kaneno, T. Myoki, T. Takasugi, *Int. J. Mater. Res.* **99,** 1229 (2008).
[6]   R.D. Cortight, R.R. Davda, J.A. Dumesic, *Nature* **418,** 964 (2002).

Mater. Res. Soc. Symp. Proc. Vol. 1128 © 2009 Materials Research Society   1128-U05-34

Catalytic Properties of Chemically Pretreated Ni3Al/Ni Two-Phase Alloy Foils for Methane Steam Reforming

Daisuke Kamikihara[1], Ya Xu[2], Masahiko Demura[2], Toshiyuki Hirano[1, 2]
[1] Graduate School of Pure and Applied Sciences, University of Tsukuba, 1-2-1 Sengen, Tsukuba, Ibaraki, 305-0047, Japan
[2] National Institute for Materials Science, 1-2-1 Sengen, Tsukuba, Ibaraki, 305-0047, Japan

ABSTRACT

We carried out methane steam reforming on the chemically pretreated Ni3Al/Ni two-phase alloy foils. The chemical pretreatment included two steps: acid leaching followed by alkali leaching. The chemically pretreated foils showed much higher activity than non-treated foils. The catalytic activity of the pretreated foils rapidly increased during the first several hours, and then reached a high-level steady state during the subsequent reaction. The surface morphology analysis revealed that many fine Ni particles were formed during the reaction. We considered that these Ni particles contributed to the increase of catalytic activity during the initial stage of the reaction.

INTRODUCTION

Ni3Al intermetallic is known as a high-temperature structural material, because of its excellent high-temperature strength and good oxidation/corrosion resistance [1-3]. The brittleness at room temperature has been a big problem which restricted its practical application. Recently we have overcome this problem and successfully developed thin foils of Ni3Al intermetallics by cold rolling of the single crystals without boron addition [4].

Up to now, the catalytic properties of Ni3Al intermetallics have attracted no attention. However, we have recently found a high catalytic activity in the single-phase Ni3Al foils for hydrogen production via methanol decomposition and methane steam reforming in spite of their low surface area [5-9]. The activity is attributed to the fine Ni particles formed on the foil surface during the reaction or by chemical pretreatment. Considering the excellent high-temperature structural properties, the foil catalysts can be used as both catalysts and structural materials of the micro reactor for hydrogen production. In addition to the single-phase Ni3Al, we recently found the catalytic property in the Ni3Al/Ni two-phase alloy foils for methane steam reforming. Since the activity is not so high compared to that in the single-phase Ni3Al foils, we have made an attempt to improve it in this study. In our previous study, it was found that a two-step chemical pretreatment of acid leaching and subsequent alkali leaching significantly enhanced the catalytic activity of atomized Ni3Al powder with a NiAl/Ni3Al two-phase structure [10]. These results show a possibility of the two-step chemical treatment for improving the catalytic activity of Ni3Al/Ni two-phase alloy foils. In this study we carried out this chemical pretreatment on the surface of Ni3Al/Ni two-phase alloy foils and studied the catalytic properties for methane steam reforming.

EXPERIMENTAL PROCEDURES

Sample preparation

Ni_3Al/Ni two-phase alloy foils of 30 μm in thickness were fabricated by cold rolling of the single crystalline plate with the composition Ni-18at%Al. The Ni_3Al/Ni two-phase structure was confirmed by microstructure analysis after cold rolling. The chemical pretreatment consisted of two steps similar to the case in atomized Ni_3Al powder [10]. The first step, acid leaching (HCl + HNO_3, vol. ratio 3:1, 60 ml) was carried out at room temperature for 20 min. The second step, alkali leaching (NaOH solution, 20 wt%, 60 ml) was carried out at 353 K for 300 min. The surface area of the foil for chemical pretreatment was 1.05 m^2.

Catalytic reaction

Methane steam reforming was performed in a conventional fixed-bed flow reactor at ambient pressure in the same way as our previous study [10]. The foils with a size of 210 mm in length and 5 mm in width were reduced by hydrogen at 873 K for 1 hour before the reaction. The reactants of CH_4 and H_2O (molar ratio of CH_4 and H_2O = 1) were introduced into the reactor at a gas hourly space velocity of 8.57 $L/h/m^2$. Catalytic reaction was carried out at 1123 K for 50 hours. The outlet gaseous composition was analyzed by gas chromatography (GL science, GC-323 equipped with thermal conductivity detectors).

Sample characterization

The surface area of foil was determined by BET analysis method using krypton adsorption (Micromeritics ASAS 2020). The surface morphology and composition of the foils were determined by scanning electron microscopy (SEM: JEOL, JSM-7000F) with an X-ray energy dispersive spectroscopy (EDS) before and after the catalytic reaction.

RESULTS & DISCUSSION

Chemical pretreatment

Table 1 shows BET surface area of the Ni_3Al/Ni two-phase alloy foils before and after the chemical pretreatment. Foil surface area after the chemical pretreatment was 6 times larger than that of the non-treated foils, indicating that the foil surface was roughed by the chemical pretreatment.

Table 1 BET surface area of Ni_3Al/Ni two-phase alloy foils.

	Non-treated	Chemically pretreated
BET surface area, (m^2/g)	0.02	0.12

Figure 1 shows the SEM images of the foil surface before and after the chemical pretreatment. Before the chemical pretreatment, the foil surface was flat, having a weak contrast of band structure as shown in Fig. 1a. This band structure was formed during cold rolling process as previously reported [11]. After the chemical pretreatment, the foil surface became rough as shown in Fig. 1b. Considering the results in the NiAl/Ni₃Al two-phase structure [10], these results suggest that the acid leaching preferentially dissolved Al-rich Ni₃Al phase compared to Ni phase, and that alkali leaching dissolved Al from both Ni₃Al and Ni phases, leading to the Ni-enriched surface.

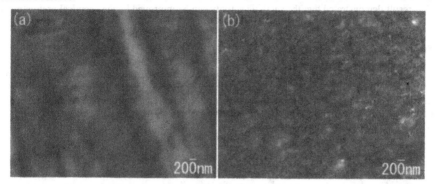

Figure 1 SEM secondary electron images of the surface of Ni₃Al/Ni two-phase alloy foils, (a) non-treated and (b) after chemical pretreatment.

Catalytic property

Fig. 2 shows conversion of methane as a function of reaction time over non-treated and chemically pretreated Ni₃Al/Ni two-phase alloy foils. For the non-treated foils, no activity was observed during the first several hours, and a sudden increase of conversion was observed after 9 hours, but it decreased subsequently. In contrast, the chemically pretreated foil showed some activity at the beginning of the reaction unlike the non-treated foils. We consider that the Ni-

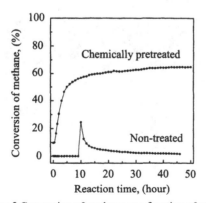

Figure 2 Conversion of methane as a function of reaction time over non-treated and chemically pretreated foil.

265

enrich surface introduced by the two-step chemical pretreatment attributed to the initial catalytic activity. Interestingly, the catalytic activity rapidly increased during the first several hours, and then tended to reach a steady state during the subsequent reaction. The maximum H_2 production rate reached about 30 L/min/m^2. This value was 40 times as high as that of the non-treated foils. This result demonstrated that the catalytic activity can be improved by the chemical pretreatment of acid leaching and subsequent alkali leaching.

Surface characterization

Figure 3 shows the SEM images of the foil surface after reaction for 50 hours at 1123 K. Some products in flake form were observed on the surface of the non-treated foil, as shown in Fig. 3a. In contrast, a large amount of spherical particles were observed on the surface of the chemically pretreated foil, as shown in Fig. 3b. The diameters of these spherical particles were found in a wide range from 50 nm to 1 μm.

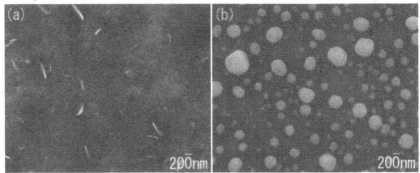

Figure 3 SEM secondary electron images of the foil surface after reaction for 50 hours, (a) non-treated and (b) chemically pretreated.

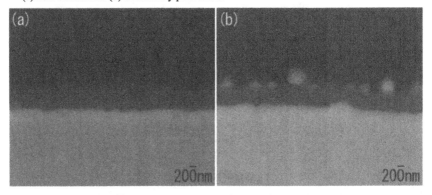

Figure 4 SEM back scattered electron images of the foil cross section after reaction for 50 hours, (a) non-treated and (b) chemically pretreated.

266

Figure 4 shows the backscattered electron images of the cross section after reaction for 50 hours. The non-treated foils showed a very flat surface, having a top layer whose thickness was about 500 nm (Fig. 4a). The chemically pretreated foils showed many spherical particles on the surface, which was consistent with the observation in Fig. 3b, being accompanied with top layer with a thickness of about 500 nm below them (Fig. 4b).

Figure 5 shows the elemental mapping on the cross section of the foil after the reaction. It shows that the spherical particles mainly consist of nickel and the thin top layer consists of aluminum and oxygen. The thin layer of aluminum oxide was determined as Al_2O_3 by XRD analysis. These results indicated that fine metallic Ni particles were produced on the surface during the reaction, and also that aluminum in the foils was selectively oxidized to form Al_2O_3 probably by steam, during the reaction.

Figure 5 SEM image and elemental mapping of the foil cross section after reaction for 50 hours, (a) SEM image, (b) Nickel, (c) Aluminum and (d) Oxygen

CONCLUSIONS

Two-step chemical pretreatment was carried out for the Ni_3Al/Ni two-phase alloy foil, and their catalytic properties for methane steam reforming were investigated at 1123 K. Main results are summarized as follow.

(1) Two-step chemical pretreatment significantly enhanced the catalytic activity of the Ni₃Al/Ni two-phase alloy foils.

(2) Ni particles were produced on the surface of the chemically pretreated Ni₃Al/Ni two-phase alloy foil during reaction, resulting in the high catalytic activity.

ACKNOWLEDGMENTS

We thank M. Takanashi at NIMS for his helpful assistance in the sample preparation, J.H. Jang and H.Y. Lee for their useful advises and comments. One of the authors (D. K.) acknowledges the National Institute for Materials Science for the provision of a NIMS Graduate Research Assistantship. This work was partly supported by the Japan Society for the Promotion of Science (Grant-in Aid for Scientific Research (KAKENHI): (C) No. 19560774, (B) No. 19360321).

REFERENCES

[1] D.P. Pope and S.S. Ezz, Int. Mater. Rev. 29 (1984) 136.
[2] N.S. Stoloff, Int. Mater. Rev. 34 (1989) 153.
[3] M. Yamaguchi and Y. Umakoshi, Mater. Sci. 34 (1990) 1.
[4] M. Demura, Y. Suga, O. Umezawa, E.P. George and T. Hirano, Intermetallics 9 (2001) 157.
[5] D.H. Chun, Y. Xu, M. Demura, K. Kishida, M.H. Oh, T. Hirano and D.M. Wee, Catal. Lett. 106 (2006) 71.
[6] D.H. Chun, Y. Xu, M. Demura, K. Kishida, D.M. Wee and T. Hirano, J. Catal. 243 (2006) 99.
[7] Y. Xu, S. Kameoka, K. Kishida, M. Demura, A.P. Tsai and T. Hirano, Intermetallics 13 (2005) 151.
[8] Y. Xu, S. Kameoka, K. Kishida, M. Demura, A.P. Tsai and T. Hirano, Mater. Trans. 45 (2004) 3177.
[9] Y. Ma, Y. Xu, M. Demura, and T. Hirano, unpublished.
[10] Y. Ma, Y. Xu, M. Demura, D.H. Chun, G.Q. Xie and T. Hirano, Catal. Lett. 112 (2006) 31.
[11] H.Borodians'Ka, M. Demura, K. Kishida and T. Hirano, Intermetallics 10 (2002) 255.

Mater. Res. Soc. Symp. Proc. Vol. 1128 © 2009 Materials Research Society   1128-U05-35

Surface modification of Ni/Ni₃Al two-phase foils by electrochemically selective etching

H.Y. Lee[1,2], M. Demura[2], Y. Xu[2], D.M. Wee[1] and T. Hirano[2]
[1]Department of Material Science and Engineering, KAIST, 335 Gwahangno, Yuseoung-gu, Daejeon, 305-701, Korea
[2]Fuel Cell Material Center, National Institute for Material Science, 1-2-1 Sengen, Tsukuba, Ibaraki, 305-0047, Japan

ABSTRACT

Surface morphology after the selective etching of the γ matrix was examined in Ni(γ)/Ni₃Al(γ′) two-phase foils with various microstructures controlled by the 98 % cold rolling and subsequent heat treatment at 873, 1073 and 1273 K for 0.5 h. In the cold-rolled state, the elongated pancake-shape γ′ precipitates were distributed in the γ matrix, and this structure was almost the same after the heat treatment at 873 K though the recrystallization partly started. These γ′ pancakes were partitioned into the fine blocky particles of 10~100 nm in the edge after the heat treatment at 1073 K and the γ′ blocks significantly became larger at 1273 K. These foils were electrochemically etched in the electrolyte of distilled water with 1 wt.% (NH₄)₂SO₄ and 1 wt.% citric acid at a constant potential of 1.75 V for 5 h. In this etching, the γ matrix was selectively dissolved and the γ′ precipitates were left behind, yielding rough and irregular surface. The surface morphologies corresponded to the γ/γ′ two-phase structures, thus demonstrating that the two-phase structure can be used as a template to make the surface area larger. The foil heat-treated at 1073 K had a number of the fine γ′ particles with 10~100 nm in size densely dispersed on the surface. Such fine surface structure was expected to improve the catalytic activity of Ni₃Al for the hydrogen production reaction.

INTRODUCTION

We have recently succeeded in fabricating thin Ni₃Al foils by cold rolling [1-3] and furthermore found that they show catalytic property for hydrogen production reaction such as methanol decomposition [4, 5]. They can be used as a plate-type catalyst for a micro-channel chemical reactor to produce hydrogen. The catalytic activity is expected to be higher on rough and irregular surface having larger surface area compared with flat surface [6, 7]. Such irregular surface can be obtained from two-phase alloys including Ni₃Al combined with the selective etching of the other phase. In this method, the Ni₃Al phase would be left behind on the surface, yielding rough and irregular surface depending on the two-phase structure. One of the possible templates is Ni(γ)/Ni₃Al(γ′) two-phase alloys, where the γ′ is coherently precipitated in the γ matrix. We have already succeeded in fabricating the two-phase foils by the cold rolling [8, 9] and found that the heavy cold rolling and subsequent heat treatment can refine the two-phase structure [10], which may yield fine surface structure after the selective etching.

In this paper, we reported the surface morphology obtained by the selective etching of the γ matrix from the γ/γ′ two-phase foils, paying attention to the effect of the microstructure which changed through the cold rolling and subsequent heat treatment processes. We here applied the electrochemical method for the selective etching, which method was expected to allow us to control the phase being etched and the extent of etching precisely.

EXPERIMENTAL

Ni(γ)/Ni$_3$Al(γ') two-phase foils with the thickness of 30 μm were fabricated by the 98 % cold-rolling of single crystalline ingots in the same way as we previously reported [1-3, 9]. The nominal composition was binary Ni-18 at.%Al without the addition of boron. They were heat-treated at the temperatures of 873, 1073 and 1273 K for 0.5 h in a flowing argon gas after evacuation to ~10^{-3} Pa.

The electrochemical etching test was conducted in the electrolyte of distilled water with 1 wt.% (NH$_4$)$_2$SO$_4$ and 1 wt.% citric acid at 1.75 V for 5 h. This condition was selected from the previous study by Mukherji [11] and we confirmed that it works for the selective etching of the γ matrix in case of the single crystalline ingot with the same composition [12]. A conventional three-electrode system was used for the electrochemical etching tests, in which the two-phase foils of 10 x 10 mm, a graphite rod and a saturated calomel electrode (SCE) were used as the working, the counter and the reference electrodes, respectively. The electrode potentials were referenced to the SCE. A potentiostat (Hokuto Denko Corp., HZ-5000) controlled with a computer was used in the electrochemical tests.

The microstructure and the composition were examined by scanning electron microscope (SEM, Jeol, JSM-7000F with a field-emission gun) and energy dispersive X-ray fluorescence spectrometer (EDS, Edax), respectively. The orientation was determined by the electron backscatter diffraction (EBSD, TSL, OIM4) method in the SEM.

RESULTS

Fig. 1 shows the back-scattered electron images obtained from the mechanically polished foils in the SEM. The darker and brighter phases correspond to the γ' precipitates and the γ matrix respectively, which correspondence was supported by the EDS analysis. The initial single crystalline ingot had the microstructure where the cuboidal type of γ' precipitates with the edge of 1~2 μm, were distributed in the γ matrix. These cuboidal precipitates were compressed and elongated along the ND and RD, respectively by the heavy cold rolling deformation, and thus changed to the elongated pancake shape [13]. The width of the elongated pancake was estimated 1~2 μm from Fig. 1(a), even though the difference in the contrast between the two phases was weak. This pancake structure kept its outer shape even after the heat treatment at 873 K (Fig. 1(b)), though the recrystallization occurred partly. However, the γ' pancakes were divided into small blocks with the edge of 10~100 nm after the heat treatment at 1073 K, resulting in the formation of very fine two-phase structure [10]. The microstructure was thus heterogeneous, as indicated by Fig. 1(c): very fine blocks were densely dispersed in the regions where the pancake existed originally, and there were relatively wide γ corridors between regions. At 1273 K, the γ' pancakes completely disappeared, and relatively coarse and blocky γ' precipitates of 0.5~1.5 μm were formed in the γ matrix instead, as shown in Fig. 1(d).

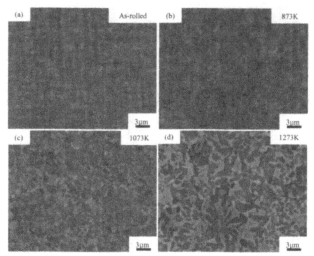

Figure 1. Back-scattered electron images fro the ND sections of Ni(γ)/Ni₃Al(γ′) two-phase foils by SEM: (a) cold-rolled, heat-treated at (b) 873 K, (c) 1073 K, (d)1273 K for 0.5 h

The EBSD measurements revealed that the cold-rolled foils had a strong {110} texture. The recrystallization occurred partly at 873 K and completed at 1273 K, which was consistent to our previous report [13]. The recrystallized grains had different orientations from the cold-rolled {110} texture.

Fig. 2 shows the surface morphologies of the foils after the electrochemical etching test. The γ matrix was selectively etched and the γ′ precipitates were left behind in all the foils and the surface morphologies strongly depended on the two-phase structures as expected. In the cold-rolled foil, there were a number of small flakes elongated along the RD and striae between them on the surface (Fig. 2(a)). They corresponded to the elongated γ′ pancakes and the γ matrix in the cold-rolled microstructure, respectively (e.g. compare Fig. 1(a) and Fig. 2(a)). In the foil heat-treated at 873 K, the flakes and striae were observed similarly to the cold-rolled foil, and in addition, a number of dimples were formed on the surface of the flakes (Fig. 2(b)). These small dimples may be related to the partial recrystallization which occurred at this heat treatment, as will be discussed later.

In the foil heat-treated at 1073 K, a number of fine particles were exposed on the surface after the etching (Fig. 2(c)). These fine particles corresponded to γ′ precipitates and had higher Al content according to the EDS analysis. Fig. 3 shows the detailed structure in a high magnification. The blocky γ′ particles were densely distributed and their size ranged from 10 to 100 nm. It should be noted that the striae or gap between the particles are narrow, ~10 nm. Thus, the present electrochemical etching condition can selectively dissolve the γ matrix in such fine structure.

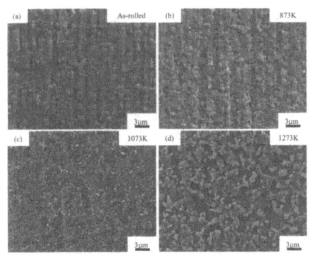

Figure 2. SEM images showing surface morphology of Ni(γ)/Ni$_3$Al(γ') two-phase foils electrochemically etched in the electrolyte of distilled water with 1 wt.% (NH$_4$)$_2$SO$_4$ and 1 wt.% citric acid at 1.75 V for 5 h: (a) cold-rolled, heat-treated at (b) 873 K, (c) 1073 K, (d) 1273 K for 0.5 h

Figure 3. SEM image with high magnification showing the surface morphology of Ni(γ)/Ni$_3$Al(γ') two-phase foil heat-treated at 1073 K for 0.5 h after the electrochemical etching

Similarly, the blocky γ' particles were clearly exposed on the surface of the foil heat-treated at 1273 K as shown in Fig. 2(d). The size of the exposed γ' precipitates was 0.5~1.5 µm, as predicted from the observed microstructure (Fig. 1(d)). The orientation of each blocky particle differed from the region to the next, making the surface irregular. Such irregularity is ascribed to the polycrystalline structure after the recrystallization; the γ′ particles were coherently precipitated in each grain, trying to keep a cube-on-cube relationship as in the case of single crystalline ingot. Thus, the large and irregular γ′ precipitates brought about the most winding surface among the present specimens.

DISCUSSION

The present results demonstrated that the rough and irregular surface can be obtained from the template of the γ/γ′ two-phase foils by the selective etching of the γ matrix as shown in Fig. 2. The electrochemical etching method used here was appropriate for the foils in the various states: 98 % cold-rolled, partly recrystallized, and fully recrystallized states. Note that the γ matrix was selectively etched from the fine structure in the foil heat-treated at 1073 K, resulting in the formation of the fine γ′ particles densely dispersed on the surface (Fig. 2(c)). Such fine surface structure is expected to show high catalytic activity since the fine particles act as a catalytic center, as we previously reported [4-7]. That is, the foil heat-treated at 1073 K can be one of the most promising templates consequently. The measurements of the surface area, roughness and catalytic properties after the surface modification are undergoing.

It was confirmed that the main factor to control the surface morphology is the template, i.e. the two-phase structure in this method (e.g. compare Fig.1 with Fig. 2). That is, the fine two-phase structure is necessary for the fine surface structure, which is preferable for our purpose as described above. A key to the refinement of the γ′ particles is the combination of the heavy cold rolling and subsequent heat treatment. First, the initial cuboidal γ′ precipitates are rolled into the elongated and thin pancakes in the heavy cold rolling process, and then they are partitioned into the fine γ′ blocks by the narrow γ corridor in the heat treatment. It is, in particular, crucial to select the appropriate heat treatment temperature, 1073 K in the present condition. In fact, no refinement or change occurred at a lower temperature of 873 K (Fig. 1(b)) because the atomic diffusion rate is too slow. On the other hand, the γ′ blocks grew rapidly at a higher temperature of 1273 K (Fig.1(d)), resulting in the coarse structure.

Another minor factor is the effect of crystallographic orientation or recrystallization. As shown in Fig. 2(b), a number of dimples were additionally formed on the foil heat-treated at 873 K, comparing with the cold-rolled foil (Fig. 2(a)). Since the γ′ pancake structure was not changed through the heat treatment at 873 K, this minor difference is ascribed to the partial recrystallization. The dissolution rate can be affected by the crystallographic orientation [14, 15] and the recrystallized regions had different crystallographic orientations from the {110} texture as mentioned in the result section. It is likely that the slower or higher dissolution rate on the recrystallized regions led to the formation of small dimples on the surface of the γ′ pancakes in the foil heat-treated at 873 K. This effect, however, became unclear in the foils heat-treated at 1073 and 1273 K. It is because the surface was much more rough and irregular due to the blocky shape of the γ′ particles, prevailing over the minor difference with the crystallographic orientation. In other words, the shape of the γ′ determined the resultant surface morphology dominantly.

CONCLUSIONS

The surface morphologies after the electrochemical etching were examined in $Ni(\gamma)/Ni_3Al(\gamma')$ two-phase foils cold-rolled and heat-treated at 873, 1073 and 1273 K for 0.5 h and the following results were obtained.

(1) In the cold-rolled foil, the γ' precipitates having a pancake shape elongated along the RD with the width of 1~2 μm were distributed in the γ matrix. After the heat treatment at 873 K, the pancake structure was not changed, even though the recrystallization started. On the other hand, the foil heat-treated 1073 K had very fine two-phase structure by partitioning of the γ' pancakes into the fine blocky particles with the edge of 10~100 nm. In the foil heat-treated at 1273 K, the recrystallization completed and the blocky γ' particles grew up to 0.5~1.5 μm in the edge.

(2) After the electrochemical etching tests in the electrolyte of distilled water with 1 wt.% $(NH_4)_2SO_4$ and 1 wt.% citric acid at 1.75 V for 5 h, the γ matrix was selectively etched and the γ' precipitates were left behind. The surface morphology after the etching strongly depended on the two-phase structure. The foil heat-treated at 1073 K had the fine surface structure where the fine γ' particles of 10~100 nm in size were densely exposed on the surface, which was considered preferable for improving the catalytic activity.

ACKNOWLEDGMENTS

The authors thank M. Takanashi for his helpful assistance in the sample preparation. This work was partly supported by the Korea Science and Engineering Foundation (KOSEF) grant funded by the Korean government (MOST) (No. R01-2007-000-10008-0).

REFERENCES
1. M. Demura, Y. Suga, O. Umezawa, E.P. George and T. Hirano, Intermetallics 9, 157 (2001)
2. M. Demura, K. Kishida, Y. Suga and T. Hirano, Metall. Mater. Trans. A 33A, 2607 (2002)
3. M. Demura, K. Kishida, Y. Suga, M. Takanashi and T. Hirano, Scrip. Mater. 47, 267 (2002)
4. D.H. Chun, Y. Xu, M.Demura, K. Kishida, M.H. Oh, T. Hirano and D.M. Wee, Catal. Lett. 106, 71 (2006)
5. D.H. Chun, Y. Xu, M. Demura, K. Kishida, D.M. Wee and T. Hirano, J. Catal. 243, 99 (2006)
6. Y. Xu, M. Demura and T. Hirano, Appl. Surf. Sci. 254, 5413 (2008).
7. Y. Xu, S. Kameoka, K. Kishida, M. Demura, A.P. Tsai and T. Hirano, Intermetallics 13, 151 (2005)
8. H. Borodians'ka, M. Demura, K. Kishida and T. Hirano, Intermetallics 10, 255 (2002)
9. D. Li, K. Kishida, M. Demura and T. Hirano, Intermetallics 16, 1317 (2008)
10. M. Demura, S. Hata, K. Kishida, Y. Xu and T. Hirano, Unpublished work.
11. D.Mukherji, G.Pigozzi, F.Schmitz, O.Nath, J.Rosler and G.Kostorz, Nanotechnology 16 2176 (2005)
12. H.Y. Lee, M. Demura, Y. Xu, D.M. Wee and T. Hirano, Unpublished work.
13. M. Nakamura, M. Demura, Y. Xu and T. Hirano, MRS proc. 0980-II05-30 (2006)
14. C.A. Schuh, K. Anderson and C. Orme, Surf. Sci. 544, 183 (2003)
15. J.J. Gray, B.S. El Dasher and C.A. Orme, Surf. Sci. 600, 2488 (2006)

Mater. Res. Soc. Symp. Proc. Vol. 1128 © 2009 Materials Research Society   1128-U05-36

Surface structure modification of Ni3Al foil catalysts by oxidation-reduction treatment

Jun Hyuk Jang[1,2], Ya Xu[2], Masahiko Demura[2], Dang Moon Wee[1] and Toshiyuki Hirano[2]

[1]Department of Materials Science and Engineering, KAIST, Daejeon 305-701, Korea
[2]National Institute for Materials Science, 1-2-1 Sengen, Tsukuba, Ibaraki 305-0047, Japan

ABSTRACT

A two-step treatment, oxidation in air followed by reduction in hydrogen, was carried out to modify the smooth Ni3Al foil surface into Ni particles supported on the oxide structure. The surface structure significantly changed depending on the oxidation temperature. A layer of granular NiO formed on the outer surface and inner oxide zone (IOZ) over Ni3Al foil surface after oxidation at 973 K. The IOZ was a mixture of Al and Ni oxides. In contrast, a large amount of faceted NiO particles formed on the outer surface after oxidation at 1173 K. Beneath the NiO particles, NiAl2O4 thin layer formed on IOZ over Ni3Al foil surface. And then, these NiO was selectively reduced to Ni after reduction treatment, constituting an oxide supported Ni particles structure. These results suggest that it is possible to modify the surface structure of Ni3Al foils simply by oxidation-reduction treatment.

INTRODUCTION

An efficient hydrogen production process from alcohol or hydrocarbon is an important part of fuel cell technologies [1-3]. Micro-channeled reactors are highly promising for this process due to their high surface-to-volume ratio and high rates of heat and mass transfer compared with conventional reactors [4,5]. At present, it is a crucial issue for micro-channeled reactors to develop efficient and inexpensive heterogeneous catalysts. Another important issue is the development of heat-resistant metal sheets and simplification of the coating process of catalyst layers in fabrication of micro-channeled reactors.

Intermetallic compound Ni3Al has been known as a promising high-temperature structural material due to its excellent oxidation/corrosion resistance at high temperatures [6,7]. So far, many researches were carried out on the oxidation behavior of Ni3Al, and it was reported that various types of Ni oxide are formed on (Ni, Al-rich) oxide layer over Ni3Al substrate depending on the oxidation conditions [8-12]. We consider that these NiO can be reduced in hydrogen, constituting Ni particles supported on the oxide structure which is catalytically active surface structure. To our best knowledge, most of the previous researches have focused on the oxidation/corrosion behavior of the Ni3Al, but no studies were carried out on the reduction behavior of the oxide layer. In this study, we examine the effects of oxidation-reduction treatment on the surface structure modification of the Ni3Al foils for the catalyst application.

EXPERIMENTAL

Ni3Al foils with 30 μm in thickness were fabricated by 98% cold-rolling of single crystalline plates of Ni3Al (Ni-24at%Al) without intermediate annealing. Details of the foil fabrication procedure were previously described [13,14]. The Ni3Al foils were oxidized at 973 and 1173 K for 1 h in air. The oxidized foils were reduced at 773 K for 1 h in flowing hydrogen and kept in nitrogen flow during cooling the samples to the ambient temperature.

The surface and cross-section of the samples were observed by means of scanning electron microscopy (SEM) with a field emission gun (JEOL JSM-7000F). The cross-sections were polished using an argon ion polishing machine (JEOL SM-09010, Cross Section Polisher). The chemical composition was measured by energy dispersive X-ray spectrometry (EDS). The crystal structures of the surface products were characterized by X-ray diffraction (XRD) using a Cu$K\alpha$ source (Rigaku RINT 2500V). The electron binding energies (BE) of nickel and aluminum in the Ni$_3$Al foils were measured by means of X-ray photoelectron spectroscopy (XPS) (VG Scientific ESCALab 200R).

RESULTS AND DISCUSSION

Surface structure after oxidation

Fig. 1 shows a surface morphology after oxidation treatment of the cold-rolled Ni$_3$Al foils which were macroscopically smooth [15]. It is noted that the surface morphology strongly depended on the oxidation temperature. When the foils were oxidized at 973 K, a continuous outer layer consisted of agglomerated granular particles formed on the foil surface (Fig. 1(a)). The XRD and SEM-EDS observation revealed that this outer layer corresponds to NiO (not shown here). On the other hand, when the foils were oxidized at 1173 K, a large amount of faceted particles ($\leq$ 600 nm) formed on the foil surface (Fig. 1(b)). These particles were also confirmed as NiO. Perez et al. reported the oxidation behavior of PM Ni$_3$Al alloy, suggesting a different type of oxidation behavior depending on the temperature [10]. They have found that the agglomerated granular NiO particles formed on the surface in low temperature region below 908 K, while faceted NiO particles formed in high temperature region above 1103 K. This report is well consistent with our results, though the temperature was slightly different. We consider that it is caused by the different chemical composition and fabricating process of Ni$_3$Al alloy used in the experiment.

Figure 1. Surface morphology after 1 h of oxidation at (a) 973 K and (b) 1173 K.

Fig. 2 shows the cross-sectional structures of the Ni$_3$Al foils after oxidation at 973 K and 1173 K, respectively. Figs. 2(a) and (b) are the secondary-electron images (SEI), and Figs. 2(c) and (d) are the corresponding back-scattered electron images (BEI). The foils formed multi-layer structure over the Ni$_3$Al substrate. The multi-layer structure strongly depended on the oxidation temperature. When the foils were oxidized at 973 K, a two-layer surface structure formed, as clearly shown in BEI (Fig. 2(c)). EDS analysis indicated that the outer layer was NiO, and the

276

internal layer ($\leq$ 100 nm) was a mixture of Al and Ni oxides which was called inner oxide zone (IOZ). When the foils were oxidized at 1173 K, a four-layer structure formed over the Ni_3Al substrate (Fig. 2(b)). The faceted NiO particles, as shown in Fig. 1(b), formed on the outer surface over the thick IOZ ($\geq$ 200 nm). In addition, discontinuous Al_2O_3 layer formed between the IOZ and Ni_3Al substrate (Fig. 2(b)). An additional thin layer, distinguished by different contrast with the IOZ in BEI (Fig. 2(d)), formed beneath the NiO particles and over the IOZ.

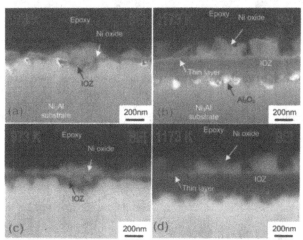

Figure 2. Cross-section of the foils. (a, b) secondary-electron images (SEI) and (c, d) the corresponding back-scattered electron images (BEI) after 1 h of oxidation at 973 K (a, c) and 1173 K (b, d).

Surface structure of oxidized foils followed by reduction

The surface morphology after reduction of the oxidized Ni_3Al foils is shown in Fig. 3. It is noted that the oxidized surface morphology was further modified with reduction. When the foils oxidized at 973 K were reduced, the granular NiO layer (Fig. 1(a)) was replaced by discontinuous porous layer, and fine particles were dispersed on the exposed foil surface (Fig. 3(a)). The porous layer and the small particles were identified to metallic Ni by XRD and SEM-EDS. On the other hand, after reduction of the foils oxidized at 1173 K, the large amount of the particles was remained (Fig. 1(b)) on the foil surface, while the faceted plane changed into dimple structure having small pores in a part. These particles were also identified to metallic Ni.

Fig. 4 shows the cross-sectional structure of the Ni_3Al foils after oxidation-reduction treatment. The layer structure after oxidation treatment was remained, except that the NiO was selectively reduced to metallic Ni. When the oxidized foils at 973 K were reduced, discontinuous Ni layer formed on the IOZ over the Ni_3Al substrate (Fig. 4(a)). These layers can be clearly recognized in BEI (Fig. 4(c)). The thickness of IOZ ($\leq$ 100 nm) was similar with that of oxidized foils (compare Figs. 2(c) and 4(c)), indicating that the IOZ was not affected by reduction treatment. The reduced foils following oxidation at 1173 K showed that Ni particles formed on the thick IOZ over the Ni_3Al foil substrate. The thickness of the IOZ was over 200 nm, similarly to that after oxidation. The results support that only outer NiO was selectively reduced to

277

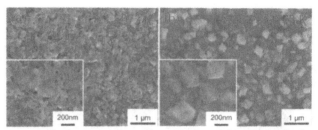

Figure 3. Surface morphology after reduction of the foils oxidized at (a) 973 K and (b) 1173 K.

metallic Ni, without affecting the IOZ. We consider that it is due to the low reduction temperature of NiO. The reduction temperature of NiO is reported to below 773 K, while that of Al_2O_3 and $NiAl_2O_4$ is over 1100 K [16]. The thin layer beneath the Ni particles and over the IOZ, formed after oxidation at 1173 K, was remained (Fig. 4(d)) even after reduction treatment.

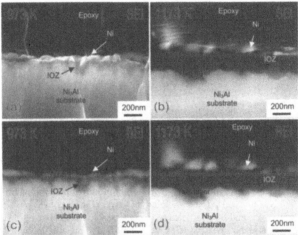

Figure 4. Cross-section after reduction of the oxidized foils. (a, b) secondary-electron images (SEI) and (c, d) back-scattered electron image (BEI). The oxidation temperatures were 973 K (a, c) and 1173 K (b, d).

The surface chemical state of the foil surface was examined by XPS. Fig. 5 shows XPS spectra for Ni $2p$, Ni $3p$ and Al $2p$ before and after oxidation-reduction. Before oxidation, the Ni $2p$ spectra showed a peak at 852.7 eV corresponding to the binding energy of metallic Ni (Ni^{met}) [17,18]. The Ni $3p$ spectra showed a peak at 66 eV corresponding to the binding energy of Ni^{met}. The Al $2p$ spectra showed two peaks at 72.2 and 74.3 eV corresponding to the binding energy of metallic Al (Al^{met}) and a native amorphous Al oxide (AlO_X), respectively [18]. After oxidation at 973 K followed by reduction, Ni $2p$ and Ni $3p$ spectra showed no obvious difference with those before oxidation-reduction treatment, indicating that metallic Ni existed on the foil surface. In contrast, for the Al $2p$ spectra, the peak of Al^{met} disappeared, and a small peak at 74.5 eV was detected, which corresponds to the binding energy of Al_2O_3 [19]. These results indicate that the metallic Ni and Al_2O_3 formed on the foil surface, showing a good consistency with the surface analyses by SEM and XRD. After oxidation at 1173 K followed by reduction, the Ni $2p$ spectra showed a small hill at 856.0 eV corresponding to the binding energy of Ni_2O_3 and $NiAl_2O_4$ [17], in addition to the main peak for Ni^{met}. Here, the formation of Ni_2O_3 can be excluded because Ni_2O_3 is unstable at this high oxidation temperature [12]. The Al $2p$ spectra showed a peak at 74.0 eV which corresponds to the binding energy of $NiAl_2O_4$ [20]. The results indicate that the thin layer, formed by the oxidation at 1173 K, was $NiAl_2O_4$ (Figs. 2(d) and 4(d)). It is reported that $NiAl_2O_4$ layer forms beneath NiO and over IOZ with the reaction between NiO and Al_2O_3 at high temperature as following [10,12];

$$NiO(s) + Al_2O_3(s) \rightarrow NiAl_2O_4(s) \quad \Delta G^0 = -1,609,036 + 284T(\text{Joule}) \qquad (1)$$

These results confirmed the SEM and EDS analyses, i.e., only the NiO was selectively reduced by reduction, and the IOZ and $NiAl_2O_4$ were not reduced.

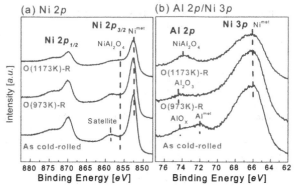

Figure 5. XPS spectra of Ni₃Al foils for (a) Ni $2p$ and (b) Al $2p/$ Ni $3p$ before and after oxidation-reduction treatment: O(973K)-R and O(1173K)-R correspond to the samples oxidized at 973 and 1173 K following by reduction, respectively.

The surface analyses reveal that the oxidation-reduction treatment successfully modified the surface structure of the Ni₃Al foil into Ni particles supported on the oxide structure. This surface structure is similar to those of the activated Ni₃Al foils in methanol decomposition, i.e., fine Ni particles dispersed on the thin Al_2O_3 layer [15,21]. The result suggests a possibility to develop the Ni₃Al foils as both heat-resistant structural sheet and catalytic material simply by the oxidation-reduction treatment.

CONCLUSIONS

This study examined surface modification of Ni₃Al foil catalysts by oxidation-reduction treatment for catalysts application. The following results were obtained.

(1) Surface morphology of Ni_3Al foils was significantly affected by the oxidation temperature. A granular NiO layer formed on IOZ over Ni_3Al foil surface after oxidation at 973 K, while a large amount of faceted NiO particles formed on $NiAl_2O_4$ thin layer over IOZ after oxidation at 1173 K.

(2) The reduction treatment following the oxidation selectively reduced the NiO to metallic Ni, modifying the oxidized surface morphology. The foils oxidized at 973 K formed a discontinuous porous Ni layer on the Al-rich IOZ after reduction treatment, while the foils oxidized at 1173 K formed a large amount of Ni particles on the $NiAl_2O_4$ thin layer.

(3) The smooth surface of the Ni_3Al foils were successfully modified into Ni particles supported on the oxide structure simply by the oxidation-reduction treatment, suggesting a high possibility for catalysts application.

ACKNOWLEDGMENTS

This work was supported partly by the Korea Science and Engineering Foundation (KOSEF) grant funded by the Korean government (MOST) (No. R01-2007-000-10008-0), and a Grant-in Aid for Scientific Research (KAKENHI) (C) (No. 19560774) and (B) (No. 19360321) from the Japan Society for the Promotion of Science.

REFERENCES

1. Brown LF. Int J Hydrogen Energy 2001;26:381.
2. Rostrup-Nielson JR. Phy Chem Chem Phys 2001;3:283.
3. Ogden JM. Phys Today 2002;55:69.
4. Janicke MT, Kestenbaum H, Hagendorf U, Schüth F, Fichtner M, Schubert K. J Catal 2000;191:282.
5. Aartun I, Venvik HJ, Holmen A, Pfeifer P, Görke O, Schubert K. Catal Today 2005;110:98.
6. Stoloff NS. Int Mater Rev 1989;34:153.
7. Stoloff NS, Liu CT, Deevi SC. Intermetallics 2000;8:1313.
8. Cao G, Geng L, Zheng Z, Naka M. Intermetallics 2007;15:1672.
9. Susan DF, Marder AR. Oxid Met 2002;57:159.
10. Perez P, Gonzalez-Carrasco JL, Adeva P. Oxid Met 1997;48:143.
11. Haanappel VAC, Pérez P, González-Carrasco JL, Stroosnijder MF. Intermetallics 1998;6:347.
12. Qin F, Anderegg JW, Jenks CJ, Gleeson B, Sordelet DJ, Thiel PA. Surf Sci 2008;602:205.
13. Demura M, Kishida K, Suga Y, Takanashi M, Hirano T. Scr Mater. 2002;47:267.
14. Demura M, Suga Y, Umezawa O, Kishida K, George EP, Hirano T. Intermetallics 2001;9:157.
15. Chun DH, Xu Y, Demura M, Kishida K, Oh MH, Hirano T, Wee DM. Catal Lett 2006;106:71.
16. Jeong JH, Lee JW, Seo DJ, Seo Y, Yoon WL, Lee DK, Kim DH. Appl Catal A Gen 2006;302:151.
17. Bolt PH, Grotenhuis E, Geus JW, Habraken FHPM. Surf Sci 1995;329:227.
18. Haerig M, Hofmann S. Appl Surf Sci 1998;125:99.
19. Paparazzo E. Surf Interface Anal 1988;12:115.
20. Ng KT, Hercules DM. J Phys Chem 1976;80:2094.
21. Chun DH, Xu Y, Demura M, Kishida K, Wee DM, Hirano T. J Catal 2006;243:99.

Mater. Res. Soc. Symp. Proc. Vol. 1128 © 2009 Materials Research Society   1128-U05-38

Effect of growth rate on microstructure and microstructure evolution of directionally solidified Nb-Si alloys

Yoshihito Sekito[1], Seiji Miura[1], Kenji Ohkubo[1], Tetsuo Mohri[1], Norihito Sakaguchi[2], Seiichi Watanabe[2], Yoshisato Kimura[3], and Yoshinao Mishima[3]
[1]Division of Materials Science and Engineering, Graduate School of Engineering, Hokkaido University; [2]Center for Advanced Research of Energy Conversion Materials, Hokkaido University; [3]Interdisciplinary Graduate School of Science and Engineering, Tokyo Institute of Technology

ABSTRACT

In the present work, Nb-18.1Si-1.5Zr alloy rods are produced with a growth rate ranging from 1.5 to about 1500 mm/h using the optical floating zone (OFZ) furnace. A part of each specimen is heat-treated at 1650 °C for 100 h. The microstructure was observed using SEM and TEM and analyzed using EPMA and EBSD. Eutectic-cells are observed in as-grown specimens with a growth rate of 150 mm/h or higher. It is found by EBSD analysis that the solidification direction of Nb is along <113> and that of Nb_3Si is along <001], and {112} of Nb and {110} of Nb_3Si are parallel. The present crystallographic orientation relationship between Nb and Nb_3Si is different from that found in previous reports by several researchers. It was also confirmed that the heat-treated microstructure in the specimen grown by OFZ with a growth rate of 150 mm/h is similar to that in the heat-treated specimen prepared by arc-melting.

INTRODUCTION

The high-temperature material is one of the keys for establishing a high efficiency energy conversion system. Currently, the turbine foils in jet engines are subjected to a temperature above 1100 °C [1], which exceeds 80 % of the melting point of Ni. Therefore the improvement of the high temperature capability of Ni based super alloys becomes very hard.

Nb has attracted attention as potential structural material for high-temperature use because of the melting point is 1000 °C higher than that of Ni and because it has a lower density than Ni and other refractory metals. Nb-Si based alloys with various additives are composed of Nb solid solution (Nb_{ss}) and Nb_5Si_3, and this alloy group attracts attention as one of the candidates for heat-resistant alloys [2]. Although as-cast eutectic Nb-Si alloys proposed by the present authors consist of Nb_{ss} and Nb_3Si, the constituent phases change to Nb_{ss} and Nb_5Si_3 through the eutectoid decomposition of Nb_3Si -> $Nb+Nb_5Si_3$. This phase transformation and the following spheroidization of Nb_5Si_3 form a microstructure in which Nb_5Si_3 is finely dispersed in the Nb matrix. These alloys are expected to show both, an excellent deformation resistance at high temperature and toughness at room temperature [3-6].

Because the Nb/Nb_3Si interfacial energy is one of the dominant factors of the eutectoid decomposition, understanding of the Nb/Nb_3Si interface structure is very important. There are several reports on the crystallographic orientation relationship between Nb and Nb_3Si in several alloys produced by different methods. Thus the aim of this study is to investigate the effect of the growth rate on the crystallographic orientation relationship between Nb and Nb_3Si in a eutectic Nb-Si alloy containing Zr and Mg [3].

EXPERIMENT

Ingot buttons of about 50 g were prepared by arc-melting of high purity Nb, Si, Zr and $MgCl_2$ powder under Ar atmosphere. The nominal composition of the alloy is Nb-18.1at. % Si-1.5at. % Zr-100ppm Mg. The ingots were unidirectionally solidified in the optical floating zone (OFZ) furnace (Crystal Systems Co., FZ-T-12000-X-S-PC) at a growth rate ranging from 1.5 mm/h to about 150 mm/h under Ar-flow atmosphere. The typical size of the rods produced by OFZ is 5 mm in diameter and 100 mm in length. Some of the rods that were prepared with a growth rate of 150 mm/h were remelted and solidified with a growth rate of 1500 mm/h in the OFZ furnace. A part of the latter specimen was heat-treated at 1650 °C for 100 h. To minimize oxidation during the heat-treatment for the eutectoid decomposition, alloys were wrapped with Ta foils.

The microstructure observation was conducted on each specimen after carefully polishing with colloidal SiO_2 powder (40 nm in diameter) using both, a scanning electron microscope (SEM, JEOL-JXA-8900M) and a field emission-SEM (FE-SEM, JEOL-JSM-6500F) for an electron backscatter diffraction pattern (EBSD) analysis with TexSEM Laboratories-orientation imaging microscopy (OIM) software (TSL OIM 4.5). Observation by a transmission electron microscope (TEM, JEM-2010F) operated at 200 kV was also carried out on as-grown specimens with a growth rate of 150 mm/h. TEM samples were prepared by mechanically grinding the specimen down to 50 μm followed by Ar-ion milling in a Gatan Precision Ion Polishing System (PIPS).

RESULTS and DISCUSSION

Microstructure

The microstructures of the cross-sections of as-grown alloys are shown in Figure 1, together with the heat-treated microstructure of the alloy with a growth rate of 1500 mm/h. The bright phase is Nb_{ss} and the dark phase is Nb_3Si. A eutectic cell structure, similar to that in as-arc melted alloys is observed in alloys with a growth rate of 150 mm/h or higher. The eutectic cell size and the size of Nb rods in the center of the cells decrease with increasing growth rate. In alloys with a growth rate lower than 15 mm/h, the Nb rods becomes rounded and no eutectic cells appear. Fig. 1(f) shows the microstructure of the 1500 mm/h specimen after the heat-treatment at 1650 °C for 100 h. Phase compositions were measured by Electron Probe Micro-Analysis (EPMA). Before the heat-treatment, the compositions of the bright and the dark phase were Nb-0.6at.%Si-0.7at.%Zr and Nb-24.2at.%Si-3.3at.%Zr, respectively. Thus bright phase is Nb and dark one is Nb_3Si. On the other hand, after the heat-treatment, compositions of bright and dark phases were Nb-1.1at.%Si-0.3at.%Zr and Nb-39.7at.%Si-1.2at%Zr, respectively. Thus the bright phase is still Nb but the dark phase is Nb_5Si_3. The results of EPMA suggests that Nb_3Si decomposed completely into Nb_{ss} + Nb_5Si_3 during the heat-treatment. After the eutectoid decomposition, the Nb phase forms a continuous microstructure, i.e. a Nb-network appeared.

Crystallographic orientation relationship analyzed by EBSD and TEM

Figure 2 shows the results of EBSD analysis of the as-grown specimens. As Nb_3Si has a tetragonal crystal structure (a=b≠c), the asymmetrical notations <...] and {...} are used to indicate all possible permutations, which are allowed within the two first indices, while the third one is set [7]. The solidification direction of Nb_3Si is <001], regardless of the growth rate. In each eutectic cell all the Nb rods have the same crystallographic orientation in alloys with a growth rate of 150 mm/h or higher. On the other hand, in the alloys with the growth rate of 15 mm/h or lower, the Nb phase has various crystallographic orientations. This is due to the effect of constitutional undercooling in multi-component alloys. Under such conditions, a higher solidification rate leads to destabilization of the solid-liquid interface which results in the formation of eutectic cells, where crystallographic orientation relationships can be seen [8]. On the other hand, at lower solidification rates, Nb has several different crystallographic orientations because it grows coarsely and independently.

Various researchers have reported on the crystallographic orientation relationship between Nb and Nb_3Si. Cockram et al. [9] studied the crystallographic orientation relationship between Nb_3Si precipitates and a Nb matrix, and proposed two kinds of relationship, $\{110\}_{Nb}//\{101\}_{Nb3Si}$, $<111>_{Nb}//<110>_{Nb3Si}$ and $\{110\}_{Nb}//\{110\}_{Nb3Si}$, $<111>_{Nb}//<111>_{Nb3Si}$, in Nb-Si binary alloys. On the other hand Drawin et al. [10] reported $\{111\}_{Nb}//\{111\}_{Nb3Si}$, in as-arc melted Nb-Si binary alloys, which is different from the result reported by Cockram et al. Sekido et al. [11] also reported $\{110\}_{Nb}//\{110\}_{Nb3Si}$, $<112>_{Nb}//<001\}$ $_{Nb3Si}$ in Nb-Si-Ti ternary alloys prepared by the OFZ method. In the present alloys with a growth rate of 150 mm/h or higher, the relationship is $\{110\}_{Nb3Si}//\{112\}_{Nb}$, $<001\}_{Nb3Si}//<113>$ $_{Nb}$, i.e. different from previous results. No orientation relationship is found in alloys with a growth rate of 15 mm/h or lower.

TEM observations were performed on the as-grown specimen with a growth rate of 150 mm/h. Figure 3 shows a bright-field and a high-resolution TEM images and corresponding selected-area electron diffraction patterns taken from Nb and Nb_3Si. The diffraction patterns indicate that $(112)_{Nb}$ is parallel to $\{110\}_{Nb3Si}$, which is consistent with the present result by EBSD. The high-resolution TEM of the Nb/Nb_3Si interface shows that a ledge structure with the $\{110)_{Nb3Si}$ is formed around there.

Atomic structure at the interface between Nb$_{ss}$ and Nb$_3$Si

To refine the possible orientation relationship on the basis of the $\{110)_{Nb3Si}//(112)_{Nb}$ planar parallelism, the atomic configurations of the two planes are considered. The refinement of the orientation relationship will be conducted by rotating the matching planes and observing the atomic row matching along the close-packed directions. Because Nb_3Si has a complex crystal structure, several planes parallel to $\{110)$ are examined. Among them, it was found that $\{440)$ satisfies the orientation relationship determined in the present study as shown in Figure 4. A reasonable coincidence of the Nb atom positions is found along $[001]_{Nb3Si}$ and $<113>_{Nb}$. With this matching model the interatomic spacing misfit is about 1.3 % along the A direction and about 5.5% along the B direction, which seems to be sufficiently small [12].

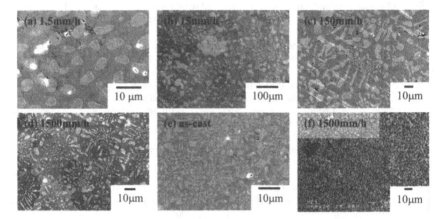

Figure 1: (a) to (e) show microstructures of the cross-sections of as-grown alloys with various growth rates. (f) shows the microstructures of the cross-section of the alloy after the heat-treatment.

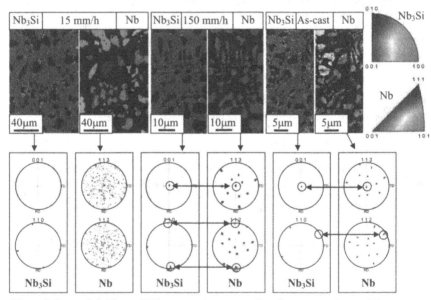

Figure 2: Invers Pole Figure (IPF) maps and corresponding discrete pole figures (PF).

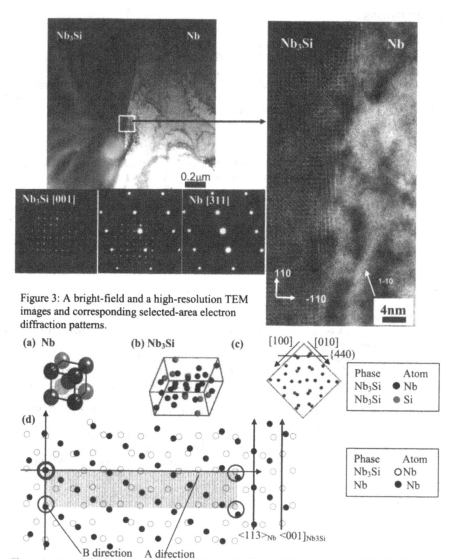

Figure 3: A bright-field and a high-resolution TEM images and corresponding selected-area electron diffraction patterns.

(a) Nb   (b) Nb₃Si   (c)

Phase	Atom
Nb₃Si	● Nb
Nb₃Si	● Si

(d)

Phase	Atom
Nb₃Si	○ Nb
Nb	● Nb

Figure 4: (a) and (b) Crystal structures of Nb and Nb₃Si, (c) [001] projection of Nb₃Si, (d) atomic configurations of the {112}$_{Nb}$ and {440)$_{Nb3Si}$ planes satisfying the orientation relationship determined in the present study. The {440)$_{Nb3Si}$ configuration includes atoms whose center positions are within 1.022 nm (one-tenth of the lattice constant along [100]) from exact {440)$_{Nb3Si}$ plane of Nb₃Si.

CONCLUSIONS

With a growth rate of 150mm/h or higher, eutectic cell structures appears. With increasing growth rate, both, the size of the Nb rods in the cell center and the cell size are decreasing. In alloys with a growth rate of 150 mm/h or higher, the solidification is <001] for Nb_3Si and <113> for Nb, and $\{110\}_{Nb3Si}$ is parallel to $\{112\}_{Nb}$. With the proposed matching model for Nb and Nb_3Si, the interatomic spacing misfit is about 1.3 % along the A direction and 5.5% along the B direction, which seems to be sufficiently small.

ACKNOWLEDGMENTS

This work was sponsored by Grant-in-Aid for Scientific Research, Ministry of Education, Culture, Sports, Science and Technology, Japan, No.19206078.

REFERENCES

[1]   G. Ghosh, G. B. Olson, Acta Mater. **55** (2007), 3281-3303 .
[2]   B. Bewlay, M. Jackson, J. Zhao P. Subramanian, M. Mendiratta, J. Lewandowski, MRS Bull. **28** (2003), 646-653.
[3]   S. Miura, M. Aoki, Y. Saeki, K. Ohkubo, Y. Mishima, T. Mohri, Metall. Mater. Trans. A, **36** (2005), 489-496.
[4]   S. Miura, K. Ohkubo, T. Mohri, Intermet., **15** (2007), 783-790.
[5]   S. Miura, Y. Murasato, K. Ohkubo, Y. Kimura, N. Sekido, Y. Mishima, T. Mohri, MRS. Symp. Proc. **980** (2006), 0980-II05-33.
[6]   S. Miura, J. H, Kim, K. Ohkubo, Y. Kimura, N. Sekido, Y. Mishima, T.Mohri, Mater. Sci. Forum, **539-543** (2007), 1507-1522.
[7]   P. Veyssiere, J. Douin, ed. J. H. Westbrook, R. L. Fleischer, Intermetallic Compounds, John Wiley & Sons Ltd, Chichester, **Vol. 1** (1995), p. 530.
[8]   B. Cockram, H. A. Lipsitt, R. Srinivasan, I. Weiss, Scr. Metall. Mater., **25** (1991), 2109-2114.
[9]   R. Elliott, Eutectic Solidification Processing, Butterworths & Co Ltd., London, (1983), p. 92.
[10]   S. Drawin, P. Petit, D. Boivin, Metall. Mater. Trans. A, **36** (2005), 497-505.
[11]   N. Sekido, Y. Kimura, S. Miura, Y. Mishima, Mater. Sci. Eng., A, **444** (2007), 51–57.
[12]   N. Sekido, F. G. Wei, Y. Kimura, S. Miura, Y. Mishima, Philos. Mag. Lett., **86**, (2006).

Mater. Res. Soc. Symp. Proc. Vol. 1128 © 2009 Materials Research Society   1128-U05-39

Solidification Microstructure Evolution Modeling in Nb-Si Based Intermetallics-Strengthened-Metal-Matrix Composites

Sujoy Kar[1], Bernard Bewlay[2] and Ying Yang[3]
[1]GE Global Research Center, Bangalore, KA, India, 560066
[2]GE Global Research Center, Niskayuna, NY, USA, 12309
[3]CompuTherm LLC, Madison, WI, USA, 53719

ABSTRACT

For higher fuel efficiency and greater thrust to weight ratios, there is a continuous drive for higher performance turbine engine components. Nb-silicide intermetallics, owing to their high melting point and high-temperature strength, are potential candidates for high temperature applications. These intermetallics when precipitated in the metal matrix of a (Nb) solid solution, result in intermetallic-strengthened metal matrix composites that have a good combination of room temperature toughness and high temperature strength. The microstructures of these in-situ composites can be complex and vary significantly with the addition of elements such as Ti and Hf. Hence an improved understanding of phase stability and the microstructural evolution of these alloys is essential for alloy optimization. In the present paper we describe binary alloy microstructural evolution modeling of dendritic and eutectic solidification obtained using phase-field simulations. The effect of parameters such as heat extraction rate, the ratio of the diffusivity of the solute in liquid to solid, and the liquid-solid interfacial energy, on microstructural evolution during solidification is discussed in detail.

INTRODUCTION

The manufacture of turbine engine components using near net shape casting gives an economically attractive option. The mechanical properties of Nb-Si alloys depend on the as-cast microstructure. The as-cast microstructure depends on the phase stability, processing parameters, and composition. Prediction of microstructural evolution during solidification in these alloys is of a very high significance in terms of understanding the mechanical behavior. Phase-field modeling [1] has been used for modeling microstructural evolution. The current paper describes modeling of microstructural evolution of a binary Nb-16at%Si alloy as a function of cooling rate and other parameters, such as the ratio of diffusivity of Si in the Nb(Si) solid solution to the liquid. The Nb-16at%Si alloy was selected, because it is the basis for more complex alloys of greater engineering relevance. The alloy is representative of the hypoeutectic binary alloys. In the current study, the phase-field modeling was performed using MICRESS™ [2] software, a multi-component, multiphase phase-field modeling tool.

THEORY

The hypo-eutectic Nb-Si alloy composition proceeds with dendritic solidification of a Nb(Si) solid solution until the eutectic temperature is reached, where the eutectic reaction gives Nb_3Si and Nb(Si). The Nb_3Si phase further undergoes eutectoid reaction into Nb_5Si_3 and Nb(Si). The eutectoid is very sluggish. The current work is limited to the binary Nb-16at%Si alloy that results in the calculated values of the eutectic temperature of 1917.8 °C, and the eutectic composition of 17.5at%Si using the ternary thermodynamic database [3]. The model has been applied to dendritic and eutectic solidification of the current alloy to study the evolution of the Nb(Si) and Nb₃Si phases in the microstructure; the results are presented in the present paper.

Phase-field models have been used and further developed in recent past to simulate complex interfacial phenomena during solidification [4, 5]. The attraction of the phase-field method is in fact that explicit tracking of the interface is completely avoided. The present model was developed using MICRESS™, a commercially available Phase field-modeling package. It is based on the multi-phase field model reported by Steinbach and co-workers [6]. MICRESS uses ThermoCalc™'s TQ interface™ with ThermoCalc™ software to calculate the thermodynamic quantities from the database that are then used for the calculation of molar Gibbs energies, and hence chemical potentials, to calculate the driving force at the diffuse interface for motion of the same. Much work on solidification modeling using phase-field is concentrated on the complex morphologies of dendritic solidification [7] in relation to the process parameters and fluid flow [8], and independently on modeling eutectic solidification [9]. MICRESS has the capability to address both dendritic and eutectic solidification processes in the same simulation owing to its versatility in modeling multiple phase transformations.

RESULTS AND DISCUSSION

Dendritic solidification in Nb – 16at%Si

The dendritic morphology obtained during solidification is affected by various physical parameters such as heat extraction from the system domain, ratio of solute diffusivity in liquid to solid, and the interfacial energy between the solid and the liquid. The effect of these parameters will be discussed in this section. Separate 400x400 and 200x200 grid-calculation-domains with grid spacing of 0.05 μm have been used for the solidification microstructure modeling.

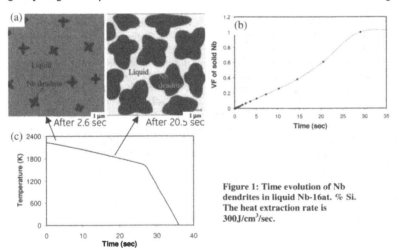

Figure 1: Time evolution of Nb dendrites in liquid Nb-16at. % Si. The heat extraction rate is 300J/cm³/sec.

Effect of heat extraction rate from the system

In the present simulation series, the latent heat of the different phases has been taken into account; the effect of temperature is evaluated by the global interaction of latent heat and heat extraction. There is no temperature gradient in the calculation domain. Figure 1 shows the time evolution of the dendrites of the Nb(Si) phase plotted as a function of Si in the system, where the starting under-cooling is specified to be 10° C and the heat extraction rate is $300*10^6$ J/m^3/sec. Figure 1 also shows the evolution of temperature and volume fraction of Nb(Si) phase during solidification.

The effect of heat extraction from the system on dendrite morphology and branching were investigated. Simulations were performed using eleven nucleation sites of the Nb(Si) phase in the melt for the three cases of heat extraction rate values ($100*10^6$, $300*10^6$ and $1000*10^6$ J/m^3/sec). Fig 2a shows the microstructures for the three cases of heat extraction values after 25% of the Nb(Si) phase has solidified. Figures 2b, 2c show the associated temperature and volume fraction evolution. Figure 2 suggests that higher heat extraction causes higher tendency of front instability leading to more secondary dendritic arm formation and higher solidification rate, which agrees with the experimental observations of dendritic solidification reported in literature.

(a)

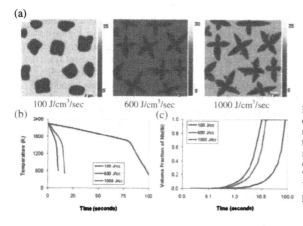

(b)   (c)

Figure 2a: Effect of heat extraction rate on dendritic morphology at fixed volume fraction. Finer structures result with the increased heat extraction rates. 2b: Temperature vs time. 2c: Volume fraction of solid Nb phase vs. time

Effect of ratio of solute diffusivity in liquid to that in solid:

The Si diffusivity in the solid Nb phase has a significant effect on secondary dendrite arm formation and also on the concentration gradient of solute from core to the periphery of the dendrite. To study this effect, two ratios of the Si diffusivity in liquid to solid Nb(Si) are considered – 10 and 100. The composition profiles along a line across the dendritic microstructures were generated from two different heat extraction rates ($200*10^6$ and $300*10^6$ J/m^3/sec), for these diffusivity ratios. Figure 3 shows a case where the ratio of diffusivities has been maintained at a value of 10. The associated composition profile does not show any coring effect (concentration gradient caused due to solidification of enriched liquid resulting in a lower Si (solute) content at the core of the dendrite) due to change in heat extraction rate. This could be due to the higher solute diffusivity in the solid. Typically, the solute diffusivity in the solid is

289

close to two orders of magnitude lower than that in the liquid phase.

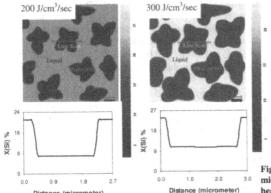

Figure 3: Dendritic solidification microstructures solidified at the heat extraction rates of $200J/cm^3/sec$ and $300J/cm^3/sec$

For the case of the ratio of diffusivities being 100, coring does occur, as there is not enough time for diffusion to occur. Figures 3 and 4 do show this effect explicitly. Figure 4 shows one dendrite column growing from the corner of the simulation domain at 45° angle. The case with solute diffusivity ratio of 10 in liquid to solid does not show a Si (color) gradient from the core of the dendrite to the interface. However, the case of slow Si diffusivity shows the Si concentration gradient, which indicates coring. The tendency of formation of side branches is also related to the solute diffusivity ratio in liquid to solid. As seen in figure 4(b), faster diffusion of the solute in solid phase causes reduced segregation which in turn reduces the tendency of the solid front instability, as a result of which faster Si diffusivity in solid phase lowers the tendency of formation of side branches on the dendrites.

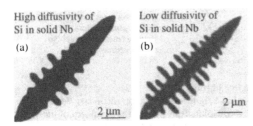

Figure 4: Effect of diffusivity of Si on dendritic morphology. (a) High (b) Low diffusivity in solid Nb(Si).

Effect of interfacial energy between liquid and solid:

Solid/liquid interfacial energy influences the evolution of the morphology of the solidifying dendritic phase. For the case of a lower interfacial energy, the system will allow an increase in the surface area that can lead to more branching of the dendrites. This is evident in the Figure 5. The growth velocity, or mobility of the interface, also has effect on side branch formation and front stability. While low mobility of the interface causes fewer tendencies to form dendritic side branches, high growth velocity can cause front instability and disintegration

290

of the solidification front. The parameters that were chosen are based on the fact that the surface energies of some of the metallic materials are in the range of 10 – 100 mJ/m².

(a) Solid-Liquid interfacial energy: $1*10^6$ J/cm²  (b) Solid-Liquid interfacial energy: $5*10^6$ J/cm²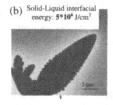

Figure 5: Effect of solid-liquid interfacial energy on dendritic morphology: (a) lower interfacial energy and (b) higher interfacial energy. More branching when interfacial energy is lower in (a).

Dendritic & Eutectic Solidification in Nb-16at%Si

In this solidification simulation of the hypo-eutectic alloy Nb-16at%Si, the Nb dendrites are formed as the primary phase until the system reaches a given under-cooling (set at 100° C) below eutectic temperature, when the remaining liquid attains the eutectic composition. At that point, the eutectic microstructure nucleates and grows from the dendrite-liquid interface. Figure 6 shows the microstructure evolution at an intermediate time, 21 sec. In this case the heat extraction rate is $45*10^6$ J/m³/sec.

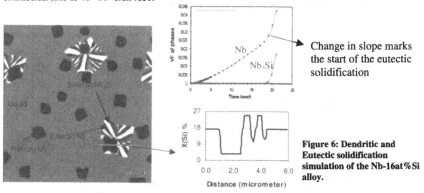

Change in slope marks the start of the eutectic solidification

Figure 6: Dendritic and Eutectic solidification simulation of the Nb-16at%Si alloy.

In the simulation, the Nb(Si) phase has been considered as an anisotropic phase (i.e. the interfacial energy is anisotropic) and the eutectic Nb₃Si particles have faceted shapes with <110> type facets, with kinetic anisotropic factor of 0.5 for each facet. This parameter represents a factor that would increase the surface energy in a preferred direction and decrease the mobility in other directions. Depending on the isotropic or anisotropic or faceted nature of the phases, and on different process parameters, there can be different morphologies of the eutectic microstructure. The typical value that would suit most alloy systems [10] would be between 0 and 0.5. For example, the eutectic can be lamellar, or one phase can be rod shaped in the matrix of the other phase. In either case, both phases grow continuously from the melt. The volume fraction of the Nb₃Si in fully solidified condition is around 50% in the current alloy – Nb-16at%Si. Experimental observation shows that for hypoeutectic Nb-Si binary alloys, the inter-dendritic eutectic structure consists of niobium rods and plates dispersed in the Nb₃Si intermetallic matrix [11]. This also confirms the phase-field results on the morphology of

291

eutectics reported by Lewis et al [9]; higher volume fractions of the second phase would result in the plate like morphologies. In the present simulation, a periodic boundary condition is maintained and the solid-liquid interface thickness is taken as 0.175 μm, which corresponds to 3.5 grid points. The phase-field methodology needs the diffuse interface thickness to be at least 3-4 grid points for numerical stability. The effect of process parameters and physical quantities on the eutectic solidification is currently being studied and will be reported subsequently.

CONCLUSIONS

Multi-component Nb-Si composites are regarded as potential future candidates for high temperature structural components. In the current study, phase-field modeling using MICRESS™ has been performed on the binary Nb-16at%Si alloy to understand the microstructural evolution of the Nb(Si) and the eutectic Nb_3Si phase. The effect of the heat extraction, reiterates that finer-scale structures are obtained with increased heat extraction rates. Finer-scale structures are also seen with lower solute diffusivity in the solid. Coring is observed for the cases of lower solute diffusivity in the solid. Using MICRESS™, it was possible to simulate dendritic solidification followed by eutectic solidification in one simulation due to the ability to represent multiple phase transformations. The modeling of subsequent eutectoid transformation after solidification that results in the formation of eutectoid Nb_5Si_3 phase will be reported subsequently.

REFERENCES

1. L.Q. Chen, *Annu. Rev. Mater. Res.*, **32**, 113 (2000).
2. http://www.micress.de
3. H. Liang and Y.A. Chang, *Intermetallics*, **7**, 561 (1999).
4. A. Karma and W.J. Rappel, *Phys. Rev. E.*, **53**, 3017 (1996).
5. R. Kobayashi, *Physica D.*, **63**, 410 (1993).
6. I. Steinbach, F. Pessolla, B. Nestler, M. Seeßelberg, R. Prieler, G.J. Schmitz, and J.L.L. Rezende, *Physica D.*, **94**, 135 (1996).
7. A. Karma, W.J. Rappel, *Phys. Rev. E.*, **60**, 3614 (1999).
8. J.H. Jeong, N. Goldenfield, J.A. Dantzig, *Phys. Rev. E*, **64**, 041602 (2001).
9. D. Lewis, T. Pusztai, L. Granasy, J. Warren, W. Boettinger, *JOM*, **35** (2004).
10. Private communication with MICRESS support group.
11. K.-M. Chang, B.P. Bewlay, J.A. Sutliff and M.R. Jackson, *JOM*, **59** (1992).

Mater. Res. Soc. Symp. Proc. Vol 1128 © 2009 Materials Research Society   1128-U05-42

Effect of Cr Addition on the Multiphase Equilibria in the Nb-rich Nb-Si-Ti System - Thermodynamic Modeling and Designed Experiments

Y. Yang[1], B.P. Bewlay[2], S.-L. Chen[1], M.R. Jackson[2] and Y.A. Chang[3]

[1]CompuTherm LLC, Madison, Wisconsin 53719, USA
[2]General Electric Global Research, Schenectady, New York 12301, USA
[3]University of Wisconsin-Madison, Madison, Wisconsin 53706, USA

ABSTRACT

Refractory Metal Intermetallic Composites (RMICs) based on the Nb-Si system are considered as candidates of next-generation high temperature materials (i.e. >1200°C). Ti and Cr have been shown to have beneficial effects on the oxidation resistance and mechanical properties of Nb-Si alloys. Phase equilibria in the Nb-Si-Ti system have been studied in detail. The present study has investigated multiphase equilibria in the Nb-Si-Ti alloys with Cr additions via an approach of integrating thermodynamic modeling with designed experiments. The alloying effects of Cr on the microstructure of the Nb-Si-Ti alloys are described using both phase equilibria and solidification paths that were calculated from the thermodynamic description of the Nb-Cr-Si-Ti system developed in the present study.

INTRODUCTION

In searching for higher temperature materials (i.e. > 1200°C) for future propulsion systems, Refractory Metal-Intermetallic Composites (RMICs) [1] based on the Nb-Si-X system have received considerable interest [2]. The typical alloying elements (X's) are Ti, Cr, Hf, and Al. Ti and Cr have been shown to have beneficial effects on the oxidation resistance and mechanical properties of Nb-Si alloys [3-6]. Phase diagrams of the Nb-Si-Ti-Cr system, are a prerequisite for the successful development of this family of materials. In the present study, a thermodynamic database that includes the Gibbs energy functions of all of the phases in the Nb-Si-Ti-Cr system was developed using the Calphad approach. The thermodynamic description was then coupled with Pandat software for the calculation of phase equilibria and solidification paths of multicomponent Nb-Si-Ti-Cr alloys. The calculated results were then compared with the results of designed experiments, and the subsequent analyses were used to refine the thermodynamic description.

THERMODYNAMIC MODELING AND EXPERIMENTAL PROCEDURES

The strategy for building the thermodynamic database for the Nb-Si-Ti-Cr quaternary system began with deriving Gibbs energy of each phase in each of the constituent binaries. There are six constituent binaries: Nb-Si, Nb-Cr, Nb-Ti, Cr-Si, Cr-Ti, and Ti-Si. After thermodynamic descriptions for all the constituent binary phases were established, the Gibbs energy of a phase in a ternary system was obtained by the weighted average of those of the same phase in the constituent binaries using geometric models, such as the Muggianu model [7] in the present study. This extrapolation method for obtaining the thermodynamic description of each ternary phase works quite well in many cases when describing thermodynamic properties and phase equilibria. However, in the present work, it was found that ternary interaction parameters are required with the thermodynamic models for the phases in all the constituent ternary systems, due to the rather strong ternary interactions. Thermodynamic descriptions were developed for all the constituent ternary systems, including Nb-Cr-Ti [8], Nb-Ti-Si [8], Nb-Cr-Si [9], and Cr-Si-Ti [10]. Details on the thermodynamic models of these individual systems are provided in the original literature.

Based on thermodynamic descriptions of phases in the constituent subsystems, a thermodynamic database of the Nb-Si-Ti-Cr quaternary was obtained using the extrapolation method. No experimental phase equilibria and thermodynamic property data for the Nb-Si-Ti-Cr quaternary were available in literature. Therefore, alloys for experimental validation were selected using phase diagrams calculated from the newly developed Nb-Si-Ti-Cr thermodynamic description. Comparisons between the calculated phase equilibria and the experimental data suggested that the thermodynamic database of the Nb-Si-Ti-Cr system obtained through extrapolation can satisfactorily describe the experimental results. There was no need to introduce the quaternary interaction parameters. Pandat software [11] was used to calculate the phase diagrams and solidification paths.

The alloys for this study were cast using cold crucible directional solidification [3, 4] after triple melting the starting charges from high purity elements (>99.99%). Samples for heat treatment were wrapped in Nb foil and heat-treated at 1350°C for 100 hours in a vacuum (<10^{-5} Torr) furnace, followed by furnace cooling.

All of the samples were examined using Scanning Electron Microscopy (SEM) Back-Scatter Electron (BSE) imaging and Energy Dispersive Spectrometry (EDS). Phase compositions were measured by Electron Probe Micro-Analysis (EPMA), which was performed on a JEOL 733 microprobe operating at 15KV, 20nA, with a ~1 µm beam diameter (Japan Electron Optics Ltd., Tokyo). High purity Nb, Cr, Ti, and Si were used as standards, and conventional matrix corrections (Z, A, and F) were used to calculate the compositions in wt% from measured X-Ray intensities. The crystal structures of the phases were identified using X-Ray Diffraction (XRD) and Electron Back Scattering Diffraction (EBSD).

RESULTS AND DISCUSSIONS

Summary of experimental results

The phase equilibria in the Nb-Si-Ti-Cr system were calculated at selected temperatures and for a range of compositions using the initial thermodynamic description. In this section, only the solid-state phase equilibria at 1350°C and liquid-solid phase equilibria for selected alloys are shown. The use of these data in the validation of the calculated results is also described.

Solid-state phase equilibria at 1350°C in the Nb-rich region

Figure 1(a) shows the calculated isothermal sections at 1350°C of the Nb-Ti-Si system with the black lines as phase boundaries and gray lines as tie-lines. Figure 1(b) shows the calculated isothermal section at the same temperature of the Nb-Cr-Ti-Si system with the concentration of Cr being fixed at 10 at%. Tie-lines are not shown in this figure as they are not

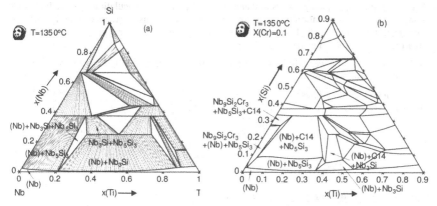

Fig. 1 Isothermal sections at 1350°C for (a) Nb-Ti-Si (b) Nb-Cr-Si-Ti (X(Cr)=0.1) systems.

inside this section. Alloy compositions are stated in atomic percent in this article. Only the Nb-rich regions are labeled because they are of the current interest. In the Nb-Ti-Si system, the phase equilibria of the Nb-rich region include (Nb)+Nb$_5$Si$_3$, (Nb)+Nb$_3$Si, and (Nb)+Nb$_3$Si+Nb$_5$Si$_3$. With Cr addition, the regions of stability of the Nb+Nb$_5$Si$_3$ and the (Nb)+Nb$_3$Si fields are significantly reduced due to the formation of the C14-Cr$_2$Nb phase. At low Ti concentrations, the Nb$_9$Si$_2$Cr$_3$ phase that emanates from the Nb-Cr-Si system is in equilibrium with (Nb) and Nb$_5$Si$_3$. The most prominent phase equilibrium region is (Nb)+Nb$_5$Si$_3$+C14-Cr$_2$Nb. Two alloys from this region were selected for experimental study and they are Nb-10Cr-23Ti-15Si and Nb-11Cr-23Ti-13Si. The

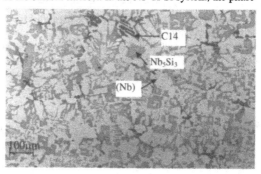

Fig.2 BSE image of the annealed (1350°C for 100h) microstructure of Nb-10Cr-23Ti-15Si.

microstructural and microchemical evidence from these two alloys (annealed at 1350°C for 100h) provided strong support for the calculated results. Fig. 2 shows a BSE image of the Nb-10Cr-23Ti-15Si alloy. Three phases can be observed in this micrograph. They are (Nb) (light gray), Nb$_5$Si$_3$ (medium gray), and C14-Cr$_2$Nb (dark gray), which is also referred to as C14. Their crystal structures were confirmed by EBSD analysis. The microstructure of the Nb-11Cr-23Ti-13Si alloy was similar to that of the Nb-10Cr-23Ti-15Si, and is not shown here. Comparisons between the calculated phase compositions (in italic font) and the EPMA measurements are shown in Table 1. In view of the fact that these experimental compositions were not used in the initial model parameter optimization, the calculated phase compositions from the extrapolated thermodynamic description agree well with the experimental measurements. The major composition discrepancy occurs in that of the C14 phase; the calculation shows a higher Cr level and lower Nb concentration than the experimental values. This discrepancy mainly comes from uncertainties in the Cr-Nb-Ti system. Additional experimental work is needed to improve the quantitative understanding for the Cr-Nb-Ti system.

Table 1. Comparisons between the EMPA measured and the calculated (italic font in parentheses) phase compositions.

Sample	Phase	Nb (at%)	Cr (at%)	Ti (at%)	Si (at%)
Nb-10Cr-23Ti-15Si	(Nb)	58.2 (60.9)	13.3 (11.9)	27.4 (26.1)	1.2 (1.1)
	Nb₅Si₃	42.2 (41.8)	0.6 (1.3)	21.2 (19.4)	36.1 (37.5)
	Cr₂Nb_C14	29.1 (21.9)	47.1 (55.6)	14.7 (12.6)	9 (9.9)
Nb-11Cr-23Ti-13Si	(Nb)	59.4 (61.2)	13.5 (12.0)	26 (25.7)	1.2 (1.1)
	Nb₅Si₃	43.7 (41.9)	0.5 (1.3)	19.5 (19.3)	36.2 (37.5)
	Cr₂Nb_C14	26 (22.1)	51 (55.5)	13 (12.5)	10 (9.9)

Liquid-solid phase equilibria in the Nb-rich region

The liquid-solid phase equilibria at the Nb-rich region are described using liquidus projections which were then validated by comparing the as-cast microstructure of selected alloys with the simulated solidification paths. The calculated liquidus projection of the Nb-Si-Ti system is shown in Fig. 3(a), which has been well validated previously. In the metal-rich region, five primary phase regions were calculated: βNb₅Si₃, αNb₅Si₃, Nb₃Si, Ti₅Si₃, and (Nb). It should be noted that the boundary between βNb₅Si₃ and αNb₅Si₃ was not investigated experimentally. In the following discussion, both βNb₅Si₃ and αNb₅Si₃ were denoted by Nb₅Si₃. The solidification reactions for the Nb-rich Nb-Ti-Si alloys are dominated by the Liquid+Nb₅Si₃→Nb₃Si and the Liquid→(Nb)+Nb₃Si reactions. There is no direct solidification reaction between (Nb) and

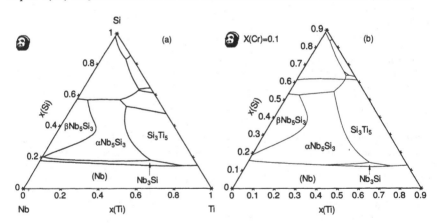

Fig. 3 Calculated liquidus projections of (a) Nb-Ti-Si (b) Nb-Cr-Si-Ti (X(Cr)=0.1).

αNb₅Si₃. The Nb₃Si was always present in the as-cast Nb-rich Nb-Si-Ti alloys investigated. The

296

3D liquidus projection of quaternary Nb-Ti-Si-Cr is very complicated and difficult to represent. Fig. 3(b) shows the calculated Nb-Ti-Si-Cr liquidus projection at 10 at% Cr, in which the Nb$_3$Si primary phase region completely disappears at the Nb-rich region. Direct eutectic solidification occurs between (Nb) and Nb$_5$Si$_3$ in the Nb-rich Nb-Cr-Si-Ti liquidus projection with the addition of 10 at% Cr. The calculation suggests that Cr addition destabilizes Nb$_3$Si primary solidification. The as-cast microstructures of four alloys Nb-10Cr-23Ti-15Si, Nb-11Cr-23Ti-13Si, Nb-21Cr-23.5Ti-15.5Si, and Nb-21Cr-28Ti-13Si were used to validate the calculated liquid-solid phase equilibria. Solidification path simulations were performed on these four alloy compositions using the Scheil [12] and lever-rule models. The Scheil model assumes no diffusion in solid and thermodynamic equilibrium only at liquid-solid interface. The lever-rule model assumes complete mixing in both liquid and solid, and that the composition of the solid and the liquid are in thermodynamic equilibrium. Both models require thermodynamic information and they are integrated into the solidification simulation module of Pandat. The solidification path and as-cast microstructure of the Nb-11Cr-23Ti-13Si alloy are described as an example. Fig. 4(a) and (b) shows the simulated solidification paths and BSE image of the as-cast microstructure of the Nb-11Cr-23Ti-13Si alloy. The solidification paths predicted using both the Scheil and lever-rule models indicate primary (Nb) solidification followed by the two-phase eutectic solidification (Nb)+Nb$_5$Si$_3$ and the ternary eutectic solidification (Nb)+Nb$_5$Si$_3$+C14. In addition to the above solidification reactions, the Scheil model predicts two more solidification reactions for the Nb-11Cr-23Ti-13Si alloy, and they are Liquid→(Nb)+Nb$_3$Si+C14 and Liquid→(Nb)+Si$_3$Ti$_5$+C14. However, the total mole fraction of Nb$_3$Si and Si$_3$Ti$_5$ is less than 0.04,

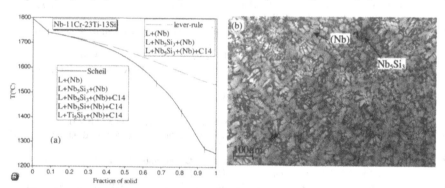

Fig. 4 (a) Simulated solidification paths using Scheil model (solid line) and lever-rule model (dash line) (b) BSE image of as-cast microstructure of Nb-11Cr-23Ti-13Si.

which means that even though these two reactions may happen, they are very difficult to detect experimentally. Fig. 4(b) shows that the predominant phases in the as-cast microstructure of this alloy are (Nb) and Nb$_5$Si$_3$. The (Nb) solidifies as dendrites, and the Nb$_5$Si$_3$ solidifies in two-phase eutectic cells with (Nb). These observations are consistent with the predicted solidification path. The comparisons between the predicted solidification paths and the as-cast microstructures of the remaining alloys also show good agreement with each other, which indicates the thermodynamic description developed in the present study is suitable for describing the liquid-solid phase equilibria of Nb-rich Nb-Cr-Si-Ti alloys.

297

SUMMARY

In summary, the phase diagram of the metal-rich region of quaternary Nb-Cr-Ti-Si system has been efficiently established using the Calphad approach. The calculated isothermal sections and liquidus projections were validated by microstructural and microchemical data. The solid-state phase equilibria at the Nb-rich region of the Nb-Cr-Ti-Si system at 1350°C shows the three-phase Nb_5Si_3+(Nb)+C14 field, as that C14 phase is stabilized by the Cr additions. The liquid-solid phase equilibria in this region involve primary solidification (Nb) and Nb_5Si_3, as well as eutectic solidification of (Nb)+Nb_5Si_3. The addition of Cr to the Nb-Ti-Si system promotes the formation of the Nb_5Si_3+(Nb) eutectic.

ACKNOWLEDGEMENTS

The authors would like to thank D.J. Dalpe for the directional solidification experiments and Dr. D. A. Wark for the EMPA measurements for the Nb-Cr-Ti-Si alloys.

REFERENCES

1. B. P. Bewlay, M. R. Jackson, J. C. Zhao, and P. R. Subramanian, *Metall Mater Trans A*, **34A**(10), 2043 (2003).
2. B. P. Bewlay, M. R. Jackson, J. C. Zhao, P. R. Subramanian, M. G. Mendiratta, and J. J. Lewandowski, *MRS Bull*, **28**(9), 646 (2003).
3. B. P. Bewlay, M. R. Jackson, and H. A. Lipsitt, *Metall Mater Trans A*, **27A**(12), 3801 (1996).
4. B. P. Bewlay, M. R. Jackson, and P. R. Subramanian, *JOM*, **51**(4), 32(1999).
5. M. R. Jackson, B. P. Bewlay, R. G. Rowe, D. W. Skelly, and H. A. Lipsitt, *JOM*, **48**(1), 39 (1996).
6. M. G. Mendiratta, J. J. Lewandowski, and D. M. Dimiduk, *Metall Mater Trans A*, **22A**(7), 1573 (1991).
7. Y. M. Muggianu, M. Gambino, and J. P. Bros, *J Chem Phys*, **72**(1), 83 (1975).
8. Y. Yang and Y. A. Chang, Thermodynamic Database of Nb Silicide Based Alloys, CompuTherm LLC, Madison, WI 53719 (2008).
9. Y. Yang, B. P. Bewlay, and Y. A. Chang, *Intermetallics*, Submitted (2008).
10. Y. Du and J. C. Schuster, *Scand J Metall*, **31**(1), 25 (2002).
11. S. L. Chen, S. Daniel, F. Zhang, Y. A. Chang, X. Y. Yan, F. Y. Xie, R. Schmid-Fetzer, and W. A. Oates, *Calphad*, **26**(2), 175-188 (2002).
12. E. Scheil, *Z Metallkd*, **34**, 242-246 (1942).

Mater. Res. Soc. Symp. Proc. Vol. 1128 © 2009 Materials Research Society   1128-U05-46

Consolidation and mechanical properties of mechanically alloyed Al-Mg powders

Mira Sakaliyska[1], Sergio Scudino[1], Hoang Viet Nguyen[2], Kumar Babu Surreddi[1], Birgit Bartusch[1], Fahad Ali[1], Ji-Soon Kim[2] and Jürgen Eckert[1,3]

[1] IFW Dresden, Institut für Komplexe Materialien, Postfach 270116, D-01171 Dresden, Germany

[2] Research Center for Machine Parts and Materials Processing, University of Ulsan, Namgu Mugeo 2-Dong, San 29, Ulsan 680-749, Republic of Korea

[3] TU Dresden, Institut für Werkstoffwissenschaft, D-01062 Dresden, Germany.

ABSTRACT

Nanostructured Al-Mg bulk samples with compositions in the range of 10 – 40 at.% Mg have been produced by consolidation of mechanical alloyed powders. Powders with composition $Al_{90}Mg_{10}$ and $Al_{80}Mg_{20}$ were consolidated into highly dense specimens by hot extrusion. Room temperature compression tests for the $Al_{90}Mg_{10}$ specimen reveal interesting mechanical properties, namely, a high strength of 630 MPa combined with a plastic strain of about 4 %. The increase of the Mg content to 20 at.% increases the strength by about 100 MPa but it suppresses plastic deformation. The $Al_{60}Mg_{40}$ powder was consolidated at different temperatures by spark plasma sintering and the effect of the sintering temperature on microstructure, density and hardness have been studied. The results reveal that both density and hardness of the consolidated samples increase with increasing sintering temperature, while retaining a nanocrystalline structure. These results indicate that powder metallurgy is a suitable processing route for the production of nanocrystalline Al-Mg alloys with promising mechanical properties.

INTRODUCTION

In recent years, nanostructured materials have been attracting much attention due to their scientific and engineering significance [1-3]. In particular, substantial increase in strength has been observed in a number of alloys with nanoscale microstructures [3,4].

Among the different processing routes, solid-state processing such as mechanical alloying (MA) have gained increasing interest as versatile non-equilibrium techniques for the production of metastable materials including amorphous alloys, quasicrystalline and nanocrystalline alloys [5-8]. However, materials in powder form, such as milled powders, have to be consolidated to achieve dense bulk specimens [4]. Consolidation of nanocrystalline powders into fully dense bulk specimen is thus of primary interest in the development of near-net shape parts for technological applications [4].

The essence of all compaction techniques is to apply high pressure for densification, and rather high temperature to soften the material so that plastic deformation allows better filling and material flow by diffusion helps to remove the remaining porosity [4]. However, the consolidation of nanocrystalline powders is a very complex procedure and the operational conditions, e.g. pressure, temperature and holding time during the powder consolidation process, are difficult to control precisely in order to obtain fully dense specimens with complete bonding between the initial particles and, at the same time, avoiding or limiting grain coarsening [4].

Among the advanced engineering materials, nanocrystalline and ultra-fine-grained Al-Mg alloys are of great interest due to their remarkable mechanical properties comprised by a beneficial combination of high-strength, good ductility and low density [9-12], which makes these alloys potential candidates for automotive and aerospace applications.

Accordingly, in this work, nanostructured Al-Mg powders with compositions in the range of 10 – 40 at.% Mg have been produced by mechanical alloying of elemental powder mixtures. Consolidation into bulk samples was done by hot extrusion and spark plasma sintering and the mechanical properties of the samples have been investigated by room temperature compression tests and hardness measurements.

EXPERIMENT

Milling experiments starting from pure elemental powder mixtures (purity >99.9wt.%) with nominal compositions $Al_{90}Mg_{10}$, $Al_{80}Mg_{20}$ and $Al_{60}Mg_{40}$ were performed using a Retsch PM400 planetary ball mill and hardened steel balls and vials. The powders were milled for 100 h with a ball-to-powder mass ratio (BPR) of 13:1 and a milling intensity of 150 rpm. To avoid or minimize possible atmosphere contamination during milling, vial charging and any subsequent sample handling was carried out in a glove box under purified argon atmosphere (less than 1 ppm O_2 and H_2O). Consolidation of the powders with composition $Al_{90}Mg_{10}$ and $Al_{80}Mg_{20}$ was done by uni-axial hot pressing followed by hot extrusion under argon atmosphere at 673 K and 500 MPa. The extrusion ratio was 10:4. The powder with composition $Al_{60}Mg_{40}$ was consolidated into cylindrical specimens of 10 mm diameter under high-vacuum by spark plasma sintering at different temperatures (463, 473, 493, 513 and 523 K) using a heating rate of 10 K/min and with an applied pressure of 500 MPa. The microstructure of the consolidated samples was studied by X-ray diffraction (XRD) using a Philips PW 1050 diffractometer (Co K_α radiation) and by optical microscopy (OM) with a Zeiss Axioskop 40. The Rietveld method was applied for the profile-fitting structure refinement using the WinPlotR software package [13].The density of the consolidated samples was evaluated by the Archimedes principle. Cylinders with a length/diameter ratio of 2.0 (8 mm length and 4 mm diameter) were prepared from the extruded samples. The specimens were tested with an Instron 8562 testing facility under quasistatic loading (strain rate of 8×10^{-4} s^{-1}) at room temperature. Both ends of the specimens were polished to make them parallel to each other prior to the compression test. Hardness measurements were done using a computer-controlled Struers Duramin 5 Vickers hardness tester. The device is equipped with a typical diamond indenter in the form of a pyramid with square base and an angle of 136° between the opposite faces. The applied load was 1.96 N for 10 seconds.

RESULTS AND DISCUSSION

The characterization of the mechanically alloyed Al-Mg powders with 10 to 40 at.% Mg has been reported elsewhere [14,15]. However, some key features have to be quoted here. During milling, the Al-Mg elemental powder mixtures transform into nanoscale supersaturated Al(Mg) solid solutions with dimensions of about 30 - 80 nm. During heating, the solid solutions display a complex thermal behavior characterized by several exothermic events, indicating a sequence of phase transformations upon heating. At low temperatures, an increasing amount of Mg is rejected from the solid solution with increasing temperature. At higher temperature, the β'-phase, a hexagonal intermediate phase with approximate composition Al_3Mg_2, is formed. The β'-phase most likely serves as precursor for the formation of the equilibrium β-Al_3Mg_2 phase, which occurs in the following exothermic event [14].

The supersaturated Al(Mg) solid solutions with compositions $Al_{90}Mg_{10}$ and $Al_{80}Mg_{20}$ were consolidated by hot pressing followed by hot extrusion, giving rise to consolidated samples with a relative density of about 98 %. For comparison purposes, a bulk specimen was produced by extrusion of pure Al powder using the same consolidation parameters as used for the Al-Mg powders. No extrusion of the $Al_{60}Mg_{40}$ powder was possible without excessive grain growth of the nanocrystalline structure.

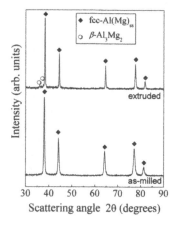

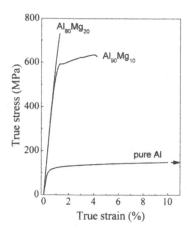

Figure 1. XRD patterns (Co K_α radiation) for the as-milled and extruded powder with composition $Al_{90}Mg_{10}$.

Figure 2. Room temperature compression true stress-true strain curves for the extruded samples: pure Al and MA powders with compositions $Al_{90}Mg_{10}$ and $Al_{80}Mg_{20}$.

As a typical example for the structure of the extruded samples, Fig. 1 shows the XRD patterns of the as-milled and extruded $Al_{90}Mg_{10}$ powder. The pattern of the as-milled powder is characterized by the presence of few broad diffraction peaks which can be identified as a nanoscale supersaturated fcc Al(Mg) solid solution with dimension of about 80 nm [15]. Beside fcc Al, the pattern of the extruded material displays the presence of a minor amount of β-Al_3Mg_2 phase, indicating that the formation of the equilibrium phase occurs during consolidation. The diffraction peaks of the extruded powder are rather broad, indicating that the phases formed are of nano or ultra-fine dimensions. Indeed, Rietveld structure refinement reveals an average grain size for fcc Al ranging between 150 and 300 nm. Smaller dimensions (< 100 nm) have been observed for the β-Al_3Mg_2 phase.

Typical room temperature uni-axial compression true stress-true strain curves of the tests under quasistatic loading for the extruded materials are shown in Fig. 2 together with the curve for the extruded pure Al. The $Al_{90}Mg_{10}$ sample displays the best compromise between strength and plastic deformation. The compressive strength reaches a value of 630 MPa combined with a fracture strain of about 4 %. On the other hand, the specimen with composition $Al_{80}Mg_{20}$ is characterized by a higher strength (730 MPa) but with no measurable plastic deformation. This is

301

most likely due to the larger amount of β-Al_3Mg_2 formed during extrusion of the $Al_{80}Mg_{20}$ sample with respect to $Al_{90}Mg_{10}$. It is worth noticing the remarkable increase of strength of the Al-Mg specimens with respect to pure Al. Although for the extruded $Al_{90}Mg_{10}$ the plastic strain is only about 4 % (60 % for pure Al), the compressive strength reaches a value exceeding the strength level of pure Al by a factor of 4.

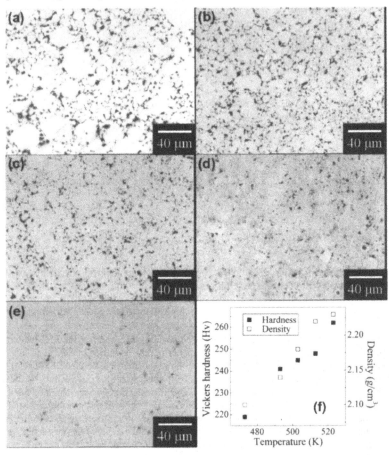

Figure 3. OM micrographs for the $Al_{60}Mg_{40}$ powder consolidated by SPS at (a) 463, (b) 473, (c) 493, (d) 513 and (e) 523 K. (f) hardness and density of the samples consolidated by SPS as a function of the sintering temperature.

Nanostructured materials with composition $Al_{60}Mg_{40}$ (corresponding to the equilibrium β-Al_3Mg_2 phase) are of extreme interest for their possible engineering applications. The β-Al_3Mg_2 phase (space group $Fd\overline{3}m$), an intermetallic compound with a giant unit cell ($a_0 = 2.824$ nm) containing about 1168 atoms [16], has been extensively investigated with particular attention to its structure as well as to its physical and mechanical properties [17-20]. The β-Al_3Mg_2 phase displays interesting properties, such as low density (about 2.25 g/cm^3 [17]) and high-temperature strength (~ 300 MPa at 573 K [20]), which makes this material an attractive candidates for structural applications. Therefore, since extrusion of $Al_{60}Mg_{40}$ has led to inadequate results, the $Al_{60}Mg_{40}$ powder was consolidated at different temperatures by spark plasma sintering (SPS) and the effect of the sintering temperature on microstructure, density and hardness have been studied. The choice of SPS as a consolidation technique was done because by this method sintering can be carried out at relatively low temperatures for a shorter time than in conventional sintering processes [21]. Therefore, the SPS process shows a large potential for achieving fast and full densification of nanostructured materials suppressing grain growth [21].

In accordance with previous structure investigations [14,15], the XRD patterns of the $Al_{60}Mg_{40}$ specimens consolidated by SPS (not shown here) reveals that the structure of the sample sintered at 473 and 493 K consists of a nanocrystalline Al(Mg) solid solution together with a small amount of γ-$Al_{12}Mg_{17}$ phase. With increasing sintering temperature to 513 and 523 K, the patterns display the formation and growth of the nanocrystalline hexagonal β'-phase.

Figs. 3(a)-(e) show the microstructure of the sintered samples investigated by optical microscopy. The porosity characterizing the samples remarkably decreases with increasing sintering temperature from 473 to 523 K. This is corroborated by the increase of the relative density with increasing sintering temperature from 94 % for the sample sintered at 473 K to 98 % for the sample sintered at 523 K [Fig. 3(f)]. Hardness measurements reveal encouraging mechanical properties. The Vickers hardness (H_v) increases with increasing sintering temperature [Fig. 3(f)] from 220 for the sample sintered at 473 K to 260 for the sample sintered at 523 K. Using the well-known relation $H_v = 3\sigma_Y$ [22], this would give a yield strength σ_Y ranging between 750 and 830 MPa, which is in good agreement with the values reported for single- and poly-crystalline β-Al_3Mg_2 [20].

CONCLUSIONS

Al-Mg powders with compositions in the range of 10 – 40 at.% Mg have been produced by mechanical alloying of elemental powder mixtures. The structure of the as-milled Al-Mg powders consist of nanoscale supersaturated Al(Mg) solid solutions with dimensions of about 30 - 80 nm. Powders with composition $Al_{90}Mg_{10}$ and $Al_{80}Mg_{20}$ were consolidated into highly dense specimens by hot extrusion. Room temperature compression tests of the extruded samples reveal a high strength ranging between 630 and 730 MPa but only limited plastic deformation. No extrusion of the $Al_{60}Mg_{40}$ powder was possible without excessive grain growth of the nanocrystalline structure. Therefore, the $Al_{60}Mg_{40}$ powder was consolidated at different temperatures by spark plasma sintering, which permits fast and full densification of nanostructured materials limiting grain growth. The results show that both density and hardness of the sintered nanocrystalline samples increase with increasing the sintering temperature. These preliminary results show that powder metallurgy methods, such as mechanical alloying followed by extrusion or spark plasma sintering, are suitable processing routes for the production of nanocrystalline Al-Mg alloys with promising mechanical properties.

ACKNOWLEDGMENTS

The authors thank M. Frey and H. Schulze for technical assistance, and M. Stoica and S. Venkataraman for stimulating discussions. This work was supported by the EU within the frameworks of the European Network of Excellence on Complex Metallic Alloys (NoE CMA) and by the German Science Foundation under grant Ec 111/16-2.

REFERENCES

1. H. Gleiter, *Prog. Mater. Sci.* **33**, 223 (1989).
2. R. W. Siegel in *Mechanical Properties and Deformation Behavior of Materials Having Ultrafine Microstructures*, edited by M. Nastasi, D. M. Parkin, and H. Gleiter (NATO ASI Series, Kluwer, Dordrecht, 1993).
3. A. S. Edelstein and R. C. Cammarata, *Nanomaterials: Synthesis, Properties and Applications*, (IOP Publishing, Bristol, 1996).
4. *Nanostructured Materials: Processing, Properties and Potential Applications*, edited by C. C. Koch, (Noyes Publications/William Andrew Publising, Norwich, NY 2002).
5. J. Eckert, *Mater. Sci. Eng. A* **226**, 364 (1997).
6. S. Scudino, J. Eckert, X. Y. Yang, D. J. Sordelet and L. Schultz, *Intermetallics* **15**, 571 (2007).
7. C. C. Koch, "Mechanical Milling and Alloying", in *Materials Science and Technology* vol. 15, edited by R. W. Cahn, P. Haasen and E.J. Kramer, (VCH Verlagsgesellschaft, Weinheim, 1991).
8. C. Suryanarayama, *Mechanical Alloying and Milling*, (Marcel Dekker, New York, 2004).
9. F. Zhou, X. Z. Liao, Y. T. Zhu, S. Dallek and E. J. Lavernia, *Acta Mater.* **51**, 2777 (2003).
10. Y. S. Park, K. H. Chung, N. J. Kim and E. J. Lavernia, *Mater Sci Eng A* **374**, 211 (2004).
11. G. J. Fan, G. Y. Wang, H. Choo, P. K. Liaw, Y. S. Park, B. Q. Han and E. J. Lavernia, *Scripta Mater.* **52**, 929 (2005).
12. K. M. Youssef, R. O. Scattergood, K. L. Murty and C. C. Koch, *Scripta Mater.* **54**, 251 (2006).
13. T. Roisnel and J. Rodríguez-Carvajal, *Mater. Sci. Forum* **378–381**, 118 (2001).
14. S. Scudino, M. Sakaliyska, K. B. Surreddi and J. Eckert, *J. Alloys Compd.* (in press).
15. S. Scudino, M. Sakaliyska, K. B. Surreddi and J. Eckert, *J. Phys.: Conf. Series* (in press).
16. *Pearson's Handbook of Crystallographic Data for Intermetallic Phases*, edited by P. Villars and L. D. Calvert, (American Society for Metals, Metals Park (Oh), 1985).
17. M. Feuerbacher, C. Thomas, J. P. A. Makongo, S. Hoffmann *et al.*, *Z. Kristallogr.* **222**, 259 (2007).
18. J. Dolinšek, T. Apih, P. Jeglič, I. Smiljanić, A. Bilušić, Ž. Bihar, A. Smontara, Z. Jagličić, M. Heggen and M. Feuerbacher, *Intermetallics* **15**, 1367 (2007).
19. E. Bauer, H. Kaldarar, R. Lackner, H. Michor, W. Steiner, E.-W. Scheidt, A. Galatanu, F. Marabelli, T. Wazumi, K. Kumagai and M. Feuerbacher, *Phys. Rev. B* **76**, 014528 (2007).
20. S. Roitsch, M. Heggen, M. Lipińska-Chwałek and M. Feuerbacher, *Intermetallics* **15**, 833 (2007).
21. V. Mamedov, *Powder. Metall.* **45**, 322 (2002).
22. L. A. Davies, in *Mechanical Behavior of Rapidly Solidified Materials*, edited by S. M. L. Sastry and B. A. MacDonald, (The Metallurgical Society, Warrendale, PA 1986).

Mater. Res. Soc. Symp. Proc. Vol. 1128 © 2009 Materials Research Society   1128-U05-47

Effect of Precipitate Depleted Zones on Precipitation Hardening in Mg-Based Alloys

D. Shepelev, A. Katsman, E. Edelshtein, M. Bamberger

Department of Materials Engineering, Technion, Haifa 32000, Israel

ABSTRACT

The formation of precipitate depleted zones (PDZ) near grain boundaries in Mg-based alloys strengthened by precipitation hardening is deemed as detrimental to the material since wide depleted zones may affect the mechanical and corrosion properties of the alloy. Experimental investigation of PDZ evolution in Mg-Zn-Sn-alloys aged at different temperatures for different times was conducted by SEM and TEM and by measuring the microhardness of near grain boundary zones at low loads. It was found that at early stages of aging (175°C, ≤1 day) the hardening is caused by formation of $MgZn_2$ needle- and T-like $MgZn_2/Mg_2Sn$ particles. The near-grain boundary zone is harder than the grain matrix, due to large round $MgZn_2$ and Mg_2Sn particles formed at grain boundaries. Increasing the aging was found to decrease the hardening in the matrix as well as in the near-grain boundary zones due to the dissolution of $MgZn_2$ needles and the coarsening of T-like particles. Substantial microhardness decrease in the near-grain boundary zone (from 70 to ~30 HV) found at low loads (10 gr) was connected with the formation of PDZ. This was confirmed by TEM and SEM studies. Further aging, at 225°C for 1-8 days, leads to the formation of "crusts" of enlarged T-like particles around depletion zones. As a result, the microhardness of PDZ's measured at higher loads (25 and 50 gr) increases up to ~ 60÷70 HV close to the ones measured in the grain matrix.

INTRODUCTION

Extensive research work has been devoted to Mg-based alloys strengthened by precipitation hardening [1-5]. The Mg-Zn-Sn system is considered a promising candidate for a creep resistant Mg-alloy for the automotive industry. However, due to overaging, these alloys exhibit poor structural stability at elevated temperatures. All age-hardening alloys contain regions adjacent to the grain boundaries which are depleted of precipitates (PDZ's) or precipitate free zones (PFZ's). This phenomenon is detrimental to the material since wide PFZs may affect the mechanical and corrosion properties of the alloy. Increasing the aging time leads to the widening of PDZ's. This is a diffusion controlled process which can be explained by near-grain boundary (NGB) coarsening [6]. The evolution of PFZ's and the development of a PDZ between the PFZ and the inner part of the grain are believed to have a considerable effect on the mechanical properties of alloys because of microcracks generated inside the PFZ due to the preferential deformation of this zone [7].
It was previously found that the aging of the solution treated specimens of Mg-Zn-Sn-alloys at elevated temperatures in the range of 175-250°C is accompanied by precipitation of the binary phases $MgZn_2$ and Mg_2Sn [8]. Plate-like Mg_2Sn-particles are often formed on the lateral faces of $MgZn_2$-needles near their tips, perpendicular to the needle's axis, forming the T-shape particles [9,10]. The precipitation of $MgZn_2$-needles at grain- and sub-grain boundaries was revealed by high-resolution TEM after aging for 1 hour at 175°C. Zones depleted of precipitates near grain boundaries were observed even after 2 hours of aging at 225°C [11]. PDZ's in Mg-Zn-Sn-alloys aged at different temperatures for different times were experimentally studied in the present work by SEM and TEM and by measuring the microhardness of near grain boundary zones at low loads.

EXPERIMENTAL DETAILS

Pure Magnesium of 99.98% was melted in a cemented graphite crucible under protective atmosphere of 1 liter/min CO_2 and 60cc/min CHF134A gas mixture. 99.8% pure zinc and 99.95% pure Sn and Y were added to the melt. The melt was poured at a temperature of 720°C into a steel disc shaped mold of 60 mm in diameter and 9 mm thick.

The composition of casting is Mg-4.01%Sn-4.47%Zn (Dirats Lab USA). Samples for solution treatment were encapsulated in a quartz tube filled with Ar at a pressure of 53.3kPa. The solution treatment includes the following steps: 96h exposure at 300°C, heating to 440°C at a rate of 1°C/h and holding at this temperature for additional 96h, followed by water quenching. Aging was conducted in a molten salt (sodium nitrate 50% + potassium nitrate 50%) in the temperature range of 175-225°C for 1 to 8 days.

EXPERIMENTAL RESULTS

SEM and TEM images of solution treated and aged samples (Figs.1-3) reveal the formation of near grain boundary zones depleted of precipitates (PDZ's). These zones were clearly observed after 1 day of aging at 175°C. The width of the depleted zones increases with temperature and aging time. It was measured by SEM EDS using the linear method of concentration non-uniformity [6] and is present in Table I. The outer borders of the depleted zones are adjoined by bands of enlarged precipitates having density smaller than that in the middle of the grains (Figs. 2c, 3). Large precipitates are located at grain boundaries.

Table I. Average width of PDZ's after different aging periods

t [days]	w [μm]		
	175°C	225°C	250°C
1	0.6	2.3	3.5
2	0.83	3.4	5.8
4	1.3	4.7	-
8	1.62	5.6	-
12	1.75	7.1	-

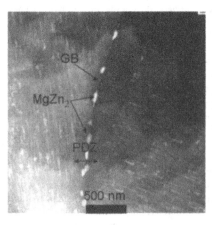

Figure 1. STEM micrographs of Mg-Sn-Zn alloy after aging for 1 day at 175°C. Different orientations of $MgZn_2$-needles across the grain boundary are seen; the boundary contains round $MgZn_2$ and Mg_2Sn particles.

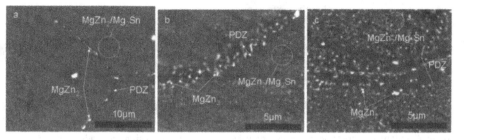

Figure 2. SEM micrographs of the Mg-Sn-Zn alloy after aging at 175°C for (a) 2 days; (b) 4 days; (c) 8 days.

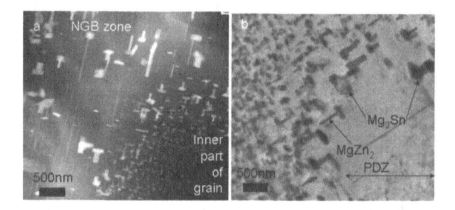

Figure 3. Microstructure of the Mg-Sn-Zn alloy: (a) STEM micrograph of the near-grain boundary zone and the adjacent inner part of grain; aging for 4 days at 225°C; (b) TEM micrograph after aging for 8 days at 175°C. Zone depleted of precipitates near the grain boundary, adjoined by enlarged T-shaped precipitates with density smaller than that in the center of the grain.

The microhardness of the aged samples was measured with different loads, ranging between 10 to 50 gr. The indents were arranged on a straight line, across a grain, perpendicular to the grain boundary so that the distance between the indents varied from ~20 μm for a load of 10 gr to ~40 μm for a load of 25 gr and to ~60 μm for a load of 50 gr (Fig. 4). The measurements were performed in 5 grains (with a size range of 100-150μm) for every load.

The microhardness measurements were made after aging at 175°C and 225°C for 1, 2, 4 and 8 days with different loads (Figs.5-7). It was found that at the low load (10gr), the microhardness value depends on the distance from the grain boundary while the values measured at higher loads (25 and 50 g) are almost independent on the position of the indents (Fig.5). Inside the body of grains, the microhardness decreases and then increases slightly with the aging time, while it substantially decreases in the NGB zones (Figs. 5, 6a, 7a).

307

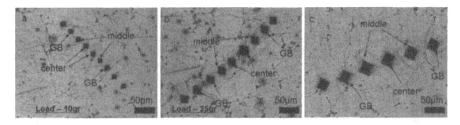

Figure 4. Indenter impression of the microhardness measurements using the loads of 10gr (a), 25gr (b) and 50gr (c).

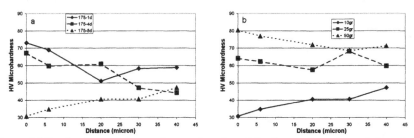

Figure 5. Microhardness of the samples as a function of the distance from the grain boundary; (a) the samples aged for 1, 4 and 8 days at 175°C measured with a load of 10gr; (b) the samples aged for 8 days at 175°C measured with different loads.

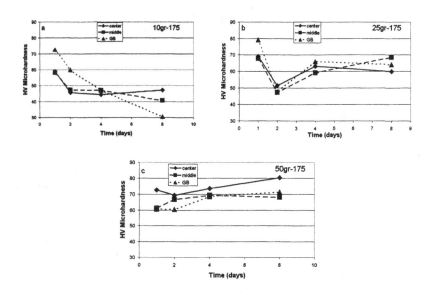

Figure 6. Microhardness of the samples aged at 175°C as a function of time measured with different loads (a) 10gr; (b) 25gr; (c) 50gr.

308

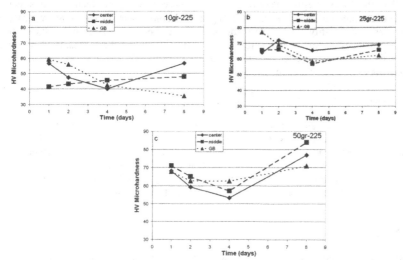

Figure 7. Microhardness of the samples aged at 225°C as a function of time measured with different loads (a) 10gr; (b) 25gr; (c) 50gr.

DISCUSSION

The aging of solution treated and quenched Mn-Sn-Zn alloys leads to homogeneous precipitation of the hcp-$MgZn_2$ needles (possibly on vacancies or vacancy-Sn clusters) in the body of grains as well as heterogeneous nucleation on grain boundaries. First $MgZn_2$ needle-shaped particles were detected after aging for 1 hour at 175°C [8]. The first observable Mg_2Sn-particles were found after 16 h aging at 175°C in the vicinity of $MgZn_2$-needle tips. After 1 day aging they are clearly seen as part of the T-shape particles (Fig.1). During further aging, both single $MgZn_2$-needles and $MgZn_2+Mg_2Sn$ T-shaped particles grow and undergo coarsening. Along with this, narrow precipitate free zones (PFZ's) are formed near grain boundaries. The formation of PFZ's is explained mainly by two theories. One is the vacancy depletion theory, that suggests that the depleted vacancy concentration near grain boundaries is smaller than a certain critical vacancy concentration that is required for nucleation [12,13]. The other is the solute depletion theory that suggests that depleted solute concentrations occur near grain boundaries, resulting in a decreased supersaturation required for precipitation [14]. The early formation of $MgZn_2$-needles at grain boundaries is compatible with the two theories. Increasing the aging time and the aging temperature leads to the development of a precipitate depleted zone (PDZ) between the PFZ and the inner part of the grain (Table 1). The development of PDZ's was explained by an NGB coarsening model taking into account the diffusion of the alloying elements from the precipitates toward the grain boundary and diffusional exchange between adjacent precipitates [6].

The microhardness values measured with 25 and 50gr loads showed weak dependence on the aging time in the range of 1 to 8 days with a slight minimum after 2-4 days of aging (Figs. 5-7) in agreement with the previous results [2, 9,10]. However, at the low load of 10gr, a substantial decrease in the microhardness, from ~70 HV to ~30 HV, was found in the NGB zone after 8 days (Figs.5, 6a, 7a). This decrease may be connected with the widening of the PDZ's near grain boundaries.

The width of the PDZ's ($< 2\mu m$) measured by SEM EDS seems to be too small, in comparison with the size of the indenter trace ($\sim 12\mu m$ for a 10gr load) in order to influence the microhardness value. However, it may testify that the plastic deformation can be concentrated on the NGB zone depleted of precipitates, in particular, due to a "crust" formed by enlarged particles. By this way the PDZ's may have a considerable effect on the mechanical properties of the alloys, in particular, due to the generation of microcracks in the NGB zone.

CONCLUSION

Zones depleted of $MgZn_2$ and Mg_2Sn precipitates near grain boundaries were observed regardless of the aging conditions used. It was found that these zones grow with time and temperature. The outer borders of the depleted zones are adjoined by bands of enlarged T-shaped particles. Large precipitates are also located at grain boundaries. Microhardness tests with low loads showed a significant decrease in the microhardness values in the NGB zone that can be related to the preferential plastic deformation of this zone followed by the formation of microcracks.

REFERENCES

1. S. Cohen, G.R. Goren-Muginstein, S. Abraham, G. Dehm and M.Bamberger in Magnesium Technology, ed. by Alan A. Luo (TMS annual meeting Proceed., San Diego, CA, 2003) pp. 301-304.
2. S. Cohen, G.R. Goren-Muginstein, S. Abraham, B. Rashkova, G. Dehm and M.Bamberger, Z. Metallkd. **96** ,1081 (2005).
3. B. Rashkova, J. Keckes, G. Levi, A. Gorny, M. Bamberger and G. Dehm, in Magnesium, edited by K.U. Kainer (7th Int.Conf. on Magnesium Alloys and Their Applications Proc., DGM, Dresden, Germany, 2006) pp. 486-489.
4. T.T. Sasaki, K. Oh-ishi, T. Ohkubo and K. Hono, Scripta Mater., **55**, 251 (2006).
5. C.L. Mendis, C.J. Bettles, M.A. Gibson, C.R. Hutchinson, Mater.Sci.Engineer., **A435-436**, 163 (2006).
6. A. Katsman, A. Gorny, D. Shepelev, M. Bamberger, presented at the DIMAT 2008 Conference, Lanzorete, Spain, 2008 (in press).
7. Ogura Tomo, Hirosawa Shoichi, Sato Tatsuo, Keikinzoku, **56**, 644 (2006).
8. A. Gorny, A. Katsman, D. Shepelev and M. Bamberger, in Thermodynamics and Kinetics of Phase Transformations in Inorganic Materials, edited by C. Ambromeit, P. Bellon, J-L. Bocquet, D.N. Seidman (Mater. Res. Soc. Symp. Proc. **979E**, Warrendale, PA) (2007), published online.
9. A. Gorny, A. Katsman, I. Popov and M. Bamberger, in Magnesium Technology 2007, edited by Randy S. Beals, Alan A. Luo, N.R. Neelameggham and M.O. Pekguleryuz (TMS Proc., Orlando, Florida, USA, 2007), pp.307-312.
10. A.Gorny and A.Katsman, J. Mater. Res., **23**, 5, 1228 (2008).
11. A. Gorny, M. Bamberger and A. Katsman, J. Mater. Sci., **42**, 10014 (2007).
12. C. R. Shastry and G. Judd, Metal.Mater.Trans., **B2**, 3283 (1971).
13. Tomo Ogura, Shoichi Hirosawa and Tatsuo Sato, Sci.Tech.Adv. Mater., **5**, 491 (2004).
14. L. S. Bushnev, Russ.Phys.Journal, **22**, 405 (1979).

Mater. Res. Soc. Symp. Proc. Vol. 1128 © 2009 Materials Research Society   1128-U05-49

Electron Irradiation Induced Crystal-to-Amorphous-to-Crystal (C-A-C) Transition in Intermetallic Compounds

Takeshi Nagase[1,2], Kazuya Takizawa[2], Akihiro Nino[3] and Yukichi Umakoshi[4]
[1]Research Center for Ultra-High Voltage Electron Microscopy, Osaka University, 7-1, Mihogaoka, Ibaraki, Osaka 567-0047, Japan
[2]Division of Materials and Manufacturing Science, Graduate School of Engineering, Osaka University, 2-1, Yamada-oka, Suita, Osaka 565-0871, Japan
[3]Department of Materials Science and Engineering, Faculty of Engineering and Resource Science, Akita University, Tegatagakuen-machi, Akita, Akita 010-8502, Japan
[4]National Institute for Materials Science, 1-2-1, Sengen, Tsukuba, Ibaraki, 305-0047 Japan

ABSTRACT

Mechanical atomic displacement in intermetallic compounds can be introduced by various processes such as severe plastic deformation, mechanical milling and electron irradiation, resulting in the crystal-to-amorphous (C-A) transition and amorphous-to-crystal (A-C) transition in some alloy systems. Recently, unique atomic displacement-induced-phase transitions containing both amorphization and crystallization was found: a crystal-to-amorphous-to-crystal (C-A-C) transition under electron irradiation and a cyclic C-A transition during mechanical milling. The features and predicting factors of alloy systems which undergo C-A-C transition under MeV electron irradiation are discussed based upon the analysis of experimental data reported to date in intermetallic compounds irradiated by high voltage electron microscope (HVEM).

INTRODUCTION

It is well known that some intermetallic compounds show crystal-to-amorphous (C-A) transition by using a mechanical process below the glass transition temperature (T_g). The C-A transition can be induced by various methods such as particle irradiation, severe plastic deformation, mechanical milling, hydrogen-absorption, inter-diffusion between multi-layers and so on. Among many C-A transition processes, electron-irradiation induced solid-state amorphization is an attractive method because in situ observation is available without oxidation, contamination, or change in chemical composition, and with negligible temperature increase during C-A transition [1-5]. Mori et al. reported that an amorphous state was realized in 37 among 70 electron irradiated crystalline alloys [4]. Such systematically experimental research works clearly indicate that electron irradiation is an effective technique for revealing the origin of C-A transition.

Recently a unique phase transition containing C-A transition was found in metallic glass former alloys during mechanical milling and electron irradiation techniques: cyclic crystalline-amorphous (Cyclic-CA) transformation during mechanical milling [6,7], electron irradiation induced crystal-to-amorphous-to-crystal (C-A-C) transitions [8,9], and electron irradiation induced quasicrystal-to-amorphous-to-crystal (Q-A-C) transition [10]. Figure 1 shows the typical example of C-A-C transition in $Nd_2Fe_{14}B$ polycrystalline intermetallic compound under 2.0 MeV electron irradiation at 298 K [9]. Changes in bright field (BF) images and selected area diffraction (SAD) patterns of the $Nd_2Fe_{14}B$ compound can be detected in detail by in situ HVEM observation. The specimen before irradiation shows a typical poly-crystalline structure with a grain size of about 50 nm in Figure 1a. The $Nd_2Fe_{14}B$ compound can not maintain its original structure under 2.0 MeV electron irradiation, C-A transition occurs after 15 s of electron irradiation. After amorphous phase formation, the amorphous area spreads with increasing irradiation time. In the central area of electron irradiation, an amorphous phase can maintain a glassy structure until the irradiation time reaches 180 s. At this time, nano-crystallization of the amorphous phase, namely, amorphous-to-crystal (A-C) transition occurs in the central area of electron irradiation as indicated by a white arrow in Figure 1g, and the nano-crystalline area spreads with increasing irradiation time. The nano-crystalline structure composed of b.c.c.-solid solution, defective crystalline compounds and residual amorphous matrix is formed through C-A-C transition.

In the present study, the features and predicting factors of metallic alloy systems which show electron irradiation induced C-A-C transition are discussed based upon the analysis of the experimental data reported to date in intermetallic compounds irradiated by HVEM.

DATA-BASE AND EXPERIMENTAL PROCEDURE

Experimental data of MeV electron irradiation in 84 intermetallic compounds are listed in Table 1 which is constructed by the previous papers [4, 11-13]. The first column indicates the intermetallic compounds, the second column indicates the Strukturbericht symbol and crystal system as the crystalline structure, the third column indicates the occurrence of electron irradiation induced amorphization, the fourth and fifth columns indicate the existence of C-A-C transition and the reference, respectively.

In most cases the irradiation experiments have been carried out by a research group at Osaka University using the ultra-high voltage electron microscope (UHVEM) of HU-2000 and H-3000 in Osaka University. Most of intermetallic compounds were irradiated with 2.0 MeV electrons at the temperature between 4.2 K and 298 K. The dose rate was fixed at approximately $1 \times 10^{24} \, m^{-2} s^{-1}$.

RESULTS AND DISCUSSION

49 of the 84 intermetallic compounds in Table 1 undergo a C-A transition under electron irradiation, while the other 35 remain crystalline. This indicates that electron irradiation induced C-A transition of intermetallic compounds is a phenomenon of wide generality in metallic materials.

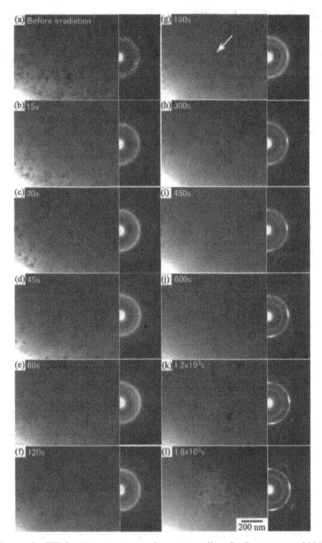

Figure 1. Change in TEM microstructure and corresponding SAD patterns of $Nd_2Fe_{14}B$ inter-metallic compound during 2.0 MeV electron irradiation at 298 K at the dose rate of $1.0 \times 10^{25} m^{-2} s^{-1}$. $Nd_2Fe_{14}B$ crystalline phase was obtained by melt-spinning method in $Fe_{82.3}Nd_{11.8}B_{5.9}$ alloy. (a) before electron irradiation, (b) 15 s, (c) 30 s, (d) 45 s, (e) 60 s, (f) 120 s, (g) 180 s, (h) 300 s, (i) 450 s, (j) 600 s, (k) 1.2×10^3 s, (l) 1.8×10^3 s. [9,14]

Table 1. Occurrence of C-A and C-A-C transition in intermetallic compounds under MeV electron irradiation. The list is constructed based on Ref. 4, 11, 12 and 13.

(a) Metal-Metal system

Compounds	Structure	C-A	C-A-C	References
Al_9Co_2	(monoclinic)	Yes		
Al_7Cr	(monoclinic)	Yes		
Al_5Cr	(monoclinic)	Yes		
Al_4Cr	(monoclinic)	Yes		
Al_3Fe	(monoclinic)	Yes		
Al_6Mn	$D2_h$	Yes		
Al_4Mn	(orthorhombic)	Yes		
$Al_{10}V$	(cubic)	Yes		
$Al_{45}V_7$	(monoclinic)	Yes		
$Al_{23}V_4$	(hexagonal)	Yes		
Al_2Zr	C14	Yes		
Al_3Zr_2	(orthorhombic)	Yes		
$AlZr$	B_f	Yes		
Al_4Zr_5	(hexagonal)	Yes		
Al_3Zr_4	(hexagonal)	Yes		
Al_2Zr_3	(tetragonal)	Yes		
$AlZr_2$	C16	Yes		
Co_2Ti	C15	Yes		
Cr_2Zr	C15	Yes		
Cu_3Ti_2	Dl_3	Yes		
Cu_4Ti_3	(tetragonal)	Yes		
$CuTi$	B11	Yes		
$CuTi_2$	$C11_b$	Yes		
$Cu_{10}Zr_7$	(orthorhombic)	Yes		
$CuZr$	B2	Yes		
$CuZr_2$	$C11_b$	Yes	Reported	8, 9
$Fe_{17}Nd_2$	(hexagonal)	Yes	Reported	14
Fe_2Ti	C14	Yes		
$FeZr_2$	C16	Yes		
$FeZr_3$	$E1_a$	Yes		
Mn_2Ti	C14	Yes		
$MoNi$	(orthorhombic)	Yes		
Nb_7Ni_6	$D8_5$	Yes		
$NiTi$	B2	Yes		
$NiTi_2$	(cubic)	Yes		
Ni_3Zr	$D0_{19}$	Yes		
$NiZr$	B_f	Yes		
$NiZr_2$	C16	Yes	Reported	15
$PdZr_2$	$C11_b$	Yes	Reported	9
Pt_3Zr_5	$D8_8$	Yes		
$ZrAlNi$	(hexagonal)	Yes		
Al_2Au	C1	No		
Al_5Co_2	$D8_{11}$	No		
Al_9Cr_4	(cubic)	No		

Continued, Metal-Metal system

Compounds	Structure	C-A	C-A-C	References
Al_8Cr_5	$D8_{10}$	No		
Al_2Cu	C16	No		
$AlCu(\eta_2)$	(monoclinic)	No		
Al_5Fe_2	(orthorhombic)	No		
Al_2Fe	(triclinic)	No		
$AlFe$	B2	No		
Al_3Mn	(orthorhombic)	No		
$Al_{11}Mn_4$	(triclinic)	No		
$AlMn(\gamma_2)$	$D8_{10}$	No		
$Al_{12}Mo$	(cubic)	No		
Al_8Mo_3	(monoclinic)	No		
Al_3Ni	$D0_{11}$	No		
Al_3Ni_2	$D5_{13}$	No		
$AlNi$	B2	No		
$AlNi_3$	$L1_2$	No		
Al_3Ti	$D0_{22}$	No		
Al_3V	$D0_{22}$	No		
Al_8V_5	$D8_2$	No		
$Al_{12}W$	(cubic)	No		
Al_3Zr	$D0_{23}$	No		
$AlZr_3$	$L1_2$	No		
$CoTi$	B2	No		
Cr_2Ti	C15	No		
Cu_4Ti	(orthorhombic)	No		
$FeTi$	B2	No		
Ni_3Ti	$D0_{24}$	No		

(b) Metal-Metalloid system

Compounds	Structure	C-A	C-A-C	References
BCo_2	C16	Yes		
BCo_3	$D0_{11}$	Yes		
BFe_3	$D0_e$	Yes		
BNi_2	C16	Yes		
B_3Ni_4 (o)	(orthorhombic)	Yes		
$Fe_{23}Nd_2B_3$	(cubic)	Yes	Reported	14
$Fe_{14}Nd_2B$	(tetragonal)	Yes	Reported	9, 14
$Fe_{81}Zr_9B_{10}$		Yes	Reported	16
BCo	B27	No		
BFe_2	C16	No		
BNi	B_f	No		
BNi_3	$D0_{11}$	No		
B_3Ni_4 (m)	(monoclinic)	No		
$Fe_4Nd_{11}B_4$		No		

The electron irradiation induced C-A-C transition was found in 7 intermetallic compounds; $C11_b$-Zr_2Cu, $C16$-Zr_2Ni, $C11_b$-Zr_2Pd, tetragonal-$Nd_2Fe_{14}B$, cubic-$Nd_2Fe_{23}B_3$, $Fe_{17}Nd_2$ and α-Mn type $Fe_{81}Zr_9B_{10}$. It can be concluded that not only C-A transition but also C-A-C transition is phenomena is of wide generality in metallic materials. C-A-C transition can be seen in various intermetallic compounds such as b.c.c.-based $C11_b$, $C16$ and other intermetallic compounds of complex structure regardless of their crystal structure. From the view points of metallic glass formation by liquid quenching, the occurrence of C-A-C transition can be seen in both metal-metal type and metal-metalloid types of compounds. These parameters therefore cannot be thought to be the predicting parameters of C-A-C transition. A map of chemical mixing enthalpy (ΔH_{chem}) versus mismatch entropy (S_σ/k_B) map [17,18] is more effective for predicting the alloy composition for which metallic amorphous phase can be obtained by liquid quenching. This map is also effective to predict the occurrence of electron irradiation induced C-A transition [11]. The relationship between the occurrence of the C-A-C transition and the ΔH_{chem} - S_σ/k_B map is shown in Figure 2. The intermetallic compounds which undergo C-A transition show the tendency to have large negative ΔH_{chem} and large positive S_σ/k_B. All intermetallic compounds with $\Delta H_{chem} <$ -35 KJ/mol show electron irradiation induced C-A transition. In contrast, the tendency of the occurrence of C-A-C transition can not be seen in Fig. 2. The features and predicting factors for unique disordering-ordering of C-A-C transition in alloy systems would need the accumulation of more systematic experimental data.

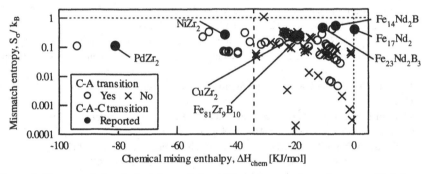

Figure 2. The map of chemical mixing enthalpy (ΔH_{chem}) versus mismatch entropy (S_σ/k_B) map. [11]

CONCLUSIONS

The features and predicting factors of alloy systems which show electron irradiation induced crystal-to-amorphous-to-crystal (C-A-C) transition are discussed based upon the analysis of the present experimental data. The following conclusions can be drawn:
(1) C-A-C transition can be observed in 7 intermetallic compounds, indicating that this transition

is of comparatively wide generality in metallic crystals. (2) Whereas a map of chemical mixing enthalpy (ΔH_{chem}) versus mismatch entropy (S_σ/k_B) map give the necessary criterion of $\Delta H_{chem} < -35$ KJ/mol for C-A transition, it cannot offer an useful information for the occurrence of C-A-C transition due to the lack of systematic experimental data.

ACKNOWLEDGEMENTS

This work was supported by "Priority Assistance of the Formation of Worldwide Renowned Centers of Research -The Global COE Program- (Project: Center of Excellence for Advanced Structural and Functional Materials Design)" from the Ministry of Education, Culture, Sports, Science and Technology of Japan.

REFERENCES

1. G. Thomas, H. Mori, H. Fujita, and R. Sinclair, *Scr. Mater.* **16**, 589 (1982).
2. H. Mori, H. Fujita, *Jpn. J. Appl. Phys.* **21**, L494 (1982).
3. A. Mogro-Camprero, E. L. Hall, J. L. Walter and A. J. Ratkowski, in *Metastable Materials Formation by Ion Implantation*, edited by S. T. Picraux and W. J. Choyke, (Mater. Res. Soc. Proc. 7, Boston, MA, 1981) pp. 203-208.
4. H. Mori, "Current Topics in Amorphous Materials," *Physics and Technology*, ed. Y. Sakurai, Y. Hamakawa, T. Masumoto, K. Shirac and K. Suzuki (Elsevier, 1997) pp. 120-126.
5. P. R. Okamoto, N. Q. Lam and L. E. Rein, "Physics of Crystal-to-glass transformations," *Solid State Physics, vol. 52*, ed. H. Ehrenreich and F. Spaepen (Academic Press, 1999) pp. 1-135.
6. M. S. El-Eskandarany, K. Aoki, K. Sumiyama and K. Suzuki, *Appl. Phys. Lett.* **70**, 1679 (1997).
7. M. S. El-Eskandarany, K. Aoki, K. Sumiyama and K. Suzuki, *Scr. mater.* **36**, 1001 (1997).
8. T. Nagase and Y. Umakoshi, *Scr. Mater.* **48**, 1237 (2003).
9. T. Nagase, A. Nino, T. Hosokawa and Y. Umakoshi, *Mater. Trans.* **48**, 1651 (2007).
10. T. Nagase, T. Hosokawa and Y. Umakoshi, *Intermetallics.* **14**, 1027 (2006).
11. T. Nagase, *Materia Japan.* **47**, 519-523 (2008).
12. T. Nagase and Y. Umakoshi, *Mater. Trans.* **47**, 1469 (2006).
13. T. Nagase, A. Nino and Y. Umakoshi, *Mater. Trans.* **48**, 1340 (2007).
14. A. Nino, T. Nagase and Y. Umakoshi, *Mater. Trans.* **48**, 1659 (2007).
15. T. Nagase, K. Takizawa, M. Nakamura, H. Mori and Y. Umakoshi, *J. Phys.* Conference series, (Inter. Conf. on ASFMD2008, Osaka, Japan, 2008) (to be submitted).
16. T. Nagase and Y. Umakoshi, *ISIJ*, **46**, 1371 (2006).
17. A. Takeuchi and A. Inoue, *Mater. Trans.* **41**, 1372 (2000).
18. A. Takeuchi and A. Inoue, *Mater. Trans.* **46**, 2817 (2005).

Mater. Res. Soc. Symp. Proc. Vol. 1128 © 2009 Materials Research Society   1128-U05-50

An Intermetallic Precipitate Strengthened Alloy for Cryogenic Applications

Ke Han, Yin Xin and Robert P. Walsh

National High Magnetic Field Laboratory, Florida State University, 1800 E. Paul Dirac Drive, Tallahassee, Florida, 32310, USA

ABSTRACT

In high field magnets, most of the structural materials operate at very low temperatures (e.g. 1.8 K) and high magnetic fields (e.g. >20 T). Such an extreme environments ask for demanding properties from the materials. Austenite stainless steels have been used for such an application for decades and provide good combination of mechanical strength and toughness at cryogenic temperatures. In the National High Magnetic Field Laboratory, we have studied intermetallic precipitate strengthened alloys for high field magnet and cryogenic applications. An examination of the data from an intermetallic precipitate strengthened alloys shows that intermetallic precipitate strengthened alloys can be used at cryogenic temperatures and have some superior properties in extreme environments. Based on our experimental data, we developed a new alloy that has even better properties than the current intermetallic precipitate strengthened alloys.

INTRODUCTION

Austenitic stainless steels have been used for cryogenic applications (mainly at 4 K or below) for decades because of their good combination of the mechanical strength and toughness at cryogenic temperatures. The structural materials in high field magnets, such as the materials in fusion magnets and high field research Series-Connected Hybrid (SCH) magnets [1], also operate in high fields. Significant numbers of stainless steels will have phase transformation at a high filed, therefore are not suitable for such applications. In addition, the structural materials, such as the conduit alloy for Cu-Nb-Sn wires in high field magnets, are exposed to a heat treatment at temperatures between 600 -700 °C for more than 24 hours in order to transform the precursor to Cu/CuSn/Nb$_3$Sn composites. At such a heat treatment condition, most of the stainless steels are sensitized and become brittle at cryogenic temperatures. Low carbon 316 LN (denoted as 316LLN in this paper) is one of the austenitic stainless steels that can work in such an extreme environment: cryogenic temperatures, high magnetic fields and experience in sensitization temperature range for a long time. The low carbon content retards the formation of the grain boundary carbides and makes the materials less vulnerable to sensitizations than commercial 316LN. However, their yield strength in annealed status of 316 LLN is limited to a range of 1000 MPa at 4 K partially due to their low carbon content. In addition, the intermetallic compounds may form at grain boundaries at certain temperature ranges and make the materials less ductile than the one heat treated at optimized heat treatment temperatures. The researchers in the national high magnetic field laboratory (NHMFL) studied other materials for cryogenic applications and focused the efforts on Haynes 242 alloy as a candidate alloy for cryogenic applications [2] [3]. The NHMFL research on the cryogenic mechanical and physical properties of Haynes 242 [14] is fruitful. The research is also expanded to evaluate the microstructure and

TABLE 1. Chemical composition of alloys investigated in wt.%

Alloy	C	N	Mn	Si	Cr	Ni	Mo	Nb+Ta	Co	Cu	Fe
316LN	0.012	0.15	1.58	0.36	17.6	13.07	2.23	0.1	0.06	0.03	Bal
H 242	0.01	---	0.3	0.15	7.82	Bal	25.13	---	0.04	0.02	1.03

mechanical property evolution as a result of the processing because of the phase transformations that occur during a special heat treatment for Nb_3Sn conductors. The processing includes both the deformation and the heat treatment at temperatures where sensitization in austenite stainless steels occurs. The materials are received at the NHMFL in either sheet or round tube forms and subsequently shaped into rectangular form with the superconductor cable inserted and then wound into coils. This fabrication procedure introduces deformation strains at room temperature. The estimated strains for shaping are about 10-15%. The maximum bending strain caused by coil winding is about 3%. The deformed materials are then heat treated in the temperature range of 650°C -700°C for 50-100 hours. A combination of deformation and heat treatment was also performed in order to study the stress accelerated grain boundary oxidation (SAGBO) [5]. The heat treatment promotes the formation of intermetallic compound precipitates in both 316LLN and Haynes 242 alloys. In 316LLN, the intermetallic compounds were found to form at the grain boundaries whereas in Haynes 242, they form inside the grains as an ordered phase. This paper reviews some of our activities on a) providing an insight to understand the possibility of occurrence of sensitizations, and SAGBO in these alloys, b) identifying the pre-stress contribution to the formation of the intermetallic compounds and to the performance of the materials at 4 K, c) microstructural examinations of the materials exposed to extreme enviroments, and d) comparison of the mechanical properties tested at room temperatures and 4 K after the materials were heat treated under the stresses.

EXPERIMENTAL

The chemistries of two alloys are listed in Table 1. However, this paper focuses on the results from samples H242.

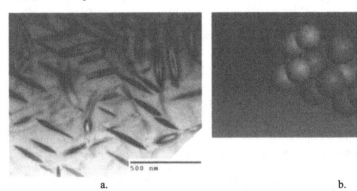

a.   b.

FIGURE 1. Microstructure and crystallographic structure of Ni_2Mo in Haynes 242 alloy: a. TEM bright field image; b. the arrangement of the atoms of Ni_2Mo.

Bending Test Simulations

Bending tests were done on Haynes 242. 150 mm (6") long flat strips were sectioned from conduit for our tests so that the pre-strain for the shaping from the round to rectangular conduit are included. Control samples (without pre-stress) from annealed Haynes 242 sheets were also used as a comparison. The thickness of the samples from the conduit and annealed sheet strips were 1.6 mm and 1.5 mm, respectively. The strips were bent around and secured to stainless steel mandrels of three different diameters to create three magnitudes of bending strains of 2.6%, 3.3% and 3.8%.

Two conduit strips were used at each strain level; one contained a butt weld through its mid-plane and one was the base metal. The Haynes 242 strip was added to the small diameter tube with the highest strain. The heat treatment tests were performed in air to examine if there is any stress accelerated grain boundary oxidization occurs. The heat treatment schedule was 210°C for 48 hour; then 400°C for 48 h and finally 650°C for 50 hours with ramping rate of 100°C/hour. These particular temperatures were designed for application in superconductor heat treatments.

A large scale test was undertaken to validate the small scale tests. The test used a one-meter sample wound around a half cylinder winding mandrel at the inner radius of 0.305m (12 inches; that is the radius of the outsert coil in our SCH design). The ends of the conduit will be fusion welded to the mandrel. The heat treatment was undertaken in a large Mellen furnace. Glass cloth was applied between the mandrel and conduit.

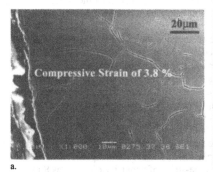 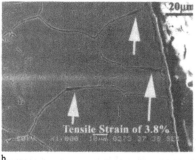

a.   b.

FIGURE 2 Scanning electron microscopy images of the samples: a) a region with a compressive strain of 3.8%; and b) a region with a tensile strain of 3.8%. The arrows indicate the large crack-like features due to the precipitation formation in tensile regions of the materials.

Tensile Strain and Heat Treatments

A special frame was designed to mount the tensile test samples. The cross bars of the frame are made with Maraging C300 steel while the axial load bearing legs were made of either 316LN SS or Haynes 242 to coincide with the test sample to be heat treated. The mount and the samples were heat treated to simulate the pre-stress + heat treatment effects on the microstructure and mechanical properties of the conduit alloys. The pre-strain was measured by a clip-on extensometer and was nominally 3%. This strain is equivalent to the maximum strain applicable to the conduit during the winding operations in the SCH at the NHMFL. Tensile tests are performed according to ASTM E1450 in displacement control mode. The pin loaded flat specimens have 38 mm gage lengths. Strain was measured with 25mm gage length extensometers for a strain below 10%. Elongation is determined from scribed marks. Engineering strength and ductility values are given in the tables for application and comparison with literature data. Scanning electron microscopy work was undertaken with a JEOL 6380 microscope operating at 15 KV. Transmission electron microscopy work was performed on a JEOL 2011 microscope operating at 200 KV.

RESULTS AND DISCUSSIONS

Microstructure in Aged Conditions

After aging at temperature range between 650C and 700 C for up to 100 hours, Ni_2Mo ordered intermetallic phase was observed in a precipitates form, as shown in Figure 1. These precipitates strengthened materials by 20-40%,

Pre-strain Effects on Microstructure and Cryogenic Properties

None of the Haynes 242 material failed during the heat treatment with a bending strain from 2.6 to 3.8%. This result is important for the material fabrications under the stresses. The materials can survive some bending strains without fracture at high heat treatment temperatures in air.

However, some microstructure changes were observed in the samples heat treated under the stresses. Figure 2 shows a comparison of the x-section SEM images of the samples heat treated at the same temperature but with the different external strains. The area in figure 2a was subjected to a compressive strain of 3.8 % whereas the one in figure 2b subjected to a tensile strain of 3.8 %. In the area of near the surface with tensile strain, some large groves were observed. Close examinations indicate large precipitates were formed within the groves. Therefore, in the experiments described in the following section, tensile samples were prepared with tensile strains.

In this experiment, Haynes 242 samples were strained in a fixture before being exposed to a high temperature heat treatment. Only tensile strain was applied to the samples. After the heat treatment (denoted as H in Figures 3 and 4) under a pre-strain of about 3% (denoted as S in Figures 3 and 4), the cryogenic mechanical properties changed, as shown in Figures 3 and 4. Heat treatment with a pre-strain of 3% reduces the tensile strength and the ductility of base metal (denoted as B in the figure) of the Hayes 242, as shown in Figure 3. From the stress-strain curves in Figure 3, it appears that the strain-hardening rates were not affected by the straining during the heat treatments. The reduction of the mechanical properties is probably due to the grain boundary precipitate formation and change of the intermetallic Ni_2Mo precipitates orientation within the grains. In the tensile side of the samples, we indeed observe relatively larger grain boundary precipitates than in compressive sides. The heat treatment with a pre-strain of 3% also reduces the yield strength and tensile strength of the welded portion of the Haynes 242, but increase the ductility slightly.

320

Residual Stress in the Conduit of Haynes 242

Full-sized conduit samples were shaped prior to heat treatment. After the heat treatment, residual stresses may be developed in the materials that introduce internal strain. The internal strain can be detected by measurement of the lattice parameter shifts that are measurable by X-ray diffraction techniques. The total strain amplitude after heat treatment is estimated to be 0.3% (the difference between 0.13% and -0.17%) [6]. This is about 1/20th of the original strain amplitude (6%; which is 2 times of the maximum anticipated winding strain) applied to the conduit. Since Young's modulus for Haynes 242 is about 230 GPa [7] one can estimate the residual stress amplitude to be 680 MPa. The strain amplitude can be divided into both tensile and compressive portions. If both the tensile and compressive strains or stresses have the equivalent absolute values, the maximum tensile residual stress is about 340 MPa. The design of the magnet should take this residual stress into account since it is a significant portion of the

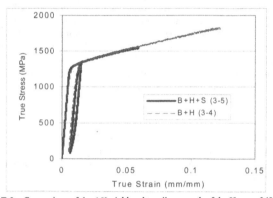

FIGURE 3a. Comparison of the 4 K yield and tensile strength of the Haynes 242 conduit materials after the materials were heat treated with (solid thick line) and without (thin broken line) 3% strain.

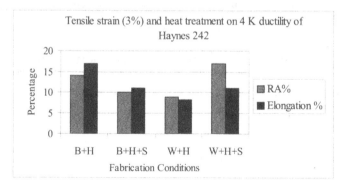

FIGURE 4. Comparison of the 4 K reduction-in-area at fracture and elongation of the Haynes 242 conduit materials after the materials were heat treated with 3% strain.

321

materials 0.2% offset yield strength (1240 MPa). In other words, one needs only to apply 900 MPa tensile stress to reach the flow stress of the material and cause permanent deformation of the conduit.

New Alloy Developments

To further enhance the strength of the materials, high modulus elements, such as Re, was added to existing Hayes alloys with adjustment of the chemistry. The new alloys need as short as 4 minutes for formation of the intermetallics to strengthen the matrix. At the same, the materials have a very high resistance to over-aging. At the same time, the materials indeed have the high modulus and high strength.

CONCLUSIONS

The Nb_3Sn reaction heat treatment of base metal of Haynes 242 shows a decrease in yield strength, tensile strength and ductility after the Nb_3Sn reaction heat treatment under an applied strain of 3%. The welded Haynes 242 samples also show decrease in yield strength and tensile strength and increases of the ductility after heat treatments under an applied strain of 3%.

For Haynes 242, tensile strain promotes the grain boundary precipitates but have no significant impact on the intermetallic compound precipitation inside the grains. The bending strain from winding of conduit up to 6% will be relaxed by a factor of 20 after heat treatment. The residual stress amplitude of up to 680 MPa can be introduced and should be taken into account in the design of CICC magnets that use Haynes 242.

ACKNOWLEDGMENTS

We thank L.T. Summers, J.R. Miller, A. Bonito-Oliva and M. Bird for helpful discussions and supports, L. Marks, and R. Stanton for performing the welding experiments. Financial supports of the National Science Foundation under grant of DMR-0084173 and the State of Florida are gratefully acknowledged.

REFERENCES

1 I. R. Dixon, M.D. Bird, and J.R. Miller, *IEEE Trans. on Appl. Supercon.*, **16,(2)**, 981 (2006).
2 M. Kumar and V.K. Vasudevan, *Acta Mater.* **44**, 4865 (1996).
3 R.P. Walsh, V.J. Toplosky, K. Han and J.R. Miller, *Advances in Cryogenic Engineering (Materials)* **52A**, 107 (2006).
4 J. Lu, K. Han, E.S. Choi, , Y. Jo, L. Balicas, and Y. Xin, *J. Appl. Phys.*, **101**, 123710 (2007)
5 L.S. Toma, M.M. Steeves, R.P. Reed, *Advances in Cryogenic Engineering*, **40B**, 1291, (1994).
[6] K. Han, R.P. Walsh, J. Toplosky, R.E.Goddard, J. Lu, and I. R. Dixon, *Advances in Cryogenic Engineering*, **54**, 84 (2008)
7 R.P. Walsh, V.J. Toplosky, and J.R. Miller, *Advances in Cryogenic Engineering*, **50**, 145, (2004).

Mater. Res. Soc. Symp. Proc. Vol. 1128 © 2009 Materials Research Society   1128-U05-52

Direct Modeling of the Simultaneous Flow of Compressible Atomizing Gas Jets and a Weakly Compressible Liquid Intermetallic Stream During Gas Atomization

Mingming Tong, David J. Browne
Engineering and Materials Science Centre, University College Dublin, Belfield, Dublin 4, Ireland

ABSTRACT

The authors have developed a new atomization model enabling direct numerical simulation of the simultaneous flow of compressible atomizing gas jets and a weakly compressible liquid metal stream. It has been used to simulate the atomization of a Ni-50wt.%Al melt stream by argon gas jets in a closed-coupled atomizer. The 2D simulation results show that the presence of the liquid intermetallic stream significantly influences the field variables, particularly the aspiration pressure. At the gas plenum pressure used, the gas nozzles are choked and hence the gas flow upstream of the tip of the liquid delivery tube is not influenced by the presence of the liquid intermetallic stream, whereas the downstream gas flow is affected. Significant differences between model predictions assuming either incompressible or compressible gas are reported. Besides the atomization of liquid intermetallic stream by argon gas, this unified atomization model is available for use to simulate a variety of different twin-fluid atomization processes.

INTRODUCTION

Due to the high catalytic activity caused by their high specific area, porous Raney nickel powders [1] have been widely used in industry. Gas atomization is a very efficient route to fine, clean, and spherical catalytic powders directly from the melt. The hydrodynamic interaction between the melt stream and impinging atomizing gas jets is a key issue. Based on the experience of the authors on front-tracking formulation [2-5], they developed an atomization model [6-7] for direct numerical simulation of the forced disintegration of a continuous Ni-Al melt stream by gas jets during the early stages of the process, based on the assumption that the gas is incompressible. In order to improve the accuracy of the model and enable more realistic simulations for the industry, the compressibility of gas needs to be considered. Although the flow of compressible atomizing gas jets has been investigated by many researchers using numerical modelling and simulation [8-15], such numerical models are only capable of handling compressible gas flow in the gas-only case, i.e. ignoring the interaction with the liquid metal.

The authors have developed a new compressibility model [16] for 2D studies of gas-only flow in a typical atomizer. This model also has the potential to simulate the simultaneous flow of compressible fluid and weakly compressible fluid. In this paper, based on front-tracking [7] and gas compressibility [16], the authors report on their development of a novel unified atomization model. This model can be used to directly simulate the interaction between the compressible atomizing gas jets and weakly compressible liquid metal stream during their simultaneous flow in the atomization process. The practical significance of the model is that it can capture the initial transients which will develop to become the quasi-steady state behavior in real industrial atomizers.

NUMERICAL MODEL

The unified atomization model uses a pressured-based projection type Eulerian-Lagrangian method. The main function of the model is to deal with the simultaneous flow of significantly compressible gas and a weakly compressible liquid intermetallic stream within a one-fluid formulation. Both gas and liquid are regarded as a single fluid with spatially variable material properties and compressibility. The gas-liquid two-phase flow is dominated by a single set of governing equations, although the properties (such as compressibility and specific heat) of respective phases are significantly different. Space limitations preclude the reproduction of the governing equations of continuity, momentum and energy conservation here.

The model can be reduced two sub-models. One is related to the compressibility of fluids, the other to the tracking of inter-fluid interfaces. Details of the models and particularly the relevant verification and validation are presented elsewhere [7, 16].

The compressibility model is used to deal with the challenge of significantly different compressibility of the simultaneous flowing fluids – here gas and liquid metal. This compressibility model is based on the combined unified procedure method originally developed by Yabe and Wang [17] and further developed by other researchers [18-20], and particularly by Xiao et.al.[21]. The basic idea is introducing compressibility into the projection procedure of the pressure based method. The front-tracking model for atomization [5-7] is used to track the moving sharp interfaces between liquid metal stream and gas, in order to facilitate the identification of some material properties (such as dynamic viscosity and specific heat) which jump in value across the interfaces and to determine the surface tension forces operating on the interfaces. In this way the interaction/coupling between compressible gas and weakly compressible liquid intermetallic stream is considered.

NUMERICAL SIMULATION RESULTS AND DISCUSSION

The target of the numerical simulation is the hydrodynamic interaction between the weakly compressible, initially continuous, Ni-50wt.%Al melt stream and compressible argon gas jets during the early stage of atomization in a close-coupled atomizer. This atomizer is of a type used by CERAM, UK - a research partner within the project. The material properties of liquid Ni-50wt.%Al were obtained from key project reports [22]. The main geometrical parameters of the nozzles are shown in Fig.1a. The configuration is a vertical liquid delivery tube surrounded by discrete inclined gas nozzles. A descending Ni-50wt.%Al melt stream enters the chamber through the central liquid delivery tube and pressurized argon gas jets enter through the surrounding gas nozzles. The initial parameters and boundary conditions used in the simulation are schematically shown by Fig.1b with relevant parameters as $P_o = 3MPa$, $P_a = 1.0 \times 10^5 Pa$ and $T_o = T_a = 293K$. The inlet velocity of the liquid phase is assigned according to the mass flow rate of the Ni-Al melt. The initial temperature of the melt stream is 1830K, around 200K above the liquidus. Non-reflecting boundary conditions are applied at the gas inlet and the outlet of the simulation domain. For the sake of illustration convenience, the simulation domain is, from here on, rotated by 90 degrees such that the gravity force acts horizontally towards the right.

Figure 2 shows the temporal evolution of the density field, in which the density is in units of kg/m^3. In the simulations shown in row (a) a continuous liquid intermetallic stream presents in the domain at the beginning of the simulation. In the simulations shown in row (b) the liquid stream is absent and this corresponds to a gas-only case study.

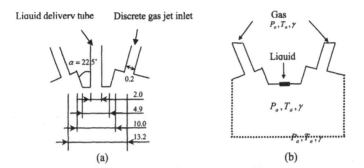

(a)   (b)

Figure 1. Schematics of (a) the atomizer nozzles and (b) initial parameters and boundary condition used; dimensions in millimeters.

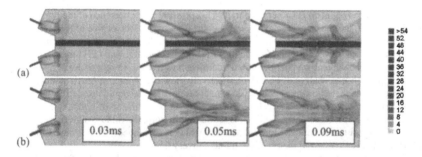

Figure 2. Density fields (a) with and (b) without the presence of liquid intermetallic stream, at different times; density in kg/m³; liquid has high density, is only very weakly compressible, and therefore appears as the black phase.

As can be seen in both Fig. 2 (a) and (b), atomizing gas jets emerge from the gas nozzles and converge towards the center line. In both systems the predicted gas pressure at the gas nozzle exit is 1.54 atm. which is very close to the value predicted by isentropic one-dimensional flow analysis. This means that, here, the atomizing gas is under-expanded, since the pressure at the nozzle exit is above atmospheric, and the gas nozzle is "choked".

The difference of the status of fluids flow in systems with and without the presence of the central liquid intermetallic stream is further analyzed in Fig.3. Detailed analysis of the contours of Mach number and pressure for the gas-only case can be found in [16]. This current paper mainly addresses the difference caused by the presence of the continuous liquid stream.

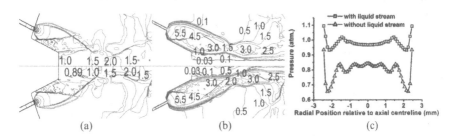

(a)   (b)   (c)

Figure 3. Contours of (a) pressure and (b) Mach number, in which the upper half is in the presence of the liquid stream, and (c) pressure profile along the tip of the liquid delivery tube; all at 0.05ms

The top and bottom halves of Fig.3(a) respectively show the systems with and without the presence of the liquid intermetallic stream, near the liquid nozzle. The pressure is in units of atm., and three features are obvious. Firstly, near the tip of the liquid delivery tube, the pressure is lower when the liquid stream is absent. Secondly, when the liquid stream is present, the contour lines of pressure are basically perpendicular to the axis of the liquid delivery tube and hence to the continuous liquid intermetallic stream. Thirdly, the pressure across the tip of the liquid delivery tube (Fig.3(c)) is sub-atmospheric. This pressure in the presence of the liquid is higher than that for the gas-only case, particularly within the range of the orifice of the liquid delivery tube (from -1mm to +1mm). The aspiration pressure of the two respective systems (with/without the presence of liquid intermetallic stream) are 0.974 atm. and 0.831 atm., respectively. i.e. subambient whether the liquid intermetallic stream is present or not.

The top and bottom halves of Fig.3(b) respectively show the contour of Mach number when the liquid intermetallic stream does or does not present, near the liquid nozzle. Without the liquid stream (bottom), the wake is enveloped by the sonic lines and the tip of liquid nozzle. However, due to the presence of the continuous liquid stream (top), the wake is significantly squeezed to be confined to a much more limited area which is enveloped by the surface of the liquid stream, the sonic line and the tip of the liquid delivery tube.

Checking the difference between the contours of pressure and Mach number for the two different cases, along the downstream gas jets, it can be seen that the gas flow dynamics upstream of the tip of the liquid delivery tube is not significantly influenced by the presence of the liquid intermetallic stream, while the downstream gas flow is obviously affected. The downstream gas flow cannot affect the gas flow inside the choked gas nozzles and hence the gas flow dynamics near the gas nozzle exits is not significantly affected by the downstream configuration (i.e. whether or not a liquid stream is present).

In the new simulations, many ripples are predicted along the surface of the melt stream. This phenomenon is very different from the simulations of atomization by incompressible gas jets [7], where the surface of melt stream was predicted to be quite smooth and flat, except for the concavity caused by the impact of gas jets. The incompressible gas jets were reflected by the melt stream, flowing away from the melt following impact. However, due to the cellular shock wave structure [16] now predicted in the current model, the atomizing gas jets flow along the surface of the melt stream even after impact, as shown in Fig. 4, which illustrates the flow at

326

later times than in Fig. 2. The attached flow of compressible gas jets causes very significant shear and fragmentation all along the melt stream. The reader should compare Fig. 4 to Fig. 2(c) and 2(d) of [7] to observe the effects of including compressibility in the model.

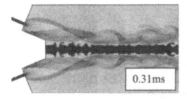 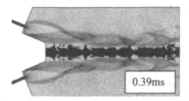

Figure 4. Temporal evolution of density field in the compressible atomizing gas at later times; density key as per Fig. 2.

CONCLUSIONS

A new model has been developed which can handle the simultaneous flow of compressible atomizing gas jets and a weakly compressible liquid metal stream during the process of gas atomization. In this unified model, the disparity in the compressibility of different phases is solved and the variation in material properties is solved using a front-tracking approach. The model was used to predict the nature of the atomization of a liquid Ni-50wt.%Al stream by argon gas jets in a closed-coupled atomizer, using direct numerical simulation.

When the liquid intermetallic stream is present, the pressure around the tip of the liquid delivery tube, particularly the aspiration pressure, is higher than that for the much-published gas-only case. So one has to be very careful when the results obtained from gas-only case studies are used to make predictions of realistic atomization processes. Although gas flow downstream of the tip of the liquid delivery tube is significantly affected by the presence of the liquid stream, the upstream flow of gas is not. This implies that, when the gas nozzles are choked, the gas flow near the gas nozzles can be regarded as a stable input to the atomization process. The design of the tip of the liquid delivery tube, such as flat tip, slotted tip or sunken tip, does not influence the upstream flow of gas jets but does influence the downstream flow of gas and liquid metal.

This paper mainly addresses the hydrodynamic interactions between the gas and continuous liquid metal stream during the very early stage of the atomization process. The analysis is relevant to the further disintegration of the intermetallic stream into ligaments and the subsequent formation of melt droplets, work yet to be completed.

Besides simulating the atomization of intermetallic melt by gas, due to its feature of "one-fluid", the new model can be used to numerically simulate a variety of different twin-fluid atomization processes, such as liquid metal atomization by water and fuel atomization by gas.

ACKNOWLEDGMENTS

This work is financially supported by European Commission (contract number NMP-CT-2004-500635) Sixth Framework Programme as the project "Intermetallic Materials Processing in

Relation to Earth and Space Solidification", which is co-funded and coordinated by the European Space Agency. The authors thank Dr. Steven P. Mates from National Institute of Standards and Technology, USA, for helpful discussions.

REFERENCES

1. M. Raney. Method of producing Finely Divided Nickel, US Patent 1628190, issued May 10,1927.
2. D. J. Browne and J. D. Hunt, *Arch. Thermodynamics* **24**, 25 (2003).
3. D. J. Browne and J. D. Hunt, *Numer. Heat Transfer, Part B* **45**, 395 (2004).
4. J. Banaszek, D.J. Browne, *Mater. Trans.* **46**, 1378 (2005).
5. M. Tong and D. J. Browne, *Commun. Numer. Meth. Engng.* **24**, 1171 (2008).
6. M. Tong and D. J. Browne, *Proc. International Symposium on Liquid Metal Processing and Casting*, ed. Peter D. Lee, Alec Mitchell, Jean-Pierre Bellot and Alain Jardy, Nancy, France, 2007, pp.255-260.
7. M. Tong and D. J. Browne, *J. Mater. Processing Tech.* **202**, 419 (2008).
8. U. Fritsching and K. Bauckhage, *J. Comp. Fluid Dynamics* **5**, 81 (1992).
9. U. Heck, Zur Zerstaubung in Freifalldusen. Dissertation, Universitat Bremen, 1998
10. J. Ting and I. E. Anderson, *Mater. Sci. Eng. A* **379**, 264 (2004).
11. Q. Xu, D. Cheng, G. Trapaga, N. Yang and E.J. Lavernia, *J. Mater. Res.* **17**, 156 (2002).
12. N. Zeoli and S.Gu, *Comp. Mater. Sci.* **43**, 268 (2008).
13. J. Mi, R.S. Figliola and I.E. Anderson, *Metall. Mater. Trans. B* **28**, 935 (1997).
14. P. I. Espina, Study of an underexpanded annular wall jet past an axisymmetric backward-facing step, Ph.D thesis, 1997, University of Maryland
15. P. I. Espina and U. Piomelli, Numerical simulation of the gas flow in gas-metal atomizers, *Proceedings of the 1998 ASME Fluids Engineering Division Summer Meeting* (Washington, DC: ASME 1998), FEDSM98-4901, pp.1-11
16. M. Tong and D.J. Browne, *Computers and Fluids*, accepted for publication, November 2008.
17. T. Yabe and P. Wang, *J. Phys. Soc. Japan* **60**, 2105 (1991).
18. K. Yokoi and F. Xiao, *Physica D* **161**, 202 (2002).
19. S.Y. Yoon and T. Yabe, *Computer Physics Commun.* **119**, 149 (1999)
20. M. Ida, *Computer Physics Commun.* **150**, 300 (2003).
21. F. Xiao, R. Akoh and S. Ii, *J. Comput. Phys.* **213**, 31 (2006).
22. Various Authors, First-Phase Report on Structure Characterisation Results for Turbine and Catalytic Materials, IMPRESS project, Deliverables D9-2 & D9-3, European Space Agency, 2006, pp.98-99.

Mater. Res. Soc. Symp. Proc. Vol. 1128 © 2009 Materials Research Society   1128-U05-53

Role of the Microstructure on the Deformation Behavior in Mg$_{12}$ZnY With a Long-Period Stacking Ordered Structure

K. Hagihara[1], N. Yokotani[1], A. Kinoshita[1], Y. Sugino[1], H. Yamamoto[1]
M. Yamasaki[2], Y. Kawamura[2] and Y. Umakoshi[3]

[1]Division of Materials and Manufacturing Science, Graduate School of Engineering, Osaka University, 2-1, Yamada-oka, Suita, Osaka 565-0871, Japan
[2]Department of Materials Science, Kumamoto University, 2-39-1, Kurokami, Kumamoto 860-8555, Japan
[3]National Institute for Materials Science, 1-2-1, Sengen, Tsukuba, Ibaraki 305-0047, Japan

ABSTRACT

The influence of a heat-treatment on the plastic deformation behavior in Mg$_{12}$ZnY with a long-period stacking ordered (LPSO) structure was investigated by using crystals grown by the Bridgman method. Annealing of the crystal at 798 K for 3 days induced the change in the crystal structure of Mg$_{12}$ZnY from the 18-fold rhombohedral structure (18R) to the 14-fold hexagonal structure (14H). The plastic behavior of those LPSO crystals showed a large variation depending on the loading axis in both crystals, because of the limitation of operative deformation modes in them. The change in the stacking sequence in the LPSO crystals did not show a large influence on the plastic deformation behavior at room temperature.

INTRODUCTION

Recently, new-types of magnesium (Mg) alloys which possess very high strength and relatively good ductility have been developed especially in the Mg-Y-Zn system, eg. see [1, 2].Some of the alloys contain the Mg$_{12}$ZnY phase which has the long-period stacking ordered (LPSO) structure of the close-packed plane, which corresponds to the (0001) basal plane in Mg (called LPSO-phase hereafter). The role of the LPSO-phase on the superior mechanical properties of these alloys has been focused, but the details are not enough clarified yet. In order to examine this, we prepared the LPSO single-phase alloys with the 18R structure, the detail on the crystal structure will be described later, and the mechanical properties of the LPSO-phase itself have been investigated by using the directionally solidified (DS) crystals [3].

On the crystal structure of the LPSO-phase, it was recently reported by some researchers [4-7] that in Mg/LPSO two-phase alloys, the stacking sequence in the LPSO-structure was varied by the heat-treatment at high temperatures. In this study we examined the influence of a heat-treatment on the plastic behavior of the Mg$_{12}$ZnY LPSO-phase. By comparing the plastic behavior between the as-DS-grown crystal and the heat-treated crystals, the effect of the change in the stacking sequence on the mechanical properties was investigated.

EXPERIMENTAL PROCEDURE

The mother alloy with the composition of Mg88Zn5Y7 (at.%) was prepared by induction melting. Using the alloy, the directional solidification was conducted by means of the Bridgman method at a rate of 5 mm/h under an Ar atmosphere. By X-ray diffraction (XRD) analysis, the gained crystal was confirmed to be almost composed of the LPSO-phase, although small amounts of second phases, $Mg_3Zn_3Y_2$ (W-phase) and Mg-phase, also existed especially at grain boundaries. Some of the gained DS crystals were subsequently heat-treated for the control of the microstructure (crystal structure) under an Ar atmosphere at 798 K for 3 days. The microstructure in alloys was observed with an optical microscope (OM) and a transmission electron microscope (TEM, JEOL JEM-3010) operated at 300 kV.

The mechanical properties of these crystals were examined in compression tests at room temperature. Rectangular specimens with dimensions of approximately $2x2x5mm^3$ were cut by electro-discharge machining from the alloys. Two different directions were selected for loading axis. One is parallel to the growth direction, and the other is the direction inclined at an angle of 45° from it, which are called the 0°- and 45°-orientations, respectively. Compression tests were performed on an Instron testing machine at a nominal strain rate of $1.67x10^{-4}\,s^{-1}$. Deformation markings were observed by the OM equipped with Nomarski interference contrast. The deformation substructure was observed in the TEM.

RESULTS AND DISCUSSION

Change in the crystal structure depending on the heat-treatment in DS crystals
Figure 1 shows the selected area electron diffraction pattern (DP) of the as-DS-grown $Mg_{12}ZnY$ LPSO-phase, taken along the [11$\bar{2}$0] direction. In the DP, the diffraction spots were asymmetrically arranged with respect to the [0001] axis at the positions divided the height between the incident-beam spot and the 0002 fundamental spot of a simple hcp unit cell into 18-fold. This indicates the LPSO-phase was made up with the 18-fold periodical stacking of the close-packed planes. In the HRTEM observation, the stacking order of the close-packed plane was determined to be ABABABCACACABCBCBC. This is in good agreement with a previous report by Luo and Zhang [8]. This complicated 18-fold stacking order leads to the change in the crystal symmetry from the hexagonal cell in pure Mg to the rhombohedral cell. In rhombohedral-based crystals, the forbidden criterion of diffraction spots "-h+k+l ≠ 3n" (n =

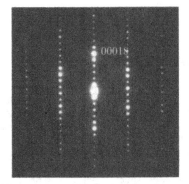

Figure 1 A [11$\bar{2}$0] selected area electron diffraction pattern observed in the as-DS-grown crystal.

integer) is applied, and hence the [11$\bar{2}$0] DP shows the asymmetrical arrangement with respect to the [0001] axis. That is, the LPSO-phase in the as-DS-grown crystal shows the "18R" structure. However, in the rhombohedral lattice system the crystal structure can also be indexed

330

by the hexagonal notation by considering the unit cell whose volume is three times larger than that of the rhombohedral cell. Therefore, in this paper the 18R crystal structure is also expressed by the hexagonal notation for the sake of simplicity.

Figure 2(a) shows the bright field image of the microstructure in the as-DS-grown crystal viewed along the [11 $\bar{2}$ 0] direction. Although the frequency was not so high, the grain was sometimes divided by the band-like regions with the interface on (0001). Figure 2(b) shows the DP taken at the boundary between the band-like region and the matrix shown by a circle in Fig.2(a). From this it can be recognized that the two separated regions show mirror relationship on the stacking sequences of close-packed planes each other. That is, the direction of the c-axis in band-like regions is anti-parallel to that in the matrix as depicted in Fig.2(c). Such a microstructure containing the twin-variants along the c-axis is often observed in the other materials having the rhombohedral crystal symmetry [9].

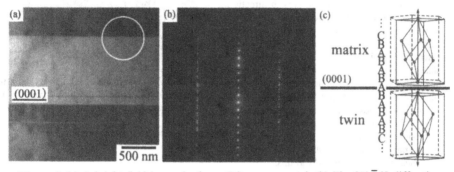

Figure 2 (a) A bright field image in the as-DS-grown crystal. (b) The [11 $\bar{2}$ 0] diffraction pattern taken from the regions shown by the circle in the Fig. (a). (c) A schematic of the twin-variants observed in the 18R LPSO-crystals.

This 18-fold stacking sequence in the as-DS-grown crystal was varied in the wide region of the crystal by the heat-treatment at 798 K for 3 days. Figure 3 shows the [11 $\bar{2}$ 0] DP gained in the heat-treated specimen. In the DP, the diffraction spots were symmetrically arranged with respect to the [0001] axis, contrary to that in the 18R-crystal, and the spots were arranged at the positions divided the height between the incident-beam spot and the 0002 fundamental spot of a simple hcp unit cell into 14-fold. This indicates the LPSO-phase in the annealed specimen was made up with the 14-fold periodical stacking of the close-packed planes with the hexagonal crystal symmetry; that is 14H-structure. The present results on the change in the stacking sequence shows in good correspondence to previous reports observed in Mg/LPSO two-phase alloys [4-7].

Figure 3 A [11 $\bar{2}$ 0] diffraction pattern observed in the crystal annealed at 798 K for 3 days.

331

Plastic deformation behavior of the LPSO-phase at room temperature; the influence of the change in the crystal structure

As previously reported [3], the microstructure and the texture in alloys composed of the $Mg_{12}ZnY$ LPSO-phase could be controlled by the DS-treatment. The grains of LPSO-phase generally show the plate-like morphology with the wide interface parallel to (0001) in solidification [4, 6]. In the DS treatment, the plate-like grains were grown as their (0001) wide interfaces became to be aligned almost parallel to the growth direction. Although the stacking sequence in the LPSO-phase was varied from the 18R to the 14H as described in the previous section, the macroscopic microstructure did not show a large change by the heat-treatment at 798 K as shown in Fig.4; the plate-like grains were remained to be aligned along the growth direction. It was confirmed by XRD analysis that the annealed crystal shows the similar texture as that in the as-DS-grown crystals; the growth direction is nearly parallel to the (0001) plane, in many cases nearly <11$\overline{2}$0> in many grains.

The mechanical properties of the annealed crystals were examined by compression tests at ambient temperature. In the compression tests, two different loading directions, which were parallel to the growth direction and the direction inclined at an angle of 45° from it, were selected. By considering the characteristic features of the microstructure, it is recognized that the 0°- and the 45°-loading orientations correspond to the direction where the Schmid factor for the basal slip is negligible, and it shows the some large values in grains, respectively. Figure 5 shows the room temperature yield stress (the average values in the several times of tests) of the as-DS-grown crystals with the 18R structure and the crystals annealed at 798K with the 14H structure. Although the data showed some scatter due to the disturbance of texture in each specimen, it is obviously seen from the figure that the yield stress shows a strong variation between at the 0°- and 45°-loading orientations

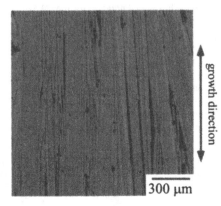

Figure 4 An OM image of the DS crystal after annealed at 798 K for 3 days, observed on the longitudinal section along the growth direction. Plate-like shapes of grains are aligned approximately along the growth direction.

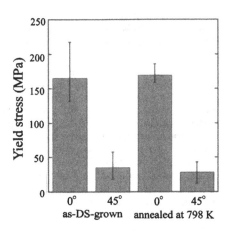

Figure 5 The room temperature yield stress of the crystals as-DS-grown and after annealed at 798 K for 3 days, at the 0°- and 45°-loading orientations.

in both crystals. On the other hand, the difference of the yield stress between the 18R and the 14H crystal was not so significantly detected at both orientations in the deformation at room temperature. The results indicate that the plastic behavior of the LPSO-crystals show the strong anisotropy at room temperature, irrespective of the stacking sequence of the close-packed plane.

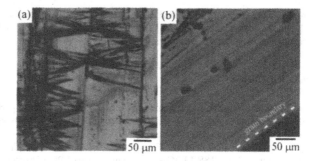

Figure 6 Deformation markings in the annealed crystals with the 14H structure deformed at the (a) 0°- and (b) 45°-loading orientations, to about 2% plastic strain.

Figure 6 (a) and (b) show the deformation markings observed in the annealed crystals with the 14H structure deformed to about 2% plastic strain at the 0°- and 45°-loading orientations, respectively. At the 45°-orientation, slip traces were very finely introduced in some of the grains as shown in Fig.6 (b). It is seen that the slip traces were all introduced parallel to the grain boundaries. Since the wide interfaces in the plate-like shapes of grains are considered to be almost parallel to the (0001), the result indicates the basal slip was operative in the 14H LPSO-crystals deformed at the 45°-orientation. The Burgers vector of the basal dislocations was identified to be parallel to $<11\bar{2}0>$ in the TEM observation (not shown here).

On the other hand, the different plastic behavior was observed at the 0°-orientation, since the operation of the basal slip was strongly limited because of the small Schmid factor at this orientation. In the deformed specimen, large band-like deformation products were abundantly introduced as shown in Fig.6 (a). In the TEM observation, the change in the crystal orientation in the band-like product was detected in many cases as the rotation nearly around the $<1\bar{1}00>$ axis, and distinct crystal orientation relationships between the band-like product and the matrix was not confirmed. This result strongly suggests the observed band-like products are deformation kinks which were formed by the explosive generation of the pair of basal dislocations in the restricted region [10].

It is to note that the gained results on the deformation microstructure were almost the same as that observed in the as-DS-grown crystals with the 18R structure [3], and as shown in Fig.5 the yield stress also exhibited similar values between them. These results demonstrate that the change in the stacking sequence from the 18R to the 14H in LPSO-crystals by the heat-treatment does not induce a strong influence on the plastic behavior at room temperature, and in both crystals the plastic deformation behavior shows the strong plastic anisotropy because of the limitation of operative deformation modes. The attention should be paid, however, that the influence of the change in the crystal structure might appear in the deformation behavior at high temperatures. The operation of non-basal slip was not observed in the deformation of the DS crystals at room temperature in this study. However their operation was confirmed in the extruded alloys which show much higher yield stress [3]. Therefore the operation of the non-basal slip is expected to occur in the deformation under much higher stress fields or at high temperatures. In that case, the operative non-basal slip systems, whether a prism slip occurs or a pyramidal slip does, may vary depending on the crystal symmetry in the LPSO-phases. Related results were previously clarified in Ni-based LPSO-compounds in our study [9]. The study on the plastic deformation behavior of the LPSO-phases at high temperatures is in progress and the results will be published elsewhere.

CONCLUSIONS

The plastic deformation behavior of $Mg_{12}ZnY$ with the long-period stacking ordered (LPSO) structure and the influence of the heat-treatment on it was investigated in this study. The gained results are summarized as follows:

[1] By the heat-treatment of DS crystals at 798 K for 3 days, the crystal structure of $Mg_{12}ZnY$ transforms from the 18R LPSO-structure to the 14H LPSO-structure.

[2] The plastic deformation behavior of the 14H crystal shows strong orientation dependence as similarly to that in the 18R crystals. The yield stress shows the low value at the loading orientation where the (0001) basal slip could be operative. In case that the basal slip could not be operative, kink deformation occurred accompanied by the relatively high yield stress.

ACKNOWLEDGMENTS

This work was supported by the project "Development of Key Technology for Next-generation Heat-resistant Magnesium Alloys, Kumamoto prefecture Collaboration of Regional Entities for the Advancement of Technological Excellence" from Japan Science and Technology Agency, and also by funds from the "Priority Assistance of the Formation of Worldwide Renowned Centers of Research - Global COE Program (Project: Center of Excellence for Advanced Structural and Functional Materials Design, Osaka University)" and a Grant-in-Aid for Scientific Research and Development from the Ministry of Education, Culture, Sports, Science and Technology of Japan.

REFERENCES

[1] Y. Kawamura, K. Hayashi, A. Inoue and T. Masumoto, *Mater. Trans.* **42** 1172 (2001).

[2] A. Inoue, M. Matsushita, Y. Kawamura, K. Amiya, K. Hayashi and J. Koike, *Mater. Trans.* **43** 580 (2002).

[3] K. Hagihara, A. Kinoshita, Y. Sugino, N. Yokotani, M. Yamasaki, Y. Kawamura and Y. Umakoshi, *Magnesium technology 2009* (Proc. of TMS annual meeting 2009), in press.

[4] T. Itoi, T. Seimiya, Y. Kawamura and M. Hirohashi, *Scripta Mater.* **51** 107 (2004).

[5] M. Matsuda, S. Ii, Y. Kawamura, Y. Ikuhara, M. Nishida, *Mat. Sci. Eng. A* **393** 269 (2005).

[6] S. Yoshimoto, M. Yamasaki and Y. Kawamura, *Mater. Trans.* **47** 959 (2006).

[7] A. Ono, E. Abe, T. Itoi, M. Hirohashi, M. Yamasaki and Y. Kawamura, *Mater. Trans.* **49** 990 (2008).

[8] Z. P. Luo and S.Q. Zhang, *J. Mater. Sci. Lett.* **19** 813 (2000).

[9] K. Hagihara T. Tanaka, T. Nakano and Y. Umakoshi, *Acta Mater.* **53** 5051 (2005).

[10] J. B. Hess and C. S. Barrett, *Trans. Ame. Ins. Min. Met. Eng.* **185** 599 (1949).

Mater. Res. Soc. Symp. Proc. Vol. 1128 © 2009 Materials Research Society   1128-U05-57

Pseudo Fine Particle Effect on Unstable Phase Transition of Gd_2Al

A.Yazdani, N. Kamali Sarvestani, R. Osaty Araghi
Tarbiat Modares University, Jalal Al Ahmad Exp. way, P.O. Box: 14115-175, Tehran, Iran.

ABSTRACT

The unusual duality of magnetic phase transition (especially high transition temperature, remarkable shoulder) is thought to be due to second phase impurities of pure Gd or Gd_3Al_2 or even both. The temperature dependence and field dependence of magnetization have been studied in the temperature range of 4.2-350 K in three different strengths of magnetic field to investigate the cause or/and the source of metamagnetic character which is a puzzle on Gd-intermetallic compounds, IMC, in order to find the critical field needed to overcome the fluctuation field, above which the magnetic behavior of the system is stabilized, and a critical turning field is defined.
However for the clearance of intrinsic instability behavior and the nature (and source) of anisotropy in the Gd_2Al system, the following experimental works has been prepared.
The study of the effect of increasing the magnetic field up to the critical "threshold" field which is exactly the same as decreasing the conduction electron concentration "c.e.c" from Gd_2Al to Gd_2Au; The contribution of short range order as well as thermal fluctuation of short range exchange "J_{sh}" through the long – range interaction "J_l" which can be induced by localized s-f (or d-f) exchange. This effect can be the cause (or the source) of internal magnetic field $H_{in}=<S_i.Sj>$ and fluctuation field $H_f=H_{in}-H_{ext}$.

INTRODUCTION

The exchange interaction between localized 4f moments in the rare earth compounds is usually mediated by conduction electrons, whose character can change by the strength and the sign of exchange. The contribution of these electrons to magnetic ordering which unequivalently links the magnetic structure to the lattice parameters [1], offer a nice opportunity to investigate the origin of the double magnetic transition on the Gd_2Al where the exchange fluctuation is known to be important. In order to understand the cause and role of this fluctuation in this system, the compounds with s-state ions (Gd), without crystal field effects (c.f.e) are investigated where, the metamagnetic character is still a puzzle. Even though the reported results of the unusual magnetic behavior of the system is still in contradiction [2-4], but, the high magnetic phase transition (especially the remarkable shoulder) is attributed to second phase impurities of Gd [5] and/or Gd_3Al_2 [6]. These suggestions have been made on the basis of metallurgical phase diagrams in which, these compounds peritectically exist at ≈870°C [7]. The main findings about these IMC are a dynamic range of Curie temperature which is magnetic field dependent as well as a thermal magnetic history without hysteresis character at low temperatures.

335

EXPERIMENT

This study is a part of our efforts to investigate the unusual magnetic properties of polycrystalline Gd_2Al IMC as a frustrated mictomagnetic system, called cluster-glass. The Gd_2Al sample was prepared by melting the stochiometric amounts of elements (4N for Gd and 5N for Al) in an argon atmosphere using an induction melting furnace. X-ray diffraction of this compound before and after annealing process reveals an orthorhombic crystal structure [8].

The isotherm magnetization and temperature dependence of susceptibility measurements on the prepared sample where performed in the temperature range of 4.2-350 K and in three different strengths of magnetic field, 0<H<20 kOe, H=20 kOe and H=40 kOe, by the quantum design "MPMS" magnetometer. In order to find the intrinsic instability and the source of fluctuation, the control parameters such as decreasing of chemical pressure, decreasing of the concentration of magnetic ions and the magnetic fields were modified.

In our previous works; (i) The conduction electron concentration was decreased to find out the effect of chemical pressure by substitution of Au^{+1} by Al^{+3}, by which the surprising mechanical shape dependence of magnetization was manifested [1] and (ii) the system was diluted to find out the critical threshold field by which the second phase can be prevented [9]. Therefore in this work the effect of magnetic field which behaves similar to the above cases are considered.

RESULTS AND DISCUSSION

Based on the isotherm magnetization in low applied fields, H=0-20 kOe (figure 1), a critical field is defined by the ratio of magnetization at 4.2 K to the value at room temperature. This ratio is field dependent and reaches to a minimum value defined as turning field ($H_c \approx 20$ kOe) where the obtained results can be divided into two distinct categories (inset of figure 1);

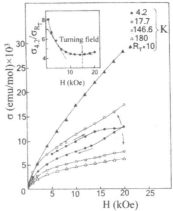

Figure 1: The isotherm magnetization at low applied fields H<20 for different temperatures.

(a) Below this field (H≤20): fluctuations are developed as $H_f = H_{in} - H_{ext}$ where H_{in} is the internal magnetic field defined by $H_{in} = <<S_i.S_j>>$ and H_{ext} is the external magnetic field. Even though the metamagnetic character is reported for the Gd_2Al IMC [2], but, the steep concave curvature (initially downward) at low applied field of H=0-20 kOe (figure 1) suggests a strong exchange interaction (weak ferromagnetic character) on which;

1) A low magnetic saturation- like is manifested at 146.6 K and 180 K (figure 1). The isotherm magnetization at room temperature behaves similar to Langevin character which shows the existence of an interaction. Irreversibility is the character of the whole process at 4.2 K in which a small hysteresis in the initial part of the curve and a shift of the sharp increase in the latter part is observed (figure 2).

2) The magnetization above and below the maximum susceptibility temperature, $T_m \approx 50$, behaves in two different ways; Above T_m the initial part rotates to a higher curvature and tends to a lower value of magnetization and eventually become flat. And bellow T_m it rotates to a lower curvature with higher values.

3) Then it increases more rapidly (sharply) as the magnetic field increases and finally reaches to the saturation state (figure 2).

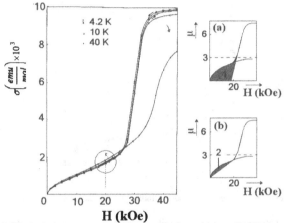

H (kOe)

Figure 2: The isotherm measurement of magnetization at high applied fields.

4) Regarding the defined turning field, bellow which the strong correlated internal field is manifested, due to intracluster interactions two points (figure 2 a) can be suggested;

i) the normal F.M behavior of $\sigma(H)$ is shifted to higher fields which can be the breaking of the intracluster exchange and then the starting of the orientation of magnetic ions from $H_{ext}>20$ (shown by shaded area in figure 2 a).

ii) the orientation of particles away from the metamagnetic alignment if it was AFM (shown by shaded area in figure 2 b).

Consequently the $\sigma(H)$ curves (figure 2 a,b) suggest that, the threshold field of cluster rotation is manifested at $H_{ext} \cong H_{in} > 20$ kOe, above which the fluctuation field is diminished ($H_f = 0$). As a result, the external field overpowers the internal field and the cluster begins to rotate (as same as the rotation of domains). The aligned clusters extend themselves by the irreversible process at 4.2 K. That means, the clustering rotated regions spread out, tunnels into each other by thermal activity. And finally at strong applied fields, all of the magnetic ions would rotate to become in the direction of field.

(b) Above this field (H≥20):

1. the magnetization depresses to a lower value below T_C in the whole range of temperature, and all transitions disappear.
2. the maximum susceptibility reaches to its stability.
3. the value of magnetization suddenly rises in all ranges of temperature with a completely paramagnetic (superpartamagnetic) behavior and follows the Longevin function in H=40 kOe (figure3).

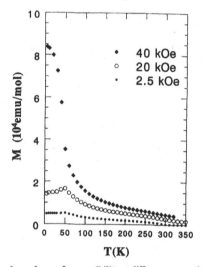

Figure 3: Temperature dependence of susceptibility at different magnetic fields.

It is shown that the inverse susceptibility in this range of field is almost a straight line passing through the origin with two strong deviations around $T_{max} \sim 50$ K and $\theta_p' \cong 295$ K. It also should be mentioned that;

- At high temperatures, above 295 K, the inverse susceptibilities follow the Curie-Wise law. At this temperature, a deviation attributed to crystal field effect is suggested [3]. By extrapolating the linear part of $\chi^{-1}(T)$, we obtain the paramagnetic Curie temperature to be $\theta_p > 220$ K at H=20 kOe applied magnetic field and $\theta_p > 150$ K at H=40 kOe with $\mu_{eff} = 8.3 \mu_\beta/Gd$. The $\theta_p \cong 150$ K is reported with no mentioned applied field [3, 10]. The nearly ionic value of μ_{eff} suggests that the superparamagnetic behavior has not been suppressed completely in this temperature range at 40 kOe.

- Bellow $T_c = 295$ K in the temperature range of 295-50 K a typical paramagnetic Curie-Wise behavior is observable and the straight line of inverse susceptibility passes through the origin ($\theta_p \cong 0$-2.2 K with $\mu_{eff} \cong 17.6 \mu_\beta$).

- A deviation around $T_m > 50$ K appears where, the maximum susceptibility stabilizes.
- And finally the maximum at 48-50 K under H≤15 kOe suggests an antiferromagnetic ordering.

CONCLUSIONS

The above consistent results strongly suggest that;
-        the anisotropy energy increases with decreasing of temperature which clearly shows the nature of exchange, as the size of cluster increase ($\mu_{eff} \cong 17.6\mu_B$), this is manifested by the very high moments observed at H≥15-20 kOe.
-        cluster glass magnetization bellow T_C is dominated by large groups of strongly correlated spins (magnetic ions) above which it behaves AF.M. This persists even at 40 kOe.
-        the fact that short range magnetic order is indeed present in this mictomagnetic character is shown by the superparamagnetic behavior at high temperatures in low and high magnetic fields, H=40 kOe while both present the Langevin behavior

Also the instability of this compound can be due to either the huge entropy of pure ground state of Gd in low applied fields in the range of interaction space $3.39 \leq R_C \leq 3.64$ Å where the crystal field effect does not exist (if Gd is in s-state), or the spin density wave type. If the latter be the case, a Fermi liquid-like behavior would play a role and the magnetic ions behave as fine particles. But if the crystal field effect does exist, which can be due to the d-like character of conduction electrons, it should be considered as the d-bond in the virtual bond state.

ACKNOWLEDGMENTS

Authors want to give their thanks to Prof. A. Tazuke Yuuchi, Hukido University, Japan, for using the apparatus and his kind helps in doing the measurements.

REFERENCES

[1] A. Yazdani, J. Stewal Gardner, Phys. Stat. Sol. (b) 208, 465, 265(1998).
[2] X. G. Li, M. Sato, S. Takahashi, K. Aoki, T. Masumoto, J. Mag. Mag. Mat. 212, 145 (2000).
[3] L. R. Sill and J. Biggers, J. Appl. Phys. 49, 1500 (1978).
[4] A. Yazdani, J. Mag. Mag. Mat. 90 & 91, 563-564(1990).
[5] V. Brédimas, H. Gamari-Seale, C. Papatriantafillou, Phys. Stat. Sol. (b) 122, 527 (1984).
[6] H. Oesterreicher, phys. Stat. Sol. (a) 39, K91 (1977).
[7] O. J. C. Runnalls and R. R. Boucher, JLCM; 13, 43 (1967).
[8] W. B. Pearson, The crystal chemistry and physics of metal alloys, Pergamon Press, 532 (1972).
[9] A. Yazdani and H. Gamari-Seale, Phys. Stat. Sol. (a) 96, 587 (1986).
[10] K. H. J. Buschow, J. Less Comm. Met. 43, 55 (1975).

Nickel/Cobalt Superalloys and Nickel Aluminides

Mater. Res. Soc. Symp. Proc. Vol. 1128 © 2009 Materials Research Society   1128-U06-01

Intermetallic B2 Bond Coats: Systems Compatibility and Platinum-Group Metal Additions

Tresa Pollock, Dan Widrevitz, Russell Pong, Fang Cao and Bryan Tryon
Department of Materials Science and Engineering, 2300 Hayward St., University of Michigan
Ann Arbor, MI 48109

ABSTRACT

Intermetallic interlayers are key to the performance of thermal barrier coating systems. The role of platinum group metal (PGM) additions to B2 bond coat interlayers has been investigated with emphasis on diffusional aspects of coating structure evolution and properties. Additions of Ru and Ir to NiAl reduce interdiffusion and the resultant thickness of the coating and interdiffusion zone. Relative to Pt and Pd, Ru and Ir exhibit slower interdiffusion coefficients. Ru additions to NiAl increase creep resistance but degrade oxidation kinetics.

INTRODUCTION

The temperature capability of turbine airfoil systems can be substantially increased with the use of multilayered thermal barrier coating (TBC) systems [1 - 3]. However, maintaining adhesion of the yttria-stabilized zirconia TBC coating to the nickel-base superalloy substrate remains a major challenge. Intermetallic interlayers with properties intermediate to the ceramic top coat and the metallic substrate can improve the durability of these systems if a balanced set of properties of can be achieved. Properties of the intermetallic that strongly influence the failure process include diffusion, high temperature strength, coefficient of thermal expansion (CTE) and oxidation resistance [3, 4]. Importantly, the intermetallic bond coat must serve as an environmental barrier, forming a thermally grown oxide (TGO) upon high temperature exposure. Coatings based on the B2 intermetallic NiAl with the CsCl structure have been effective interlayers for thermal barrier systems [1, 2, 5]. Platinum group metals, with significant solubility in the B2 NiAl have the potential to alter the properties of the intermetallic interlayer and thus improve adhesion and system life. This study focuses on additions of Ru, Ir, Pd and/or Pt to NiAl-based systems and their influence on coating structure and properties. The influence of PGM additions on interdiffusion has been investigated via diffusion couples. The structure and properties of PGM-containing coatings fabricated by vapor deposition techniques have also been investigated.

EXPERIMENTAL APPROACH

Diffusion couples were formed by joining stoichiometric single crystal NiAl to (Ni,PGM)Al polycrystalline alloys. Additions of Ir, Os, Ru, Pd and Pt were investigated with 10% of the PGM substituted for Ni. PGM-containing alloys were prepared by melting in an induction levitation melting system or by arc melting. The joining procedure for the diffusion couples, outlined in detail by Kulkarni et. al [6], consisted of loading the two halves of the couples under 10 - 20MPa compressive loads at 1000°C or 1050˚C for 13 – 30 hours in a Centorr vacuum furnace operatiang at ~10^{-7} torr. Bonded specimens were sealed in Ar-backfilled quartz tubes and annealed for either 300 or 100 hours at either 1100 or 1200°C. No evidence of oxidation or specimen degradation was found following the diffusion cycle. After heating, specimens were sectioned and prepared by traditional metallographic means and examined via scanning electron microscopy (SEM) in backscattered electron mode (BSE). All couples maintained a single phase B2 structure across the interdiffusion zone. A Cameca SX 100 electron probe microanalysis (EPMA) system was used to determine concentration profiles across diffusion zones and to image the diffusion interfaces in BSE imaging mode. The EPMA was calibrated via stoichiometric NiAl, and 99.99% pure Ir, Os, Pt, Ru and Pd standards and accelerating voltage and beam current were set to 20 kV and 10 nA, respectively to generate the profiles.

To fabricate coatings, PGM metallic layers 3 – 4µm in thickness were deposited on the surfaces of single crystal superalloy CMSX-4 substrates by either electron beam physical vapor deposition (EB-PVD) or electroplating processes. The coupons were then annealed in an inert atmosphere at 1100 °C for 4 hours. Final coating structure was established by low activity aluminization conducted with standard commercial procedures [7]. After cleaning in acetone, coated samples were placed on curved alumina platens in a conventional bottom-loading cyclic oxidation furnace. Cyclic oxidation studies were conducted with a cycle that consisted of ramping to 1100 °C in 10 minutes, followed by an isothermal dwell period of 45 minutes at 1100 °C before cooling to 100 °C in 10 minutes.

The compressive creep properties of (Ru,Ni)Al samples fabricated by induction levitation melting were investigated. Creep specimens were sectioned by wire electrodischarge machining (EDM) to approximate dimensions of $4 \cdot 4 \cdot 8$ mm^3. Constant load compression creep tests were carried out in a vacuum creep frame with a TZM creep compression fixture in the temperature range of 950˚C to 1050˚C.

RESULTS

The influence of additions of Ir, Os, Ru, Pd and Pt on interdiffusion in NiAl were investigated at 1100˚C and 1200˚C. All diffusion couples maintained a single phase B2 structure following extended annealing in this temperature range, Fig. 1. While the full interdiffusion matrix was analyzed using the MultiDiFlux approach [6], only the main PGM interdiffusion coefficients are considered here. Fig. 1 shows the interdiffusion coefficients at 1100˚C and 1200˚C. At both temperatures Pd and Pt display the highest interdiffusion coefficients among the group of PGMs investigated. Ir, Ru and Os have interdiffusion coefficients that are one to two orders of magnitude lower than the Pd and Pt in this temperature range.

Following aluminization, all PGM-containing coatings possessed a PGM-rich B2 structure with a multiphase intermetallic interdiffusion zone between the B2 phase and the superalloy substrate. Fig. 2 shows the layered structure and the distribution of elements in as-processed coatings containing, Pt, Ru Ir or Ru+Pt rich B2, respectively. The thickness of the B2 layer plus the

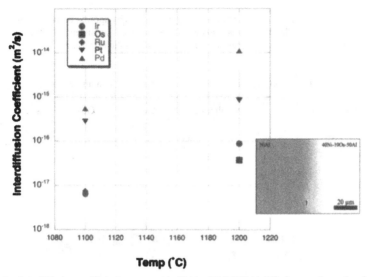

Figure 1 – Interdiffusion coefficients measured in NiAl – (Ni,PGM)Al diffusion couples and an image of the interdiffusion zone for the NiAl-(40Ni10Os)50Al couple.

interdiffusion layer varied with PGM addition, with total thicknesses of 57μm, 40 μm and 35 μm in Pt, Ir and Ru coatings, respectively. Aluminization without the presence of PGMs resulted in a total thickness of the B2 and interdiffusion layer of 75 μm.

Cyclic oxidation experiments were conducted simultaneously on 3mm thick disks of the Pt- and Ru-containing coatings, as well as a hybrid coating containing both Ru and Pt [8]. Thermal barriers of 7YSZ were applied to all coatings by an EBPVD process [8]. The 1100°C cycle outlined in more detail above was utilized. Cyclic oxidation resulted in the formation of a thermally grown oxide (TGO) of Al_2O_3 between the YSZ top coat and the B2 interlayer in all coatings. This layer serves the important function as an environmental barrier, since the YSZ is permeable to oxygen. However, as this layer thickens, the coating system becomes prone to spallation [3, 4]. The Ru-containing coating displayed the most rapid oxidation kinetics, with a TGO layer 4 – 5 μm in thickness after 60 cycles to the maximum temperature of 1100°C, Fig. 3. The coating spalled at this point in the cycling process. The TGO layer in Pt-containing coating was only 2 – 4 μm in extent after 936 cycles and spallation ultimately occurred after 1568 cycles. The hybrid Ru+Pt coating had a longer cyclic life than the Pt coating, in spite of the fact that a thicker TGO layer formed, Table 1.

Considering that Ru additions were detrimental to oxidation kinetics but beneficial to the overall system life in the hybrid Ru+Pt coating, the influence of Ru on the high temperature creep properties of NiAl was investigated across the B2 phase field in the Ni-Al-Ru system [9, 10]. As Ru is substituted for Ni across the B2 phase field, there is a dramatic increase in the creep strength. Fig. 4 shows the rise in creep strength measured experimentally for creep rates fixed in the range of 1×10^{-7}/s to 3×10^{-7}/s. The creep strength increases by approximately a factor of 50 moving from the NiAl side of the ternary toward the RuAl side. Consistent with this rise in creep strength is an increase in the melting temperature across the ternary, Fig. 4.

Table 1 – Growth of Al_2O_3 TGO layer in Ru and Pt containing coatings

Bond Coating	Number of 1100°C Cycles	TGO Thickness Range (μm)
Ru	60	4 - 5
Ru+Pt Hybrid	936	6 - 8
Pt	936	2 - 4

Pt-modified aluminide

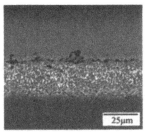
Straight aluminide

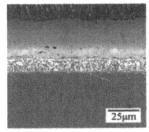
Ru/Pt-modified aluminide

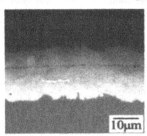
Ru-modified aluminide

Ir-modified aluminide

Figure 2 - PGM-containing B2 bond coatings on single crystal superalloy substrates. Note the higher magnification of the Ru-modified coating.

346

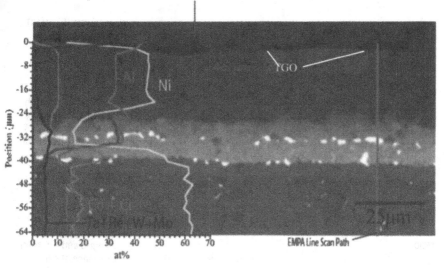

Figure 3 - Discontinuous TGO following spallation after 60 cycles to 1100°C in a Ru-containing coating.

DISCUSSION

Optimization of thermal barrier coating systems for superalloys requires a balance of properties to be achieved in the bond coat interlayer. Platinum group metals provide an opportunity for "systems optimization" through their influence on the properties of the bond coat. Since PGMs generally have some degree of solubility in NiAl, they can influence interdiffusion. Ru, Os and Ir were more effective at lowering interdiffusion coefficients, compared to Pd and Pt.
For a fixed processing path, the lower interdiffusion coefficients resulted in a thinner coating plus interdiffusion zone in the Ru- and Ir-modified coatings, compared to Pt-containing or PGM-free coatings. PGM additions such as Ru and Ir with low interdiffusion coefficients are thus desirable from the diffusional point of view, since they will inhibit migration of superalloy substrate elements through the B2 coating to the TGO. They will also minimize loss of the load-bearing superalloy due to coating application (important for thin-walled turbine airfoils) and in subsequent coating repair operations.

While an important function of the bond coat is to minimize interdiffusion, oxidation resistance is an equally important property. As the TGO layer thickens, the tendency for "rumpling" of the thermally-mismatched TGO and bond coat increases [3, 4, 11]. This drives formation of cracks at the TGO-TBC interface that ultimately link and cause spallation of the TBC. Micromechanical modeling [4, 12] demonstrates that the failure process can be suppressed by improving the high temperature strength of the bond coat. Ru additions to NiAl are shown here to substantially improve the creep resistance of the B2 structure. This higher creep resistance apparently permits the system to sustain a thicker TGO prior to the onset of spallation. Hemker

347

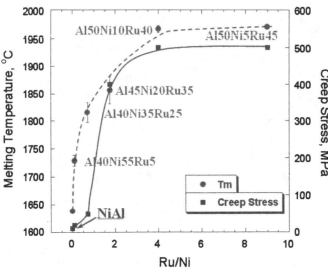

Figure 4 - Creep strength and melting temperature of ternary Ru-Ni-Al alloys.

and co-workers [13] have shown that Pt and other refractory additions to NiAl are much less effective at improving the high temperature strength of NiAl, compared to Ru. Conversely, Pt is much more beneficial than Ru in terms of reducing oxidation kinetics and growth of the TGO layer. As a result of the need to achieve a balance of high temperature strength and oxidation resistance, the hybrid Ru/Pt coatings have better cyclic lives, compared to coatings containing only Pt or Ru. This, in turn, suggests that specific mixtures of PGMs might be even more effective at improving coating life. Thus, overlay coating approaches may provide a more convenient path to improving the overall TBC coating life in these complex, multi-layered systems.

CONCLUSIONS

1. PGM additions to NiAl reduce interdiffusion coefficients. Ru, Ir and Os additions reduce interdiffusion to a greater degree than Pt and Pd.
2. PGM additions influence the growth kinetics of the Al_2O_3 TGO at the bond coat – TBC interface. Pt is more beneficial to cyclic oxidation kinetics than Ru additions.
3. Additions of Ru to NiAl dramatically improve the creep resistance, providing a benefit to the TBC system life.

ACKNOWLEDGMENTS

The authors gratefully acknowledge H. Murakami for providing an Ir-coated superalloy sample and K. Murphy for aluminization of coated samples. This research was supported by the National Science Foundation under a GOALI grant # DMR 0605700.

REFERENCES

1. B. Gleeson, AIAA J. Prop. Power **22**, 2, 375, (2006).
2. R. Schafrik, R. and R Sprague, Advanced Materials & Processes, **162,** p. 27, (2004).
3. A.G. Evans, D.R. Mumm, J.W. Hutchinson, G.H. Meier, F.S. Pettit, Prog. Mater. Sci. **46** 505, (2001).
4. A.M. Karlsson, T. Xu and A.G. Evans, Acta Mater. **50**, 1211, (2002).
5. J.R. Nicholls, MRS Bull., **9**, 659, (2003).
6. K. N. Kulkarni, B. Tryon, T.M. Pollock, and M.A. Dayananda, J. Phase Equilibria and Diffusion, **28**, 503, (2007).
7. K.S. Murphy: U.S. Patent No. 5,856,027, 1999.
8. B. Tryon, K.S. Murphy, C.G. Levi, J. Yang, and T.M. Pollock, Surf. Coatings Tech. **202**, 349, (2007).
9. B. Tryon and T.M. Pollock, Mater. Sci. Eng. A **430**, 266, (2006).
10. F. Cao and T.M. Pollock, Acta Materialia **55**, 2715, (2007).
11. V.K. Tolpygo and D.R. Clarke, Acta Materialia **48**, 3283, (2000).
12. D.S. Balint and J.W. Hutchinson, J. Mech. Phys. Sol. **53**, 4, 949, (2005).
13. K.J. Hemker, unpublished research, Johns Hopkins University.

Mater. Res. Soc. Symp. Proc. Vol. 1128 © 2009 Materials Research Society   1128-U06-04

Properties and Application of Two-Phase Intermetallic Alloys Composed of Geometrically Close Packed Ni₃X(X:Al and V) Structures

Takayuki Takasugi[1,2] and Yasuyuki Kaneno[1]

[1]Department of Materials Science, Osaka Prefecture University, 1-1 Gakuen-cho, Naka-ku, Sakai, Osaka, 599-8531, Japan

[2]Osaka Center for Industrial Materials Research, Institute for Materials Research, Tohoku University, 1-1 Gakuen-cho, Naka-ku, Sakai, Osaka, 599-8531, Japan

ABSTRACT

The two-phase intermetallic alloy, which is composed of Ni₃Al(L1₂) and Ni solid solution (A1) at high temperature annealing and is additionally refined by a eutectoid reaction at low temperature aging, according to which the Al phase is decomposed into Ni₃Al+ Ni₃V(D0₂₂), was recently developed. The two-phase intermetallic alloys exhibit highly coherent interface structure between the constituent phases of Ni₃Al and Ni₃V not only at the micron scale but also at the sub-micron scale, and also display high microstructural and phase stability at high temperature. The two-phase intermetallic alloys showed high tensile (creep) strength and creep rupture life, accompanied with high tensile elongation and fracture toughness over a broad temperature range, and also superior oxidation/corrosion resistance and surface hardening capability. The attempts to develop several applications were shown. The two-phase intermetallic alloys are promising as a new-type of high-temperature structural materials.

INTRODUCTION

A number of Ni-based superalloys currently in use are composed of a two-phase microstructure consisting of γ (Ni solid solution:A1 phase) and γ' (Ni₃Al:L1₂ phase). A1 phase is most densely packed structure when alloy exists as a solid solution. On the other hand, L1₂ phase is categorized in geometrically close packed (GCP) structure. It has been believed that attractive properties of Ni-based superalloys are attributed to not only the presence of γ' but also fine microstructure with structural coherency between Ni solid solution and γ'. Like superalloys, multi-phase intermetallic alloy composed of the GCP Ni₃X intermetallic phases is expected to have coherent interfaces and fine microstructure because of their closely related crystal structures. In the present article, alloy composition, microstructure, physical properties, mechanical properties, chemical properties (such as oxidation and corrosion resistance), surface hardening capability and application of the two-phase intermetallic alloys composed of geometrically close packed Ni₃Al and Ni₃V (**Fig. 1**) are presented.

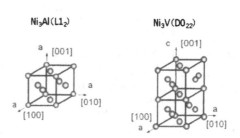

Ni₃Al(L1₂)   Ni₃V(D0₂₂)

Figure 1. Crystal structures of geometrically close packed (GCP) phases Ni₃Al(L1₂) and Ni₃V(D0₂₂).

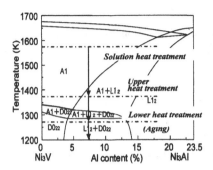

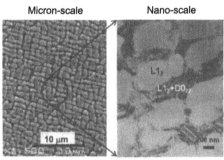

Micron-scale   Nano-scale

Figure 2. Longitudinal section along a pseudo-binary Ni₃V-Ni₃Al line at Ti = 1.5 at.% [4].

Figure 3. Microstructure of the two-phase intermetallic alloy composed of geometrically close packed (GCP) structures Ni₃Al(L1₂) and Ni₃V(D0₂₂) [3]. SEM image (left) and TEM image (right).

ALLOY COMPOSITION AND MICROSTRUCTURE

Typical alloy compositions studied are shown in **Table 1**. Essential alloy elements resulting in the two-phase microstructure are Ni, Al and V, and other minor elements such as Nb, Ti, Cr and Co are added at levels below several atomic percents if necessary. Addition of Cr, Nb and Co may be required to improve oxidation property while addition of Ta, W and Nb may be required to improve high temperature strength and wear property. Also, boron doping is necessary to suppress the propensity of intergranular fracture of this alloy.

The two-phase intermetallic alloys are generally solid-solution treated (homogenized) in Al phase region, annealed in the duplex phase region of (A1+L1₂) at high temperature, and then additionally refined (aged) by a eutectoid reaction at low temperature (see **Fig. 2**), according to which the *prior* Al phase comprising the channel-like region is decomposed into the Ni₃Al+Ni₃V [1-4], as shown in **Fig. 3**. Here, it is interesting to note that the two-phase intermetallic alloys can be forged and recrystallized in the A1 phase with a simple structure. The detailed TEM observation showed that the two- phase intermetallic alloys exhibit highly coherent interface structure between the constituent phases of Ni₃Al and Ni₃V not only at the micron scale but also at the sub-micron scale, as shown in **Fig. 3**. In the channel-like region, D0₂₂ domains are alternately aligned with the

Table 1 Typical chemical compositions of the two-phase intermetallic alloys (at.%)

Alloys	Ni	Al	Ti	Nb	V	Co	Cr
#011	75.0	7.5		2.5	15		
#032	70.5	10		3	10.5	3	3
#053	75.0	8	2.5		14.5		

Figure 4. Back-scattering (BS)-SEM images of the microstructure of the alloy Ni₇₅Al₈.₇₅Nb₃V₁₃.₂₅ aged for (a) 5 h, (b) 10 h, (b) 168 h and (c) 840 h at 1273 K after annealing for 10 h at 1373 K [5].

352

Table 2 Ambient-temperature physical properties of the two-phase intermetallic alloy($Ni_{70.5}Al_{10}V_{10.5}Nb_3Co_3Cr_3$), Inconel X750 and Inconel 718.

Alloys	Young's modulus (GPa)	Shear modulus (GPa)	Density (g/cm³)	Thermal expansion coefficient ($\times 10^{-6}$/K)	Thermal conductivity (W/mK)
The present alloy	185	114	8.01	10.41	10.9
Inconel X750	214	-	8.25	12.6	12.0
Inconel 718	200	-	8.22	12.8	11.5

different type of DO_{22} domain or with $L1_2$ phase along <110> direction, and also with habit plane of {110}.Also, it was found that the microstructural stability is extremely high even by a prolonged annealing, e.g., one month annealing at 1273 K, as shown in **Fig. 4** [4,5]. The coarsening of the two-phase microstructure essentially does not take place although the morphological change somehow takes place.

PHYSICAL AND MECHANICAL PROPERTIES

Some ambient-temperature physical properties of the two-phase intermatallic alloy are listed in **Table 2** in comparison with Inconel X750 and 718. Interesting points are that the physical parameters of the typical two-phase intermatallic alloy are smaller than those of most superalloys (Inconel). Particularly, density and thermal expansion coefficient are lower in the two-phase intermetallic alloy than in most superalloys. This feature may be advantageous to application such as a turbine blade or high temperature tools.

For the mechanical properties [3,5-7], **Fig. 5** shows yield strength, tensile strength and tensile elongation as a function of temperature for alloys $Ni_{75}Al_{8.75}V_{13.25}Nb_3$ and $Ni_{75}Al_{7.5}V_{15}Nb_{2.5}$ that were aged at 1273 K for 10h after annealing for 10 h at 1373 K [7]. In both alloys, the yield strength as well as the tensile strength increased with increasing temperature, and showed a broad maximum at intermediate temperature (around 900K), followed by a decrease at high temperature. At low temperature, both alloys showed a relatively large strain-hardening rate and then ruptured. With increasing temperature, the strain-hardening

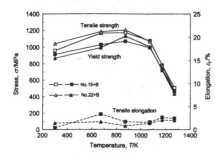

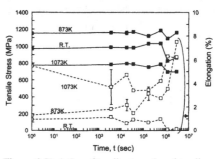

Figure 5. Tensile strength, yield strength and tensile elongation as a function of test temperature for alloys $Ni_{75}Al_{8.75}V_{13.25}Nb_3$ and $Ni_{75}Al_{7.5}V_{15}Nb_{2.5}$ with the two-phase microstructure [7].

Figure 6. Variation of tensile strength and tensile elongation with aging time for alloy $Ni_{75}Al_{8.75}V_{13.25}Nb_3$ that was aged at 1273 K, and then deformed at room temperature, 873 K and 1073 K, respectively [5].

rate decreased. At high temperature, the alloys exhibited steady-state flow after yielding and subsequent yield drop. The tensile elongation ranged between 0.4 % and 3 % at the tested whole temperatures. The observed fracture pattern of the samples deformed at room temperature reveals transgranular fracture mode accompanied with dimple-like and cleavage-like patterns. For the samples deformed at high temperature, the observed fracture pattern reveals transgranular fracture mode accompanied with more intergranular fracture patterns than those observed in the samples deformed at room temperature.

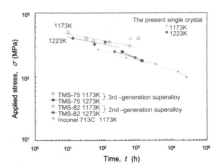

Figure 7. Applied stress vs. creep rupture time curves of the single crystal with the two-phase microstructure crept at 1173 K and 1223 K [3]. Note that the data of some advanced superalloys are included as reference.

Fig. 6 shows variations of mechanical properties with aging time for alloy Ni$_{75}$Al$_{8.75}$V$_{13.25}$Nb$_3$ aged at 1273 K [5]. Data were collected at room temperature, 873 K and 1073 K. It appears that the tensile strengths were primarily insensitive to aging time up to 10^6 s, but tended to decrease in aging time range beyond 10^6 s, especially for the samples deformed at high temperature. On the other hand, the tensile elongations were similarly insensitive to aging time up to 10^6 s. At aging time range beyond 10^6 s, the tensile elongation of the samples deformed at room temperature decreased while the tensile elongation of the samples deformed at intermediate (873 K) and high (1073 K) temperatures steadily increased.

The creep curves showed that steady (or minimum) creep region is well defined and prevails the major part of the whole creep rupture time [6]. It was shown that the minimum creep rate decreases more steeply with decreasing applied stress, therefore indicating the existence of the threshold stress. It was demonstrated that the evaluated minimum creep rate is much lower in the two-phase intermetallic alloys than that e.g., in the Ni-20Cr superalloy dispersed by ThO$_2$ oxide [8]. **Fig. 7** shows applied stress vs. creep rupture time curves of the single crystal of Ni$_{75}$Al$_{7.5}$V$_{15}$Ti$_{2.5}$with the two-phase microstructure that were crept at 1173 K and 1223 K [3]. It is emphasized from this figure that the data of the two-phase intermetallic alloys are comparable with or superior to the data of most advanced superalloys.

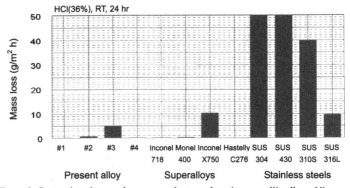

Figure 8. Comparison in mass loss among the two-phase intermetallic alloys, Ni superalloys and stainless steels in 36 % HCl solution for 24 h at room temperature.

354

CHEMICAL PROPERTY

The corrosion behavior of the two-phase intermetallic alloys was largely examined in various types of corrosive solution. **Fig. 8** shows that the corrosion resistance of some two-phase intermetallic alloys in 36 % hydrochloric acid is almost equivalent to or better than those of most Ni-based superalloys, and rather superior to those of many stainless steels. Similarly, fairly good corrosion resistance of the two-phase intermetallic alloys was observed in nitric acid, sulfuric acid and acetic acid, selecting an adequate alloy composition corresponding to each corrosive liquid.

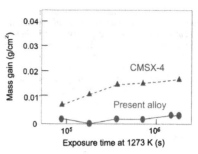

Figure 9. Comparison in mass gain between the two-phase intermetallic alloy and CMSX as a function of exposure time in air at 1273 K.

To understand oxidation behavior in air was particularly important to develop high-temperature structural materials. V element was shown to be a detrimental element for the oxidation property. Consequently, the oxidation resistance primarily decreases with increasing V, in other words, with decreasing Al content. Therefore, there is a trade-off relation between the oxidation resistance and density. On the other hand, the addition of Cr, Nb and Co to the two-phase intermetallic alloys was shown to be useful for improving the oxidation resistance.

The development of the two-phase intermetallic alloys with good oxidation resistance accompanied not only with good high-temperature strength but also with a relatively low density was currently undertaken in collaboration with industries. As one of the efforts, the two-phase intermetallic alloy with fairly better oxidation resistance at 1273 K in air than CMSX-4 has been developed, as shown in **Fig. 9**.

SURFACE HARDENING CAPABILITY

It was found that the surface layer of the two-phase intermetallic alloy was effectively hardened by both plasma-nitriding and -carburizing, depending on treating temperature and time, as shown in **Fig. 10**. The maximum surface hardness of the alloys was obtained by nitriding at ~850 K and by carburizing at ~1025 K, respectively. The XRD analysis revealed that vanadium

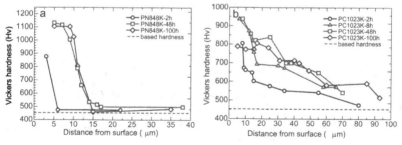

Figure 10. Hardness-depth curves of (a) the plasma-nitrided and (b) -carburized two-phase intermetallic alloy ($Ni_{70.5}Al_{10}V_{10.5}Nb_3Co_3Cr_3$).

355

nitrides (VN) and vanadium carbides (VC) were formed in the surface layer of the nitrided and carburized alloys, respectively, suggesting that the enhanced surface hardness is attributed to the dispersion hardening due to the nitrides or carbides although the role of the solid solution hardening due to nitrogen or carbon atoms in the constituent phases cannot be excluded from possible hardening mechanisms.

Figure 11. Some products or parts fabricated by the two-phase intermetallic alloys: (a) turbine blade for jet engine, (b) ball bearing, (c) friction stirring welding tool for high melting point materials, and (d) bolt and nut.

APPLICATION

In collaboration with industries, some parts such as turbine blade for jet engine, ball bearing, friction stirring welding tool for joining high melting point materials (such as steels and titanium alloys), and bolt and nut used at high temperature and/or in corrosive solution are currently developed. **Fig. 11** indicates examples of some products and parts actually fabricated from the two-phase intermetallic alloys. Here, it should be emphasized that various kinds of primary fabrication (i.e., conventional melting and casting) and secondary fabrication (conventional working, machining, heat-treatment and surface modification) procedures are applicable to the two-phase intermetallic alloys, thereby suggesting that inexpensive, reliable and environmentally harmonized production of the present novel materials is possible.

SUMMARY

For the two-phase intermetallic alloys with highly coherent interface structure between the constituent phases of Ni_3Al and Ni_3V, and also with high microstructural and phase stability at high temperature, physical, metallurgical, mechanical, chemical properties and surface modification processing have been critically evaluated. Based on these knowledge, the attempts of the compositional selection and microstructural control to develop jet-engine and gas turbine blade, high-temperature ball bearing, friction stirring welding tool for high melting point materials, and bolt and nut used at high temperature and/or in corrosive liquids have been conducted. The obtained results are promising for the development of the two-phase intermetallic alloy as a new-type of high-temperature structural materials.

REFERENCES

1. Y. Nunomura, Y. Kaneno, H. Tsuda, T. Takasugi, *Intermetallics* **12**, 389 (2004).
2. Y. Nunomura, Y. Kaneno, H. Tsuda, T. Takasugi, *Acta Mater.* **54**, 851 (2006).
3. S. Shibuya, Y. Kaneno, M. Yoshida, T. Shishido, T. Takasugi, *Intermetallics* **15**, 119 (2007).
4. S. Shibuya, Y. Kaneno, H. Tsuda and T. Takasugi, *Intermetallics* **15**, 338 (2007).
5. Y. Kaneno, W. Soga, H. Tsuda and T. Takasugi, *J. Mater. Science* **43**, 748 (2008).
6. S. Shibuya, Y. Kaneno, M. Yoshida and T. Takasugi, *Acta Materialia* **54**, 861 (2006).
7. W. Soga, Y. Kaneno, M. Yoshida and T. Takasugi, *Materials Sci. and Eng. A* **473**, 180 (2008).
8. R.W. Land and W.D. Nix, *Acta Metall.* **24**, 469 (1976).

Mater. Res. Soc. Symp. Proc. Vol. 1128 © 2009 Materials Research Society   1128-U06-06

Intermetallic Compounds in Co-Base Alloys–Phase Stability and Application to Superalloys

K. Ishida

Department of Materials Science,

Graduate School of Engineering, Tohoku University,

6-6-02 Aoba-yama, Sendai 980-8579, Japan

ABSTRACT

The phase stability of intermetallic compounds of Co_3X has been discussed with a focus on the γ' phase of Co_3Al, Co_3W and $Co_3(Al, W)$ with the $L1_2$ structure. The critical temperatures of the γ' phase of Co_3Al, Co_3W and $Co_3(Al, W)$ compounds are estimated to be about $870\,°C$, $980\,°C$ and $1076\,°C$, respectively. The effect of alloying elements on the Co-Al-W -base alloys was found to be very similar to that of Ni-base superalloys, where Ti, Ta, Nb and V are the γ' stabilizing elements, while Mn, Fe and Cr are the γ forming elements. The mechanical properties of Co-Al-W-base alloys were also found to be similar to those of Ni-base alloys. In particular, the flow strengths of Co-Al-W-base alloys at temperatures above $800\,°C$ were comparable or higher than those of Ni-base superalloys, which implies that the Co-base superalloys strengthened by the γ' phase have great potential as a new type of high-temperature alloys.

INTRODUCTION

Nickel-base superalloys have been widely used for high-temperature applications such as aircraft engines, industrial gas turbines, reactors, chemical industry, etc., where the intermetallic compound of Ni_3Al with the $L1_2$ structure plays a key role in the contribution to the strength at higher temperature [1]. As shown in figure 1 (a) of the phase diagram of the Ni-Al binary system [2], the γ' phase of the Ni_3Al compound appears stably up to $1395\,°C$, while no Co_3Al compound is formed, as shown in figure. 1 (b). As discussed on the GCP (Geometrically Close-Packed) phases which form A_3B of Co-base alloys [3], two types of GCP phase have been reported in Co_3X : Co_3Ti with the $L1_2$ structure [4,5] and ordered fcc Co_3Ta [6,7]. Although the effect of various alloying elements on the stability and morphology of the γ' Co_3Ti phase has been investigated, the usefulness of the γ' phase is restricted to temperatures below $750\,°C$. In the

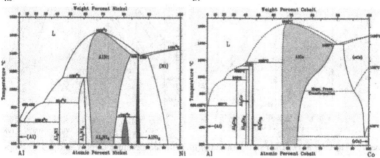

Figure 1   Phase diagrams of (a) Ni-Al and (b) Co-Al binary systems.

case of Co₃Ta, the ordered fcc phase is metastable and ready converts to the hexagonal closed-packed structure of Co₃Ta [8,9]. Therefore, GCP phases, Co₃X with the L1₂ structure, do not form easily in Co-base alloys, so other ordered phases such as hcp Co₃W and Co₃Mo are expected as strengthening phases. On the other hand, CoAl compound with the B2 structure is very stable as well as in the case of NiAl phase as shown in figure 1, which suggests that the enthalpy of formation of the metastable Co₃Al γ' phase is not a small negative value. In fact, Ohtani et al. [10] have recently reported that the enthalpy of formation of metastable Co₃Al with the L1₂ structure is – 19900 J/mol by the first principle calculation.

In this paper, the phase stability of intermetallic compounds in Co-base alloys is discussed with a focus on the Co₃X phase compared with those of Ni-base alloys. Furthermore, the phase stability of the Co₃ (Al, W) ternary compound with the L1₂ structure, which was recently discovered by our group [11], is shown and the potential for a new-type of Co-base superalloys using this ternary γ' phase is presented.

PHASE STABILITY OF CoX and Co₃X

According to the phase diagrams of Co-X binary systems [2], many compounds are formed. In view of the relative stability of intermetallic compounds, the enthalpy of formation Δ H and the melting temperature of compound (MP) are important. In the case of CoX compound, these data are shown in figure 2 compared with those of the NiX compound, where Δ H of CoX and the NiX is taken from Miedema's compilation [12]. The MP of most CoX compounds is higher than that of the NiX phase, but the Δ H of NiX shows more negative values. Moreover, no intermetallic compound is formed for Fe, Ru and Os of the VIII group, while MP of CoX and NiX for Al and Ga are almost the same and high, but Δ H of these elements does not have such large negative values. It is noted that the crystal structure of CoX and NiX for IIIb elements is the B2 ordered phase although no compound is formed in the Co-In binary system. Figure 3 shows the β (B2)-phase field in the Co-Al-X system compared with that of the Ni-Al-X system [13]. In the Co-Al-X system with X=Mn, Fe, Ni or Cu the β -phase field extends over a wide range of composition, which is quite similar to that of the Ni-Al-X system.

Similar data on Δ H and M.P for Co₃X- and Ni₃X-type compounds are shown in figure 4. Although the enthalpy of formation for Ni₃X has a larger negative value than that for Co₃X, the difference is small. Particularly noteworthy is the formation of Co₃X and Ni₃X of IIIb and IVb elements. Ni₃X compounds with the L1₂ or DO₁₉ structure are formed while no compound of Co₃X for IIIb elements appears as shown in Table 1. Since Δ H of Co₃X for IIIb elements shows values similar to those for Ni₃X, the metastable compound with the L1₂ structure of the Co₃X compound for the IIIb element would be expected to appear.

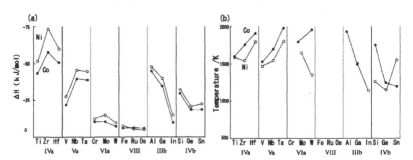

Figure 2   (a) Enthalpy of formation and (b) melting temperature of NiX and CoX compounds.

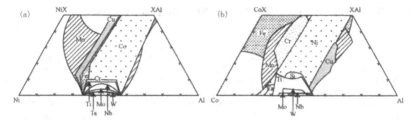

Figure 3 β (B2)-phase field in the (a) Ni-Al-X and (b) Co-Al-X systems.

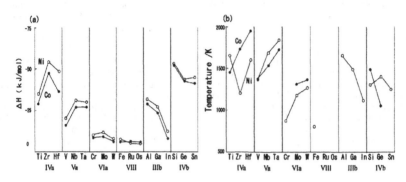

Figure 4   (a) Enthalpy of formation and (b) melting temperature of Ni₃X and Co₃X compounds.

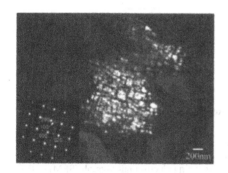

Figure 5   Transmission electron micrograph of metastable Co₃Al phase with L1₂ structure heat-treated at $600\,°C$.

Table 1   Ni₃X and Co₃X compounds for IIIb and IVb elements.

	X	Ni₃X	Co₃X
	Al	L1₂	–
IIIb	Ga	L1₂	–
	In	DO₁₉	–
	Si	L1₂	tetragonal
IVb	Ge	L1₂	Cr₃Si ?
	Sn	DO₁₉	–

PHASE STABILITY OF L1₂ COMPOUND IN THE Co-Al-W SYSTEM

Metastable Co₃Al and Co₃W phases

Although no stable Co₃Al compound is formed in the Co-Al binary system as shown in figure 1 (b), the formation of an ordered Co₃Al phase has been suggested [14,15]. Recently, we found direct evidence of a metastable Co₃Al phase with the L1₂ structure, which is formed in the Co-14 at % Al alloy annealed at 600 °C for 24 hours as shown in figure 5 [16]. This fact suggests that the critical temperature of the metastable γ' Co₃Al phase is above 600 °C. The thermodynamic information of the γ' Co₃Al phase is very limited. First principle calculations suggest that both Co₃Al and CoAl₃ have a strong ordering tendency to form an L1₂ structure [10,17]. Based on experimental and thermodynamic data on the metastable Co₃Al phase and the phase equilibria on the γ' phase in the Co-Al-W ternary system, the thermodynamic calculation of the Co-Al binary system including the Co₃Al phase has been performed [18]. Figure 6 shows the calculated phase diagram of Co-Al binary system, where the metastable fcc phase relation is also included. The critical temperatures of Co₃Al and CoAl₃ are estimated to be the same, i.e. about 870 °C , a little higher than that of L1₀ CoAl.

On the other hand, metastable Co₃W with the L1₂ structure has been reported in the precipitation of Co-W alloys aged at 600 °C and 700 °C [19], which suggests that the metastable Co₃W phase has the similar stability similar to that of Co₃Al. The only study of first principle calculation indicates that Co₃W has a certain ordering tendency to form the L1₂ structure, while CoW₃ cannot [17]. According to this information and recent experimental data [20], the thermodynamic assessments of Co-W binary system were carried out [18]. Figure 7 shows the calculated Co-W phase diagram as well as the fcc metastable phase equilibrium. Since only the ordering contribution of L1₂ Co₃W is taken into consideration, the fcc Co-W phase diagram is asymmetric. The critical temperature of Co₃W is calculated to be about 980 °C .

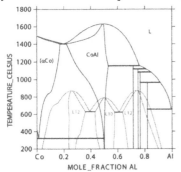

Figure 6   Calculated Co-Al binary phase diagram and metastable fcc phase relation.

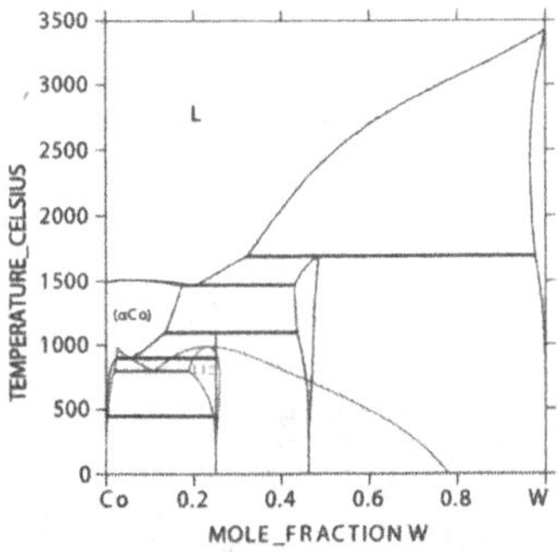

Figure 7   Calculated Co-W binary phase diagram and metastable fcc phase relation.

γ' phase in the Co-Al-W ternary system

Recently, we found a new ternary γ' phase of Co₃ (Al, W), where the composition of Al and W has an almost equiatomic ratio [11]. A typical scanning electron micrograph (SEM) of the $\gamma + \gamma'$ structure is shown in figure 8, where cuboidal γ' phase homogeneously precipitates in the γ (Al) matrix, which is very similar to the morphology observed in Ni-base superalloys.

Based on the thermodynamic analysis of Co-Al, Co-W and Al-W binary systems, CALPHAD-type assessment of the Co-Al-W ternary system was performed [18]. Figure 9 shows the calculated Co-Al-W isothermal section diagram at 900 °C compared with the experimental data, where good agreement is obtained between calculated and observed ones. Figure 10 shows the metastable γ / γ' phase equilibrium in the Co-rich corner at 950 °C, 1000 °C and 1050 °C. The critical temperature of the γ' phase is 1076 °C occurring at a composition of 10 at % Al and 15 at % W. The details of thermodynamic analysis of Co-Al-W system will be reported before long [18].

γ' phase in other Co-ternary systems

As shown in figure 7, the critical temperature of the metastable γ' Co_3W phase is estimated to be rather high at about 980 °C, which suggests that there is a great possibility for the formation of a ternary $L1_2$ compound of the Co-W-X system. Accordingly, we have found a new γ' phase in the Co-W-Ge ternary system [21]. Figures 11(a) and (b) show on an SEM image of a γ / γ' two-phase structure heat-treated at 800 °C for 15 hours and the differential scanning calorimetry (DSC) curve of Co-5.7W-11.5Ge (at %) alloy, respectively. Figure 11(a) shows that the distribution of precipitates of the γ' phase with a mean size of about 100 nm, is homogeneous and figure 11(b) shows that the solvus temperature of the γ' phase is about 907 °C. The composition of the γ' phase is Co-8.2W-12.6Ge (at. %). Other examples of the γ' phase of Co-W base alloys are shown in figure 12 [22]. The curboidal γ' phase is formed in the Co-W-Ga and Co-W-Ta ternary systems. It should be noted that the composition of the Co-W-Ga ternary γ' phase is Co-11.2W-11.4Ga (at. %) with an almost equiatomic ratio of W and Ga, which is quite similar to that of the Co_3(W, Al) γ' phase. Further investigation is required to clarify the nature of the ternary $L1_2$ compound of Co-W-X alloys.

EFFECT OF ALLOYING ON STABILITY OF γ' Co_3(Al, W) PHASE

The partition behavior of alloying elements between the γ and γ' phases is one of the important factors for alloy design in Co-base alloys as well as in Ni-base superalloys. Since the details of the effect of alloying elements on the partition behavior between the γ and γ' phases in Co-Al-W base alloys will be reported in a separate paper [23], a brief outline is shown here.

Figure 8   Scanning electron micrograph of γ + γ' structure of Co-9.2Al-10.2W(at.%) at 1000 °C.

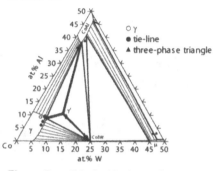

Figure 9   Calculated phase diagram of Co-Al-W ternary system at 900 °C.

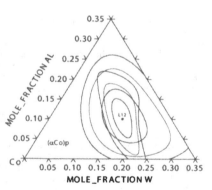

Figure 10   Metastable γ / γ' phase equilibrium in the Co-rich corner at 950 °C, 1000 °C and 1050 °C.

361

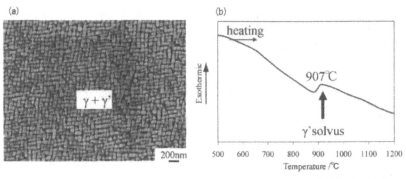

Figure 11 SEM image of a γ / γ' two-phase structure heat-treated at 800 °C for 15 hours and the DSC curve of Co-5.7W -11Ge (at. %) alloy.

Figure 12 SEM images of γ + γ' structure (a) Co-9.3W-9.5Ga (at. %) heat-treated at 900 °C for 72 hours and (b) Co-5W-2.5Ta (at. %) heat-treated at 800 °C for 8 hours.

Figure 13 shows the partition coefficient for various elements X between the γ and γ' phases of Co-9.8W-8.8Al-2X (at. %) alloy at 900 °C. Ti, Ta, Nb and V are the strong γ' stabilizing elements, while Mn, Fe and Cr are the γ forming elements. This behavior is very similar to the case of Ni-base superalloys [24].

The effect of Ni on the phase equilibria and the γ' phase stability of Co-Al-W alloy is very interesting because it is expected that the γ' Co₃ (Al, W) phase changes continuously to the γ' phase of the Ni-Al-W system. Figure 14 shows the isothermal section diagram of the Co-Ni-Al-W quaternary system at 900 °C [25]. Since the equilibrium compositions of Ni in the γ and γ' phases are not so different between the two phases, the phase relations are given in (Co + Ni)- Al-W two-dimensional

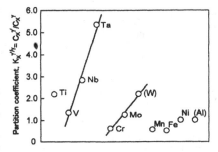

Figure 13 Partition coefficients of alloying elements between the γ and γ' phases of Co-9.8W-8.8Al-2X (at. %) alloy at 900 °C .

362

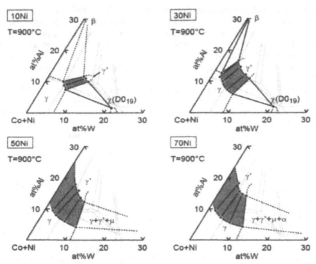

Figure 14 Isothermal section diagram of Co-XNi-Al-W (X=10, 30, 50, 70) system at 900 °C . Grey lines and shaded regions indicate the phase boundaries in Co-Al-W ternary system and the γ / γ' two-phase regions in Co-Ni-Al-W quaternary system, respectively.

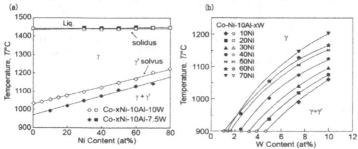

Figure 15 (a) γ' solvus and solidus temperatures of Co-XNi-10Al-7.5W and Co-XNi-10Al-10W alloys. (b)Vertical section diagram of Co-Ni-10Al-W alloys with various Ni and W contents, showing the γ / γ +γ' phase boundary, where the solid and open symbols indicate the data obtained by DSC and EPMA, respectively.

figures assuming that the Ni contents are equal in the γ and γ' phases and also in the χ (DO₁₉ : Co₃W), β (B2 : CoAl), μ (D8₅ : Co₇W₆) and α (A2 : W) phases. The grey lines and shaded regions indicate the phase boundaries in the Co-Al-W ternary system and the γ / γ' two-phase region in Co-Ni-Al-W quaternary system, respectively. While the composition range of the stable γ' is small in the Co-Al-W ternary system, the γ' phase expands and shifts to lower W and higher Al content regions with increasing Ni content. Moreover, the γ' phase exists even without W in the composition region over 50% Ni, which can be confirmed by the Ni-Co-Al ternary phase diagram [26], although the W is necessary to form the γ'-L1₂ compound in the region less than 40% Ni. It is also seen that the solubility of W in the γ phase increases with increasing Ni content.

363

The vertical section diagrams were determined by DSC and electron probe microanalysis (EPMA). The γ' solvus as a function of Ni and W is shown in figures 15 (a) and (b), respectively. The γ' solvus temperature increases with increasing Ni content, while the solidus temperature is hardly affected by the Ni content. It is also observed that the γ' solvus temperature increases with increasing W content from 7.5 at% to 10 at%. It is also seen from figure 15(b), that the γ / γ + γ' boundary monotonously increases with increasing Ni and W contents and that the substitution of Ni for Co is effective for increasing the γ' solvus temperature, at least at the fixed Al content of 10 at%, except for the 70Ni alloys with low W content. Figure 16 shows FE-SEM images of the microstructure of (82.5-X) Co-XNi-10Al-7.5W alloys (X=10, 40, 50 and 60) annealed at 900 °C for 168 hours. The 10Ni alloy shows cuboidal γ' precipitates 200-300 nm in size, and the morphology becomes spherical as Ni content increases. It is known that the morphology is affected by the lattice mismatch between the matrix and the precipitate; the γ' precipitates are formed as spheres at a 0%-0.2% mismatch, as cubes at about 0.5%-1.0% and as plates at above about 1.25% in Ni-based superalloys [1]. It was experimentally confirmed that the lattice parameters of both the γ and γ' phases decrease with increasing Ni content and that the lattice mismatch also decreases from 0.4184% to 0.1958%. Therefore, the variation of the morphology of the γ' phase in figure 16 is due to the change of the mismatch between the γ matrix and γ' precipitate. The change of morphology from cubic to spherical is also observed in the case of Mo addition to Co-Al-W ternary alloy [11]. Since variation in the γ' mismatch affects the mechanical properties and the environmental susceptibility in Ni-base and Fe-base superalloys [1], further investigations on this effect are required in Co-base superalloys.

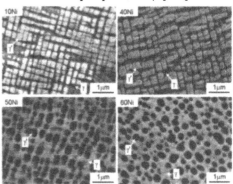

Figure 16 SEM images of Co-10Ni-10Al-7.5W, Co-40Ni-10Al-7.5W, Co-50Ni-10Al-7.5W and Co-60Ni-10Al-7.5W annealed at 900 °C for 168 hours.

MECHANICAL PROPERTIES

Co₃ (Al, W) γ' phase

The elastic properties of the γ' Co₃ (Al, W) phase have been recently reported as shown in Table 2 [27,28]. According to the first principle calculation [27], the γ' phase has a strengthening effect on the γ matrix due to the larger modulus differences between the γ and γ' phases but the γ' phase is brittle in nature. On the other hand, the experimental data on single-crystal elastic constants of the γ' phase by Tanaka et al. [28] are considerably smaller than those calculated values as shown in Table 2. The values of bulk, shear and Young moduli are also smaller than those of calculation and are 15%-25% larger than those for the Ni₃ (Al, Ta) compound. They concluded that the ductility of γ' Co₃ (Al, W) phase is expected to be sufficiently high, leading to excellent mechanical properties at high temperatures [28]. The mechanical properties of the γ' Co₃ (Al, W) polycrystalline phase were recently measured by compression testing at various temperature ranging from room temperature to 900 °C [29]. Figure 17 shows the temperature dependence of the 0.2% flow stress of the Co₃ (Al, W) phase compared with other L1₂ compounds. The 0.2% flow stress is 382 MPa and the compressive ductility is about 10% at 900 °C. A weak positive temperature dependence of γ' Co₃ (Al, W) phase is recognized. The small strain rate dependence of strength in the high temperature range is also observed. It is

364

Table 2 Elastic constants of γ' phase

(a) Single crystal

	Temperature (K)	C_{11} (Gpa)	C_{12} (Gpa)	C_{44} (Gpa)
Co₃ (Al, W) Experiment [28]	5	271	172	162
Co₃ (Al, W) Calculation [27]	0	363	190	212
Ni₃ (Al, Ta) Experiment [28]	5	238	154	130

(b) Poly crystal

	Temperature (K)	Bh (Gpa)	Gh (Gpa)	Eh (Gpa)	ν
Co₃ (Al, W) Experiment [28]	5	205	101	260	0.289
Co₃ (Al, W) Calculation [27]	0	248	148	370	0.251
Ni₃ (Al, Ta) Experiment [28]	5	182	82.8	216	0.303

Bh: bulk modulus   Gh: shear modulus   Eh: Young modulus   ½Poisson's ratio

noted that even at room temperature, γ' Co₃ (Al, W) shows ductile deformation higher than 10% plastic strain in compression, which is contrary to the theoretical calculation [27].

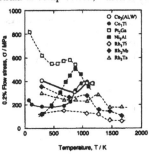

Figure 17 The temperature dependence of 0.2% flow stress of Co₃ (Al, W) and other L1₂ alloys.

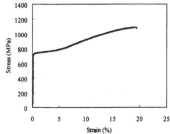

Figure 18 Tensile stress-strain curve of Co-9.2Al-9W (at. %) alloy at room temperature, where the sheet thickness and width of specimen are 1.15mm and 3.23mm, respectively. Gage length is 25mm and strain rate is 3.3×10^{-4} /sec.

γ + γ' Two-phase

Figure 18 shows the tensile stress-strain curve of Co-9.2Al-9W (at.%) alloy heat-treated at 900 °C for 1 hour at room temperature, where 0.2% proof stress, tensile strength and elongation are 737 MPa, 1085 MPa and 21%, respectively. These values are comparable to the tensile properties of Ni-base superalloys such as Waspaloy, with 0.2% proof and tensile strength of 795 and 1275 MPa, respectively, and 25% elongation [30].

The hardness of Co-base alloys together with that of the conventional Ni-base superalloy (Waspaloy) aged at various temperatures for 1 week is shown in figure 19. The peak hardness of Co-Al-W base alloys is obtained at an aging temperature of about 800 °C, while that of Co-Cr-Ta alloys is obtained at about 700 °C. This difference in peak temperature may be due to the phase stability of the γ' phase of γ' Co₃ (Al, W) and Co₃ Ta, where β -Co₃ Ta with a rhombohedral structure precipitates at 800 °C [9].

Recently, high-temperature strength and deformation behavior of the γ + γ' two-phase structure of Co-Al-W base alloys have been studied [31,32]. Figure 20 shows the temperature dependence of the compression 0.2% flow stress for Co-9Al-9W (at. %) and Co-9Al-10W-2Ta (at. %) alloys together with those of conventional Ni-base and Co-base superalloys [32,33]. It is seen that the positive temperature dependence of flow stress is observed above 600 °C in Co-Al-W-base alloys, which is very similar to that of the Co₃ (Al, W) γ' phase as shown in figure 17. Compression tests on the single-crystal specimens have revealed that the anomalous peak temperature of both Co-Al-W and Co-Al-W-Ta alloys is about 800 °C [32]. The Ta-containing alloy sustains its strength above 800 °C due to the activation of 1/3 (112) partial dislocations that shear the γ' precipitates and form a high density of stacking faults in the precipitates, while the Co-Al-W ternary alloy shows a rapid decrease in the flow stress above the peak temperature, due to 1/2 (110) dislocations by-passing the precipitates.

Figure 21 shows the most recent work on the 0.2% flow stress of Co-Al-W ternary alloys as a function of the volume fraction of the γ' phase [34]. The positive temperature dependence above 600 °C was confirmed. The γ + γ' phase with the volume fraction of the 60-80% γ' phase shows higher strength, which is similar to that of Ni-Al base alloys [1]. In these alloys, grain boundary fracture was observed, as has already been reported [32]. Since B, Zr and Hf have a remarkable effect on the suppression of grain boundary brittleness in Ni-base superalloys [1], the effect of a very small addition of B on the high-temperature compression strength of Co-Al-W alloy was studied. Figure 22 shows the 0.2% flow stress of this alloy with and without B, where the results for a single crystal of Co-9.4Al-10.7W and Co-8.8Al-9.8W-2Ta alloys are

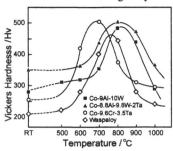

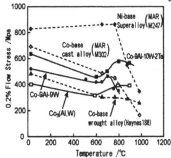

Figure 19 Vickers hardness of Co-base alloys aged at various temperatures for 1 week.

Figure 20 Temperature dependence of the compression 0.2 % flow stress for Co-9Al-9W (at. %) and Co-9Al-10W-2Ta (at. %) alloys together with conventional Ni-base and Co-base superalloys.

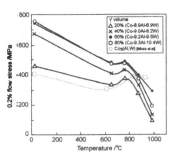

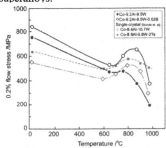

Figure 21 Compression 0.2% flow stress for Co-Al-W ternary alloys as a function of volume fraction of the γ' phase.

Figure 22 Compression 0.2% flow stress for Co-Al-W alloy with and without B. The data on single crystal is also shown [32].

366

also included for comparison [32]. The positive temperature dependence of flow stress of polycrystal Co-Al-W alloy without B is very weak, while that of alloy with B has a high positive one. The addition of B has a drastic effect on the grain boundary strengthening. Systematic studies for other elements such as C, Zr, Hf etc. are required.

The reduction of area and flow tensile stress of Co-19.9Ni-6.4W-9.1Al-1.2Ta-5Cr (at. %) alloy measured by the Gleeble test is shown in figure 23 in comparison with those of conventional Ni-base wrought alloy, Udimet 520 [35]. This alloy with a volume fraction of 40% of the γ' phase exhibits high ductility and higher flow stress at higher temperatures as compared with Udimet 520, and is expected to have great potential as a Co-base wrought superalloy.

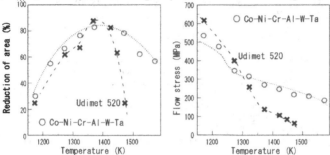

Figure 23 Reduction of area and deformation resistance on strain-rate tensile test of Co-19.9Ni-6.4W-9.1Al-1.2Ta-5Cr (at. %) alloy, where the dimension of specimen is 6mm diameter and 55mm length and crosshead speed is 50.8mm/s.

SUMMARY

The phase stability of Co_3X compounds was discussed with a focus on Co_3Al-, Co_3W- and $Co_3(Al, W)$-base phases. It was shown that the critical temperatures of the metastable γ' phase of Co_3Al and Co_3W with the $L1_2$ structure are estimated as 870 °C and 980 °C, respectively, while that of the $Co_3(Al, W)$ ternary compound is about 1076 °C. The partition behavior of alloying elements between the γ and γ' phases in Co-Al-W base alloys were shown to be very similar to that of Ni-base alloys. The results on the mechanical properties of Co-base alloys strengthening γ' $Co_3(Al, W)$ phase were similar to those on Ni-base superalloys such as tensile properties at room temperature, aging behavior, high temperature strength indicating positive temperature dependence, etc. The small addition of the microalloying element of B showed a drastic effect on the suppression of grain boundary brittleness. In particular, the flow strength at temperatures higher than 800 °C was found to be rather high compared with that of Ni-base superalloys, which implies that the Co-base superalloys have great potential as a new type of high temperature alloys. In this sense, the thermodynamic database for the prediction of phase constituent, the volume fraction of the γ' phase and other phases, solidification sequence, etc., should be constructed. Needless to say, much data such as creep and fatigue properties, oxidation and corrosion resistance, welding, coating, etc., is required.

Acknowledgement

The author would like to thank Drs. R. Kainuma, K. Oikawa, M. Jiang, I. Ohnuma, T. Omori and J. Sato for their help and useful discussions. This work was supported by a Grant-in Aid for Scientific Research from the Japan Society for the Promotion of the Science. Support from CREST, Japan Science and Technology Agency and from the Global COE Program is also acknowledged.

REFERENCES

1. R. F. Decker and C. T. Sims, in The Superalloys, C. T. Sims, W. C. Hagel, Eds., Wiley, New York, 33 (1972).
2. Binary Phase Diagrams, Vol. 1, 2nd ed., by T.B. Massalski, H. Okamoto, P. R. Subremanian and L. Kacprzak, ASM Int. Materials Park, OH 136, 181 (1990).
3. C. T. Sims, in The superalloys, C. T. Sims, W. C. Hagel, Eds., Wiley, New York, 145 (1972).
4. J. M. Blaise, P. Viatour and J. M. Drapier, Cobalt 49, 192 (1970).
5. P. Viatour, J. M. Drapier and D. Coutsouradis, Cobalt 3, 67 (1973).
6. M. Korchynsky and R. W. Fountain, Trans. Met. Soc. AIME 215, 1033 (1959).
7. R. D. Dragsdorf and W. D. Forgeng, Acta Cryst. 15, 531 (1962).
8. J. M. Drapier, J. L. de Brouwer and D. Coutsouradis, Cobalt 27, 59 (1965).
9. J. M. Drapier and D. Coutsouradis, Cobalt 39, 63 (1968).
10. H. Ohtani, M. Yamano and M. Hasebe, Calphad 28, 177 (2004).
11. J. Sato, T. Omori, K. Oikawa, I. Ohnuma, R. Kainuma and K Ishida, Science 312, 90 (2006).
12. F. R. de Boer, R. Boom, W. C. M. Mattens, A. R. Miedema and A. K. Niessen, Cohension in Metals, Transition Metal Alloys, Noth-Holland, (1989).
13. K. Ishida, R. Kainuma and T. Nishizawa, Symp. Proc. on Mechanical Properties and Phase Transformations of Multi-phase Intermetallic Alloys, TMS, 77 (1995).
14. A. J. Bradley and G. C. Seager, J. Inst. Met. 64, 81 (1939).
15. O. S. Edwards, Inst. Met. 67, 67 (1941).
16. T. Omori, Y. Sutou, K. Oikawa, R. Kainuma and K. Ishida, Mater. Sci. Eng. A438-440, 1045 (2006).
17. G. H. Johannesson, T. Bligaard, A. V. Ruban, H. L. Skriver, K. W. Jacobsen and J. K. Norskov, Phys. Rev. Lett. 88, 255506 (2002).
18. M. Jiang, J. Sato, K. Oikawa, I Ohnuma, R. Kainuma and K. Ishida, To be submitted.
19. J. Dutkiewicz and G. Kostorz, Acta metal mater. 38, 2283 (1990).
20. J. Sato, K. Oikawa, R. Kainuma and K.Ishida, Mater. Trans. 46, 1199 (2005).
21. H. Chinen, J. Sato, T. Omori, K. Oikawa, I. Ohnuma, R. Kainuma and K. Ishida, Scripta Mater. 56, 141 (2007).
22. H. Chinen, T. Omori, K. Oikawa, I. Ohnuma, R. Kainuma and K. Ishida, To be submitted.
23. T. Omori, J. Sato, K. Oikawa, I. Ohnuma, R. Kainuma and K. Ishida, To be submitted.
24. C. C. Jia, K. Ishida and T. Nishizawa, Metall. Mater. Trans. A 25A, 473 (1994).
25. K. Shinagawa, T. Omori, J. Sato, K. Oikawa, I. Ohnuma, R. Kainuma and K. Ishida, Mater. Trans. 49, 1474 (2008).
26. R. Kainuma, M. Ise, C. C. Jia, H. Ohtani and K. Ishida, Intermetallics 4, S151 (1996).
27. Q. Yao, H. Xing and J. Sun, Appl. Phys. Lett., 89, 161906 (2006).
28. K. Tanaka, T. Ohashi, K. Kishida and H Inui, Appl. Phys. Lett. 91, 181907 (2007).
29. S. Miura, K. Ohkubo and T. Mohri, Mater. Trans. 48, 2403 (2007).
30. N. S. Stoloff., in Metals Handbook, ASM International ed. 10, vol. 1, 950 (1990).
31. A. Suzuki, G. C. DeNolf and T. M. Pollock, Scripta. Mater. 56, 385 (2007).
32. A. Suzuki and T. M. Pollock, Acta Materialia 56, 1288 (2008).
33. A. Suzuki, G. C. DeNolf and T. M. Pollock, MRS Proc. 980, 499 (2007).
34. K. Shinagawa, T. Omori, J. Sato, K. Oikawa, I. Ohnuma, R. Kainuma and K. Ishida, Presented at 2008 MRS Fall Meeting.
35. M. Osaki, S. Ueta, T. Shimizu, T. Omori and K. Ishida, Denki-Seiko 79, 197 (2008).

Mater. Res. Soc. Symp. Proc. Vol. 1128 © 2009 Materials Research Society   1128-U06-07

Physical and Mechanical Properties of Single Crystals of Co-Al-W Based Alloys with L1₂ Single-Phase and L1₂/fcc Two-Phase Microstructures

Haruyuki Inui, Katsushi Tanaka, Kyosuke Kishida, Norihiko L. Okamoto and Takashi Oohashi
Department of Materials Science and Engineering, Kyoto University
Sakyo-ku, Kyoto 606-8501, Japan

ABSTRACT

Single-crystal elastic constants of $Co_3(Al,W)$ with the cubic L1₂ structure have been experimentally measured by resonance ultrasound spectroscopy at liquid helium temperature. The values of all the three independent single-crystal elastic constants and polycrystalline elastic constants of $Co_3(Al,W)$ experimentally determined are 15~25% larger than those of $Ni_3(Al,Ta)$ but are considerably smaller than those previously reported. Two-phase microstructures with cuboidal L1₂ precipitates being well aligned parallel to <100> and well faceted parallel to {100} are expected to form very easily in Co-base superalloys because of the large value of E_{111}/E_{100} and c_{ij} of $Co_3(Al,W)$. This is indeed confirmed by experiment. Values of yield stress obtained for both [001] and [123] orientations of L1₂/fcc two-phase single crystals moderately decrease with the increase in temperature up to 800°C and then decrease rapidly with temperature above 800°C without any anomaly in yield stress. Slip on {111} is observed to occur for both orientations in the whole temperature range investigated.

INTRODUCTION

Many intermetallic compounds such as TiAl have been investigated in the last two decades because of ever-increasing demands for structural materials that can withstand severe oxidizing environments and high operating temperatures [1]. The most widely used high-temperature structural materials in aircraft engines and power generation systems are at present Ni-base superalloys containing the L1₂-ordered intermetallic phase. The excellent high-temperature mechanical properties of these Ni-base superalloys have been believed to be closely related to the two-phase microstructure consisting of the solid-solution based on Ni with a face-centered cubic (fcc) structure and the stable L1₂-ordered intermetallic compound based on Ni_3Al [2-4]. The L1₂ phase forms coherent cuboidal precipitates in the fcc matrix so that the L1₂ cuboidal precipitates align parallel to <100> and facet parallel to {100}. The extent of the alignment and faceting of L1₂ cuboidal precipitates is directly related to the extent of coherent strains caused by lattice mismatch between the two phases, significantly affecting the mechanical properties of Ni-base superalloys at high temperatures [5,6]. Although Co-based alloys exhibit hot corrosion, oxidation and wear resistance better than Ni-base superalloys, their mechanical properties at high temperatures are usually inferior to those of Ni-base superalloys [7,8]. Since there is no stable binary L1₂ phase coexisting with the solid-solution based on Co with a fcc structure, it had been believed to be impossible to develop 'Co-base superalloys' with L1₂ cuboidal precipitates aligned parallel to <100> and faceted parallel to {100}. However, the recent discovery of the stable L1₂-ordered intermetallic compound, $Co_3(Al,W)$ coexisting with the solid-solution based on fcc-Co [9] has opened up a pathway to the development of a new class of high-temperature structural material based on cobalt, 'Co-base superalloys'.

Two-phase microstructures that resemble those in Ni-base superalloys have been proved to form through higher-order alloying additions to Co-Al-W based alloys [9,10]. However, almost nothing is known about mechanical properties of these Co-based two-phase alloys [9-11]. This is especially the case for the constituent $L1_2$ phase, $Co_3(Al,W)$ [12,13]. The $L1_2$-ordered intermetallic compound, $Co_3(Al,W)$ is expected to exhibit physical properties similar to those exhibited by Ni_3Al-based compounds that are the constituent phase of Ni-base superalloys. However, the result of the recent first principles calculation of the elastic constants by Yao *et al.* [14] indicates that unlike Ni_3Al, the $L1_2$-ordered intermetallic compound, $Co_3(Al,W)$ is extremely brittle. If this is indeed the case, Co-base superalloys containing the 'brittle' $L1_2$ compound, $Co_3(Al,W)$ may exhibit low toughness especially at low temperatures and their practical use will be significantly limited when compared to Ni-base superalloys. In the present study, we experimentally investigate elastic constants of single crystals of $Co_3(Al,W)$ with the $L1_2$ structure and compression deformation behaviors of single crystals with the $L1_2$/fcc two-phase microstructures, in order to see if the compound is indeed brittle or ductile enough to make further efforts for the development of 'Co-base superalloys'.

EXPERIMENTAL PROCEDURES

Single crystals of the Co-Al-W ternary system with $L1_2$ single-phase and $L1_2$/fcc two-phase microstructures were grown by the optical floating-zone method from rods respectively with nominal compositions of Co-10 at.%Al-11 at.%W and Co-9 at.%Al-9 at.%W prepared by Ar arc-melting. Specimens with a rectangular parallelepiped shape having three orthogonal faces parallel to {001} were cut from the single crystals with a $L1_2$ single-phase microstructure by spark-machining for measurements of elastic constants. Measurements of the elastic constants were carried out by ultrasound resonance spectroscopy (RUS) at liquid helium temperature [15,16]. In order to compare elastic constants of the $L1_2$ compound, $Co_3(Al,W)$ with those of the $L1_2$ compound based on Ni_3Al, measurements of the elastic constants at liquid helium temperature were carried out also for a single crystal of $Ni_3(Al,Ta)$ with a nominal composition of Ni-22 at.%Al-2 at.%Ta. Specimens with a rectangular parallelepiped shape having dimensions of 2 x 2 x 5 mm^3 were cut from the single crystals with $L1_2$/fcc two-phase microstructures by spark-machining for compression tests. Compression tests were made with an Instron-type testing machine at a strain rate of 1 x 10^{-4} s^{-1} in the temperature range of room temperature to 1000°C. The compression axes investigated are [001] and [$\bar{1}$23].

RESULTS AND DISCUSSION

Single-crystal elastic constants of the $L1_2$ compounds

Values of single-crystal elastic constants experimentally determined at liquid helium temperature for $Co_3(Al,W)$ and $Ni_3(Al,Ta)$ are tabulated in Table 1 together with those previously calculated for $Co_3(Al,W)$ by Yao *et al.* [14]. The values of all three independent elastic constants of $Co_3(Al,W)$ experimentally determined in the present study are 15~25% larger than those of $Ni_3(Al,Ta)$ experimentally determined also in the present study. This may come from the fact that the melting point for $Co_3(Al,W)$ is higher than that for $Ni_3(Al,Ta)$. The values of all three independent elastic constants of $Co_3(Al,W)$ experimentally determined are considerably smaller than those calculated by Yao *et al.* [14]. The reason for this discrepancy is not clear yet. Although

Yao *et al.* [14] never specified the chemical composition of the $L1_2$-ordered compound $Co_3(Al,W)$ they used in their calculation, the melting point for their $L1_2$-ordered compound may be considerably higher than that for the present $Co_3(Al,W)$ single crystal with the chemical composition of Co-10 at.%Al-11 at.%W. One point to be noted is that the solid-solubility range for the ternary $L1_2$-ordered compound $Co_3(Al,W)$ is reported to be quite small [9] and that we could get single crystals only in a very small composition range as reported in [9].

Polycrystalline elastic constants evaluated from the single-crystal elastic constants in Table 1 by the Hill's method are tabulated in Table 2. The values of bulk (B_h), shear (G_h) and Young (E_h) moduli experimentally determined for $Co_3(Al,W)$ are 15~25% larger than those experimentally determined for $Ni_3(Al,Ta)$ but are considerably smaller than those of $Co_3(Al,W)$ calculated by Yao *et al.* [14], as in the case of single-crystal elastic constants. The value of Poisson's ratio experimentally determined for $Co_3(Al,W)$ is smaller than that experimentally determined for $Ni_3(Al,Ta)$, indicative of more significant directionality of atomic bonding in $Co_3(Al,W)$ than in $Ni_3(Al,Ta)$. This implies that $Co_3(Al,W)$ is more brittle than $Ni_3(Al,Ta)$. Indeed, the increase in the value of shear moduli is more significant than that of bulk moduli for $Co_3(Al,W)$ than for $Ni_3(Al,Ta)$. The value of Poisson's ratio calculated for $Co_3(Al,W)$ is considerably smaller than that experimentally determined for $Co_3(Al,W)$. When judged from the values of Poisson's ratio, however, the $L1_2$-ordered compound, $Co_3(Al,W)$ is not as brittle as Yao et al. [14] expected from their calculation.

Anisotropic parameters evaluated from the single-crystal elastic constants in Table 1 are tabulated in Table 3. The value of Cauchy pressure (defined as c_{12}-c_{44} for crystals the cubic symmetry) experimentally determined for $Co_3(Al,W)$ is smaller than that experimentally determined for $Ni_3(Al,Ta)$. This again indicates the development of directional atomic bonding in $Co_3(Al,W)$ [17]. Of importance to note, however, is that the directionality of atomic bonding in

Table 1: Single-crystal elastic constants experimentally determined in the present study [13].

	Temperature (K)	c_{11} (GPa)	c_{12} (GPa)	c_{44} (GPa)
$Co_3(Al,W)$ experiment	5	271	172	162
$Co_3(Al,W)$ calculation [14]	0	363	190	212
$Ni_3(Al,Ta)$ experiment	5	238	154	130

Table 2: Polycrystalline elastic constants calculated by the Hill's method [13].

	Temperature (K)	B_h (GPa)	G_h (GPa)	E_h (GPa)	ν
$Co_3(Al,W)$ experiment	5	205	101	260	0.289
$Co_3(Al,W)$ calculation [14]	0	248	148	370	0.251
$Ni_3(Al,Ta)$ experiment	5	182	82.8	216	0.303

Table 3: Elastic anisotropy parameters [13].

	Temperature (K)	A	E_{111}/E_{100}	c_{12}-c_{44} (GPa)	G_h/B_h
$Co_3(Al,W)$ experiment	5	3.26	385/137=2.80	10	0.493
$Co_3(Al,W)$ calculation [14]	0	2.45	495/232=2.13	-22	0.585
$Ni_3(Al,Ta)$ experiment	5	3.09	315/117=2.69	24	0.455

Co₃(Al,W) is not as strong as that described with the large 'negative' value of Cauchy pressure, as calculated by Yao *et al.* [14]. Indeed, Co₃(Al,W) should be regarded as a ductile material, since the value of G_h/B_h for Co₃(Al,W) is far less than 0.57, above which the material of concern is regarded as a 'brittle' material according to the Pugh's criterion [18]. The smaller G_h/B_h value of Ni₃(Al,Ta) implies, of course, that Ni₃(Al,Ta) is more ductile than Co₃(Al,W). However, when judged from the values of Poisson's ratio, Cauchy pressure and G_h/B_h, the ductility of Co₃(Al,W) is sufficiently high so that Co₃(Al,W) can be practically used as the constituent phase of 'Co-base superalloys'.

The value of anisotropic factor ($A = 2c_{44}/(c_{11}-c_{12})$) experimentally determined for Co₃(Al,W) is a little larger than that experimentally determined for Ni₃(Al,Ta) but is considerably larger than that calculated for Co₃(Al,W). The value of E_{111}/E_{100} is also tabulated in Table 3, as deduced from the orientation dependence of Young's modulus of Fig. 1. The value of Young's modulus is the smallest in the <100> direction and is the largest in the <111> direction for all the three cases. This is consistent with the common trend observed for cubic crystals with the anisotropy factor value larger than unity [15,16]. The value of E_{111}/E_{100} experimentally determined for Co₃(Al,W) is larger than that experimentally determined for Ni₃(Al,Ta). The large E_{111}/E_{100} value of Co₃(Al,W) is beneficial to the formation of fcc (γ)/L1₂ (γ') two-phase microstructures in 'Co-base superalloys'. If the crystal is softer in the <100> directions than in the <111> directions, the interface between γ/γ' two phases tends to be parallel to {100}, resulting in cuboidal L1₂ precipitates. In addition, the driving force to form cuboidal precipitates aligning parallel to <100> is proportional to the magnitude of c_{ij}. Since Co₃(Al,W) has the large values of E_{111}/E_{100} and c_{ij}, two-phase microstructures with cuboidal L1₂ precipitates well aligned parallel to <100> and well faceted parallel to {100} are expected to form more easily in Co-base superalloys containing Co₃(Al,W) than in conventional Ni-base superalloys, when the lattice misfit between the two phases is assumed to be identical. This is expected to lead to the excellent mechanical properties at high temperatures.

Yield stress of single crystals with a L1₂/fcc two-phase microstructure

A SEM (scanning electron microscope) image of a single crystal with a L1₂/fcc two-phase microstructure is depicted in Fig. 2. As expected from the values of single-crystal elastic constants, cuboidal L1₂ precipitates (bright areas) are observed to be well aligned parallel to <100> and well

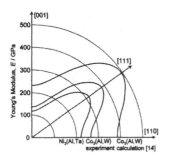

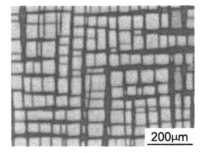

Fig. 1: Orientation dependence of Young's modulus [13].

Fig. 2: SEM image of a single crystal with a L1₂/fcc two-phase microstructure.

372

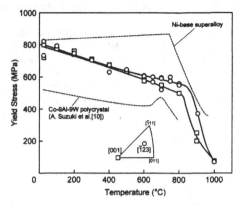

Fig. 3: Yield stress obtained for L1$_2$/fcc two-phase single crystals with [001] and [123] orientations. Those obtained for a similar two-phase polycrystalline Co-9 at.%Al-9 at.%W alloy by Suzuki et al. [10] are also indicated.

faceted parallel to {100} in the fcc matrix (dark area). The edge length of most cuboidal L1$_2$ precipitates is in the range of 50-100 nm and their volume fraction is about 60-70%. Yield stresses obtained for two-phase single crystals with [001] and [123] orientations are plotted in Fig. 3 as a function of temperature, together with those obtained for a similar two-phase polycrystalline Co-9 at.%Al-9 at.%W alloy by Suzuki et al. [10]. Slip lines observed for both orientations are exclusively those corresponding to slip on {111} in the whole temperature range investigated. Values of yield stress obtained for both orientations moderately decrease with the increase in temperature up to 800°C and then decrease rapidly with temperature above 800°C. Although a similar temperature dependence of yield stress is observed up to 650°C for the polycrystalline alloy [10], the yield stress values for the polycrystalline alloy is much lower than those obtained for the present single crystals. The difference in size for cuboidal L1$_2$ precipitates arising from different heat treatment may be the reason for that. On top of that, while Suzuki et al. [10] observed the small anomalous increase in yield stress with the peak temperature at 700°C for the polycrystalline alloy, no anomaly in yield stress is observed for both orientations of the present single crystals. The size of most cuboidal L1$_2$ precipitates in our single crystals seems to be too small to form superkinks on APB(anti-phase boundary)-coupled <110> dislocations with the relevant heights for the anomaly to appear. The dissociation scheme for <110> dislocations (of the APB-type or SISF (superlattice intrinsic stacking fault)-type) has yet to be determined.

CONCLUSIONS

(1) The values of all the three independent single-crystal elastic constants and polycrystalline elastic constants of L1$_2$-Co$_3$(Al,W) experimentally determined are considerably smaller than those previously calculated by Yao et al. [14]. Unlike the calculation result by Yao et al. [14], Co$_3$(Al,W) exhibits a positive value of Cauchy pressure and a G_h/B_h value less than 0.57. When judged from the values of Poisson's ratio, Cauchy pressure and G_h/B_h, the ductility of Co$_3$(Al,W) is expected to

be sufficiently high so that $Co_3(Al,W)$ can be practically used as the constituent phase of 'Co-base superalloys'.
(2) Because of the large values of E_{111}/E_{100} and c_{ij} of $Co_3(Al,W)$, two-phase microstructures with cuboidal $L1_2$ precipitates being well aligned parallel to <100> and well faceted parallel to {100} are expected to form very easily in Co-base superalloys containing $Co_3(Al,W)$, leading to the excellent mechanical properties at high temperatures. This is indeed confirmed by experiment.
(3) Values of yield stress obtained for both [001] and [123] orientations of $L1_2$/fcc two-phase single crystals moderately decrease with the increase in temperature up to 800°C and then decrease rapidly with temperature above 800°C. Slip on {111} is observed to occur for both orientations in the whole temperature range investigated. No anomaly in yield stress is observed for both orientations of the present single crystals.

ACKNOWLEDGMENTS

This work was supported by Grant-in-Aid for Scientific Research (A) (19656179) from the Ministry of Education, Culture, Sports, Science and Technology (MEXT), Japan and and in part by the Global COE (Center of Excellence) Program of International Center for Integrated Research and Advanced Education in Materials Science from the MEXT, Japan. The authors would like to thank Dr. Seiji Miura of Hokkaido University for supplying the single crystal of Ni-22 at.%Al-2 at.%Ta.

REFERENCES

1. M. Yamaguchi, H. Inui, and K. Ito, Acta Mater. 48, 307 (2000).
2. D.P. Pope, and S.S. Ezz, Intl. Metals Rev. 29, 136 (1984).
3. T.M. Pollock, and R.D. Field, in F.R.N. Nabarro, and M.S. Duesbery (Eds.), Dislocations in Solids, Vol. 11, Elsevier, Amsterdam, 546 (2002).
4. F.R.N. Nabarro, and M.S. Duesbery (Eds.), Dislocations in Solids, Vol. 10, Elsevier, Amsterdam, (1997).
5. H. Harada, K. Ohno, T. Yamagata, T. Yokokawa and M. Yamazaki, Superalloys 1988, 513 (1988).
6. T. Ichitsubo, and K. Tanaka, Acta Mater. 53, 4497 (2005).
7. A.M. Beltran, in C.T. Sims, N.S. Stoloff, and W.C. Hagel (Eds.), Superalloys II, Wiley, New York, 135 (1987).
8. T.C. Du Mond, P.A. Tully, K. Wikle, Metals Handbook, 9 th edition, Vol. 3, American Society for Metals, Metals Park, OH, 589 (1980).
9. J. Sato, T. Omori, I. Ohnuma, R. Kainuma, and K. Ishida, Science 312, 90 (2006).
10. A. Suzuki, G.C. DeNolf and T.M. Pollock, Scripta Mater. 56, 385 (2007).
11. A. Suzuki and T.M. Pollock, Acta Mater., 56, 1288 (2008).
12. S. Miura, K. Ohkubo and T. Mohri, Mater. Trans. JIM, 48, 2403 (2007).
13. K. Tanaka, T. Ohashi, K. Kishida and H. Inui, Appl. Phys. Let., 91, 181907 (2007).
14. Q. Yao, H. Xing, and J. Sun, Appl. Phys. Lett. 89, 161906 (2006).
15. K. Tanaka, and M. Koiwa, Intermetallics 4, S29 (1996).
16. K. Tanaka, and M. Koiwa, High Temp. Mater. Processes 18, 323 (1999).
17. D.G. Pettifor, Mater. Sci. Tech. 8, 345 (1992).
18. S. F. Pugh, Phil. Mag. 45, 823 (1954).
19. A. Couret, Y.Q. Sun and P.B. Hirsch, Phil. Mag. A, 67, 29 (1993).

Mater. Res. Soc. Symp. Proc. Vol. 1128 © 2009 Materials Research Society 1128-U06-08

Phase Stability of the L1$_2$ Compound and Microstructural Changes in Co-(W or Mo)-Ta Ternary Alloys

Hibiki Chinen[1], Toshihiro Omori[1], Katsunari Oikawa[1], Ikuo Ohnuma[1], Ryosuke Kainuma[2] and Kiyohito Ishida[1]
[1]Department of Materials Science, Graduate School of Engineering, Tohoku University, 6-6-02 Aoba-yama, Sendai 980-8579, Japan
[2]Institute of Multidisciplinary Research for Advanced Materials, Tohoku University, 2-1-1 Katahira, Sendai 980-8577, Japan

ABSTRACT

Microstructural investigations of Co-5W-2.5Ta (at.%) and Co-4Mo-4.5Ta (at.%) ternary alloys were conducted. Fine coherent precipitates were observed in these alloys annealed at 800°C, and the crystal structures of precipitates (γ') and matrix (γ) phase were identified as the L1$_2$ and A1 structure, respectively, by transmission electron microscopy. Cellular precipitation with a $\gamma+\chi$(D0$_{19}$) lamellar structure also proceeded at grain boundaries, and the alloys aged for a longer time only showed the $\gamma+\chi$ two-phase microstructure instead of the γ' phase. With aging at around 800°C, the peak hardness of these alloys with a $\gamma+\gamma$' two phase structure was about 580 Hv.

INTRODUCTION

Co-based alloys strengthened by solid solution hardening and carbide precipitation are commonly employed as high-temperature materials. On the other hand, Ni-base superalloys strengthened by the γ'-Ni$_3$Al precipitates with the L1$_2$ crystal structure are the most widely used alloys in high-temperature environments such as aircraft engines and industrial gas turbines [1]. The precipitation of the γ'-L1$_2$ phase in the disordered face-centered cubic (fcc) matrix phase is considered to be one of the key factors for high-temperature materials because of its superior high temperature strength and creep properties. Recently, new stable ternary compounds, i.e., Co$_3$(W,Al) and Co$_3$(W,Ge), have been found in the Co-W-Al [2] and Co-W-Ge [3] ternary systems. The Co-W-Al alloy strengthened by the γ' phase has been found to show superior strength at elevated temperatures. As well as in the Co-W base ternary systems, there have been some reports on the precipitation of geometrically close-packed (GCP) Co$_3$X phases in Co-based systems such as Co$_3$Ti [4-6], Co$_3$Al [7-9], Co$_3$W [10-11] and Co$_3$Ta [12-15] with the L1$_2$ structure. Co$_3$Ti is the only stable L1$_2$ phase in Co-based binary systems, and titanium-modified alloys such as the CM-7 alloy show good mechanical properties due to the homogeneous distribution of the γ'-(Co,Ni)$_3$(Ti,Al) phase. The γ' phase is, however, replaced by the η phase with the D0$_{24}$ structure in the temperature region above 750°C in the CM-7 alloy. In the Co-Al and Co-W binary systems it is known that the metastable γ' phase appears besides the stable β (B2) or χ (D0$_{19}$) phases in the fcc-γ Co solid solution [7-11]. Only a small amount of the γ' phase exists in a limited temperature range in these binary systems, and the stability is rather low. Also, in the Co-Ta based system, the γ'-Co$_3$Ta phase is metastable and transforms to the stable hexagonal β-Co$_3$Ta phase [12-13]. However, the γ' phase in the Co-Ta binary system is more stable compared to the Co-Al and Co-W systems, and the γ' phase exists after annealing at 975°C for 5 h [13]. It has also been reported that the metastable γ' phase exists after annealing at 700°C for 16 h in the Co-Cr-Ta ternary system [14]. Thus, it is interesting to investigate the phase stability of the γ' phase in the Co-W-Ta and Co-Mo-Ta systems where W and Mo belong to the same group VI as Cr. In this study, the phase stabilities of intermetallic compounds, including the L1₂ phase, appearing in the Co-W-Ta and Co-Mo-Ta ternary systems were investigated and are herein reported.

EXPERIMENTAL DETAILS

Co-5W-2.5Ta (at.%) and Co-4Mo-4.5Ta (at.%) alloys were prepared by arc melting and induction melting under an argon atmosphere, respectively, the starting materials being Co (99.9%), W (99.9%), Mo (99.9%) and Ta (99.9%). The Co-5W-2.5Ta alloy was alternately melted more than ten times during arc melting to obtain homogeneous ingots.

The obtained ingots were hot-rolled at 1200°C. Small specimens cut from the hot-rolled ingots were sealed in quartz capsules evacuated and backfilled with argon gas, and then homogenized at 1300°C for 4 hours, followed by quenching in ice water. The specimens were subsequently aged in the temperature range from 600°C to 1000°C for periods between 1 hour and 400 h, followed by water-quenching.

After the heat treatment, the quenched specimens were mechanically polished and etched to observe the microstructure. Microstructural observation was carried out using an optical microscope, a field emission scanning electron microscope (FE-SEM) and a transmission electron microscope (TEM), thin foil specimens for TEM observation being prepared by jet polishing in a solution of 20% perchloric acid and 80% ethanol.

The phase transition temperatures were determined by differential scanning calorimetry (DSC) under high-purity argon gas at a heating rate of 10°C /min.

The hardness was measured by the Vickers hardness test with loading of 500 g, and determined by a mean value for five data.

DISCUSSION

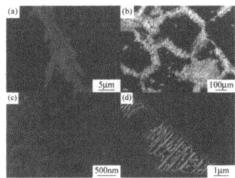

Figure 1 Secondary electron images of (a) Co-5W-2.5Ta (at.%) alloy, (b) Co-4Mo-4.5Ta (at.%) alloy and (c)(d) high magnification images of Co-5W-2.5Ta (at.%) alloy annealed at 800°C for 2 h.

Microstructral observation

Figures 1(a) and (b) show FE-SEM micrographs of Co-5W-2.5Ta and Co-4Mo-4.5Ta alloys annealed at 800°C for 2 h, lamellar structures accompanied by cellular precipitation being observed at grain boundaries in both alloys. Figures 1(c) and (d) show high magnification FE-SEM micrographs within a grain and near a grain boundary in the Co-5W-2.5Ta alloy. In addition to the lamellar domain at the grain boundaries, fine precipitates are seen in Figure 1(c). A

376

similar microstructure was also observed in the Co-4Mo-4.5Ta alloy. To investigate the microstructure of these alloys in detail, TEM observation was carried out. Figures 2(a) and (b) show the dark field image obtained using a $(100)L1_2$ super lattice reflection and the selected area diffraction pattern taken from the Co-5W-2.5Ta alloy annealed at 800°C for 1 h, respectively. There are fine coherent precipitates with a mean size of about 10-20 nm, whose morphology is similar to that observed in Ni-based superalloys. From the diffraction pattern shown in Figure 2(b), the crystal structures of the precipitate and matrix phase are identified as the $L1_2$ and A1 structures, respectively. As shown by the micrographs of the Co-4Mo-4.5Ta alloy annealed at 800°C for 2 h in Figures 2(c) and (d), fine precipitates with the $L1_2$ structure are formed in the A1 matrix with a size of about 20 nm.

Figures 3(a) and (b) show the microstructures of the Co-5W-2.5Ta alloy annealed at 800°C for 8 and 400 h, respectively. The area covered by the lamellar structure increases and the coarsening of the precipitates occurs with annealing time, as shown in Fig. 3(b). Here, the lamellar structure is identified with the γ and $\chi(Co_3W:DO_{19})$ phases by XRD examination, and no γ' phase was found in this specimen. It is considered from Figure 1(d) and Figures 3(a) and (b) that the lamella of $\gamma + \chi$ proceed to intragranular $\gamma + \gamma'$, resulting in the microstructural change, and the γ' phase transformed to the neighboring χ or γ phases at the boundary. Figure 3(c) shows the microstructure of the Co-4Mo-4.5Ta alloy annealed at 800°C for 8 h. The very fine lamellar structure with the $\gamma + \chi(Co_3Mo:DO_{19})$ phase covers almost the entire area, and a coarse lamellar region appears in the vicinity of the grain boundaries. This microstructure means that discontinuous coarsening of the lamellar structure occurs [16]. Since the γ' phase observed

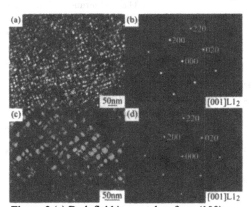

Figure 2 (a) Dark field image taken from $(100)L1_2$ reflection, (b) selected area diffraction pattern in Co-5W-2.5Ta (at.%) alloy annealed at 800°C for 1 h, (c) dark field image taken from $(100)L1_2$ reflection, and (d) selected area diffraction patterns in Co-4Mo-4.5Ta (at.%) annealed at 800°C for 2 h.

Figure 3 Back-scattering electron images of the Co-5W-2.5Ta (at.%) alloy annealed at 800°C for (a) 8 h and (b) 400 h. (c) Secondary electron image of the Co-4Mo-4.5Ta (at.%) alloy annealed at 800°C for 8 h.

in Co-5W-2.5Ta and Co-4Mo-4.5Ta alloys disappears by lengthy aging as demonstrated in Figure 3, the γ' phase in these systems is metastable at least at 800°C.

DSC measurement

Figure 4 shows the DSC heating curves obtained from the Co-5W-2.5Ta and Co-4Mo-4.5Ta alloys quenched from 1300°C. One or two endothermic peaks are detected at about 850°C in the Co-5W-2.5Ta alloy. It was confirmed by X-ray diffraction in the Co-5W-2.5Ta alloy annealed at 900°C for 3 h that this peak corresponds to the dissolution reaction of the metastable γ' phase. On the other hand, an exothermic peak at around 550°C and two endothermic peaks at

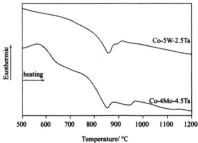

Figure 4 DSC heating curves of Co-5W-2.5Ta (at.%) and Co-4Mo-4.5Ta (at.%) alloy annealed at 1300°C for 4 h.

around 850°C and 950°C are detected in the Co-4Mo-4.5Ta alloy. The exothermic peak at 550°C seems to correspond to the precipitate reaction of the γ' phase, and at least one of the two endothermic peaks may be due to the dissolution of the γ' phase. In the Co-3.5Ta(at.%) binary alloy, it has been reported that the γ' phase exists at 900°C and 950°C for 1 h and disappears at 900°C and 950°C for 10 h [13]. The effect of W and Mo on the stability of the γ' phase was not accurately evaluated in this study; nevertheless, it is apparent that the phase stability of the L1$_2$ phase in the Co-Ta alloy is not drastically enhanced by the addition of W or Mo.

Hardness

Hardness measurements were carried out for the Co-5W-2.5Ta and Co-4Mo-4.5Ta alloys aged for 8 and 2 h, respectively. The obtained results are shown in Figure 5, where the hardness

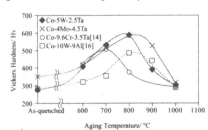

Figure 5 Vickers hardness of Co-5W-2.5Ta (at.%) alloy annealed for 8 hours and of Co-4Mo-4.5Ta (at.%) alloy annealed for 2 hours at each temperature.

was measured for the γ + γ' two phase region of the large grains. The hardness values of the as-quenched alloys are also plotted on the vertical axis for comparison with those in the aged alloys. It is seen that the Co-5W-2.5Ta and Co-4Mo-4.5Ta alloys are noticeably hardened by aging and that the Vickers hardness of each alloy reaches 584 Hv and 589 Hv, respectively. This hardening effect is clearly attributed to the precipitation of the γ' phase, as expected from the results of the microstructural investigations shown in Figure 2. It should be noted that the hardness values in the Co-W-Ta and Co-Mo-Ta alloys are higher than those of Co-9.6Cr-3.5Ta [14] and the Co-10W-9Al alloy [17], for almost the whole range of aging temperature.

378

CONCLUSIONS

(1) The existence of the γ' phase with the $L1_2$ structure was confirmed in Co-5W-2.5Ta and Co-4Mo-4.5Ta alloys aged at 800°C by TEM observation.

(2) The γ' phase disappears with lengthy aging due to the progress of the lamellar structure with $\gamma + \chi(Co_3W$ or $Co_3Mo)$, which means that the γ' phase is metastable at 800°C in these systems.

(3) The phase stability of the γ' phase in the Co-Ta alloy is not drastically heightened by the addition of W and Mo.

(4) The Co-5W-2.5Ta and Co-4Mo-4.5Ta alloys with $\gamma+\gamma'$ structure show a higher Vickers hardness than those of Co-Cr-Ta and Co-W-Al alloys.

ACKNOWLEDGMENTS

This work was supported by a Grant-in-Aid for Scientific Research from the Japan Society for the Promotion of the Science. Support from CREST, Japan Science and Technology Agency and from the Global COE Project is also acknowledged.

REFERENCES

1. C. T. Sims, N. S. Stoloff and W. C. Hagel, "Superalloys (2nd edition)", John Wiley & Sons, New York (1987) pp. 135-140.
2. J. Sato, T. Omori, K. Oikawa, I. Ohnuma, R. Kainuma and K. Ishida, Science **312**, 90 (2006).
3. H. Chinen, J. Sato, T. Omori, K. Oikawa, I. Ohnuma, R. Kainuma and K. Ishida, Scripta Mater. **56**, 141 (2007).
4. J. M. Blaise, P. Viatour and J. M. Drapier, Cobalt **49**, 192 (1970).
5. P. Viatour, J. M. Drapier, D. Coutsouradis and L. Habraken, Cobalt **51**, 67 (1971).
6. P. Viatour, J. M. Drapier and D. Coutsouradis, Cobalt 3, 67 (1973).
7. A. J. Bradley and G. C. Seager, J. Inst. Met. **64**, 81 (1939).
8. O. S. Edwards, J. Inst. Met. **67**, 67 (1941).
9. T. Omori, Y. Sutou, K. Oikawa, R. Kainuma and K. Ishida, Mat. Sci. Eng. A **438-440**, 1045 (2006).
10. J. Dutkiewicz and G. Kostorz, Acta Metall. Mater. **38**, 2283 (1990).
11. J. Dutkiewicz and G. Kostorz, Mater. Sci. Eng. A **132**, 267 (1991).
12. R. D. Dragsdorf and W. D. Forgeng, Acta Cryst. **15**, 531 (1962).
13. M. Korchynsky and R. W. Fountain, Trans. A.I.M.E. **215**, 1033 (1959).
14. J. M. Drapier, J. L. de Brouwer and D. Coutsouradis, Cobalt **27**, 59 (1965).
15. J. M. Drapier and D. Coutsouradis, Cobalt **39**, 63 (1968).
16. J. D. Livingston and J. W. Cahn, Acta Met. **22**, 495 (1974).
17. J. Sato, Ph.D. thesis, Tohoku University, Japan, (2006).

Mater. Res. Soc. Symp. Proc. Vol. 1128 © 2009 Materials Research Society   1128-U06-09

Catalytic Properties of Atomized Ni3Al Powder for Methane Steam Reforming

Yan Ma[1,2], Ya Xu[2], Masahiko Demura[2] and Toshiyuki Hirano[1,2]

[1]Graduate School of Pure and Applied Science, University of Tsukuba, 1-2-1, Sengen, Tsukuba, Ibaraki 305-0047, Japan
[2]National Institute for Materials Science, 1-2-1, Sengen, Tsukuba, Ibaraki 305-0047, Japan

ABSTRACT

The catalytic activity of Ni_3Al for methane steam reforming was investigated using its atomized powder. Taking advantage of the fine duplex microstructure, we made an attempt to increase the activity by the combined pretreatment of acid and subsequent alkali leaching. The powder surface became porous structure, having fine Ni particles on the porous surface. The pretreated powder exhibited high activity for methane steam reforming. The reason of the activity enhancement was attributed to the fine Ni particles on the surface.

INTRODUCTION

Ni_3Al has been known as an excellent high-temperature structural material and never attracted as a catalyst. We recently found that it has a potential as catalyst, which was previously unknown. Methanol decomposition was highly enhanced over Ni_3Al, producing hydrogen and carbon monoxide [1-3]. Interestingly, the catalytic activity was observed both in powder and foil of Ni_3Al. The catalytic activity was found to be due to the fine Ni particles on the surface which are produced by chemical pretreatment [1] or spontaneously produced during the catalytic reaction [2, 3]. Namely, Ni_3Al is regarded as a precursor alloy for catalyst.

Our next concern is to examine whether Ni_3Al exhibits catalytic activity for methane steam reforming or not. In this study we examined it using atomized powder. Since the atomized powder is rapidly solidified from liquid, it has a fine lamellar β'-NiAl in the Ni_3Al matrix [4, 5]. Taking advantage of this fine duplex microstructure, we made an attempt to form porous structure with Ni-enriched surface by the combined chemical pretreatment of acid and subsequent alkali leaching. Porous surface is, needless to say, advantageous to enhancing catalytic activity because of its high surface area. In our previous papers, we partly reported the microstructure change of atomized Ni_3Al powder due to various chemical pretreatments and the resultant catalytic properties for methane steam reforming [6,7]. In this study, we present the detailed microstructure change which leads to high catalytic activity.

EXPERIMENT

A stoichiometric Ni_3Al (Ni-25 at.% Al) powder was prepared by a gas atomizing process. The as-received powder was sieved for less than 150 μm in size. It was chemically pretreated in two steps, acid leaching and subsequent alkali leaching in the same way as our previous report [6,7]. First, it was dipped in 2 vol.% aqueous HNO_3 at 298 K for 15 min (acid leaching), then rinsed in deionized water and dried at 323 K for 8 h. Subsequently, it was dipped in a stirred 20 wt% aqueous NaOH solution at 366 K for 300 min (alkali leaching), then again rinsed in deionized water and dried at 323 K for 8 h. After each step of the pretreatments, the solution was subjected to an inductively coupled plasma (ICP) analysis in order to measure the amounts of aluminum and nickel leached from the powder. The surface area of the powder was determined by the Brunauer-Emmett-Teller (BET) surface area analysis method using krypton adsorption. Characterization of the powder was carried out by scanning electron microscopy (SEM), transmission electron microscopy (TEM) coupled with an X-ray energy dispersive spectroscopy (EDS) system, and X-ray diffraction (XRD) using a Cu Kα source.

Catalytic experiments were carried out in a conventional fixed-bed flow reactor at ambient pressure in the same way as described in our previous report [1-3, 6,7]. A 0.4 g sample of the pretreated Ni_3Al powder was used for each test. The height of the powder was about 2 mm. Prior to the reaction, the powder was reduced at 873 K for 1 h in a flowing hydrogen atmosphere. The hydrogen flow was then stopped and filled with pure nitrogen to flush out the hydrogen. Subsequently, mixed reactants of CH_4 and H_2O (mole ratio of $H_2O/CH_4=3$) were introduced into the reaction chamber at the gas hourly space velocity (GHSV) of 12000 h^{-1} (defined as the volume of CH_4 passed over the unit volume of catalyst per hour). The H_2O was fully evaporated in a thermal evaporator before introducing it into the reactor. Finally temperature was stepwise increased from 873 to 1232 K at an interval of 50 K. The composition and the flow rate of the outlet gas products were analyzed using a gas chromatograph and a soap bubble meter, respectively.

RESULTS AND DISCUSSION

As-received powder

The as-received powder has the overall stoichiometric composition of Ni_3Al. However, since it was rapidly solidified during the atomizing process, it has a second phase in the Ni_3Al matrix (shown as dark strip in Fig.1). The second phase is mostly lamellar (100-200 nm thick) and partly dendritic. The EDS analysis indicated that the second phase is slightly enriched in aluminum compared to the Ni_3Al matrix. Fig. 2, TEM bright field image, shows that the second phase consists of a lamellar structure with many planar faults. It was identified as $L1_0$–type β'-NiAl by the selected-area diffraction (SED) pattern, the inset of Fig. 2, which agrees with the result observed on the rapidly solidified Ni_3Al powder by Baker et al. [4, 5].

Fig.1 SEM image of the cross section of the as-received powder.

Fig. 2 TEM bright filed image of the second phase with lamellar structure. The inset is the SED pattern from the lamellar phase,

Chemically pretreated powder

Table 1 summarizes the ICP analysis results of the solutions used for each step of the chemical pretreatment. The amounts of leached aluminum and nickel are shown by the percentage to the chemical composition of the as-received powder. In the first step, aluminum and nickel atoms were almost equally leached form the powder into the HNO_3 solution. In the second step, a small amount of aluminum atom was leached out in the NaOH solution but no nickel was leached, indicating aluminum was selectively leached. The results suggest that the surface was enriched in nickel after the treatment.

Table 1 Amount of leached elements and BET surface area of the powder

Pretreatment	Amount of leached elements (wt%)		BET surface area $(m^2 \, g^{-1})$
	Al	Ni	
Step 1: Acid leaching	7.86	7.66	0.12
Step 2: Alkali leaching	3.47	BDL	0.41

BDL: below detection limit

The as-received powder was round in shape, having smooth surface. The smooth surface drastically changed into porous one after the first step of the pretreatment as shown in Fig. 3(a). The higher magnification SEM image, Fig. 3(b), shows that the lamellar β'-NiAl preferentially dissolved in the HNO_3 solution over the Ni_3Al matrix, leaving a porous structure on the powder surface. The preferential dissolution of β'-NiAl is reasonable from the previously reported fact that β'-NiAl is more easily etched by an acid solution than Ni_3Al [5]. In the second step, the surface became more complicated (Fig. 4(a)). The higher magnification SEM observation revealed that fine platelet or needle-like precipitates were formed on the surface as shown in Fig. 4(b)). TEM analysis, Fig. 5, revealed that many fine particles are present on the surface. The diffraction pattern (Fig. 5(b)) shows that the particles have an fcc-base structure. The EDS spectra were taken at positions A, B, C and D in Fig. 5(a). As shown in Fig. 5(c), all the spectra similarly show that the particles are composed of almost pure Ni. An aluminum peak was hardly detected. It is thus clear that fine Ni particles were formed on the surface of the porous structure.

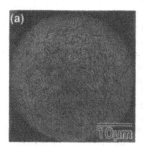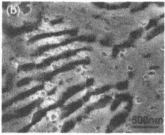

Fig. 3 SEM images of the atomized Ni_3Al powder surface after the first step of the chemical pretreatment. (a) Lower and (b) higher magnification images.

383

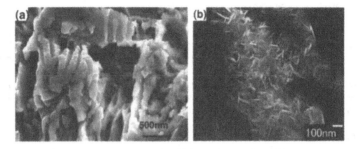

Fig. 4 SEM image of the atomized Ni₃Al powder surface after the second step of the chemical pretreatment. (a) Lower and (b) higher magnification images.

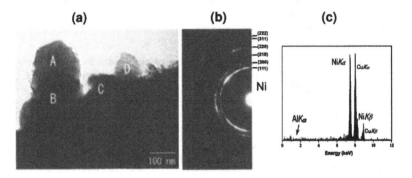

Fig. 5 (a) TEM bright field image of the chemically pretreated powder, showing the formation of fine particles on the surface, (b) the diffraction pattern, and (c) EDS spectra taken from position B showing that the particles are composed of pure Ni.

The powders were examined by XRD before and after the pretreatment. All the diffraction peaks were assigned as $L1_2$–type γ'-Ni₃Al before the pretreatment, though β'-NiAl ($L1_0$–type) was detected in the TEM analysis (Fig. 2). It is probably due to peak broadening by the strain generated during the rapid solidification. As described above, the ICP results showed that both aluminum and nickel was leached in the first step of the pretreatment, and that aluminum was selectively leached in the second step. However, no change was detected in the XRD pattern. The results indicate that the leaching is limited to the thin surface layer of the powder, leaving the bulk as is. It is thus concluded that porous structure were formed on the Ni₃Al powder surface, having fine Ni particles on the porous surface.

The surface areas measured by the BET method are also listed in Table 1. Though the surface area was as low as that of the as-received powder after the step 1 of the pretreatment, it increased after the second step. The results indicate that the combination of the present two-step chemical pretreatment is effective in increasing the surface area of the atomized Ni₃Al powder.

Catalytic activity

The catalytic activity of the chemically pretreated powder was isochronally measured from 873 to 1223 K. Three gaseous products, H_2, CO and CO_2, were detected far below 873 K at the reactor outlet by gas chromatography. From the species of product gases and their molecule composition, it is clear that the following two reversible reactions, methane steam reforming (1) and water-gas shift reaction (2), catalytically occurred similar to the conventional Ni catalysts [8].

$$CH_4 + H_2O \leftrightarrow 3H_2 + CO \qquad \Delta H = +206 \text{ KJ/mol} \qquad (1)$$
$$CO + H_2O \leftrightarrow H_2 + CO_2 \qquad \Delta H = -41 \text{ KJ/mol} \qquad (2).$$

Fig. 6(a) shows the conversions of methane and steam as a function of reaction temperature. The methane conversion rapidly increases with the increasing temperature up to 1073 K, consistent with its endothermic reaction. Above 1073 K it reaches a steady-state with a value of about 95%, which means that almost all amounts of the feed methane are converted to H_2 and CO via the forward reaction of (1). Namely, the chemically pretreated powder is catalytically active in this temperature range. Steam conversion shows the same temperature tendency as the methane conversion below 1023 K, while it decreases at temperatures above 1023 K, indicating that the forward reaction of (2) occurred less and less with temperature. This negative temperature dependence is due to the exothermic reaction of (2).

In contrast to the chemically pretreated powder, both the reactions (1) and (2) were not detected over the as-received powder until temperature increased above 1173 K. It clearly shows a remarkable enhancing effect of the present chemical pretreatment on the catalytic activity of Ni_3Al. From the characterization of the powder described above, the observed high catalytic activity can be attributed to the fine Ni particles formed on the porous surface. As is well known, fine Ni particles are catalytically active for methane steam reforming [8].

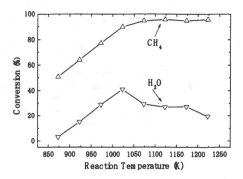

Fig. 6 Conversions of CH_4 and H_2O, and (b) selectivities of CO and CO_2 for methane steam reforming over the chemically pretreated powder as a function of reaction temperature.

385

CONCLUSIONS

The catalytic activity of Ni$_3$Al for methane steam reforming was investigated using its atomized powder. The powder was chemically pretreated by acid and subsequent alkali leaching in order to enhance the activity. The following results were obtained.
(1) After the combined chemical pretreatment, porous structure was formed on the powder surface, having fine Ni particles on the porous surface.
(2) The pretreated powder exhibited high catalytic activity for methane steam reforming. The reason of the activity was attributed to the fine Ni particles on the porous surface.
(3) The results demonstrate that the Ni3Al powder can serve as a catalyst precursor for hydrogen production by methane steam reforming.

ACKNOWLEDGMENTS

One of the authors (Y. Ma) acknowledges the National Institute for Materials Science for the provision of a NIMS Graduate Research Assistantship. This work was supported by a Grant-in Aid for Scientific Research (KAKENHI) (C) (No. 19560774) and (B) (No. 19360321) from the Japan Society for the Promotion of Science.

REFERENCES

1. Y. Xu, S. Kameoka, K. Kishida, M. Demura, A.P. Tsai, and T. Hirano, Intermetallics, **13**, 151(2005).
2. D. H. Chun, Y. Xu, M. Demura, K. Kishida, M. H. Oh, T. Hirano, and D. M. Wee, Catal. Lett., **106**, 71(2006).
3. D. H. Chun, Y. Xu, M. Demura, K. Kishida, D. M. Wee, and T. Hirano, J. Catal., **243**, 99(2006).
4. I. Baker, F. S. Ichishita, V. A. Supprenant, and E. M. Schulson, Metallography, **17**, 299(1984).
5. I. Baker, J. A. Horton, and E. M. Schulson, Metallography, **19**, 63(1986).
6. Y. Ma, Y. Xu, M. Demura, D. H. Chun, G. Xie, and T. Hirano, Catal. Lett., **112**, 31(2006).
7. Y. Ma, Y. Xu, M. Demura, and T. Hirano, Appl. Catal. B, **80**, 15(2008).
8. J.R. Rostrup-Nielsen, J. Sehested, and J. K. Nørskov, Adv. Catal. **47**, 65(2002).

Mater. Res. Soc. Symp. Proc. Vol. 1128 © 2009 Materials Research Society 1128-U06-10

Thin-Film Synthesis and Cyclic Oxidation Behavior of B2-RuAl

Karsten Woll, Rama K.S. Chinnam and Frank Mücklich
Functional Materials, Dept. for Materials Science and Engineering, Saarland University, Campus, 66123 Saarbrücken, Germany

ABSTRACT

B2-RuAl as a potential intermetallic for thin-film applications at high temperatures is studied with respect to thin-film synthesis and cyclic oxidation behavior. Using the multilayer approach, single phase RuAl thin films were fabricated. The phase sequence from the elements Ru and Al goes through $RuAl_6$ to the final product RuAl. To understand the reaction mechanism calorimetric as well as kinetic experiments were performed. The cyclic oxidation behavior is characterized at 1200 °C up to 47 h. The morphology of the grown alumina shows no cracks or regions of spallation which indicates the good cyclic oxidation behavior. Compressive stresses in the oxidation scale of about 1.2 GPa at maximum were determined.

INTRODUCTION

One group of materials that are used for protection purposes from extreme environments are intermetallic compounds based on single phase B2 aluminides [1]. Crucial factors for the application at high temperature, e.g. in thermal barrier systems [2] or protective coatings which avoid material interactions [3,4], are good oxidation as well as high creep resistance [5]. Recent experiments indicated that RuAl shows excellent oxidation properties due to the formation of a thin but compact protective Al_2O_3 scale [6]. During oxidation, only the thermodynamic stable α-Al_2O_3 layer forms on the surface with a parabolic kinetic. Tryon et al. additionally revealed that the thermal expansion coefficient of RuAl is nearly equal to that of Al_2O_3, whereas the coefficient of NiAl exceeds that of alumina about one order of magnitude [7]. Thermal induced stresses at the interface between alumina and RuAl are therefore expected to be lower and crack initiation due to thermal loads should be significantly reduced. Moreover, RuAl with its extraordinary high melting point of 2060 °C appears as a candidate material that exhibits high creep resistance. Therefore, the RuAl intermetallic compound is a potential material for applications at high temperatures [8,9].

RuAl obviously finds prospective application as a thin film or coating. However, there are only few systematic studies concerning the thin-film fabrication of RuAl [10]. Therefore, the detailed study of its synthesis is the first purpose of this paper. For this, the multilayer approach was chosen where individual Ru and Al layers were staggered. The thickness of each layer is in the order of tens of nanometers. Annealing such multilayers results in the formation of intermetallic phases.

The application of intermetallics at high temperatures requires a good cyclic oxidation resistance which is the second purpose of this paper. Good isothermal oxidation behavior is an essential precondition which is given for RuAl [6, 11]. Cao et al. studied the cyclic oxidation behavior of near single-phase RuAl and two-phase RuAl-based alloys in air between 1000 °C and 1300 °C [12]. The samples were synthesized by ingot metallurgy and a minimal volume fraction of 1% of second phase Ru was observed. This intergranular Ru gave rise to preferential intergranular oxidation [6]. Similar effects are expected in cyclic experiments. Thus, oxidation experiments on fully single-phase RuAl have to be performed to characterize the cyclic oxidation behavior.

EXPERIMENT

Magnetron physical vapor deposition was used to deposit Al-Ru multilayers. The Ar-flow rate was 80 sccm and the Ar-pressure was 0.003 mbar. To obtain the Ru/Al composition ratio of 1:1 individual Al and Ru layers with a thickness of 48 and 40 nm, respectively, were deposited. Thus, the periodicity, i.e. the bilayer thickness, is 88 nm. The total thickness of the multilayers was chosen to be 2 μm. Differential scanning calorimetry (DSC) with free standing multilayers was performed. Multilayers were deposited on mica substrates and peeled off. The peeled films were loaded into Al pans for DSC measurements which are performed under pure argon flow. Scanning experiments with heating rates between 20 and 100 °C/min were done by Perkin Elmer DSC 7. Calibration for each heating rate was run with Zn, In and Pb standards. Individual sample weights were > 5 mg. Two sequential heating runs under the same conditions were performed with one sample to construct the baseline.

Annealing experiments at 380, 515, 550 and 630 °C in vacuum (< 5·10^{-5} mbar) were done with multilayers (88 nm periode) on glass substrates. The heating rate was chosen to be 20 °C/min. Using Cu Kα radiation, X-ray diffraction under a glancing angle was then performed on a Phillips X'Pert diffractometer to study phase formation.

Experiments concerning the cyclic oxidation behavior were done using bulk RuAl samples. Powder metallurgical concepts provide best possibilities for fabricating single-phase microstructures. Details can be found in [8]. The specimens were cylindrical in shape with a thickness of > 3 mm and a diameter of 8 mm. Before oxidation, the surface was metallographically polished to 1 μm. Cyclic oxidation tests were conducted in a mirror furnace under still air up to 47 oxidation cycles. Each cycle consist of a fast ramp up to 1200 °C with a heating rate of about 300 °C/min. Here, the temperature was kept constant (within 2 °C) for 45 min. Afterwards, the sample was cooled to room temperature with 300 °C/min. A new cycle was started after 15 min.

To study the scale morphology, SEM analysis of the oxide scale surface and FIB-cross sectioning was performed in a FIB Dual Beam Workstation (Strata DB 235, FEI). To assess the stress state of the growing oxide as a function of oxidation time, stress measurements according to the $\sin^2(\Psi)$-method were done after 10 and 47 cycles. A Phillips X'Pert diffractometer was used.

DISCUSSION

Synthesis of RuAl-Thin Films

Figure 1 a) shows a typical DSC trace (here for a heating rate of 40 °C/min). Two strong exothermic events are observable at 370 and 500 °C, respectively. In addition, a smaller but broader exothermic signal is superimposed to that at 500 °C. To study the occurring reactions in the multilayer system, annealing experiments were done at 380, 515, 550 and 630 °C.

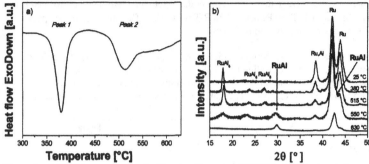

Figure 1: a) DSC trace measured with a heating rate of 40 °C/min. The inset shows results of the b) Phase analysis of annealed multilayers

Figure 1 b) summarizes the diffractograms. At 25 °C, only the reflexions of pure Al and pure Ru are visible. Thus, no reaction of the Al and Ru multilayer at 25 °C can be supposed. Annealing the sample at 380 °C, results in the formation of the Al-rich phase $RuAl_6$. Consequently, it has to be assumed that in this reaction stage Al mainly diffuses into Ru. This is corroborated by the reducing peak at $\approx 38°$ with increasing the temperature. This peak has contributions from both Al (38.472°) and Ru (38.385°). As the Ru peaks at 42 an 44° are nearly constant with increasing temperature up to 380 °C and a small fraction of the peak at 38° is left up to 380 °C, the peak reduction at 38° is associated with the consumption of Al, i.e. the diffusion of Al into Ru. Therefore, a solid solution formation is assumed at temperatures < 350 °C. This is in agreement with the studies on Ru-Al thin films metastable phases [13]. At 350 °C, a decomposition of the solid solution was experimentally verified, and it transformed to the intermetallic phase $RuAl_6$. These results are in perfect agreement with our calorimetric studies which exhibit an exothermic peak with an onset temperature of 350 °C.

Annealing at 515°C, i.e., at the maximum of Peak 2, no new phases can be observed. Annealing at 550 °C, i.e., the temperature at the end of Peak 2, it is observed that there is a decrease in Ru and $RuAl_6$ intensity in the XRD measurements. On the other side, strong evidence for the formation of RuAl is given. Characteristic peaks at 30° and 43.5° arise. Thus, the reaction of Ru and $RuAl_6$ to RuAl can be assumed. Annealing at 630 °C finally results in a mostly single-phase RuAl microstructure.

To study this phase formation in more detail, Johnson-Mehl-Avrami (JMA) analysis was applied. Nucleation and growth processes which can be modeled using the equation $x = 1-\exp(-K \cdot t^n)$ are assumed. Here, x is the fraction transformed, K is a rate constant with dimensions of time, t is time, and n is a dimensionless parameter known as the Avrami exponent. Generally, the rate constant shows an Arrhenian temperature dependence with an activation energy E_a. Thus, E_a as well as the n values for every transformation are needed. The Kissinger analysis allows the calculation of E_a by plotting $\ln(\beta/T_p^2)$ against $1/T_p$ where β is the heating rate and T_p is the peak temperature [14]. The slope of the curve defines E_a. Results for both the peaks are shown in Figure 2 a). The activation energy of Peak 1 and Peak 2 is determined to 213 kJ/mol and 235 kJ/mol, respectively. To check that these E_a values lies within the acceptable application boundaries of the classical nucleation theory, i.e. between 15 R·T and 60 R·T, a comparison of

389

the activation energy to the average thermal energy at the onset temperature was performed [15]. It is found that E_a equals 41 R·T for Peak1 and 38 R·T for Peak 2. Therefore, these two peaks seem to represent nucleation and growth reactions.

To calculate the n values within the theory of Johnson-Mehl-Avrami, the results of scanning DSC experiments were analyzed according to Criado and Ortega [16]. Therefore, the fraction transformed x has to be evaluated as a function of temperature (basing on DSC results). The value of ln(-ln(1-x)) was plotted against 1/T. Figure 2 b) shows a typical result for the analysis of Peak 1.

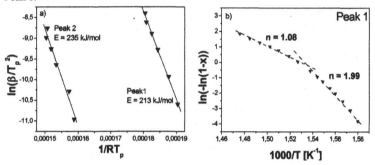

Figure 2: a) Kissinger analysis for Peak 1 and Peak 2 to determine the activation energies. b) JMA plot for the first peak in Figure 1 a).

The curve is characterized by two linear branches indicating two simultaneously occuring processes in this reaction stage. The two individual slopes defines the n values of each process. The n values of 1.08 and 1.99 were determined which characterizes nucleation and one dimensional growth. As the first peak is related to the formation of RuAl$_6$, it can be concluded that in the first reaction stage nucleation (n = 1.08) and growth along the interfaces (n = 1.99) is occuring. Peak 2 is characterized by n values of 2.5 and 1. A three dimensional growth as well as a nucleation process occurs. Based on the XRD results, it is RuAl$_6$ which grows three dimensionally, i.e. perpendicular to the interfaces. Additionally, phase nucleation of B2-RuAl is assumed. In the diffractogram at 515 °C, there is no evidence of that phase. The amount of RuAl in this stage is probably under the detection limit of the XRD technique. A further analysis of the RuAl growth process is not possible due to the strong overlapping signals at higher temperatures.

Cyclic Oxidation Behavior of RuAl

The surface analysis of the grown oxide after 47 cycles reveals a crack-free oxidation product without regions where spallation occurs. Figure 3 a) and b) shows the cross-sectional view of the oxide scales after 10 and 47 cycles, respectively. Generally, the oxide consists of two layers which are in agreement with the isothermal studies already discussed in the literature [6, 11]. On top, there is a porous alumina subscale consisting of equiaxed grains whereas the subscale next to the substrate is fully dense and columnar grained.

Comparing the relative thickness of both the subscales brings the conclusion that the porous as well as the dense subscale grow during oxidation. It is therefore assumed that new oxide forms within the alumina scale and not at the alumina/air or alumina/substrate interface. This type of oxide growth implies the evolution of compressive growth stresses during oxidation [17].

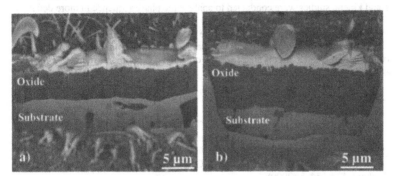

Figure 3: Cross-sectional view the oxide scale after a) 10 and b) 47 cycles. The oxide scale owns a porous (on top) as well as a dense (next to the substrate) oxidation subscale.

Figure 4 demonstrates that indeed compressive stresses in the alumina scale of about 1 GPa are measurable. Increasing the number of cycles results in a small increase of the stresses. Commonly, stresses in grown oxide scales can arise from the growth process itself or during thermal cycling due to the mismatch in the thermal expansion behavior of the substrate and the oxide. Here, it is assumed that the latter effect does not dominate the stress evolution as the coefficients of thermal expansion of alumina and RuAl are nearly equal in the considered temperature range [7]. Additionally, the stress value of 1.1 – 1.2 GPa suggests a domination of growth stresses. An exemplary study of the two effects can be found in [18]. Therefore, the increase of the stresses with the number of cycles is associated to the growing oxide. Finally, it can be concluded, that cyclic oxidation on single-phase RuAl result in a crack-free oxidation scale with small compressive stresses. It is assumed, that the latter favorably influence the resistance of the alumina scale against cracking or spallation in later oxidation stages.

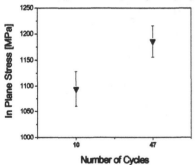

Figure 4: In plane stress state of the grown oxide after 10 and 47 oxidation cycles

CONCLUSIONS

Thin-Film synthesis of RuAl using the multilayer approach was performed for the first time. Before the formation of RuAl at temperatures > 550 °C, $RuAl_6$ is firstly developed. Calorimetric and kinetic studies were conducted to analyze the phase sequence in more detail. The experiments concerning the cyclic oxidation behavior at 1200 °C reveal a good stability of the oxide scale on the RuAl substrate. Tthe oxide is characterized by compressive stresses of 1.2 GPa at maximum. It is assumed, that these arise mainly from the growth of the scale and that thermal mismatch effects are of second order. The experiments indicate a good oxidation resistance under cyclic thermal load.

REFERENCES

[1] C.G. Levi, *Curr. Opin. Solid State Mater. Sci.* **8**, 77 (2004)
[2] J. Shi, S. Darzens, A.M. Karlsson, *Mater. Sci. Eng., A* **392**, 301 (2005)
[3] C.Y. Kang, Y.I. Chen, C.H. Lin, J.G. Duh, *Appl. Surf. Sci.* **253**, 6191 (2007)
[4] D. Zhong, E. Mateeva, I. Dahan, J.J. Moore, G.G.W. Mustoe, T. Ohno, J.Disam, S. Thiel, *Surf. Coat. Technol.* **133-134**, 8 (2000)
[5] G. Sauthoff, *Intermetallics*, VCH Verlagsgesellschaft, Weinheim, 1995
[6] F. Soldera, N. Ilic, S. Brännström, I. Barrientos, H.A. Gobran, F. Mücklich, *Oxid. Met.* **59**, 529 (2003)
[7] B. Tryon, T.M. Pollock, M.F.X. Gigliotti, K. Hemker, *Scr. Mater.* **50**, 845 (2004)
[8] F. Mücklich, N. Ilic, *Intermetallics* **13**, 5 (2005)
[9] F. Mücklich, N. Ilic, K. Woll, *Intermetallics* **16**, 593 (2008)
[10] I.M. Wolff, *JOM* **1**, 34 (1997)
[11] P. Bellina, A. Catanoiu, F.M. Morales, M. Rühle, *J. Mater. Res.* **21**, 276 (2006)
[12] F. Cao, T.K. Nandy, D. Stobbe, T.M. Pollock , *Intermetallics* **15**, 34 (2007)
[13] A. Zariff, C. Chaudhury, C. Suryanarayana, *J. Mater. Sci.* **17**, 3158 (1982)
[14] H. E. Kissinger, *J. Res. Natl. Bur. Stand.* **57**, 217 (1956)
[15] P.F. Ladwig, Y.A. Chang, E.S. Linville, A. Morrone, J. Gao, B.B. Pant, A.E. Schlutz, S. Mao, *J. Appl. Phys.* **94**, 979 (2003)
[16] J. M. Criado and A. Ortega, *Acta Metall.* **35**, 171 (1987)
[17] M. Schütze, *Protective Oxide Scales and Their Breakdown*, John Wiley & Sons (1997)
[18] V.K. Tolpygo, D.R. Clarke, *Oxid. Met.* **49**, 187 (1998)

Niobium and Molybdenum
Silicide-Based Alloys

Mater. Res. Soc. Symp. Proc. Vol. 1128 © 2009 Materials Research Society

Fracture and Fatigue of Niobium Silicide Alloys

David M. Herman[1], B.P. Bewla, y[2], L. Cretegny[2], R. DiDomizio[2], and John J. Lewandowski[1]
[1]Case Western Reserve University, Department of Materials Science and Engineering, Cleveland, OH 44106-7204
[2]General Electric Global Research, Schenectady, NY 12301, USA

ABSTRACT

The fracture and fatigue behavior of refractory metal silicide alloys/composites is significantly affected by the mechanical behavior of the refractory metal phase. This paper reviews some of the balance of properties obtained in the alloys/composites based on the Nb-Si system. Since some of the alloy/composite properties are dominated by the behavior of the refractory metal phase, the paper begins with a review of data on monolithic Nb and its alloys. This is followed by presentation of results obtained on Nb-Si alloys/composites and a comparison to behavior of some other high temperature systems.

INTRODUCTION

Because of their high melting point, low density, high fracture toughness and very good properties at ultrahigh temperatures, Niobium-Silicon alloys have attracted recent attention [1] [2]. This class of alloys has very good strength retention, low creep rate at ultrahigh temperature and very good oxidation resistance up to 1300-1400°C. These alloys continue to be considered as one of the candidates for airfoils in gas turbines/aircraft engines to replace the currently used nickel-base superalloys which are limited to temperatures less than 1150°C. [1] [2]

While the oxidation resistance of the monolithic refractory metals are generally poor, the intermetallic compounds of refractory metals with silicon(e.g. Nb,Mo) are of interest because of their high melting points, their increased oxidation resistance, and their relatively low densities [1,2]. In particular, alloys/composites in the systems Mo-Si-B-X (X=W,V,Cr,Nb,Al,Ge,Re) and Nb-Ti-Hf-Cr-Al-Si-Ge combine many of the attractive features of refractory metals and intermetallic compounds. Despite the potential of silicides for use as high temperature materials such as turbine airfoils or some hypersonic applications for service temperatures ranging from 1000-1400°C, relatively little is known about their structure and properties(in monolithic and composite form) at ambient or elevated temperatures. Some of the additional property requirements for the hottest parts of aircraft are provided elsewhere [3].

Predominant issues in the mechanical behavior of these advanced materials are their ambient and high temperature properties, including the strength, ductility, toughness, creep, fatigue performance, and oxidation resistance at intermediate and high temperatures. Recent works have successfully demonstrated the potential of improving the toughness of niobium silicide systems via compositing with ductile Nb-phases utilizing ingot casting and extrusion, directional solidification, as well as lamination via PVD or powder metallurgy techniques [3] [4] [5] [6]. Parallel efforts are being developed to improve the oxidation and creep resistance of Nb via alloying without compromising toughness.

In the Nb-Si system, there is present the Nb solid solution (denoted as Nb(ss) with a BCC crystal structure) and several possible intermetallics which are very brittle, such as Nb_3Si (tetragonal with a space group $P4_2/n$), Nb_5Si_3 (tetragonal with a space group $I4/mcm$) and Nb_5Si_3

(hexagonal with a space group P6₃/mcm). Hardened by Nb silicides and toughened by the Nb(ss) phase, the Nb-Si alloys may exhibit excellent mechanical properties at elevated temperatures [1] [2] [5] [7] [8]. Early studies on the mechanical properties and microstructure of Nb-Si alloys focused on the Nb-10 a/o Si alloy as a model alloy. Later alloys with a wide range of Si contents (12-25 a/o) have also been investigated [1]. In the equilibrated Nb-10 a/o Si alloy, there are essentially two phases in the alloy, i.e. Nb(ss) and Nb₅Si₃. The alloy contains 50 vol. % primary Nb and 25 vol. % secondary Nb [8]. It was found that the flow, fracture and fatigue behavior of the Nb-10 a/o alloy was primarily dependent on the properties of the Nb(ss) phase [5] [9] and that extrusion significantly reduced the microhardness of the Nb(ss) phase and produced an increase in the fracture toughness from 6-9.6 MPa$\sqrt{m}$ to ~21 MPa$\sqrt{m}$ [5]. In cyclic testing it was also found that increasing the load ratio, R, decreased the threshold stress intensity of the Nb-10 a/o Si alloy, consistent with the fatigue behavior of metals [8]. However, the Nb(ss) phase in the Nb-10 a/o Si alloy exhibited a distinct fracture mode transition from ductile tearing at low ΔK to cleavage fracture at high ΔK essentially triggered by K$_{max}$ in the fatigue cycle, thereby increasing the Paris Law exponent. In contrast, a bulk Nb-1.24 a/o Si solid solution did not show any evidence of cleavage fracture [8]. The fracture resistance of the Nb-10 a/o Si alloy is considerably higher than that of monolithic Nb₅Si₃, since crack initiation in the Nb₅Si₃ was followed by the propagation of these microcracks bridged by the tougher Nb (ss) ligaments [9].

The flow, fracture, and fatigue performance of the toughened Nb silicide base alloys/composites are significantly affected by the flow, fracture and fatigue performance of the toughening phase(i.e. Nb(ss). In most of the alloy/composite systems studied to date, the Nb toughening phase contained additional alloying elements, either due to contamination(e.g. O,C, etc) or due to alloying(e.g. Si, Hf, Ti, Cr, Ge, etc) to improve the alloy/composite oxidation resistance and/or other properties. In some cases, Cr additions produced NbCr₂ [10]. In order to be able to document the effects of toughening phase(s) on the composite behavior, it is necessary to understand the effects of microstructural changes on the flow, fracture, and fatigue performance of the toughening phase itself. As such, this review paper will first cover the available literature on the factors which control the flow, fracture, and fatigue behavior of polycrystalline Nb, followed by recent fracture and fatigue studies over temperatures ranging from 77K-773K on Nb toughened Nb Silicide alloys/composites prepared either by arc casting/extrusion or directional solidification (DS). The presence of the brittle silicide intermetallic in the alloy/composite may produce constrained flow and elevated tensile stresses in the Nb phase(s), thereby changing the fracture behavior in the Nb toughening phase. As such, the behavior of unconstrained and constrained Nb will be presented for a range of temperatures prior to the discussion of the alloy/composite behavior. Test temperatures as low as 77K have been utilized in order to illustrate the effects of changes in stress state and flow behavior on the flow/fracture of the toughening phase.

Polycrystalline Nb Flow and Fracture Behavior
        The flow stress of polycrystalline Nb and its alloys are relatively insensitive to changes in grain size over the range 5 μm to roughly 500 μm. The Hall-Petch slope obtained for changes in grain size ranging from 20 μm to roughly 165 μm was 8.7 X 10⁴ N/m$^{-3/2}$ [11], consistent with previous work on polycrystalline Nb [9] [12] [13] [14]. The Ultimate Tensile Stress (UTS) is also relatively insensitive to changes in grain size [11]. The strength of commercial purity Nb at 298 K is 150-300 MPa and at 77K can exceed 1 GPa, while still exhibiting significant ductility

and ductile fracture [9]. Nb with solid solution additions of Si is significantly strengthened, with strengths at 298 K exceeding 350 MPa while still possessing some ductility [9] [15] [16]. Reductions in test temperature to 77K further increase the yield strength to near 1 GPa, although the fracture at 77 K occurs via transgranular cleavage with low ductility [9] [15] [16]. The strength of commercial purity Nb is quite low at high temperatures, while Figure 1 shows the increase in strength for the Nb-Ti-Cr-Al and a number of other Nb-based alloy/composites system over a range of temperatures [17] [18] [19]. Some of the Nb-Si based alloys/composites are also shown in Figure 1 and illustrate that high strengths are possible, with high temperature strengths well in excess of that of the monolithic toughening phase due to the constraint and strengthening provided by the other intermetallic particles and phases [5] [17] [18] [19], of which Nb_5Si_3 and $NbCr_2$ are the most commonly employed.

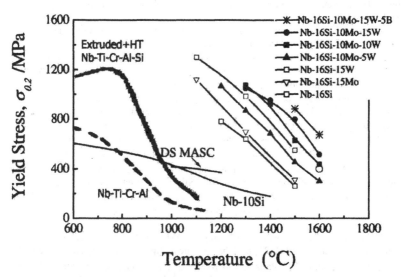

Figure 1. Example of the potential in yield stress improvements with modification of the Nb-10 a/o Si model alloy/composite system [31]. HT indicates heat treat at 1500°C/100 hrs.

The Ductile to Brittle Transition Temperature (DBTT) of polycrystalline Nb tested in impact is dependent on processing conditions and chemistry as well as grain size. As expected, decreases in grain size reduce the DBTT and increase the impact energy, while high purity Nb prepared via electron beam melting (EBM) exhibits the lowest DBTT and highest notched impact energy [11]. Work conducted at CWRU on commercial purity Nb heat treated to different grain sizes similarly revealed an effect of grain size on the DBTT. For example, at 298 K, coarse grained Nb exhibited 75% cleavage fracture while the fine grained Nb exhibited 0% cleavage.

The competition between the flow and cleavage fracture in polycrystalline Nb and Nb alloys (including Si-solid solution strengthened Nb) was investigated [9] [12] [15] where the

397

cleavage fracture stress was measured via notched specimens. The result of these works [9] [12] [15] revealed that cleavage fracture in polycrystalline Nb and Nb alloys occurs by reaching a temperature-independent critical cleavage fracture stress. The magnitude of the cleavage fracture stress was shown to depend solely on the grain size in the materials studied, with decreases in grain size producing increases to the cleavage fracture stress [13]. The temperature-independent values for cleavage fracture stress varied from 1000-1500 MPa depending on the grain size [9] [12] [15]. Extensive fractography of the notched bend specimens revealed that cleavage fracture initiated ahead of the notch in the region of peak tensile stress [9] [12] [15], entirely consistent with classic theories of tensile stress controlled cleavage fracture. The high constraint and stress levels produced by the silicide particles and other phases present in the Nb-Si-X alloys/composites indicate that cleavage of the refractory phase may occur in such systems. This will be particularly true for Nb-Si systems heavily alloyed with other elements(e.g. Hf, Ti, Cr, Ge, etc.).

The static fracture toughness of polycrystalline Nb has been determined on fatigue precracked bend specimens tested over temperatures ranging from 77 K to 298 K [20]. In addition, dynamic fracture toughness tests, K_{ID}, have been determined on fatigue precracked Charpy impact specimens test in an instrumented impact machine [11]. The fracture toughness obtained under dynamic loading conditions is in the range of 30-40 MPa $\sqrt{m}$ from 77K to above 298K, while fracture surface examination revealed the fracture to consist of 100% cleavage fracture [11]. The values obtained for the static fracture toughness [20] over a more limited temperature range were in close agreement with the data obtained under impact conditions [11], suggesting that the minimum fracture toughness of the high purity polycrystalline Nb samples is quite large(i.e. 30-40 MPa $\sqrt{m}$), despite the appearance of 100% cleavage. The fracture toughness of the Nb-Si solid solution alloy was approximately 30 MPa $\sqrt{m}$ [9] [15] at 298K, while Niobium solid solution containing 0.8 a/o Silicon revealed a "pop-in" at only 17 MPa $\sqrt{m}$ at room temperature [21]. This lower toughness and crack "pop-in" is most likely due to oxygen and/or C pickup during processing in addition to the absence of deformation processing.

The fatigue crack growth behavior of polycrystalline Nb has not been investigated to any great extent. Recent work [8] [22] [23] has revealed fatigue thresholds in the range 10-12 MPa $\sqrt{m}$ at R=0.1 at 298 K. Paris Law slopes were in the range 2-4, consistent with most metallic materials. Increases in stress ratio decreased the fatigue threshold somewhat without significantly changing the Paris law exponent, also consistent with most metallic materials. This is shown in Figure 2 and includes data for some Nb-Si-X alloys/composites to be discussed below.

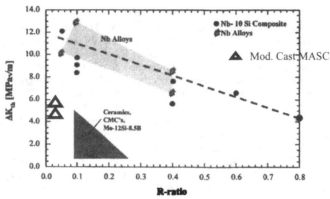

Figure 2. Effects of changes in stress ratio on fatigue threshold of a variety of high temperature materials tested at 298K. Data for ceramics, CMC's, Mo-Si-B from [29,33,34].

Fracture Behavior of Nb-Si-X Alloys/Composites

In-situ composites based on the binary Nb-10 a/o system shown in Figure 3 have been extensively investigated. The majority of the work on this system has been conducted on arc cast/extruded material, subsequently heat treated 1500C/100 hours in order to equilibrate the structure to contain Nb_5Si_3 and Nb(ss-Si). The structure shown in Figure 3 exhibits elongated primary Nb(ss-Si) present at about 50 volume %. The primary Nb has a grain size of roughly 25-50 μm, with the size of the primary Nb roughly 50-100 μm. Secondary Nb is present at about 25 volume % with a size of about 1 μm. The remaining 25% of the structure is the brittle Nb_5Si_3. Included in Figure 3 is a Nb-Si-X alloy prepared by casting containing approximately 55 vol. % of a Nb(ss) matrix with grain size of approximately 30 μm. The remaining 45 volume % consists of smaller Nb_5Si_3 and $NbCr_2$ Laves phase particles.

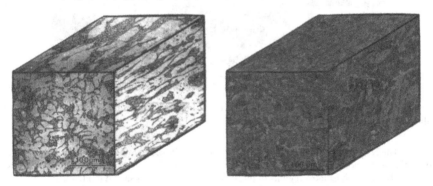

Figure 3. Typical microstructure of Arc Cast/Extruded Nb-10 a/o Si shown on the left with the conventional casting/HIP modified MASC material shown on the right. For both microstructures the solidification/extrusion direction is perpendicular to the plane of the page.

399

Fracture toughness tests conducted on the monolithic Nb silicide revealed that it possessed a fracture toughness of only $1\text{-}2\,\text{MPa}\sqrt{m}$ [16]. Fracture toughness tests on the Nb 10 a/o Si alloy/composite shown in Figure 3 conducted so that the crack growth occurred perpendicular to the extrusion direction exhibited R-curve behavior. Significant increases in toughness were exhibited by the Nb-Si alloy/composite due to the toughening provided by the Nb(ss-Si). Figure 4 presents a sequence of photographs taken at increasingly higher load (i.e. A-B-C-D) in a test conducted at 298K on the Nb 10a/o Si alloy/composite [9]. Cracks which have impinged on the Nb are blunted and the Nb deforms significantly before fracture. The toughness, roughly $24\,\text{MPa}\sqrt{m}$ is relatively insensitive to large changes in loading rate and test temperature. The very high levels of toughness exhibited at 77K at the fastest loading rates occurred with a preponderance of cleavage of the primary Nb in the alloy/composite [9]. Despite the presence of cleavage fracture of the primary Nb, the fracture toughness far exceeds that of the Nb silicide. This appears to be consistent with the high toughness of the polycrystalline Nb presented earlier despite the appearance of 100% cleavage. Testing at 773K does not significantly change the fracture toughness of the Nb-10 a/o Si alloy/composite or that of a variety of Nb-Ti-Si directionally solidified composites [24] despite a change in fracture mode of the toughening phase over that range of temperatures.

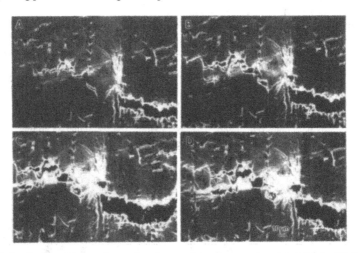

Figure 4. In-situ fracture experiment showing crack impingement on tough Nb in Nb-10 a/o Si alloy/composite. Increasing load from A-D

The Nb(ss) in multicomponent Nb-Si-X heavily alloyed to improve the high temperature oxidation resistance and high temperature strength typically displayed cleavage fracture at 298 K and 848 K; although isolated regions of plasticity were observed as shown in Figure 5. Regardless, the toughness of the more heavily alloyed Nb-Si-X systems, determined through notch toughness testing as well as with fatigue precracks display reasonable levels of toughness(i.e. K_q= 15-20 MPa√m). The fracture surface appearance of the Nb(ss) in the multicomponent Nb-Si-X within the temperature range of 298 K and 848 K is nearly identical.

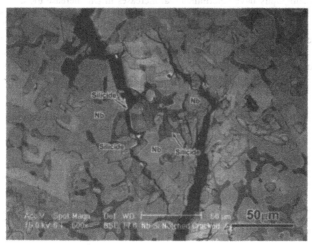

Figure 5. In-situ fracture experiment showing crack impingement on heavily alloyed Nb in Nb-Si-X alloy/composites

Fatigue Behavior of Nb-Si-X Alloys/Composites

Few studies of the fatigue crack growth behavior of Nb-Si alloys composites have been conducted. Early work [25] [26] [27] on Nb toughened intermetallics suggested that the fatigue performance of such toughened systems might be as poor as that of the monolithic brittle matrix, with Paris Law slopes approaching 100 in some cases. In those cases, the toughening constituent failed prematurely via fatigue [25] [26] [27]. In order to address the relevance of these issues for the Nb-Si system, extensive testing over a range of stress ratios and test temperatures has been conducted ([8] [22] [23] [28]), followed by quantitative fractography in order to document the type of fracture mechanism(s) operating in the different fatigue regimes for the Nb-Si alloys/composites shown in Figure 3.

Fatigue crack growth experiments at 298K on the Nb-10 a/o Si system ([8] [22] [23] [28]) are summarized in Figure 6. Fatigue crack growth experiments were conducted on bend bar specimens tested at stress ratios ranging from 0.05 – 0.8 on a closed loop MTS servo-hydraulic testing machine. Crack growth was monitored via the use of foil resistance gages (i.e. KRAK gages) bonded to the outer surfaces of the specimens. Multiple tests were conducted for each stress ratio and representative results are shown in Figures 2 and 6. The fatigue threshold and Paris Law slope at R=0.1 is similar to that of metallic materials as shown in Table 1 and Figure 2. Figures 2, 6, and Table 1 also reveal that increasing the stress ratio decreases the fatigue threshold and increases the Paris Law slope. While the former observations(i.e. decreased fatigue threshold) is consistent with the behavior of metals [29], the increase in Paris Law slope with increasing stress ratio is typically not observed in monolithic metals. Initial work [8] [22] revealed that cleavage fracture of the primary Nb occurred in the Paris Law regime of the Nb-10a/o Si alloys/composites. Tests conducted over a wide range of stress ratios combined with quantitative fractography have determined that increases in ΔK increase the amount of cleaved primary Nb. However, as shown earlier, cleavage of Nb is typically considered a static fracture mode controlled by reaching a critical value of the cleavage fracture stress [13]. If this is the case, cleavage during fatigue crack growth should be controlled by K_{max} in the fatigue cycle. The use of K_{max} appears to normalize all of the data obtained at different stress ratios at 298K for the Nb-10 a/o Si alloy/composite and suggests strongly that increases in the Paris Law slope exhibited in Figure 6 and Table 1 with increasing stress ratio for Nb-10 a/o Si is due to the intervention of static modes of fracture (i.e cleavage) in the primary Nb [8] [22]. This further suggests that microstructure manipulation and/or testing at different temperatures could affect the tendency for cleavage of the primary Nb, and here hence significantly change the fatigue crack growth characteristics. Preliminary work of this nature has produced fatigue crack growth characteristics and fractographic observations which are consistent with these predictions [28].

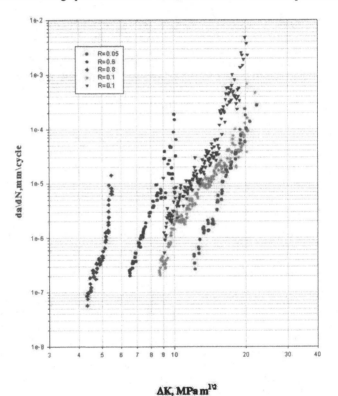

$$\Delta K, MPa\ m^{1/2}$$

Figure 6. Effects of Changes in stress ratio on Fatigue of Nb-10 a/o Si Composites.

Higher alloy variants of the Nb-10 a/o Si system have been prepared in order to improve the oxidation resistance [4]. Both arc-casting/extrusion as well DS materials have been prepared. Table 1 also summarizes the fatigue threshold and Paris Law slope of the materials prepared and tested to date [23] [28]. It is again clear that these toughened Nb alloys/composites can exhibit fatigue thresholds, Paris Law slope exponents, and values for fracture toughness which are metallic-like in character. In order to demonstrate this, Figure 2 and 6 were prepared from the data obtained presently as well as that taken from the literature [30] [31]. Included in Figure 2 is data for ceramics, toughened ceramics, and ceramic composites, and preliminary work on Mo based intermetallic composites. The fatigue thresholds are as low as 1.5 MPa$\sqrt{m}$ at R=0.1, with Paris Law slopes for these systems approaching 60-100 in some cases [29] [30] [31] [32], in contrast to the high threshold and low Paris Law slope(i.e. <20) values generally exhibited by Nb-Si systems at R=0.1. Although the Nb toughened Nb silicide alloys/composites exhibit an increase in Paris Law slope with increased stress ratio, this is apparently due to the intervention of static modes of fracture during the fatigue test. Reducing the tendency for the Nb phases(s) to cleave via microstructure changes, or testing at high temperatures should further improve the Paris Law slope of the Nb-Si systems. Fatigue fracture of the heavily alloyed Nb-Si-X samples often exhibits cleavage like fracture over a range of ΔK [32]. This also appears to be partly responsible for the higher fatigue Paris slopes reported in Table I at R=0.1 for these alloys.

SUMMARY

The data obtained to date on Nb-Si alloys indicates that it continues to be important to investigate the behavior of the monolithic toughening phase(s) in order to understand their role in the toughening and fatigue behavior of alloys/composites based on the Nb-Si system. It is clear that reasonable levels of toughness can be obtained in the most heavily alloyed Nb-Si-X systems despite the appearance of cleavage fracture of the Nb(ss) phase, although further improvements would be expected if this fracture mode was avoided/minimized. In this regard, further work is necessary in order to demonstrate that further alloying additions to the high temperature Nb-Si alloys/composites are effective in improving the oxidation resistance without negatively impacting the mechanical properties. The size scale, spacing, and orientation of the toughening phase(s) have not been optimized in the systems studied to date. The fatigue crack growth behavior of the Nb-Si-X system has been investigated over a number of different alloys. In general, the fatigue threshold behavior is similar to that reported for metals and responds to changes in load ratio, R, similar to most metallic systems. The Paris Law slopes are somewhat higher than that obtained in conventional metallic alloys, with increases in the Paris slope obtained at higher load ratios. This behavior is unlike that reported for most metallic systems and apparently relates to an increase in the amount of cleavage fracture of the Nb(ss) toughening phase with increases in alloy content and/or load ratio. However, the Paris slopes were consistently lower than that reported on various ceramics, ceramic matrix composites, and some other refractory metal systems. Additional fatigue tests at higher temperatures and stress ratios are also needed to determine the generality of the arguments presented here.

ACKNOWLEDGMENTS

The authors would like to acknowledge the work and help of a number of former and present students: JD Rigney, J Kajuch, J Short, WA Zinsser, A Samant, A Awadallah, PM Singh, L Leeson, L Ludrosky, D Li, and J Larose. Useful discussions with D Dimiduk, M Jackson, M Mendiratta, and J Larsen, RO Ritchie, and AW Thompson are acknowledged. Supply of materials by B Bewlay and M Mendritta are appreciated as is funding by Reference Metals Co., GE-CRD, MTS Systems, AFOSR-F49620-96-1-0164 and AFOSR-F49620-96-1-0167, GE Global.

REFERENCES

[1] B.P. Bewlay, M.R. Jackson, J.C. Zhao, P.R. Subramanian, M.G. Mendiratta, and J.J. Lewandowksi. *MRS Bulletin.* **28**, 646(2003).

[2] B.P. Bewlay, M.R. Jackson, J.C. Zhao, and P.R. Subramanian. *Metall. Mater. Trans. A* **34A**, 2043(2003).

[3] B. P. Bewlay, J.J. Lewandowski, and M. R. Jackson. *JOM*, **49**, 44 (1997).

[4] M.G. Mendiratta, J. J. Lewandowksi, and D.M. Dimiduk. *Metall. Trans. A*. **22A**, 1573(1991).

[5] M.G. Mendiratta, J.J. Lewandowski, and D.M Dimiduk. *Metall. Trans. A*, **24A**, 501(1993).

[6] P. J'ehanno, M. Heilmaier, H. Kestler, M. B"oning, A. Venskutonis, B. Bewlay, M. Jackson, *Metall. Mater. Trans A*. **36A**, 515(2005).

[7] Y. Kimura, H. Yamaoka, N. Sekido, and Y. Mishima. *Metall. Mater. Trans. A*, **36A**, 483(2005).

[8] W. A. Zinsser and J. J. Lewandowski, *Metall. Mater. Trans. A*. **29A**, 1749 (1998).

[9] J. D. Rigney and J. J. Lewandowski, *Metall. Trans. A*. **27A**, 3292 (1996).

[10] B. P. Bewlay, Y. Yang, R. L. Casey, M. R. Jackson, Y. A. Chang in *Advanced Intermetallic-Based Alloys*, edited by Jörg Wiezorek, Chong Long Fu, Masao Takeyama, David Morris, and Helmut Clemens (Mater. Res. Soc. Symp. Proc. **980**, 2006) pp. 0980-II05-34.

[11] D. Padhi and J. J. Lewandowski, *Metall. Mater. Trans. A*. **34A**, 1 (2003).

[12] A. Samant and J. J. Lewandowski, *Metall. Trans. A*. **28A**, 389 (1997).

[13] M. A. Adams, A. C. Adams, and R.E. Smallman, *Acta Metall.*, **8**, *328 (1960)*.

[14] J. J. Lewandowski, D. Padhi, and S. Solv'yev, *Flow, Fracture, and Fatigue of Nb and Nb Silicide Intermetallic Compossites. in Structural Intermetallics 2001*, TMS (The Minerals, Metals and Materials Society) pp. 371.

[15] M.G. Mendiratta, R. Goetz, D.M. Dimiduk, and J.J. Lewandowski. *Metall. Trans. A*. **26A**, 1767 (1995).

[16] J. Kajuch, J.W. Short, and J.J. Lewandowski, *Acta Metall. Mater.* **43**, 1955 (1995).

[17] P.R. Subramanian, M.G. Mendiratta, D.M. Dimiduk, and M.A. Stucke, *.Mater. Sci. Eng.* **A239-240**, 1 (1997).

[18] B. P. Bewlay, M. R. Jackson, and H. A. Lipsitt, *Metall. Trans. A* . **27A**, 3801(1996).

[19] M.R. Jones, and K.D. Jackson, *Refractory Metals: Extraction, Processing, and Applications,(KC Liddell, DR Sadoway, and RG Bautista, eds), TMS, Warrendale, PA,* pp.335(1990).

[20] A.V. Samant and J.J. Lewandowski. *Metall. Trans.* **A 28**, 2297 (1997).

[21] R. N. Nekkanti and D. M. Dimiduk, *Ductile-phase toughening in niobium-niobium. Mat. Res. Sac. Symp. Proc.* **Vol.194**, pp.175 (1990).

[22] W. A. Zinsser and J. J. Lewandowski, *Scripta Metall Mater.* **38**, 1775 (1998)

[23] J. J. Lewandowski, *in Fatigue '99, (XR Wu and ZG Zhang, eds.), Volume III, EAMS,* p.147 (1999) .

[24] J. D. Rigney, et al. *Unpublished Rsearch,* (1998).

[25] KTV Rao and R.O. Ritchie, *Mater. Sci. Eng.* **A153**, 479 (1992).

[26] KTV Rao, W.O. Soboyejo, and R.O. Ritchie, *Metall. Trans. A*, **23A**, 2249 (1992).

[27] W. O. Soboyejo, and Sastry, SML *Mater. Sci. Eng.* **A171**, 95 (1993)..

[28] S. Solv'yev and J. J. Lewandowski, Unpublished Research, Case Western Reserve University, Cleveland, OH (2000).

[29] R.O. Ritchie, *Inter J. Fracture.* **100**, 55(1999)

[30] H.Choe, et al. *in Fatigue and Fracture of High Temp. Matl's, (PK Liaw, eds), TMS Warrendale, PA,* p. 17 (2000).

[31] C.L. Ma, et al, *Mater. Sci. Eng.* **A384**, 377 (2004).

[32] D.M. Herman, M.S. Thesis, CWRU (2008).

[33] R.O. Ritchie, R. H. Dauskardt, and K.T. Rao, *Mater. Sci.* **30**, 277 (1994).

[34] J.J.Kruzic, J.H. Schneiber, and R. O. Ritchie, *Scripta Mater.* **50**, 459(2004).

Table 1. Fatigue crack growth behavior of Nb-Nb silicide composites (*Kmax at fatigue overload **Without precracking)

Materials	Processing Conditions	R-Ratio	ΔK_{th}, MPa√m	m-Paris Law Slope	K_{IC}, MPa√m
Nb-10Si	Extr. + HT	0.05	12.0	6.1	23.4*
		0.1	9.0	6.6	24.1*
		0.1	8.4	6.5	24.3*
		0.1	9.7	6.3	22.3*
		0.6	6.5	8.9	25.4*
		0.6	6.0	9.9	24.8*
		0.8	3.2	21.6	25.2*
		0.8	4.4	16.9	27.2*
Nb-15Si	Extr. + HT	0.1	4.4	6.4	20.6*
		0.4	4.1	--	--
		0.6	4.0	5.6	22.5*
		0.8	3.7	--	24.9*
Nb-12Si	DS	0.1	13.2	16.7	18.1
Nb-18.2Si	DS	0.1	2.5	--	3.3
Nb-30Ti-8Cr-10Al-14Si	DS	0.1	--	--	8.3**
Nb-42.5Ti-15Si	DS	0.1	5.5	9.7	10.2
Nb-Ti-Hf-Cr-Al-Si	DS	0.1	8.5	2.9	24.2
Nb-22Ti-3Hf-2Cr-2Al-17Si	Extr. + HT (1500°C/100hr)	0.1	7.1	4.8	24.4
		0.4	5.6	4.1	19.8
Nb-22Ti-3Hf-2Cr-2Al-17Si	Extr. + HT (1400°C/100hr)	0.1	7.2	4.8	17.8
		0.4	3.9	4.2	13.9
Nb-7Si-9Al	Arc Melting + HT	0.1	3.8	--	--
Nb-10Si-9Al-10Ti	Arc Melting + HT	0.1	3.5	--	--
Nb-6Si-9Al-10Ti	Arc Melting + HT	0.1	5.4	--	--
Mod. MASC	Cast + HIP	0.1	6-8	15-20	10-20

Mater. Res. Soc. Symp. Proc. Vol. 1128 © 2009 Materials Research Society 1128-U07-02

Modeling of Microstructure Evolution in Nb-Si Eutectic Alloy Using Cellular Automaton Method

Kenichi Ohsasa[1] and Seiji Miura[1]
Graduate School of Engineering, Hokkaido University
N13, W8, Kita-ku, Sapporo 060-8628, Japan

ABSTRACT

A numerical model was developed for the simulation of microstructure evolution during the solidification of Nb-Si eutectic alloy. In this model, the cellular automaton method was used to simulate the eutectic growth of Nb solid solution and Nb_3Si intermetallics. Diffusion in liquid, mass conservation at the solid/liquid interface and local equilibrium at the solid/liquid interface with consideration of curvature undercooling were solved to determine the positions of the Nb/liquid and Nb_3Si/liquid interfaces. In the alloy with eutectic composition of 17.5 at%Si, irregular eutectic growth morphology was observed in relatively low undercooling region. On the other hand, in high undercooling region over 50 K, dendrite morphology of Nb_3Si was observed. In the simulation for the alloy with hypo-eutectic composition of 16.0 at%Si, Nb solid solution grew with cell morphology in low undercooling region, while coupled eutectic morphology was formed in higher undercooling region over 8 K. The growth velocity of the coupled growth increased with increase in the degree of undercooling. In the alloy with hypo-eutectic composition of 12.0 at%Si, cell and dendrite morphology of Nb was formed in whole undercooling region,

INTRODUCTION

Eutectic alloys are widely used as industrial materials and the eutectic materials have remarkable feature as in-situ composite. Namely, the minor phase covers some functions and the major phase keeps the entire strength of the material, compensating for the lack of strength of the minor part. In order to develop a new functional structure of a eutectic alloy as a composite material, control of the eutectic solidification structure would be very important. Nb-Si eutectic alloy is a candidate for high temperature materials. In order to predict the optimum conditions to obtain a desired eutectic structure of Nb-Si alloy, it is important to clarify the effects of the process parameters on the eutectic solidification structure of the Nb-Si alloy. However, many experimental works would be required to obtain optimum conditions. Hence, it is desirable to develop a mathematical model to simulate eutectic structure formation.

In the past two decades, several models have been developed for predicting the evolution of micro and macrostructures of alloys during solidification. Among those models, the phase-field method (PF) is a very powerful tool for simulating the evolution of a dendrite structure. However, a drawback of the PF method is a very small computational domain, and the use of the PF method for the simulation of eutectic formation in a real casting will be difficult. The cellular automaton (CA) method[1,2] is another way to simulate the evolution of solidification structures. The CA method has the advantage of relatively small computational load in comparison with the phase-field method. The aim of this work was to develop a numerical model based on the CA method to simulate the formation of the eutectic solidification structure of the Nb-Si binary alloy.

METHOD

Basic procedure

Basic procedure of the cellular automaton method is as follows:
1. a calculation domain was divided into uniform square cells
2. a cell can possess the kind of states
3. there are rules of transition for a cell that determine the state of each cell in the next time step depending on the state of intended cell, and the state of the cells in its neighborhood.

In the present model, a 2D calculation domain with square shape of 250×500μm was divided into square grids of 1μm sides as CA cells. Each cell can possess one of the following five states.

    1:Liquid: (L)
    2:Interface(Nb/L)
    3:Interface(Nb$_3$Si/L)
    4:Solid(Nb)
    5:Solid(Nb$_3$Si)

Assumptions of the present model are as follows:
1. Diffusion occurs between following cells.
    1) Liquid cell-Liquid cell
    2) Liquid cell-Interface cell
    3) Interface-cell-Interface cell
2. In an interface cell, local equilibrium at the solid/liquid interface is maintained.
3. Molar volume of Nb and Nb$_3$Si is same.

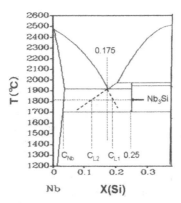

Figure 1. Nb-Si binary phase diagram. Liquidus and solidus lines for Nb and Nb$_3$Si were extrapolated to the undercooled region and local equilibrium compositions at the solid/liquid interfaces were calculated from the extrapolated lines.

Diffusion calculation was carried out by numerically solving following governing equation for the diffusion.

$$\frac{\partial C}{\partial t} = D_L \left(\frac{\partial^2 C}{\partial x^2} + \frac{\partial^2 C}{\partial y^2} \right) \quad (1)$$

Periodic boundary condition was used at the side of the calculation domain
Figure 1 shows Nb-Si binary phase diagram. In the simulation, a sample was isothermally held with undercooled state below ΔT from the eutectic temperature. As the initial condition, seed crystals of Nb and Nb_3Si were set at the bottom of the calculation domain with certain interval. Liquidus and solidus lines for Nb and Nb_3Si were extrapolated to the undercooled region and local equilibrium compositions at the solid/liquid interfaces were calculated from the extrapolated lines.

Growth rule

In the interface cells, solid and liquid coexist and the fraction solid of the cell, fs, is changed due to the diffusion through liquid according to following equation.

$$f_S = \frac{C - C_L}{C_S - C_L} \quad (2)$$

where C is average solute content of the cell, C_L and C_S are liquid and solid compositions at the solid/liquid interface at corresponding temperature. If the fraction solid of the cell becomes unity, the state of the cell is changed to solid state (Nb or Nb_3Si), and surrounding liquid cells were changed to the interface cell. Similar rule proposed by Spittle et al.[3] was used for the transition. The procedure is as follows. When a liquid cell transforms to a solid cell, the numbers of α cells and β cells existed around the liquid cell. were counted, and if the $N_\alpha > N_\beta$ then the liquid cell will transform to a α/L cell, and if $N_\beta > N_\alpha$ then the liquid cell will transform to a β/L cell.

Liquidus temperature

The liquidus composition at corresponding temperature was calculated from the phase diagram shown in the Fig.1 and then the composition was modified due to "curvature undercooling". The meaning of the "curvature undercooling" is the change in the melting temperature of curved S/L interface due to the Gibbs-Thomson effect. The liquidus temperature is expressed as follows.

$$T = T_{LP} - \Gamma \cdot K \quad (3)$$

where T_{LP} is liquidus temperature of planar S/L interface, Γ is Gibbs-Thomson coefficient, K is curvature of the S/L interface. The curvature was calculated by using Box Counting Method[4]. The properties used in the simulation were listed in Table 1.

Table 1 Properties used in the simulation

$D_L = 3.0 \times 10^{-9} (m^2/s)$
$\Gamma Nb = 2.0 \times 10^{-7} (mK)$
$\Gamma Nb_3Si = 2.0 \times 10^{-7} (mK)$

Because we could not find accurate values for the Gibbs-Thomson coefficients for Nb and Nb$_3$Si in literatures , the same values were used for both phases.

The size of a unit cell must be small enough to express the curved S/L interface. Experimentally observed microstructure of a solidified Nb-Si alloy showed that the spacing between Nb and Nb$_3$Si phases is from several microns to about 10 microns[5]. On the basis of this observation, we judged that the cell of 1 square micron is enough for estimating the curvature of the S/L interface. We also judged that the domain of 250x500 microns is enough to express the lamellar eutectic microstructure

RESULTS and DISCUSSION

Eutectic composition

Figure 2 shows simulated eutectic structures of the alloy with eutectic composition of Nb-17.5 at%Si. In the early stage of the solidification with low undercooling state (Fig.2 (a)), coupled growth of Nb and Nb$_3$Si phases was observed. However, the coupled growth becomes unstable and irregular, non-coupled eutectic morphology was formed. In the present simulation with eutectic composition, stable coupled growth was not observed in whole undercooled region. When the degree of undercooling becomes higher than 50 K, then Nb phase could not grow, and only the growth of Nb$_3$Si phase with dendrite morphology was observed as shown in Fig.2(b).

In the Nb-Si binary system, the shape of the phase diagram at the part of the Nb-Nb$_3$Si eutectic reaction is not symmetrical as shown in the Fig.1. In such case, it seems that the coupled zone does not exist at the eutectic composition.

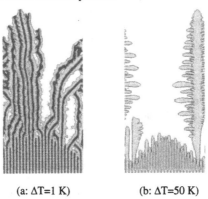

(a: ΔT=1 K)   (b: ΔT=50 K)

Figure 2. Simulated structures of Nb-Si binary alloy with eutectic composition. Dark phase is Nb and the bright phase is Nb$_3$Si.

Hypo-eutectic composition

Figure 3 shows simulated eutectic structures of the alloy with hypo-eutectic composition of Nb-16.0 at%Si. In relatively low undercooling state (Fig.3(a)), Nb$_3$Si phase could not grow, and only the growth of Nb phase with cell morphology was observed as shown in Fig.3(a). On the other hand, coupled growth of Nb and Nb₃Si phases was observed with higher undercooling state. (Fig.3(b)). The simulated result demonstrates that the coupled zone exists at the hypo-eutectic region in the Nb-Nb₃Si eutectic system. In the region over the degree of undercooling of 8 K, coupled growth was always observed in the present simulation. Figure 4 shows the relationship between the degree of undercooling and coupled growth velocity in hypo-eutectic Nb-16.0 at%Si alloy. The growth velocity of the coupled growth increased with increase in the degree of undercooling. Figure 5 shows simulated eutectic structure of the alloy with hypo-eutectic composition of Nb-12.0 at%Si. It was observed that Nb phase always grew with cell and dendrite morphology in whole undercooling region.

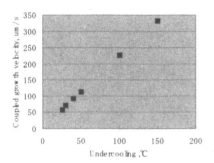

(a: ΔT=7 K)

(b: ΔT=10 K)
Figure 3 Simulated structures of Nb-Si binary alloy with hypoeutectic composition. Dark phase is Nb and the bright phase is Nb₃Si.
(Nb-16.0 at%Si)

(ΔT=50 K)

gure 4. Relationship between degree of dercooling and coupled growth velocity Nb-16.0 at%Si)

Figure 5 Simulated structure of Nb-Si binary alloy with hypoeutectic composition. Dark phase is Nb. (Nb-12.0 at%Si)

411

It was considered that the reason why coupled growth did not occur in the eutectic composition alloy is ascribed to the non-symmetrical shape of the phase diagram shown in the Fig.1. In order to grow with same velocity, Nb and Nb$_3$ phases must be received Nb and Si atoms by balanced diffusion in liquid. However, in the non-symmetrical system, the concentration gradient in liquid ahead of Nb and Nb$_3$Si phases is quite different and results in irregular eutectic growth. On the other hand, it will be expected that the balanced diffusion of Nb and Si atoms can be set up in the Nb-16.0 at%Si alloy with hypo-eutectic composition. Following fact supported this discussion. In the simulation in a hypothetical alloy with symmetrical phase diagram, the coupled growth occurred just at the eutectic composition.

CONCLUSIONS

A numerical model using cellular automaton method was developed for the simulation of microstructure evolution during the solidification of Nb-Si eutectic alloy. Coupled growth was observed in the simulation of hypo-eutectic Nb-Si alloy. The developed model demonstrated the availability of numerical simulation for analyzing the solidification mechanism of eutectic alloys.

REFERENCES

1) X.L.Yang, H.B.Dong,W.Wang and P.D.Lee: Mater. Sci. Eng., **A386,** 129 (2004).
2) M.Yamazaki, J.Satoh, K.Ohsasa and K.Matsuura : ISIJ Int., **48,** 362 (2008).
3) J.A.Spittle and S.G.R.Brown: Acta Metall. Mater. **42,** 1811 (1994).
4) L.Nastac : Acta Mater. 47, 4253 (1999).
5) Y.Sekito et al. : :2008 MRS Fall Meeting, paper No.1128-U05-38 in this volume.

Mater. Res. Soc. Symp. Proc. Vol. 1128 © 2009 Materials Research Society   1128-U07-05

Deformation Behavior of Niobium Silicides During High Temperature Compression

N. Sekido[1], S. Miura[2], Y. Yamabe-Mitarai[1], Y. Kimura[3], and Y. Mishima[3]

[1] National Institute for Materials Science, Tsukuba 305-0047, Japan

[2] Division of Materials Science and Engineering, Hokkaido University, Sapporo 060-8628, Japan

[3] Department of Materials Science and Engineering, Tokyo Institute of Technology, Yokohama 226-8502, Japan

ABSTRACT

Deformation behavior of (Nb)/α–Nb$_5$Si$_3$ two-phase alloys is examined by high temperature compression tests. The alloys exhibit brittle fracture behavior at temperatures up to 1473 K, while reasonable compressive deformability at 1673 K. Upon high temperature compression of the alloys, the flow stress gradually decreases after the peak stress due to the recrystallization/recovery in the (Nb) phase, as well as the increase in the density of mobile dislocations within the α–Nb$_5$Si$_3$ phase. Two types of slip systems that operate in α–Nb$_5$Si$_3$ have been identified as $\{011\}<1\bar{1}1]$ and $\{001\}<100]$ in the present study.

INTRODUCTION

Performance improvement of aircraft engines and land-based gas turbine engines necessitates the development of a new class of heat resistant materials that possess superior temperature capability to Ni-based superalloys. Last decades alloys based on Mo- and Nb-silicides have been attracting attention for potential high temperature applications [1-3] because of their high melting points, low densities, and ability to form protective oxide layers upon high temperature exposure. In general, these silicides are brittle at room temperature, which requests toughening by introduction of a second ductile phase for structural applications. However previous studies have demonstrated that, although (Nb) solid solution phase in the Nb-Si system is capable of plastic deformation at room temperature [4,5], multiphase alloys comprised of (Nb) and Nb-silicides possess low compressive ductility [6] and inadequate toughness at ambient temperatures [7]. This means that, even though the (Nb) phase is capable of some local deformation in multiphase alloys at room temperature, crack initiation and propagation within the silicide phases diminish the global deformability of the alloys.

The present authors have made some effort to establish effective alloy design for Nb–Si alloys to attain balanced properties of high temperature strength and room temperature toughness. It has been demonstrated that microstructural control through directional solidification onto (Nb)/Nb$_3$Si eutectic alloys is beneficial to enhancing the material performance in terms of ambient temperature toughness and high temperature compressive strength [8]. However, little information is available for the deformation behavior of the Nb-silicides at high temperatures, particularly for defect structures induced by deformation. Such knowledge may yield deeper understanding of the mechanisms that control high temperature tensile/compressive strength and creep strength of multi-phase alloys based on the α–Nb$_5$Si$_3$ phase. Thus in the present study the focal point is the identification of defect structures developed in the α–Nb$_5$Si$_3$ phase during compressive deformation at high temperatures.

EXPERIMENTAL PROCEDURES

The nominal composition of the alloy is Nb–10at%Ti–17.5at%Si in this study. Rod ingots (ϕ11 mm × 70 mm) were preliminary prepared by arc-melting of high purity raw materials under an Ar atmosphere. These ingots were unidirectionally solidified in an optical floating zone furnace with the solidification rate of 10 mm/h under an Ar atmosphere. The directionally solidified alloys were annealed at 1673 K for 500 h, by which retained high temperature phase Nb$_3$Si transforms completely into (Nb) and α–Nb$_5$Si$_3$ via a eutectoid reaction [9,10]. High temperature compression tests were conducted under vacuum at temperatures up to 1673 K with a nominal strain rate of 1.0×10^{-4} s^{-1}. The specimens (2×2×5 mm^3) were cut by EDM to tailor the loading axis to the solidification direction. Microstructures of the alloys were characterized by scanning electron microscopy and transmission electron microscopy. TEM foils are mechanically thinned to 50 μm, followed by Ar-ion milling with the acceleration voltage of 4 KV.

RESULTS

An as-directionally-solidified Nb–10Ti–17.5Si alloy exhibits a two-phase microstructure comprised of the (Nb) phase and the high temperature phase Nb$_3$Si, of which decomposition into (Nb) and α–Nb$_5$Si$_3$ requires prolonged annealing at high temperatures [11,12]. After annealing at 1673 K for 500 h, the Nb$_3$Si phase completely transforms into (Nb)/α–Nb$_5$Si$_3$ lamellae, as shown in Fig. 1a. The (Nb)/α–Nb$_5$Si$_3$ lamellae (encircled area in Fig. 1a, for instance) is found to contain some dislocations in the (Nb) phase, but few in the α–Nb$_5$Si$_3$ phase, as shown in Fig. 1b.

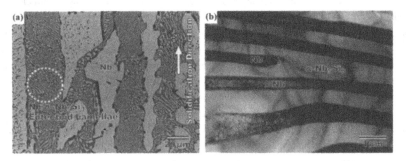

Figure 1: (a) SEM microstructure of Nb-10Ti-17.5Si alloy directionally solidified at 10 mm/h and subsequently annealed at 1673 K for 500 hrs, and (b) TEM micrograph showing the Nb/α-Nb$_5$Si$_3$ lamellae formed by the decomposition of Nb$_3$Si.

The annealed alloys were subjected to compression tests at temperatures from R.T. to 1673 K. The stress-strain curves and the specimens after compression tests are shown in Fig. 2 and 3. In Fig. 2, engineering flow stress is plotted against compressive plastic strain, where the contribution of elastic deformation is subtracted. The stress-strain curves typically exhibit the decrease in flow stress after the peak stress at around 1% of plastic strain. The 0.2% flow stress at 1673 K is above 200 MPa, indicating that the alloy possesses reasonable high temperature strength. At lower temperatures (up to 1473 K), the alloys fail in a brittle manner, as can be seen in Fig. 3. On the other hand, the alloy shows good compressive deformability at 1673 K, where no macroscopic cracks are visible on the surface of the specimen after 15% of compressive deformation.

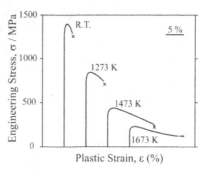

Figure 2: Stress-strain curves by the compression tests at RT, 1273 K, 1473 K, and 1673 K.

The substructure change of the alloy after deformation at 1673 K was characterized by transmission electron microscopy. Many dislocations are observed in the (Nb) phase of the deformed alloy as shown in Fig. 4a. However, the dislocation density in the (Nb) phase does not seem to be very high, and some (Nb) grains clearly exhibit a trace of recovery/recrystallization, as depicted in Fig. 4b. This suggests that recovery and/or recrystallization have occurred during compression or during cooling after the compression tests. Although the α–Nb_5Si_3 phase in the undeformed alloy possesses few dislocations as shown in Fig. 1b, a number of dislocations are found to develop after compressive deformation at 1673 K. Planar faults or dissociation of dislocations in α–Nb_5Si_3 was observed in the present study.

The Burgers vectors of the dislocations that developed in α–Nb_5Si_3 (D8$_f$: tP32, I4/mcm, Cr_5B_3 type) were determined by the weak-beam thickness fringe method[13]. The reflection vector, g, the Burgers vector of dislocation, b, and the number of fringes terminated at the dislocation end, n, satisfy the following equation:

Figure 3: The specimens after compression tests; (a) undeformed, (b) RT, (c) 1273 K, (d) 1473 K, and (e) 1673 K. The solidification direction is in accordance with the loading axis (longitudinal direction).

415

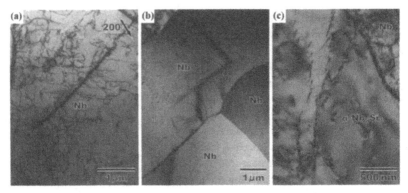

Figure 4: TEM micrographs taken from the specimen deformed at 1673 K showing (a) dislocations in (Nb), (b) recovered/recrystallized area in (Nb), and (c) dislocations in α-Nb₅Si₃.

$$\mathbf{g} \cdot \mathbf{b} = n \qquad (1).$$

Thus the Burgers vector of a dislocation can be determined from the number of excess thickness contours terminating at the end of each dislocation line imaged in weak-beam dark-field mode under, at least, three non-coplanar reflection vectors. Figure 5 shows the weak-beam dark-field images of dislocations imaged under the reflection vectors of $2\bar{2}0$, $\bar{1}2\bar{3}$, and $1\bar{2}\bar{3}$. Here, the dislocations labeled as A and B are the focus of this investigation, of which results are summarized in Table 1. Note that dislocation-A in Fig. 5c is not visible since the invisibility criterion is satisfied under $\mathbf{g} = 1$ $\bar{2}\bar{3}$. By solving the Eq. 1 with the numbers of excess thickness fringes and the reflection vectors, two Burgers vectors have been identified: $\mathbf{b} = 1/2[1\bar{1}1]$ for dislocation-A, and $\mathbf{b} = [0\bar{1}0]$ for dislocation-B. From trace analysis based on tilting experiments, the line vector, ξ, of these dislocations are determined as [16 9 8] and [23 $\bar{5}$ $\bar{2}$], respectively. Since the Burgers vector and the line vector of a mobile dislocation lie on its glide plane, the slip plane can be deduced from the cross product be-

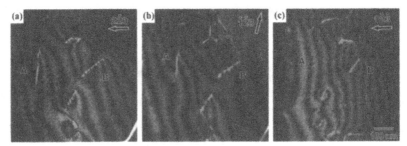

Figure 5: Weak-beam dark field images of dislocations in the α-Nb₅Si₃ phase under the reflection vectors of (a) $\mathbf{g} = 2\bar{2}0$, (b) $\mathbf{g} = \bar{1}2\bar{3}$, and (c) $\mathbf{g} = 1\bar{2}\bar{3}$.

416

Table 1: Summary of the Burgers vector and line vector analysis on the dislocations shown in Fig. 5.

| Dislocation | $g \cdot b$ | | | Burgers Vector | Line vector | Cross product |
	$g=2\bar{2}0$	$g=\bar{1}23$	$g=1\bar{2}3$	b	ξ	$b \times \xi$
A	2	1	0	$1/2[1\ \bar{1}\ 1]$	$[16,9,-8]$	$(-1,24,25)$
B	2	2	2	$[0\ \bar{1}\ 0]$	$[23,-5,-2]$	$(2,0,23)$

tween the two. The $b \times \xi$ product for dislocation-A is ($\bar{1}$ 24 25), which is almost parallel to the (011) plane. On the other hand, the $b \times \xi$ product for dislocation-B is (2 0 23), which is only 8° tilted away from the (001) plane. These results indicate that the slip systems that can operate in α–Nb$_5$Si$_3$ are {011}<1$\bar{1}$1] and {001}<100].

DISCUSSION

Upon compressive deformation, the Nb–10Ti–17.5Si alloy typically exhibit a yield-drop like behavior, where flow stress decreases with strain after showing the peak as depicted in Fig. 2. At lower temperatures, the alloys exhibit brittle fracture behavior, where growth of large cracks leads to catastrophic failure. This implies that the major reason for this behavior at lower temperatures is the stress relief caused by generation and propagation of micro-cracks during the compression tests. TEM observations have revealed that recovered or recrystallized grains are present in the (Nb) phase of the alloy deformed at 1673 K. This indicates that dynamic recovery/recrystallization in the (Nb) phase during compression may also be responsible for the yield drop phenomenon. In addition, many dislocations have been found to develop in the α–Nb$_5$Si$_3$ phase after 15% of compression at 1673 K, while few in the undeformed sample. The increase in the density of mobile dislocations in the α–Nb$_5$Si$_3$ phase would be another reason for this behavior. It is also possible that buckling is partly responsible for the reduced flow stress during compression, which is evidenced from the specimen after the test as shown in Fig. 3e.

The present study reveals that, at least, two types of slip systems are operative in the α–Nb$_5$Si$_3$ phase: {011}<1$\bar{1}$1] and {001}<100]. Development of dislocations with these Burgers vectors seems to be reasonable, since they are the shortest and second-shortest translation vectors in the α–Nb$_5$Si$_3$ phase. Information on the slip systems that are operative in the α–Nb$_5$Si$_3$ phase is limited; however, some is available for Mo$_5$SiB$_2$ [14-17] which has the same crystal structure as α–Nb$_5$Si$_3$. The slip systems reported in Mo$_5$SiB$_2$ are {001}<100] [14,15], {112}<11$\bar{1}$] [14], and {100}<001] [16]. The slip system {001}<100] is consistent with the present result. However, the slip plane of a dislocation with b = 1/2<11$\bar{1}$] in the α–Nb$_5$Si$_3$ phase is {011} rather than {112}. In terms of elastic line energy of dislocation, a dislocation with b = <001] is unlikely to develop, because the line energy of a <001] dislocation can be lowered by dissociating it into two 1/2<111] dislocations, or other multiple partial dislocations, as has been discussed [16]. However, no dissociation of dislocations was observed in the present study. Further study is necessary for slip system activation in the α–Nb$_5$Si$_3$ phase by taking critical resolved shear stresses into account.

CONCLUSIONS

A Nb–10Ti–17.5Si alloy with a (Nb)/α–Nb$_5$Si$_3$ two-phase microstructure possesses good compressive deformability at 1673 K, where no macroscopic cracks are visible on the surface of the specimen after 15% of compression. Upon high temperature compression, the alloys exhibit a yield-drop like behavior due to the dynamic recrystallization or recovery of the Nb phase, and the increase in the density of mobile dislocations in the α–Nb$_5$Si$_3$ phase. The slip systems that operate in the α–Nb$_5$Si$_3$ phase have been identified as {011}<1$\bar{1}$1] and {001}<100].

REFERENCES

1. Bewlay BP, Jackson MR, Zhao JC, Mendiratta MG, Lewandowski JJ, Subramanian PR. MRS Bull 2003;28:646.
2. Dimiduk DM, Perepezko JH. MRS Bull 2003;28:639.
3. Yamaguchi M, Inui H, Ito K. Acta Mate 2000;48:307.
4. Mendiratta MG, Goetz R, Dimiduk DM, Lewandowski JJ. Metall Mat Trans A 1995; 26A:1767.
5. Samant A, Lewandowski JJ. Metall Mat Trans A 1997;28A:389.
6. Sekido N, Miura S, Mishima Y, in: Proc. Third Pacific Rim Intnl. Conf. on Advanced Materials and Processing (TMS 1998), 2393.
7. Bewlay BP, Lipsitt HA, Jackson MR, Reeder WJ, Sutliff JA. Mater Sci Eng A 1995; A192-193:534.
8. Sekido N, Kimura Y, Wei FG, Miura S, Mishima Y. J Alloys Compds 2006;425:223.
9. Sekido N, Kimura Y, Miura S, Mishima Y. Mater Trans 2004;45:3264.
10. Sekido N, Kimura Y, Miura S, Mishima Y. Mater Sci Eng A 2007;444:51.
11. Mendiratta MG, Dimiduk DM. Scripta Metall Mater 1991;25:237.
12. Sekido N, Kimura Y, Wei F-G, Miura S, Mishima Y. J Japan Inst Metals 2000;64:1056.
13. Ishida Y, Ishida H, Kohra K, Ichinose H. Philos Mag 1980;42A:453.
14. Field RD, Thoma DJ, Cooley JC, Chu F, Fu CL, Yoo MH, Hults WL, Cady CM. Intermetallics 2001;9:863.
15. Meyer MK, Kramer AJ, Akinc M. Intermetallics 1996;4:273.
16. Ito K, Ihara K, Tanaka K, Fujikura M, Yamaguchi M. Intermetallics 2001;9:591.
17. Sekido N, Sakidja R, Perepezko JH. Intermetallics 2007;15:1268.

Mater. Res. Soc. Symp. Proc. Vol. 1128 © 2009 Materials Research Society   1128-U07-07

Current Status of Mo-Si-B Silicide Alloys for Ultra-High Temperature Applications

Martin Heilmaier[1], Holger Saage[2], Manja Krüger[2], Pascal Jehanno[3], Mike Böning[3] and Heinrich Kestler[3]

[1]Materials Science Department, TU Darmstadt, Petersenstr. 23, D-64287 Darmstadt, Germany
[2]Institute for Materials & Joining Technology, Otto-von-Guericke University Magdeburg, D-39106 Magdeburg, Germany
[3]Technology Centre, Plansee SE, A-6600 Reutte/Tyrolia, Austria

ABSTRACT

We review the current development status of molybdenum borosilicide (Mo-Si-B) alloys for ultra-high temperature applications in excess of 1100°C in air. The assessment of several ingot and powder metallurgy approaches revealed that (i) the presence of a continuous Mo solid solution matrix is mandatory for adequate low temperature toughness and (ii) wrought processing of such alloys at temperatures established for refractory metals requires the presence of an ultrafine (sub-micron) microstructure. Both the prerequisites could be fulfilled using mechanical alloying (MA) as the crucial processing step , however, values for the ductile-to-brittle transition temperature (DBTT) below 800°C could not be obtained due to grain boundary embrittlement by Si segregation. First results on the effect of different microalloying additions (e.g. Zr) on the reduction of this segregation will be presented and discussed.

INTRODUCTION

The development of structural materials for ultra-high temperature applications is a major challenge for the materials science and engineering community. In the case of energy producing gas turbines the establishment of Nickel based superalloys in the 1950's opened an Era of 50 years of continuous growth and development. Today's engines expose these superalloys to temperatures approaching 1150°C, corresponding to homologous temperatures above 0.8 of the melting point, while the turbine inlet temperature was raised to about 1500°C. However, when looking at the maximum service temperature of superalloys over the last 20 years, only a few major improvements, i.e. the introduction of single-crystals (including the addition of Re) and the application of thermal barrier coatings, respectively, have been achieved [1,2]. New metallic materials which could withstand surface temperatures higher than 1200°C would be desirable in order to increase the (thermodynamic) efficiency. Amongst other metallic materials systems such as alloys based on precious metals [3] which will be disregarded here for weight reasons, multiphase silicide alloys with Nb or Mo as a base metal have shown promise as novel structural materials for applications in excess of 1100°C in air [1,4-6].

In this overview we report on the current status of Mo-Si-B alloys consisting of Mo solid solution (Mo_{ss}) and of the intermetallic phases Mo_3Si and Mo_5SiB_2 which could take advantage of (i) the beneficial oxidation resistance of the silicide phases and (ii) the outstanding mechanical properties of molybdenum. A typical alloy composition that showed balanced properties with respect to oxidation resistance and mechanical properties was given as Mo–9Si–8B (unless stated otherwise all compositions in at.%) [5] which, thus, has been taken as a baseline composition in what follows. However, Berczik [5] reported several difficulties to reach this goal making manufacturing on an industrial scale hardly feasible: (i) to obtain materials with a metallic matrix mandatory for adequate low temperature toughness rapid solidification techniques were utilized (rotating electrode process with He gas atomization) and (ii) wrought processing of such alloys needed temperatures as high as 1760°C. To circumvent these limitations, we chose an alternative powder-metallurgical route and employed mechanical alloying (MA) as the crucial processing step [7]. We will discuss the current status and future trends of alloys based on the ternary Mo-Si-B system with a special emphasis placed on the key properties (i) BDTT as a representative for ambient temperature strength and toughness, (ii) creep and (iii) oxidation resistance. The impact of suitable micro-alloying additions on BDTT of the Mo solid solution is finally addressed and reveals the potential for further improvement of the mechanical properties.

EXPERIMENT

While details on the mechanical alloying, the degree of supersaturation and the possible formation of amorphous phases are described in [7], the manufacturing of consolidated material consists of the following elementary processing steps which were followed by X–ray diffractometry (XRD, Siemens D5000) with CuK_α radiation. Elemental Mo, Si and B powders of 99.95, 99.6 and 98% purity respectively, were used to obtain three-phase Mo_{ss}-Mo_3Si-Mo_5SiB_2 alloys as described above. Comparatively, a single phase Mo solid solution with about 1.5 at.% Si in Mo was extrapolated from the isothermal section of the respective ternary phase diagram at 1600°C [8] at the targeted application temperature of at least 1200°C. About 0.7 vol.% of Y_2O_3 and La_2O_3 powders, respectively, or about 1 wt.% ZrH_2 powder which is reduced to about 1 at.% Zr upon sintering were added in selected cases. Mechanical alloying (MA, a high energy milling process [9]) was carried out for 20h under protective (argon) atmosphere in a planetary ball mill with a rotational speed of 200 rpm and a powder to ball weight ratio of 1:13. In some cases PM is used as prefix instead which refers to alloys prepared from solely mixed powders (not mechanically alloyed). The detailed processing of these alloys is described elsewhere [10]. After milling the powders were cold isostatically pressed at 200 MPa and sintered in hydrogen atmosphere at 1500°C to reduce the impurity content (mainly oxygen, nitrogen, hydrogen). Depending on the preparation technique the powders respond differently to this treatment. Simple mixtures of the elements Mo and Si have impurity levels of below 5 wt. ppm as measured by hot gas carrier extraction (see [10]), whereas the MAed powders studied here possess 200 to 500 wt. ppm oxygen after sintering which remains at this level upon further consolidation. As a final step the sintered samples were further consolidated by hot isostatic pressing (HIP) at 1500°C and 200 MPa.

Besides XRD, scanning electron microscopy (SEM) (FEI ESEM XL30 FEG equipped with EDX and EBSD) was employed to analyse the microstructure of the powders, consolidated samples and fracture surfaces. Fresh fracture surfaces, which revealed a mixed trans- and intergranular fracture mode in SEM, of samples with the composition Mo-1.5Si and Mo-1.5Si-1Zr were made by fracturing under high vacuum and subsequently analysed in an Auger microscope for possible segregation of the alloying elements towards grain boundaries and grain interior, respectively.

Electro-discharge machining was used to prepare bending samples with a squared cross section of about 3.8 x 3 mm and a length of 45 mm from the HIP billets and final surface quality was attained by grinding. The mechanical properties were assessed by 3-point bending tests with a span of 40 mm in a temperature range of RT-1093°C using a Zwick electromechanical testing device equipped with a Maytec furnace at a crosshead speed of 0.01 mm/min. For each temperature and material two to three samples were tested. They showed negligible scatter. At temperatures above 538°C a gas mixture of argon with 2% hydrogen was used for the bending tests to prevent oxidation. The bend deformation was continuously monitored by Al_2O_3 rods attached to the centre of the lower (tensile) side of the samples and to the lower pushrod, respectively. The stresses and (elastic) strains on the tensile side of the bend sample were calculated utilizing elastic beam theory [11] which is sufficient to detect BDTT, i.e., the temperature above which plasticity occurs. With this configuration even small deviations from the elastic behaviour could clearly be resolved. The same procedure was applied for preparing and testing of compressive creep samples with rectangular shape (same cross-section as the 3-point bend samples, but 6 mm in height) and polished compression sides. Creep testing was carried out at constant true stress in a temperature range between 1050 and 1200°C. This involves a continuous monitoring of creep strain throughout the test.

Interrupted oxidation tests were carried out at the temperatures of 650°C (the pesting regime), 1100°C which is a typical service temperature for Nickelbase superalloys and 1315°C (an anticipated service temperature for Mo-Si-B alloys). For each alloy three round samples with 10 mm diameter and 3 mm thickness were simultaneously introduced in a pre-heated furnace at the given temperature and the mean value of weight loss after 1, 3, 10, 30 and 100 hours (weighted after cooling) was recorded.

DISCUSSION

Microstructure

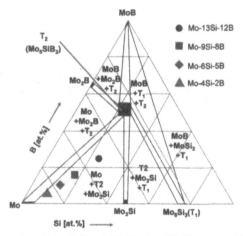

Figure 1. Isothermal section of Mo-Si-B at 1600°C, after [8]. The compositions chosen in this study are marked.

421

Alloy compositions which showed a reasonably good balance of room temperature toughness and high temperature creep and oxidation resistance were given as Mo-9Si-8B [5,12-14]. This composition which corresponds to about 55% Mo_{ss} in the three phase field (see Fig. 1) has been taken as a reference in this study. Additionally, three phase alloys with about 30% (Mo-13Si-12B), 70% (Mo-6Si-5B) and 85% (Mo-4Si-2B) Mo_{ss} were chosen to investigate the effect of different ratios of the intermetallic phases to the α-Mo solid solution on the microstructure.

The microstructures of the Mo-Si-B alloys after MA for 20 h and subsequent consolidation as described above are shown in Fig. 2. A continuous α-Mo matrix is obtained for Mo-4Si-2B (not shown), Mo-6Si-5B and for Mo-9Si-8B. In contrast, the Mo-13Si-12B alloy shows a continuous intermetallic matrix with islands of α-Mo. In Fig. 2d the microstructure of an atomized and HIPed Mo-8.9Si-7.7B alloy with a non continuous α-Mo is shown for comparison [7]. The average sizes of the three phases present were quantified on micrographs such as the ones in Fig. 2a to c and were found to lie all in the sub-micron range. Therefore, materials in the asHIP condition may be considered possessing an ultrafine-grained (UFG) microstructure.

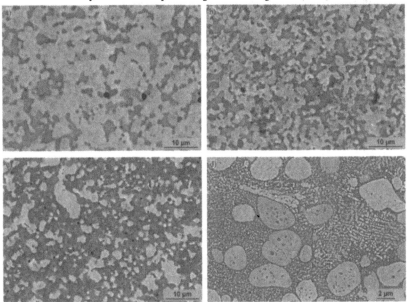

Figure 2. Microstructures of the consolidated MA samples with a composition of a) Mo-6Si-5B, b) Mo-9Si-8B and c) Mo-13Si-12B. Figure d) shows the microstructure of a gas atomized and HIP consolidated alloy with a composition of Mo-8.9Si-7.7B [7,14], all SEM in BSE mode.

Brittle to ductile transition temperature

In Figure 3 the influence of the microstructure on BDTT is revealed. Values for the atomized and as HIPed Mo-8.9Si-7.7B and extruded Mo-8.9Si-7.7B samples were taken from tensile tests [6] and designated AH-T and EX-T, respectively. In contrast to the alloys prepared from mechanically alloyed powders alloys consolidated from the gas atomized powders do not show a continuous α-Mo matrix after HIPing (Fig. 2(d)). In fact, hot extrusion is required to convert this microstructure to a (mostly) continuous α-Mo matrix [6]. From Fig. 3 it is obvious that BDTT decreased by more than 150 K by switching from a discontinuous to a continuous α-Mo matrix.

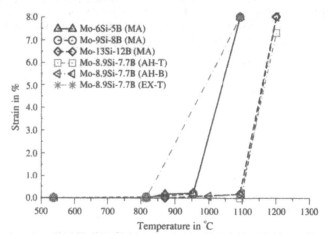

Figure 3. Outer fiber tensile strain of selected Mo-Si-B alloys versus test temperature showing the decrease of BDTT for alloys with a continuous α-Mo matrix. Tensile data for the gas atomized and HIPed (AH-T) and the extruded (EX-T) Mo-Si-B alloys were taken from [6]. All MA alloys were tested in 3-point bending with the outer tensile fibre strain shown for comparison. Tests were terminated after 8% plastic strain without cracking or even failure.

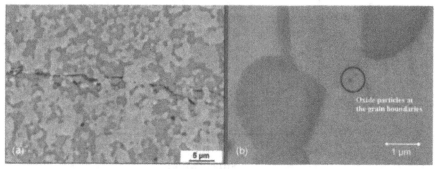

Figure 4. SEM BSE images showing (a) the path of a crack in a Mo-9Si-8B bend specimen after fracture at room temperature and (b) lanthana particles decorating the grain/phase boundaries.

On the one hand, the decrease of the BDTT due to the establishment of a continuous α-Mo matrix is promising, Fig. 3. The main reason for this behavior is the potential of crack

423

trapping and bridging by the α-Mo areas in these alloys (see Fig. 4a): propagating crack tips cannot avoid the relatively ductile α-Mo. Interestingly, in spite of different intermetallics volume fractions the curves for the alloys with a continuous α-Mo matrix coincide. Furthermore, the same behaviour is observed for the alloys with intermetallic matrix. Hence, the BDTT seems to be characteristic for the respective matrix phase. On the other hand, it is obvious that a three-phase ternary Mo-Si-B alloy appears rather brittle at RT, especially tensile tests show zero ductility [6]. However, values for the fracture toughness as high as 20 MPa√m have been reported in literature [15] for alloys with a continuous Mo matrix and comparable chemical composition. Also, Berczik [5] found a BDTT of about 500°C for a two-phase Mo-6Si-8B alloy. We believe that the key for a further performance increase lies in the ductility enhancement of the Mo solid solution through further micro-alloying. An outlook on the current state of progress will be given in the last subsection.

Creep resistance

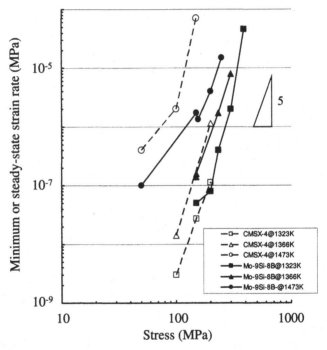

Figure 5. Double-logarithmic representation of strain rate vs. applied stress (Norton plot) for lanthana doped Mo-9Si-8B alloys (full symbols) and for the single-crystalline Nickelbase superalloy CMSX-4 (open symbols); temperatures as indicated.

The compressive creep behaviour of a MA processed Mo-9Si-8B batch doped with Lanthana was comparatively assessed against a state of the art Nickelbase single-crystalline alloy CMSX-4. The typical microstructure is displayed in Fig. 4b which shows La_2O_3 particles decorating grain and phase boundaries and, thus, stabilizing the grain and phase sizes, respectively, in the low micron range. In case of the Mo-silicide alloy, after attaining a constant, i.e. steady-state creep rate during a reasonable amount of strain the stress was increased stepwise until a new steady-state was established. Finally, a stress reduction towards the initial stress level was carried out in order to check the validity of the steady-state assumption. In doing so it was possible to obtain three data points in the Norton plot, Fig. 5, with a single specimen. However, due to the well-known thermal instability of the γ/γ'-microstructure in Nickelbase superalloys which leads to the directional coarsening of the γ' phase upon creep loading (the so-called rafting phenomenon [16-18]), the creep curves of CMSX-4 exhibit a pronounced minimum creep rate which has been taken for comparison. Consequently, a separate compressive creep specimen of CMSX-4 was tested for each single stress level for comparison in Figure 5.

From Figure 5 it is obvious that the creep performance of the lanthana doped Mo-9Si-8B alloy is comparable to CMSX-4 at the lower test temperatures of 1050 and 1093°C, respectively. At 1200°C, however, it is slightly superior exhibiting a reduced creep rate by a factor of approximately 4 over the whole stress region investigated. However, two points are noteworthy: first, the temperature of 1200°C is already beyond the capability of the single-crystalline CMSX-4 alloy because of its close proximity to the γ' solvus temperature. This is accompanied by a substantial loss in γ' strengthening due to, both, the directional coarsening and the reduction in volume fraction of the γ' phase [18,19]. Second, whereas the stress exponent $n = \Delta \log \dot{\varepsilon} / \Delta \log \sigma$ is between 5 and 7 at the lower test temperatures typically observed in particle strengthened alloys, it decreases to <3 at 1200°C and low applied stresses. While the latter value can again be rationalized for the Nickelbase CMSX-4 alloy by the above thermal instability of the γ/γ' microstructure, the microstructure of the Mo-Si-B alloys is still stable at these temperatures. Hence, even superplastic deformation due to grain boundary sliding processes can be expected [20] which quite naturally explains the observed reduced values of the stress exponent between 2 and 3. In fact, for a significant coarsening of this ultrafine triplex microstructure [20] annealing temperatures above 1600°C are needed [21]. Consequently, the annealed material shows a promising reduction of creep rate by at least one order of magnitude when the grain size of the Mo solid solution is increased to about 7 µm [21].

Oxidation resistance

The oxidation resistance of three differently processed Mo-Si-B alloys was assessed in two temperature regions according to the oxidation mechanisms described in the literature [13]: the comparison of the Ar-gas atomized Mo-8.9Si-7.7B alloy [6,14] with two MA processed Mo-9Si-8B batches doped with either Lanthana or Yttria [22] should gain insight about the effect of (a) the continuity and type of the matrix of the alloy and (b) the fineness and homogeneity of the microstructure on oxidation behavior.

The first temperature range (pesting regime) extends from 450°C to 700°C, where a B-Si-Mo-O porous oxide scale forms followed by the diffusion of oxygen in the oxide scale, resulting in a weight gain. We selected the temperature of 650°C, just below 700°C in order to characterize the oxidation rate within this temperature domain. Tests in the second range above 700°C usually result in a weight loss because of sublimating MoO_3 which decays with increasing exposure time due to the formation of a Si-B-O passivating scale. The kinetics of the formation of this boro-silicate glass scale as well as the kinetics of the oxygen diffusion through the oxide scale are the key parameters for the oxidation resistance of the material.

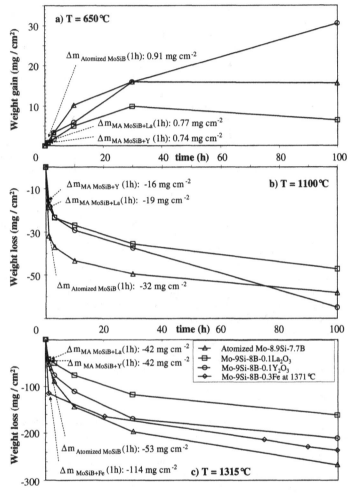

Figure 6. Oxidation tests in air on MA lanthana (open squares) and yttria (open circles) doped Mo-9Si-8B, respectively, at a) 650°C, b) 1100°C and c) 1315°C. For comparison data for Ar-gas atomized Mo-8.9Si-7.7B [6,14] and Fe-doped Mo-9Si-8B [23] are plotted as open triangles and open diamonds, respectively.

The results of the oxidation trials reveal (as expected) two different mechanisms characterized by a weight gain at the low temperature of 650°C (Fig. 6a) and weight losses at the higher temperatures of 1100 (Fig. 6b) and 1315°C (Fig. 6c), respectively. At 650°C all alloys have a very similar weight gain within the first hour exhibiting an appoximately parabolic weight gain. Further oxygen penetration through the oxide scale occurs at a lower rate for the lanthanum containing alloy compared to the other alloys. Due to the extremely volatile MoO_3 scale spallation occurs at longer testing times which makes not only the determination of a kinetic rate constant impossible but also reveals that the alloys under consideration here are still prone to pesting. By contrast, testing at both higher temperatures results in a pronounced weight loss within the first hour followed by an approximately linear weight loss. Not only within the first hour the mechanically alloyed and doped variants are superior to their Ar-gas atomized counterpart, compare for example weight losses at 1100°C between only 16 and 19 mg cm^{-2} against 32 mg cm^{-2}. The linear weight loss is still about one orders of magnitude faster than the (parabolic) weight gan obtained for alumina forming Nickelbase single crystals [24]. At the targeted temperature of 1315°C, further oxygen diffusion through the oxide scale occurs and the weight loss is increased with a rate which is nearly one order of magnitude lower for both MA alloys than for the gas atomized variant. The faster formation of the oxide scale of the MA alloys may be attributed to the homogeneously distributed intermetallic phases with an ultra-fine grain size effectively pinned by the La_2O_3 oxide particles, Figure 4b. This has also been observed in alumina forming steels [25]. The fine microstructure obviously facilitates the formation of a dense, protective boro-silicate glass layer on the material surface at high temperatures. Micro-alloying with Fe (or Ni) may yield a further slight improvement in oxidation resistance as the curve adopted from Woodard et al. [23] essentially reveals the identical course of weight loss as our MA alloys, however, at a higher test temperature of 1371°C. At present, the obtained values of the rate constants for the weight losses are still too high to be acceptable for any uncoated high temperature long-term structural application in air. Therefore, further effort has to be directed on the development and investigation of protective coatings, see e.g. [14,26].

Future perspectives

In Figure 3 we demonstrated that the BDTT of three-phase Mo-Si-B alloys could be reduced to values as low as 800°C by establishing a continuous Mo_{ss} matrix. As it is commonly accepted that the two other (intermetallic) phases are brittle at low temperatures [27,28] we focused our development trials on the Mo solid solution in order to further reduce BDTT. In previous work [10] we studied the mechanical behaviour of high purity Mo(Si) solid solutions with Si contents ranging between 0.1 wt.% to 1 wt.% as a reference for the matrix phase of the above mentioned three phase Mo-Si-B alloys. As an important outcome we could demonstrate that while Si is one of the most effective solid solution strengthener [29], it also strongly promotes the tendency towards brittle intergranular failure [10].

It is well known in bcc metals and alloys that ductility can be improved by reducing the grain size since this reduces the effective oxygen concentration at grain boundaries [30]. Thus, one of the approaches to ductilize Mo(Si) solid solutions was to add a thermodynamically stable oxide Y_2O_3 in order to stabilise a small grain size during powder metallurgical (PM) manufacturing taking into account that this will have the above demonstrated detrimental impact on the creep performance at higher. Therefore, the second and possibly more viable way is to find suitable micro-alloying additions which improve the ductility by reducing the tendency of solute atoms to segregate to the grain boundaries. Among the possible candidates which have been identified through a theoretical approach [31] Zr has been shown recently to improve the fracture toughness of Mo-12Si-8.5B (in at.%) [32]. According to Miller and Bryhan [33], who measured the distribution of elements at and close to the grain boundaries by atom probe tomography, the basic mechanism of the improvement of the grain boundary fracture stress of a welded Mo-base alloy containing small amounts of Zr, C and B was found to be the replacement of oxygen and possibly nitrogen by these additions at the grain boundaries. Therefore, in our second trial we used ZrH_2, which disproportionates into its elements during manufacturing at high temperatures, to ductilize Mo(Si) solid solutions prepared by PM.

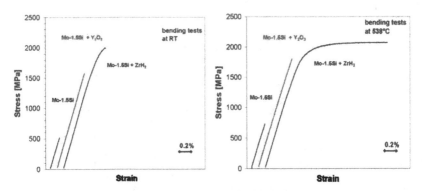

Figure 7. Stress-strain curves (outer tensile fibre) of pure Mo-1.5Si and Mo-1.5Si with Y_2O_3 and ZrH_2 addition at room temperature (left) and at 538°C (right).

Stress – strain (calculated for the outer tensile fibre) curves for the alloys tested at room temperature and at 538°C are compared in Figure 7. Accordingly, the MAed Mo-1.5 Si solid solution and the alloy with additional Y_2O_3 dispersoids behave brittle at temperatures up to 538°C. However, both alloys show some plastic deformability at 816°C (not shown here, see [34]) which coincides nicely with the BDTT values found for the 3-phase Mo-Si-B alloys with continuous Mo matrix in Fig. 3. In contrast, the alloy with ZrH_2 addition reveals already at room temperature limited plastic strain at a very high strength level of around 2 GPa. Moreover, this alloy yields a limited plastic strain of about 1.4% at 538°C (test was terminated without failure) while exceeding the strength level of 2 GPa.

Fracture of the Mo(Si) solid solution occurs predominately along grain boundaries with a fraction of intergranular failure of about 90% with the remainder showing transgranular fracture. In contrast, the Zr containing alloy only partly (≈60%) fractures intergranularly (see Fig. 8), whereas the fracture of the Y_2O_3 containing alloy occurs almost 100% along grain boundaries at room temperature. Tensile tests have revealed that the addition of only 0.3 at.% Si leads already to brittle intergranular fracture at room temperature while its Si free counterpart almost reaches 15% plastic strain to failure and shows a transgranular fracture mode [10].

The Auger analyses of both, the grain interior and grain boundaries freshly broken in ultrahigh vacuum reveal that there is a competition between Zr and Si segregating towards the grain boundaries. Mo-1.5Si alloys containing Zr show significantly lower amounts of Si at the grain boundaries which have the observed beneficial effect on grain boundary strength, cf.

Figures 8a and b respectively. Consequently, the full potential of the strength and ductility of the Mo solid solution can be exploited, see Fig. 7.

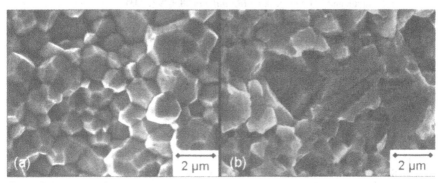

Figure 8. Fracture surfaces after deformation at room temperature of (a) Mo-1.5Si and (b) Mo-1.5Si-1Zr (at.%) showing the positive effect of Zr addition on grain boundary fracture strength.

CONCLUSIONS

On the basis of the present experimental results the following conclusions on enhancing the performance of multi-phase Mo-Si-B alloys can be drawn:

1. In Mo-Si solid solutions Si segregates to grain boundaries resulting in a reduced grain boundary strength and, hence, brittle behaviour at low temperatures.
2. The addition of Y_2O_3 particles results in a reduced grain size with improved low temperature strength but does not result in any ductility up to the testing temperature of 538°C.
3. Zr increases grain boundary cohesion which even allows limited plastic deformation during room temperature bend testing at a strength level of about 2 GPa. Auger analyses reveal that the grain boundaries of this alloy show a lower level of Si segregation.
4. In multi-phase Mo-Si-B alloys the change from a continuous intermetallic matrix with isolated islands of Mo solid solution towards a continuous Mo_{ss} matrix with finely dispersed intermetallic particles results in a reduction of BDTT of more than 150 K.
5. Ultrafine microstructures have a beneficial effect on the oxidation resistance of Mo-Si-B alloys due to a faster establishment of a protective boro-silicate glass scale.

ACKNOWLEDGMENTS

Financial support by the German Science Foundation (DFG) in the frame of the research unit 727 "Beyond Nickelbase Superalloys" is gratefully acknowledged. Partial funding by the European Union through the ULTMAT project (contract AST3-CT-2003-502977) is appreciated. One of the authors (M.H.) acknowledges support from the Japan Society for the Promotion of Sciences (JSPS) through a short-term fellowship.

REFERENCES

1. D.M. Dimiduk and J.H. Perepezko: *MRS Bulletin* **28**, 639 (2003).
2. J.R. Nicholls, *MRS Bulletin* **28**, 659 (2003).
3. Y. Yamabe-Mitarai, Y. Gu, C. Huang, R. Völkl, and H. Harada, *JOM* **56**, 34 (2004).
4. B.P. Bewlay, M.R. Jackson, J.-C. Zhao, P.R. Subramanian, M.G. Mendiratta, and J.J. Lewandowski, *MRS Bulletin* **28**, 646 (2003).
5. D.M. Berczik, US Patent 5,595,616 (1997).
6. P. Jéhanno, M. Heilmaier, H. Kestler, M. Böning, A. Venskutonis, B. Bewlay, and M. Jackson, *Metall. Mater. Trans.* **36A**, 515 (2005).
7. M. Krüger, S. Franz, H. Saage, M. Heilmaier, J.H. Schneibel, P. Jéhanno, M. Böning, and H. Kestler, *Intermetallics* **16**, 933 (2008).
8. J.H. Perepezko, R. Sakidja, and S. Kim, in (Mat. Res. Soc. Proc. **646**, Pittsburgh, PA, 2001) pp. N4.5.1.
9. C. Suryanarayana, *Prog. Mater. Sci.* **46**, 1 (2001).
10. D. Sturm, M. Heilmaier, J. H. Schneibel, P. Jéhanno, B.Skrotzki, and H. Saage, *Mater. Sci. Eng.* **A463**, 107 (2007).
11. A. Guha, in *Metals Handbook, 9th edition, vol. 8, Mechanical Testing and Evaluation* (American Society for Metals, 1985) pp. 132-136.
12. M. Mendiratta, T.A. Parthasarathy, and D.M. Dimiduk, *Intermetallics* **10**, 225 (2002).
13. T.A. Parthasarathy, M. Mendiratta, and D.M. Dimiduk, *Acta Mater.* **50**, 1857 (2002).
14. P. Jéhanno, M. Heilmaier, and H. Kestler, *Intermetallics* **12**, 1005 (2004).
15. J.J. Kruzic, J.H. Schneibel, and R.O. Ritchie, *Metall. Mater. Trans.* **36A**, 2393 (2005).
16. A. Pineau, *Acta Metall.* **24**, 559 (1976).
17. P. Caron and T. Khan, *Mater. Sci. Eng.* **61**, 173 (1983).
18. H. Mughrabi, M. Ott, and U. Tetzlaff, *Mater. Sci. Eng.* **A234-236**, 434 (1997).
19. P. Jéhanno, M. Heilmaier, H. Saage, H. Heyse, M. Böning, H. Kestler, and J. Schneibel, *Scripta Mater.* **55**, 525 (2006).
20. P. Jéhanno, M. Heilmaier, H. Saage, M. Böning, H. Kestler, J. Freudenberger, and S. Drawin, *Mater. Sci. Eng.* **A463**, 216 (2007).
21. R.C. Reed, D.C. Cox, and C.M.F. Rae, *Mater. Sci. Eng.* **A448**, 88 (2007).
22. P. Jéhanno, M. Heilmaier, and H. Kestler, European Patent EP 1 718 777 B1 (2006).
23. S. Woodard, R. Raban, J.F. Myers, and D.M. Berczik, US Patent 6,652,674 (2003).
24. M.H. Li, X.F. Sun, T. Jin, H.R. Guan, and Z.Q. Hu, *Oxid. Met.* **60**, 195 (2003).
25. N. Nomura, T. Suzuki, K. Yoshimi, S. Hanada, *Intermetallics* **11**, 735 (2003).
26. J. Disam, H.-P. Martinz, and M. Sulik, Austrian Patent AT001251 (1996).
27. K. Ito, K. Itahara, K. Tanaka, M. Fujikura, M. Yamaguchi, *Intermetallics* **9**, 591 (2001).
28. I. Rosales and J.H. Schneibel, *Intermetallics* **8**, 885 (2000).
29. L. Northcott, *Molybdenum* (Academic Press Inc., New York, 1956).
30. O.D. Sherby and J. Wadsworth, *Prog. Mater. Sci.* **33**, 169 (1989).
31. C.B. Geller, R.W. Smith, J.E. Hack, P. Saxe, and E. Wimmer, *Scripta Mater.* **52**, 205 (2005).
32. J.H. Schneibel, R.O. Ritchie, J.J. Kruzic, and P.F. Tortorelli, *Metall. Mater. Trans.* **36A**, 525 (2005).
33. M. K. Miller and A.J. Bryhan, *Mater. Sci. Eng.* **A327**, 80 (2002).

Mater. Res. Soc. Symp. Proc. Vol. 1128 © 2009 Materials Research Society   1128-U07-08

Effect of Particle Size of Mo Solid Solution on Hardness of Mo$_5$SiB$_2$/Mo-Based Alloys

Kyosuke Yoshimi, Yusuke Kondo and Kouichi Maruyama
Graduate School of Environmental Studies, Tohoku University, Sendai, Miyagi 980-8579, Japan

ABSTRACT

Three kinds of Mo-Si-B ternary alloys and a 1 at.% Al added Mo-Si-B alloy with the compositions near Mo-8.7 at.% Si-17.4 at.% B that is in the Mo$_5$SiB$_2$ and Mo two-phase compositional region were produced by Ar arc-melting followed by the heat treatment at 1800 °C for 24 h. These alloys have the characteristic fine microstructure composed of small Mo solid solution (Mo$_{ss}$) particles in the Mo$_5$SiB$_2$ (T$_2$) matrix with the primary phase (Mo$_{ss}$ or T$_2$ depending on composition). The volume fraction of the Mo$_{ss}$ particles ranges from 25.5 to 30.5 % and its average size from 3.0 to 6.4 μm in the fine microstructure of the alloys. Micro cracks were introduced by Vickers hardness tests into the microstructures, and their propagation is disturbed by the small Mo$_{ss}$ particles. Thus, each hardness value seems to relate to the cracking behavior around each indent. On the other hand, Vickers hardness values do not show correlation with the volume fraction of the Mo$_{ss}$ particles, but clearly decrease with increasing the average particle size of Mo$_{ss}$. Therefore, it should be concluded that the increase in the particle size of Mo$_{ss}$ could enhance the toughness of the Mo$_5$SiB$_2$/Mo-based alloys effectively by ductile phase toughening.

INTRODUCTION

Mo$_5$SiB$_2$ intermetallic compound, so-called T$_2$ phase, has attracted great attention as an ultra-high temperature structural material because of its high melting temperature over 2100°C [1], extremely high strength in the ultra-high temperature range [2], relatively low density close to Ni [3] and so on. It has been reported that Mo$_5$SiB$_2$ coexists with Mo solid solution (Mo$_{ss}$) from room to the melting temperature [1, 4]. The Mo$_5$SiB$_2$/Mo$_{ss}$ two-phase alloys have the characteristic fine microstructure composed of small Mo$_{ss}$ particles distributing uniformly in the matrix with the primary phase (Mo$_{ss}$ or T$_2$ depending on composition) [4]. Since the Mo$_5$SiB$_2$/Mo two-phase alloys show excellent ultra-high temperature yield strength [4], they are also one of the most promising material systems for ultra-high temperature structural use beyond Ni-base superalloys. The fracture toughness of monolithic Mo$_5$SiB$_2$ is as low as that of ceramics [5]. However, its fracture toughness should be improved by the incorporation of the small Mo$_{ss}$ particles in the Mo$_5$SiB$_2$/Mo two-phase alloy system. In our previous work [4], it was found that the coefficient of thermal expansion (CTE) in the Mo$_5$SiB$_2$/Mo two-phase alloys monotonously decreases with increasing the volume fraction of Mo$_{ss}$ because of CTE of Mo$_{ss}$ lower than that of Mo$_5$SiB$_2$. As a result, in the compositional range where Mo$_{ss}$ solidifies as the primary phase, residual tensile stress generated upon cooling by the difference in CTE introduced numbers of micro-cracks in Mo$_5$SiB$_2$ phase [4]. On the other hand, in the compositional range where Mo$_5$SiB$_2$ is dominant, micro-cracking was suppressed rather than in the Mo-rich compositional range. Those micro-cracks would reduce not only the strength but also the fracture toughness of Mo$_5$SiB$_2$/Mo-based alloys. Thus, in order to improve fracture toughness without

Table 1 Chemical compositions of four Mo$_5$SiB$_2$/Mo-based alloys.

Alloy	Nominal Composition (at.%)	Analyzed Composition							
		Element (at.%)				Impurity (wt. ppm)			
		Mo	Si	B	Al	H	C	N	O
A	Mo-8.7Si-17.4B	74.5	8.6	16.9	–	< 2	79	< 11	59
B	Mo-8.5Si-13.2B	78.7	8.3	13.0	–	2	40	< 11	24
C	Mo-10.3Si-20.5B	70.1	9.8	20.0	–	< 2	86	< 11	21
D	Mo-8.4Si-13.0B-1.0Al	77.7	8.2	12.9	1.2	< 2	34	< 11	27

deteriorating high temperature strength, microstructure controlling is necessary for the Mo$_5$SiB$_2$/Mo-based alloys so that ductile phase toughening effectively works. The objectives of this work are to investigate the correlation between microstructure and hardness in the Mo$_5$SiB$_2$/Mo-based alloys, particularly focusing on the characteristic fine microstructure of small Mo$_{ss}$ particles and the Mo$_5$SiB$_2$ matrix. Furthermore, the toughness of the fine microstructure is roughly estimated by the indentation fracture method.

EXPERIMENTAL PROCEDURE

Four different kinds of Mo$_5$SiB$_2$/Mo-based alloys were produced by the conventional Ar arc-melting method. Table 1 shows the nominal and chemically analyzed compositions of these alloys. Alloy A has the almost uniform, fine microstructure consisting of small Mo$_{ss}$ particles in the Mo$_5$SiB$_2$ matrix as reported in the refs. [1, 4]. Alloys B and C contain the primary phase of Mo$_{ss}$ and Mo$_5$SiB$_2$ with the microstructure, respectively. Alloys D has the composition of the alloy B plus 1 at.% Al. These alloys were heat-treated at 1800 °C for 24h in an Ar atmosphere. After the heat treatment, their microstructures were observed by SEM. Constituent phases were examined by XRD, and their composition was quantitatively analyzed by EPMA. For the fine microstructures of small Mo$_{ss}$ particles with the Mo$_5$SiB$_2$ matrix, Vickers hardness was measured, and the hardness of each constituent phase was also measured by the nano-indentation method using an ELIONIX ENT-1100a.

RESULTS AND DISCUSSION

Figure 1 shows the SE micrographs of four Mo$_5$SiB$_2$/Mo-based alloys. Alloy A has mainly the fine microstructure composed of the matrix (dark phase) of Mo$_5$SiB$_2$ and small particles (bright phase) of Mo$_{ss}$. Thin, elongated primary Mo$_{ss}$ is occasionally seen as shown in Fig. 1(a). In alloys B and D (figs. 1(b) and (d)), the fine microstructure exists between largely elongated, dendritic primary Mo$_{ss}$. In contrast, in alloy C, the fine microstructure exists between largely elongated primary Mo$_5$SiB$_2$ as shown in fig. 1(c). Thin, plate-like precipitates of Mo$_{ss}$ are observed in primary Mo$_5$SiB$_2$ for alloy C. Figure 2shows XRD profiles of these alloys. The XRD results indicate that the constituent phases of the alloys are mainly Mo$_5$SiB$_2$ and Mo$_{ss}$, though a

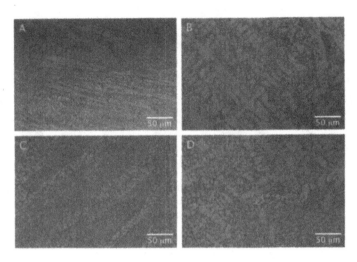

Figure 1 SE micrographs of four Mo$_5$SiB$_2$/Mo-based alloys after the heat treatment at 1800 °C for 24 h.

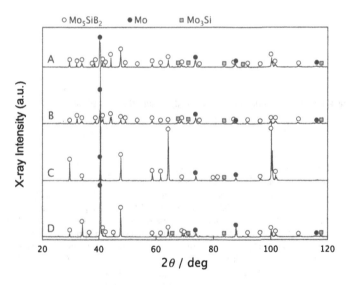

Figure 2 XRD profiles of four Mo$_5$SiB$_2$/Mo-based alloys after the heat treatment at 1800 °C for 24 h.

433

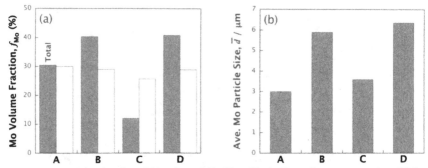

Figure 3 (a) Volume fraction of Mo_{ss} in total (dark) and in the fine microstructure of small Mo_{ss} particles with Mo_5SiB_2 matrix (bright). (b) Average particle size of Mo_{ss} in the fine microstructure.

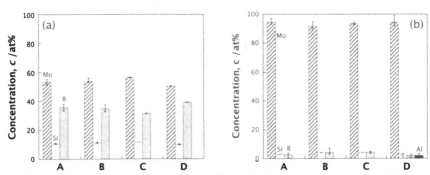

Figure 4 Chemical composition of (a) Mo_5SiB_2 and (b) Mo_{ss} phases analyzed by EPMA.

small amount of Mo_3Si is contained.

Figure 3(a) shows the volume fraction of Mo_{ss} in the total and in the fine microstructure composed of small Mo_{ss} particles and the Mo_5SiB_2 matrix of these alloys. Since alloy A has mainly the fine microstructure, the volume fraction of Mo_{ss} in total is almost equal to that in the microstructure, which are approx. 30.5 %. Alloys B and D have the primary phase of Mo_{ss} and alloy C does that of Mo_5SiB_2, so that the volume fraction of Mo_{ss} in total is quite different between the alloys but that in the fine microstructures ranges from approx. 25.5 to approx. 29 %. Figure 3(b) shows the average particle size of Mo_{ss} in the fine microstructure. The average particle size of Mo_{ss} is approx. 3.0μm for alloy A, 5.9 μm for alloy B, 3.6 μm for alloy C and 6.4 μm for alloy D, not corresponding to the order of Mo_{ss} volume fraction.

Figure 4 shows the composition of (a) Mo_5SiB_2 and (b) Mo_{ss} analyzed by EPMA. It should be noted that Al is partitioned into Mo_{ss} but not into Mo_5SiB_2. Al is also partitioned into

Mo₃Si. However, the hardness of constituent phases measured by the nano-indentation method is not apparently changed with alloys as shown in figure 5. Therefore, the solid-solution strengthening of Mo_{ss} and Mo_3Si by Al seems to be not remarkable in the Al-added Mo_5SiB_2/Mo-based alloy.

Figure 6 shows the SE micrograph of an indent introduced into the fine microstructure by a Vickers hardness test. As shown here, micro-cracking frequently occurred from the indent, and those cracks were initiated randomly not only at the corners of the indent. It is obvious that the propagation of all the cracks is disturbed (stopped or deflected) by small Mo_{ss} particles. That is, Mo_5SiB_2 phase is toughened by small Mo_{ss} particles in the fine microstructure. However, the Vickers hardness values were overestimated since the deformation around the indents was

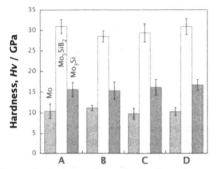

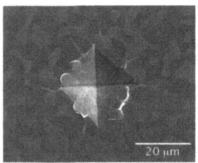

Figure 5 Hardness of constituent phases in the fine microstructure obtained by the nano-indentation.

Figure 6 SE micrograph of an indent introduced by a Vickers hardness test for alloy A.

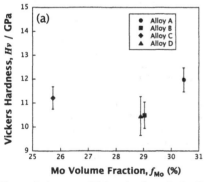

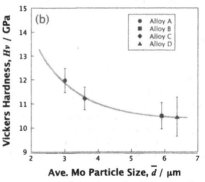

Figure 7 Apparent Vickers hardness in the fine microstructure vs. (a) volume fraction of small Mo_{ss} particles and (b) average particle size of Mo_{ss}.

compensated by crack opening. In other words, each hardness value largely relates to the cracking behavior around each indent. On the other hand, though they do not exhibit correlation with the volume fraction of small Mo_{ss} particles (figure 7(a)), it is clear that they decrease with increasing the average particle size of Mo_{ss}, as shown in figure 7 (b). Alloy A shows the highest Vickers hardness values at the smallest average particle size of Mo_{ss} in spite of the highest volume fraction of Mo_{ss}. Alloys B and D show Vickers hardness values lower than that of alloy A irrespective of the Al addition due to the average particle sizes of Mo_{ss} about two times larger than that of alloy A. This relation should result from micro-cracking generated from indents during Vickers hardness tests, meaning that the fracture toughness in the fine microstructure of Mo_5SiB_2/Mo-based alloys may change with the average particle size of Mo_{ss}.

Ihara et al. [5] reported that the fracture toughness of Mo_5SiB_2 single crystal evaluated by the indentation method is approx. 2 MPa√m. Comparing with this value, the fine microstructure of the Mo_5SiB_2/Mo-based alloys would be significantly toughened by the incorporation of small Mo_{ss} particles. Controlling the particle size of Mo_{ss} may be one of the keys to improve the fracture toughness of the fine microstructure in Mo_5SiB_2/Mo-based alloys. The results obtained in this work suggest that larger Mo_{ss} particles work more effectively for toughening even in the same volume fraction of Mo_{ss}. This idea is in good agreement with the results reported by Schneibel *et al.* [8].

CONCLUSIONS

In this study, the correlation between microstructure and hardness was studied for the fine microstructure in Mo_5SiB_2/Mo-based alloys. The volume fraction and average particle size of Mo_{ss} in the fine microstructure are slightly changed with composition. Al is preferentially partitioned into Mo_{ss} and Mo_3Si but not into Mo_5SiB_2 within 1 at.% addition. However, solid solution strengthening by Al is not remarkable in Mo_{ss} and Mo_3Si phases. Apparent hardness increases with decreasing the average particle size of Mo_{ss} because of micro-cracking at and around indents. Therefore, Mo_5SiB_2 is toughened by the incorporation of small Mo_{ss} particles in the Mo_5SiB_2/Mo-based alloys. It is suggested from the obtained results that larger Mo_{ss} particles work more effectively for toughening the fine microstructure of Mo_5SiB_2/Mo-based alloys.

REFERENCES

1. C.A. Nunes, R. Sakidja and J.H. Perepezko: Structural Intermetallics 1997, 831 (1997).
2. K. Ito, K. Ihara, K. Tanaka, M. Fujikura and M. Yamaguchi: Intermetallics, **9**, 591 (2001).
3. R. Mitra: Int. Mater. Rev., **51**, 13 (2006).
4. K. Yoshimi, S. Nakatani, N. Nomura and S. Hanada: Intermetallics, **11**, 787 (2003).
5. K. Ihara, K. Ito, K. Tanaka and M. Yamaguchi: Mater. Sci. Eng. A, **A329 – 331**, 222 (2002).
6. Testing methods for fracture toughness of fine ceramics, **JIS-R-1607**, 6 (1995).
7. Kinzoku Data Book 3rd Edition, ed. by JIM, Maruzen, 31 (1993).
8. J.H. Schneibel, M.J. Kramer, O. Unal and R.N. Wright: Intermetallics, **11**, 787 (2003).

Mater. Res. Soc. Symp. Proc. Vol. 1128 © 2009 Materials Research Society 1128-U07-10

Stacking Faults and Dislocation Dissociation in MoSi2

Miroslav Cak[1,2], Mojmir Sob[2,3], Vaclav Paidar[4] and Vaclav Vitek[1]
[1]Department of Materials Science and Engineering, University of Pennsylvania, 3231 Walnut Street, Philadelphia, PA 19104, USA.
[2]Institute of Physics of Materials, Academy of Sciences of the Czech Republic, CZ-616 62 Brno, Czech Republic.
[3]Department of Chemistry, Faculty of Science, Masaryk University, CZ-611 37 Brno, Czech Republic.
[4]Institute of Physics, Academy of Sciences of the Czech Republic, CZ-182 21 Praha 8, Czech Republic.

ABSTRACT

The intermetallic compound $MoSi_2$ crystallises in the body-centred-tetragonal $C11_b$ structure and while it is brittle when loaded in tension, it deforms plastically in compression even at and below the room temperature. The ductility of $MoSi_2$ is controlled by the mobility of $1/2\langle 331]$ dislocations on $\{013)$ planes but the critical resolved shear stress for this slip system depends strongly on the orientation of loading and it is the highest for compression along the $\langle 001]$ axis. Such deformation behaviour suggests that the dislocation core is controlling the slip on the $\{013)\langle 331]$ system. Since the most important core effect is dissociation into partial dislocations connected by metastable stacking faults the first goal of this paper is to ascertain such faults. This is done by employing the concept of the γ-surface. The γ-surfaces have been calculated for the (013) and (110) planes using a method based on the density functional theory. While there is only one possible stacking fault on the (110) plane, three distinct stacking faults have been found on the (013) plane. This leads to a variety of possible dislocation splittings and the energetics of these dissociations has been studied by employing the anisotropic elastic theory of dislocations. The most important finding is the non-planar dissociation of the $1/2\langle 331]$ screw dislocation that is favoured over the planar splittings and may be responsible for the orientation dependence of the critical resolved shear stress for the $\{013)\langle 331]$ slip system.

INTRODUCTION

Mechanical properties of molybdenum silicides have been investigated extensively in recent years owing to their high melting temperature, low density and high creep strength [1, 2]. The mechanical behaviour of $MoSi_2$ was studied most exhaustively since it deforms plastically in compression even at and below the room temperature [3-7]. The tetragonal $C11_b$ structure of $MoSi_2$ is shown in Figure 1. The c/a ratio, where a and c are defined in this figure, is 2.452 [8], which is very close to the ideal ratio $\sqrt{6}$ when all the separations between Mo and Si nearest neighbours are the same. Many slip systems have been found depending on orientation of the loading axes and temperature. The slip systems operating at and below room temperature are $\{013)\langle 331]$, $\{110)\langle 111]$, $\{101)\langle 010]$. In all cases the yield stress increases steeply with decreasing temperature.

The Schmid law is valid for the $\{110)\langle 111]$ and $\{101)\langle 010]$ slip systems but not for the $\{013)\langle 331]$ system. The critical resolved shear stress (CRSS) for the latter system is strongly dependent on orientation of the compressive axis and it is the highest for orientations close to $\langle 001]$ [2-5, 9]. Moreover, for this orientation of the compressive axis the Schmid factors for $\{110)\langle 111]$ and $\{101)\langle 010]$ slip systems approach zero and thus the $\{013)\langle 331]$ system is the only one that can produce plastic flow and accommodate strain components parallel to the ⟨001] axis. Consequently, the room temperature ductility of MoSi₂ is strongly impeded by the low

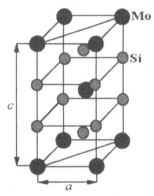

mobility of 1/2⟨331] dislocations on {013} planes for compression near ⟨001]. In fact no plastic flow occurs for loading along the ⟨001] axis and the material breaks in a brittle manner. As first noted by Rao et al. [10], such deformation properties are akin to those of BCC metals (see e. g. [11-13]), which suggests that the dislocation core is controlling the critical resolved shear stress of the {013}⟨331] system. Since the most important dislocation core feature is possible splitting into partial dislocations, we first investigated in this paper what are, if any, available stacking faults on (013) and (110) planes. This was carried out employing the concept of the γ-surface. When the displacement vectors and energies of the possible stacking faults are known the energetics of available dislocation dissociations has been analysed using the anisotropic elastic theory of Volterra dislocations. Implications of the corresponding dislocation dissociations are then discussed.

Figure 1. C11ᵦ structure of

γ -SURFACES FOR (013) AND (110) PLANES

γ-surface is a theoretical construct defined as follows. A crystal is cut along a chosen crystal plane and the upper part displaced with respect to the lower part by a vector **u**, parallel to the plane of the cut. The energy per unit area of the planar fault created in this way, γ(**u**), can be evaluated when an appropriate description of atomic interactions is available. Relaxations perpendicular to the fault have to be allowed but no relaxations parallel to the fault are permitted. Repeating this procedure for various vectors **u** within the repeat cell of the selected crystal plane, an energy-displacement surface is obtained that is commonly called the γ-surface [11, 13, 14].

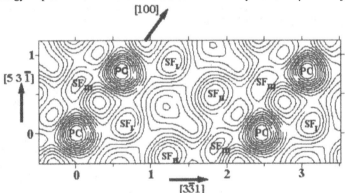

Figure 2. Contour plot of the (013) γ-surface. PC corresponds to the perfect crystal position and SF to stacking faults. Axes are in units of *a*.

438

Local minima on this surface determine the displacement vectors of possible metastable single-layer stacking faults and the values of γ at these minima are the energies of these faults. In the present study all the γ-surfaces were calculated using the pseudopotential method based on the density functional theory as implemented in the VASP (Vienna Ab-initio Simulation Package) code [15, 16]. The γ-surfaces for the (013) and (110) planes in MoSi₂ are shown in Figs. 2 and 3, respectively.

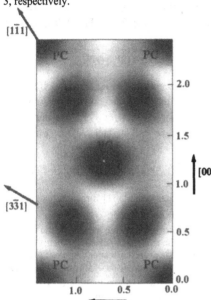

Figure 3. (110) γ-surface plotted using grey shading. Dark regions represent minima and bright regions maxima. PC corresponds to the perfect crystal position and SF to stacking faults. Axes are in units of a.

There are three minima on the (013) γ-surface, marked SF$_I$, SF$_{II}$ and SF$_{III}$ in Figure 2, which correspond to the three distinct stacking faults. The energies of these faults are 1.22, 1.06 and 1.12 J/m^2, respectively. In earlier calculations [17, 18], in which only the cross-section of the (013) γ-surface along the [3$\overline{3}$1] direction was studied, minima were found for the displacements 1/8[3$\overline{3}$1] and 3/8[3$\overline{3}$1]. These correspond closely to the components parallel to the [3$\overline{3}$1] direction of the displacements of the faults SF$_I$ and SF$_{III}$, respectively. However, displacements of these faults also have components in the [5 3 $\overline{1}$] direction that is perpendicular to the [3$\overline{3}$1] direction. In [18] a minimum was also found for the displacement 1/4[3$\overline{3}$1]. This minimum corresponds most likely to the fault SF$_{II}$ whose displacement vector has a larger component parallel to [5 3 $\overline{1}$] than do the other two faults. Only one metastable stacking fault was found for the (110) plane. Its displacement is 1/4[1$\overline{1}$1] or, equivalently, 1/4[3$\overline{3}$1] (see Figure 3). This is in agreement with earlier calculations for the [1$\overline{1}$1] cross-section [17, 18].

SPLITTING OF 1/2[3$\overline{3}$1] DISLOCATIONS

Since there are three possible stacking faults on the (013) plane and they all have very similar energies, the 1/2[3$\overline{3}$1] dislocation may dissociate in a number of different ways. However, the splittings that are likely to lead to a decrease of the energy are those for which the Burgers vectors of the partials make obtuse angles. The dissociations presented in the following are of this type. The first is the splitting composed of two partials separated by the stacking fault SF$_{II}$ that may take place according to the reaction

$$1/2[3\overline{3}1] = [PC \rightarrow SF_{II}] + SF_{II} + [SF_{II} \rightarrow PC]. \tag{1}$$

439

Here $[PC \rightarrow SF_{II}]$ is used to represent the vector between the position corresponding to the perfect crystal and the stacking fault SF_{II}; $[SF_{II} \rightarrow PC]$ has analogous meaning (see Figure 2). This notation is used since these vectors cannot be expressed in terms of crystallographic directions with low indices. If these vectors had no components perpendicular to the $[3\overline{3}1]$ direction then this reaction would be well described as $1/2[3\overline{3}1] = 1/4[3\overline{3}1] + 1/4[3\overline{3}1]$. The second dissociation is composed of three partials involving stacking faults SF_I and SF_{III} and may occur according to the reaction

$$1/2[3\overline{3}1] = [PC \rightarrow SF_I] + SF_I + [SF_I \rightarrow SF_{III}] + SF_{III} + [SF_{III} \rightarrow PC], \qquad (2)$$

where the meaning of the terms in square brackets is the same as in the previous case. If the stacking fault vectors had no components perpendicular to the $[3\overline{3}1]$ direction this reaction would be described approximately as $1/2[3\overline{3}1] = 1/8[3\overline{3}1] + 1/4[3\overline{3}1] + 1/8[3\overline{3}1]$. The third dissociation is composed of four partials and involves all three stacking faults. It may occur according to the reaction

$$1/2[3\overline{3}1] = [PC \rightarrow SF_I] + SF_I + [SF_I \rightarrow SF_{II}] + SF_{II} + [SF_{II} \rightarrow SF_{III}] + SF_{III} + [SF_{III} \rightarrow PC]. \qquad (3)$$

If the stacking fault vectors had no components perpendicular to the $[3\overline{3}1]$ direction this reaction would closely correspond to the splitting into four $1/8[3\overline{3}1]$ partials.

The widths of these three splittings were calculated as usual from the balance of the forces between the partials and the stacking fault tensions using the anisotropic elasticity theory to evaluate the dislocation stress fields (see [19]). The energies of the corresponding configurations were calculated using the same approach. For the screw orientation of the dislocation the energy gains due to the splittings (1) and (2) were found to be 5.96×10^{-9} J / m and 6.04×10^{-9} J / m, respectively. On the other hand, the splitting (3) does not lead to any energy gain, principally because in this case the separations of the partials are very small, comparable with the lattice spacing.

However, in the case of the screw dislocation a non-planar dissociation can occur into the planes (013), $(\overline{1}03)$ and (110) that all intersect along the $[3\overline{3}1]$ direction. This situation is analogous to that arising in the BCC structure if we identify the $[3\overline{3}1]$ direction in C11$_b$ with the $[1\overline{1}1]$ direction in BCC and the planes (013) and $(\overline{1}03)$ in C11$_b$ with the planes (011) and $(\overline{1}01)$ in BCC, respectively. While no stacking faults exist in the BCC lattice the dislocation core spreads into three $\{101\}$ type planes of the $[1\overline{1}1]$ zone [11-13]. In MoSi$_2$ an analogous core configuration may be a true splitting into partials, for example according to the reaction

$$1/2[3\overline{3}1] = [PC \rightarrow SF_I]_{(013)} + [PC \rightarrow SF_I]_{(\overline{1}03)} + 1/4[3\overline{3}1]_{(110)} + \vec{b}_0. \qquad (4)$$

This splitting is shown schematically in Figure 4. Here $\vec{b}_0$ is the Burgers vector of the partial positioned at the intersection of the (013), $(\overline{1}03)$ and (110) planes. If the Burgers vectors of the partials on (013) and $(\overline{1}03)$ planes were $1/8[3\overline{3}1]$ there would be no dislocation at the intersection and the reaction (4) would then be the same as that proposed in [20]. However, since the Burgers vectors of these partials deviate from $1/8[3\overline{3}1]$ there is a remainder at the intersection. The energy gain associated with the non-planar splitting (4) is 11.5×10^{-9} J / m and thus this splitting of the screw dislocation is more favourable than any of the planar splittings on the (013) plane.

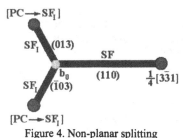

Figure 4. Non-planar splitting according to equation (4).

The energetically favourable non-planar splitting of the $1/2[3\bar{3}1]$ screw dislocation may be the reason why the CRSS for the $\{013\}\langle 331]$ system becomes very high for orientations of the compressive axis close to $\langle 001]$. In this case the Schmid stress on the (110) plane is approaching zero and thus there is virtually no Peach-Koehler force acting on the $1/4[3\bar{3}1]_{(110)}$ partial. Hence the splitting is very difficult to constrict and the $1/2[3\bar{3}1]$ screw dislocation is thus sessile and acts as a strong obstacle for the motion of other dislocations. This is somewhat analogous to the Lomer-Cottrell lock [19, 21]. However, there are other non-planar splittings that may play the same role and need further investigation. One example is the dissociation according to the reaction

$$1/2[3\bar{3}1] = [PC \rightarrow SF_I]_{(013)} + [PC \rightarrow SF_I]_{(\bar{1}03)} + 1/4[1\bar{1}1]_{(110)} + \vec{b}_1,\qquad(5)$$

where $\vec{b}_1 = 1/4[5\bar{5}1] - 2[PC \rightarrow SF_I]$ is the Burgers vector of the partial positioned at the intersection of the planes (013), ($\bar{1}03$) and (110). The partial on the (110) plane is now $1/4[1\bar{1}1]$ that bounds the same stacking fault as the $1/4[3\bar{3}1]$ partial but its Burgers vector is shorter.

CONCLUSIONS

In this paper we have shown that there are three possible stacking faults on the (013) plane in $MoSi_2$. These faults are not symmetry dictated and may, but need not, be found in other C11$_b$ compounds. Nevertheless, displacement vectors of these faults are close to the vectors $1/8[3\bar{3}1]$, $1/4[3\bar{3}1]$ and $3/8[3\bar{3}1]$ that can be anticipated on the basis of the hard-spheres model [22] but they have significant components perpendicular to the [3$\bar{3}$1] direction. However, no antiphase boundaries identified with displacements that bring Mo atoms into the positions of Si atoms and vice versa were found. Since the three stacking faults posses similar energies a number of planar dissociations of the $1/2[3\bar{3}1]$ dislocation may occur on the (013) plane. The present calculations indicate that the energetically most favourable planar splitting is into three partials with stacking faults SF$_I$ and SF$_{III}$.

Only one stacking fault, which can be foreseen geometrically, was found on the (110) plane. However, the most important finding is that the $1/2[3\bar{3}1]$ screw dislocation may dissociate in a non-planar way on the three crystallographic planes intersecting in the [3$\bar{3}$1] direction. This dissociation, described by equations 4 or 5, involves stacking faults on (013), ($\bar{1}$03) and (110) planes and may be very significant since it renders this screw dislocation sessile and may thus be responsible for the strong dependence of the yield stress on the orientation of the loading axis, in particular for compression axis close to the [001] direction.

ACKNOWLEDGEMENTS

This research was supported by the Department of Energy, BES Grant no. DE-PG02-98ER45702 (MC and VV) and by the Academy of Sciences of the Czech Republic Research projects Nos. AV0Z20410507, AV0Z10100520 and IAA100100920 (MS, MC and VP).

REFERENCES

1. D. M. Dimiduk and J. H. Perepezko, *MRS Bull* **28**, 639 (2003).
2. A. A. Sharif, A. Misra and T. E. Mitchell, *Scripta Mater.* **52**, 399 (2005).
3. K. Ito, H. Inui, Y. Shirai and M. Yamaguchi, *Philos. Mag. A* **72**, 1075 (1995).
4. K. Ito, H. Yano, H. Inui and M. Yamaguchi in *High-Temperature Ordered Intermetallic Alloys VI*, edited by J. A. Horton, I. Baker, S. Hanada, R. D. Noebe and D. S. Schwartz (Pittsburgh, Materials Research Society), Vol. 364, p. 899 (1995).
5. K. Ito, K. Matsuda, Y. Shirai, H. Inui and M. Yamaguchi, *Mat. Sci. Eng. A* **261**, 99 (1999).
6. U. Messerschmidt, S. Guder, L. Junker, M. Bartsch and M. Yamaguchi, *Mater. Sci. Eng. A* **319**, 342 (2001).
7. C. Dietzsch, M. Bartsch and U. Messerschmidt, *Phys Stat. Sol. A* **202**, 2249 (2005).
8. P. Villars and L. D. Calvert, *Pearson's Handbook of Crystallographic Data for Intermetallic Phases*, (ASM, Metals Park, OH, 1985).
9. H. Inui, T. Nakamoto, K. Ishikawa and M. Yamaguchi, *Mat. Sci. Eng. A* **261**, 131 (1999).
10. S. I. Rao, D. M. Dimiduk and M. G. Mendiratta, *Philos. Mag. A* **68**, 1295 (1993).
11. M. S. Duesbery in *Dislocations in Solids*, edited by F. R. N. Nabarro (Amsterdam, Elsevier), Vol. 8, p. 67 (1989).
12. V. Vitek in *Handbook of Materials Modeling Part B: Models*, edited by S. Yip (New York, Springer), p. 2883 (2005).
13. V. Vitek and V. Paidar in *Dislocations in Solids*, edited by J. P. Hirth (Amsterdam, Elsevier), Vol. 14, p. 439 (2008).
14. V. Vitek, *Philos. Mag. A* **18**, 773 (1968).
15. G. Kresse and J. Hafner, *Phys. Rev. B* **47**, 558 (1993).
16. G. Kresse and J. Furthmüller, *Comp. Mat. Sci.* **6**, 15 (1996).
17. U. V. Waghmare, V. Bulatov, E. Kaxiras and M. S. Duesbery, *Philos. Mag. A* **79**, 655 (1999).
18. T. E. Mitchell, M. I. Baskes, S. P. Chen, J. P. Hirth and R. G. Hoagland, *Philos. Mag. A* **81**, 1079 (2001).
19. J. P. Hirth and J. Lothe, *Theory of Dislocations*, (Wiley-Interscience, New York, 1982).
20. T. E. Mitchell, M. I. Baskes, R. G. Hoagland and A. Misra, *Intermetallics* **9**, 849 (2001).
21. A. H. Cottrell, *Philos. Mag.* **43**, 645 (1952).
22. V. Paidar and V. Vitek in *Intermetallic Compouds: Principles and Practice*, edited by J. H. Westbrook and R. L. Fleisher (New York, John Wiley), Vol. 3, p. 437 (2002).

Mater. Res. Soc. Symp. Proc. Vol. 1128 © 2009 Materials Research Society   1128-U07-11

The European ULTMAT Project: Properties of New Mo- and Nb-Silicide Based Materials

Stefan Drawin
ONERA, Metallic Structures and Materials Department, 92320 Châtillon, France

ABSTRACT

The development, in the frame of the European ULTMAT project, of alloys offering at least 150°C surface temperature increase above the Ni-based superalloys' capability is presented. The expected achievement of the project is a thorough evaluation of the capability of refractory metal (Nb and Mo) silicide based multiphase materials to withstand enhanced temperature turbine service conditions (up to 1300°C). This is based on microstructural, mechanical, physical and environmental investigations in close connection with industrial scale material processing and component fabrication technologies. The paper presents an overview of the project results. Base materials are the metal/intermetallic ductile/brittle composites in the Nb-Nb$_5$Si$_3$ and Mo-Si-B systems. Improvements in high temperature creep resistance (up to 1300°C) as well as oxidation resistance (700°C to 1300°C) have been obtained. Processing routes have been developed (ingot and powder metallurgy) that allowed the manufacture of complex shaped parts.

INTRODUCTION

Developing cleaner aero-engines, with low specific fuel consumption (SFC) and emissions, is an objective of industry and government agencies [1] and one of the drivers of the "Aeronautics and Space" thematic priority within the European 6th Framework Programme (FP6). This issue has been addressed by advanced engine architectures and cycle designs, novel combustor designs, optimised aerodynamics and new cooling concepts, while hot section materials development has clearly contributed to performance gains. Increased turbine airfoil materials temperature capability would allow significant performance enhancements.

In this context, the ULTMAT (ULtra high Temperature MAterials for Turbines) FP6 project (2004-2008) aims at providing a sound technological basis for the introduction of new materials, namely Mo- and Nb-silicide based multiphase alloys, which will allow a significant increase (ca. 150°C) in airfoil material operating temperature in aircraft/rotorcraft engines over those possible with Ni-based single-crystal superalloys. The increased capability will allow reduction of SFC, CO$_2$ emissions, and cooling air requirement, the latter leading to a further increase in efficiency and reduction in component weight. The consortium (Table I) has been built up to bring together highly skilled companies, universities and research institutes involved in research and development of high temperature (HT) materials for turbine engines. This paper presents the main project achievements, focusing on alloy development for improved HT creep and oxidation resistance, and alloy processing and fabrication.

Table I. Participants in the ULTMAT project.

▪ ONERA (coordinator), F	▪ Avio S.p.A., I	▪ Electricité de France, F
▪ Plansee SE, A	▪ Rolls-Royce plc, UK	▪ Snecma, F
▪ Turboméca, F	▪ University of Birmingham, IRC in Materials, UK	
▪ University of Magdeburg, D	▪ University of Nancy, F	▪ University of Sheffield, UK
▪ University of Surrey, UK	▪ Walter Engines a.s., CZ	

ALLOY SYSTEMS

Refractory metal (RM) and RM alloys are used in HT applications, however they are restricted to inert atmospheres because of their bad oxidation resistance. By contrast, RM-silicides can be used in air above 1500°C, as in furnace construction or glass industry. But these compounds also suffer from catastrophic oxidation ("pesting") at intermediate temperatures (600°C to 800°C) and exhibit brittleness at ambient as well as, for some of them, poor creep strength at elevated temperatures. The objective of current research on structural RM silicide alloys is to manufacture a composite material that takes advantage of the oxidation resistance of the silicides and the outstanding mechanical properties of the RM.

RM based metal-intermetallic composites have been developed in the late 1980's and two systems show the greatest potential [2], Nb-Si and Mo-Si, as metallic phase-toughened intermetallic-strengthened materials with volume fraction of metallic phase of 35% - 60%. Melting points of these multiphase alloys are *ca.* 1750°C (Nb-Si based) and 1950°C (Mo-Si based), thus substantially higher than those of Ni-based superalloys. Both systems have been studied within the ULTMAT project. In a first screening step, alloy development focused on the simultaneous improvement of three key properties, namely the HT creep strength and oxidation resistance, and the room temperature toughness, to reduce the amount (together with duration and cost) of extensive characterisations. These have been performed in a second step on a selection of alloys showing the best compromise in the above mentioned properties.

Mo-silicide based alloys

Berczik [3] patented compositions for Mo-rich Mo-Si-B alloys and a manufacturing route which in essence comprises a rapid solidification step. This yields a matrix of Mo_{ss} solid solution, providing fracture toughness and ductility below 600°C, and embedded intermetallics Mo_3Si and/or Mo_5SiB_2 (T2 phase) for enhancing the creep resistance as well as for providing a Si and B reservoir for the formation of a borosilicate glass surface layer. The studied alloys exhibit typical volume fractions of *ca.* 55% Mo_{ss}, 30% T2 and 15% Mo_3Si for the ternary Mo-3Si-1B (wt.%), quaternary Mo-3Si-1B-3Nb [4] and Mo-3Si-1B-0.1%La_2O_3 [5] compositions.

Nb- silicide based alloys

Binary Nb-Si alloys exhibit excellent creep resistance, although poor toughness and oxidation resistance. This can be improved by alloying, leading *e.g.* to the NbTiHfCrAlSi Metal And Silicide Composites family developed by General Electric [6,7]. The Nb-based materials consist basically of a metal solid solution M_{ss} (M = Nb+Ti+Hf+...) with M_5Si_3 and/or M_3Si silicide phases [8]. M_{ss} is the softest phase and the silicide the most creep resistant. Improvement of the HT mechanical properties (mainly creep resistance) can thus be obtained by enhancing the properties of the soft phase (M_{ss}) and/or increasing the silicide volume fraction, the latter by increasing the Si content. High silicide contents simultaneously improve oxidation resistance, but decrease toughness and ductility at low and intermediate temperatures. Actually, alloys with Si content in a wide range have been studied in this project. With constituting phases being modifications of M_{ss}, M_5Si_3 and M_3Si, depending on their heavy element content, the effect of composition and microstructure on oxidation and creep resistance has been established. Two creep resistant alloy compositions have been selected; one patent is currently being filed.

MANUFACTURING

Two processing routes, *i.e.* ingot metallurgy (IM) and powder metallurgy (PM), have been selected, partly using industrial scale facilities able to produce materials of the required quantity, in order to compare the properties of various types of microstructures.

Ingot metallurgy

Vacuum arc melting has been used to manufacture small buttons (< 500 g) for micro-structural, mechanical and oxidation characterisations. Bigger alloy batches, up to 50 kg, are manufactured by plasma melting, thus enabling a second set of more extensive characterisations.

Investment casting, currently used for production of Ni-based superalloy blades and vanes, has been successfully applied to Nb-silicide based materials to manufacture shaped parts. Figure 1 shows as-cast blades with length up to 320 mm [9], exhibiting a very low defect level.

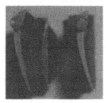 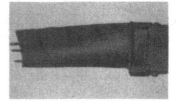

Small blade, with wax on the left (~120 mm)   *Large blade (~320 mm)*
Figure 1. Investment cast Nb-base silicide blades.

Powder metallurgy

Mechanical alloying (MA) of elemental powders has been used to produce homogeneous powders with a supersaturated Mo_{ss}. After sintering and hot isostatic pressing (HIP) (for details on processing conditions, see ref. [4]), the materials exhibit small size Mo_{ss} particles in an intermetallic matrix (Figure 2-a) and show improved properties compared to the microstructure obtained when gas-atomised (argon quenched) powders are used. Heat treatment at 1700°C for 10 h allows grain coarsening and evolution towards a continuous Mo_{ss} matrix (Figure 2-b) [10]. NbTiHfSi based materials have also been manufactured (after sintering and HIPing) from elemental powder blends. Throughout the project, about 350 kg material were produced using the industrial PM route; part has been used to investigate the effect of thermo-mechanical treatments (forging, but mainly extrusion). Turbine blades could be machined out of Mo-Si-B and Nb-Si based billets (Figure 3).

MECHANICAL AND THERMO-PHYSICAL PROPERTIES

High temperature mechanical properties are of prime importance for the targeted applications in turbo-engines and some examples are reported hereafter. Table II shows the ultimate compressive strength (UCS) and the maximum plastic strain of the HIPed Mo-Si-B-La_2O_3 alloy: very high strength levels were measured at all temperatures, showing the large potential of this ultrafine grained material.

a – As-HIPed	b – After heat treatment (1700°C, 10 h)

Figure 2. Micrographs of a Mo-Si-B-Nb alloy. Mo$_{ss}$ (bright), intermetallics (grey) and undesired silica (dark).

Figure 3. Machined blades (Mo-Si-B-La$_2$O$_3$ alloy).

Table II. UCS and maximum plastic strain for the Mo-Si-B-La$_2$O$_3$ alloy.

650°C		900°C		1200°C	
UCS (MPa)	ε_{max}	UCS (MPa)	ε_{max}	UCS (MPa)	ε_{max}
1700	2.8%	1540	>10%	710	>10%

Tensile tests were performed on Mo-Si-B-Nb alloys up to 1600°C (Figure 4-a) [4]. While no significant ductility and yield and ultimate tensile strengths (YS, UTS) of about 500 MPa were measured below 1100°C, a total elongation to fracture of about 5% could be observed at 1200°C. The significant plasticity at temperatures above 1100°C indicates a substantial potential for the application of hot deformation processes such as extrusion, rolling and forging and for component manufacturing. The compressive creep behaviour (steady-state rate) of this alloy is presented in Figure 4-b. It shows that a heat treatment (1700°C, 10 h), through grain coarsening (Figure 2), can reduce the creep rate by more than one order of magnitude.

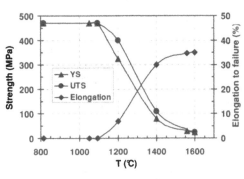
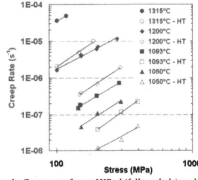

a- YS, UTS, and fracture elongation	b- Creep rate for as-HIPed (full symbols) and heat-treated (open symbols) states

Figure 4. Properties of the HIPed Mo-Si-B-Nb alloy.

Strength and creep behaviour of Nb-Si alloys have also been measured, as well as other properties, such as fracture toughness, Young's and shear moduli, coefficient of thermal expansion, thermal diffusivity and conductivity and, to a lesser extent, LCF and HCF behaviour, for both Nb-Si and Mo-Si systems. A property database has been built up that can now be used by engine designers to integrate RM-silicide based components in future engine, and by researchers as a basis for further improvements of the alloys.

OXIDATION RESISTANCE

Insufficient oxidation resistance at medium (pesting regime) and high temperature is one of the weaknesses of the RM-based silicide materials. Basically, the work performed within ULTMAT consisted in the investigation of the oxidation behaviour of uncoated alloys to identify the related mechanisms and in the design, deposition and testing of oxidation resistant coatings, with emphasis on the Nb-silicide based materials. The oxidation tests were performed in the 700°C-1300°C range in cyclic and isothermal conditions.

The coating deposition technique used was pack-cementation. The operational conditions (masteralloy composition, deposition and annealing temperatures, etc.) have been adjusted to co-deposit complex silicide coatings on silicide-based alloys, with the help of gaseous phase composition modelling. A complementary study of the deposition process has been performed in the simplified ternary system Nb-Cr-Si, which has also been investigated through thermodynamic modelling (CALPHAD) and experimental work. The control of the composition (Si, Ti, Fe, Cr, B content, etc.) of the successive layers during the optimisation steps has led to outstanding oxidation performances at both intermediate and high temperature thanks to the formation of protective silica-based scales (Figure 5).

Finally, several Mo- and Nb-silicide based alloys were tested in simulated service conditions. Several coated and uncoated specimens (with uncoated CMSX-4 as a reference) were tested in a burner rig for over 500 one hour cycles, facing burned gas containing Na_2SO_4. For some of them, an oxidation-corrosion resistance comparable or better than CMSX-4 has been demonstrated.

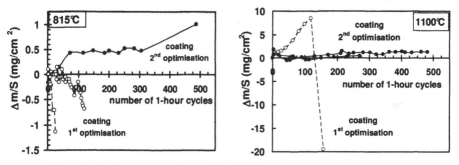

Figure 5. Cyclic oxidation of coated NbHfTiSi based alloy at 815°C and 1100°C.

CONCLUSIONS

The development of new gas turbine hot section materials with increased HT capabilities is crucial for the design of future efficient turbines with low CO_2 and NO_x emission levels. Refractory metal silicide based materials are promising candidates for such applications. A European consortium bringing together industry, university and research centres was built up in the EU-FP6 funded ULTMAT project to create a synergy that addresses the scientific and technological challenges and aims at providing a breakthrough solution.

The main scientific and technological outcomes of this fast-track programme are a significant advance in the understanding of synergistic effects of alloying elements on *(i)* phase selection, *(ii)* phase stability, *(iii)* segregation phenomena, *(iv)* phase transformations, *(v)* phase equilibria, *(vi)* oxidation behaviour, *(vii)* microstructure architecture, *(viii)* mechanical properties and *(ix)* oxidation resistance of both alloy systems, and the development of processing routes (IM and PM) and fabrication technologies at industrial scale.

ACKNOWLEDGMENTS

Financial support under the European FP6 Project ULTMAT, contract No. AST3-CT-2003-502977, as well as all project partners for fruitful cooperation and discussions are gratefully acknowledged.

REFERENCES

1. Strategic Research Agenda, Advisory Council For Aeronautics Research in Europe, October 2002 (www.acare4europe.org).
2. Zhao J.C. and Westbrook J.H., MRS Bull. 28, 622-627 (2003).
3. Berczik D.M., US Patents 5,595,616 and 5,693,156 (1997).
4. Jéhanno P., Heilmaier M., Saage H., Böning M., Kestler H., Freudenberger J. and Drawin S., Mater. Sci. Eng. **A463**, 216-223 (2007).
5. Jéhanno P., Böning M., Kestler H., Heilmaier, M., Saage, Krüger M., Powder Metallurgy **51**, 99-102 (2008).
6. Bewlay B.P., Jackson M.R. and Lipsitt H.A., Metall. Mater. Trans. **27A**, 3801-3808 (1996).
7. Bewlay B.P., Jackson M.R., Zhao J.C., Subramanian P.R., Mendiratta M.G. and Lewandowski J.J., MRS Bull. 28, 646-653 (2003).
8. Schlesinger M.E., Okamoto H., Gokhale A.B. and Abbaschian R. J., Phase Equil. **14**, 502-509 (1993).
9. Hu D., Wickins M., Harding R.A., Li Q., Loretto M.H. and Wu X, presented at the Aeromat 2008 conference, 23-26 June 2008, Austin, TX, USA (unpublished).
10. Jéhanno P., Heilmaier M. and Kestler H., Intermetallics **12**, 1005-1009 (2004).

Laves Phases—Structure
and Properties

Mater. Res. Soc. Symp. Proc. Vol. 1128 © 2009 Materials Research Society   1128-U08-01

Chemical Bonding in Laves Phases Revisited: Atom Volumina in Cs-K System

Yu. Grin[1], A. Simon[2], A. Ormeci[1]

[1]Max-Planck-Institut für Chemische Physik fester Stoffe, Nöthnitzer Str. 40, 01187, Dresden, Germany.
[2]Max-Planck-Institut für Festkörperforschung, Heisenberg Str. 1, 70569, Stuttgart, Germany.

ABSTRACT

Laves phases comprise a large group of binary and ternary intermetallic compounds with general composition AB_2. The crystal structures of Laves phases are often regarded as closest packing of differently sized spheres. This observation, beginning with very early work on Laves phases, has led many researchers over the years, to emphasize the role of geometrical factors in the formation of Laves phases. In order to develop a firm understanding of chemical bonding in Laves phases and assess the importance of geometrical factors, we undertake a first-principles-electronic structure-based chemical bonding analysis for several representatives. As a first step towards this goal we concentrate on the K-Cs system which contains the Laves phase CsK_2 and the hexagonal compound Cs_6K_7. In such alkali-metal-only compounds it is generally expected that chemical bonding-caused energy effects are minimal. Atom volumina and charge transfer investigations reported here, however, suggest that even in alkali metal-alkali metal Laves phases chemical bonding plays a non-negligible role.

INTRODUCTION

Laves phases form one of the largest groups among the intermetallic compounds. More than 1000 binary and ternary representatives are known [1]. The membership in this group is defined by a special structural feature: three-dimensional framework of the majority component (B) based on the so-called Kagome pattern with cavities in form of a truncated tetrahedron where the minority atoms (A) are located. The Laves phases are found in three main types of crystal structure: $MgCu_2$ (C15, cubic), $MgZn_2$ (C14, hexagonal), $MgNi_2$ (C36, hexagonal). The ideal stoichiometry is AB_2, and the constituent elements come from almost any part of the periodic table. This large variety of Laves phases makes them very interesting from a chemical bonding viewpoint, because one expects different chemical bonding patterns depending on the particular A and B elements. It seems to be not obvious for this group of compounds taking into account the fact that they all have the same structural features. The crystal structures of Laves phases are often interpreted as closest packing of differently sized spheres, giving rise to a demand to explain how the expected different chemical bonding patterns derived from the different chemical nature of the components yield closest packing of atoms at the end. Another related question is: are factors responsible for formation of more than 1000 Laves phase compounds the same for all, or are there sub-categories with different dominant factors?

Two rules, one geometric, other electronic, are discussed in regard to stability of Laves phases. The geometric rule is based on a hard sphere model. Close-packing condition gives sphere radii with ratio $r_A/r_B = \sqrt{(3/2)} \sim 1.225$, so that $A(B)$ spheres touch only $A(B)$ spheres. The electronic rule suggests the valence electron concentration (electrons per atom) as responsible factor for formation of the main structural pattern of the Laves phases and its variation in different crystal structure types. Recent analysis of the available information reveals that neither rule is obeyed [1]. In particular, the radius ratio is found to vary between 1.05 and 1.70 for the known Laves phases. Already this finding alone, suggests that chemical bonding effects may play an important role so that the tabulated values of atomic radii get effectively modified in the distinct compound as a result of chemical interactions and, thus, cannot be used directly as a parameter of a field for description of the chemical bonding. The idea, then, as originally put forth in 1983 by one of us [2], is to investigate the compounds for which chemical bonding effects are expected to be minimal. Good candidates for this purpose are the Laves phases formed between the alkali metals [2]. There are three known examples: KNa_2 [3], $CsNa_2$ [4] and CsK_2 [5], all forming in the hexagonal C14 structure. As a first step towards a more comprehensive study, here we focus on the Cs-K system only.

In addition to the Laves phase CsK_2 the compound Cs_6K_7 is formed [5]. Electronic structures of these compounds were first studied by pseudopotential methods in late 1970's [6, 7]. Here, especially the question was addressed, how far the Laves phase possesses in some sense a unique position or whether it fits the general tendency of the change of crystal chemical characteristics in this system depending on the composition. For this purpose, we will focus on charge transfer and atom volumina in the Cs-K binary compounds. First-principles electronic structure calculations based on the local density approximation to the density functional theory were carried out for Cs_6K_7 and CsK_2. The obtained self-consistent total charge densities were used to calculate atomic volumes and atomic charges within the framework of the quantum theory of atoms in molecules (QTAIM) [8]. The importance of the geometric factor in the formation of Cs_6K_7 and CsK_2 will be discussed according to the results of these calculations.

CALCULATION METHODS

The all-electron full-potential local orbital (FPLO) method [9] is used to perform the electronic structure calculations reported here. Perdew-Wang parametrization [10] of the exchange-correlation energy is employed. Brillioun zone integrations are done by the linear tetrahedron method. The basis set includes the valence (s,p) states as well as the d states. In addition, the (s,p) states of the penultimate shell have to be treated as valence (semicore states). In these calculations, the crystal structure data taken from the reference [5] are used directly.

The theoretical framework for obtaining atomic charges is provided by Bader's quantum theory of atoms in molecules [8]. According to QTAIM the boundaries of an atom in a molecule or crystal is defined by the zero-flux surfaces of the total charge density gradient field. The zero-flux surfaces divide the real space into mutually exclusive and space-filling regions (called basins). These basins are identified as atoms, so that the basin volumes are regarded as atomic volumes, and the charge density integrated inside the basins gives the atomic charge. Since this topological analysis depends on the total charge density, in principle, the results are independent of the particular functions (atom-centered orbitals or plane waves, etc.) employed in the basis set. Consequently, QTAIM-based charge transfer analyses are bias-free and thus more reliable than those based on empirical parameters, *e.g.* electronegativities.

The topological analysis of the well-converged self-consistent total charge densities obtained from the FPLO calculations are performed by the program Basin [11]. The margin of error in atomic charges is less than 0.05 electrons.

452

RESULTS AND DISCUSSIONS

Attempts for using structural data from known compounds to predict the lattice parameters or unit cell volumes of solid solutions, alloys, or ordered compounds have a long history. These attempts may be divided into two groups. One group uses lattice parameters or interatomic distances as basis, usually referred to as Vegard's rule [12]. For the other group, volume is the essential quantity (Biltz's rule [13]). The final expression used in either approach can be written for an $A_x B_{1-x}$ binary compound as $Q_{AB} = Q_A *x + Q_B * (1-x)$, where Q stands for either a lattice parameter or unit cell volume. The underlying assumption, here, is that the elemental contributions (increments), Q_A and Q_B, remain constant in the given system (but, see also [14]). These ideas, although quite old and not much supported by actual evidence, still form the subject matter in recent investigations [14, 15, 16]. In particular it was shown, that in the system Al-Pt the atomic volumina clearly change depending on the composition and reflecting changes in the chemical interactions [14].

The above mentioned phenomenological rules can also be used for analyzing existing crystallographical data for purposes of finding general tendencies or common aspects in a given crystal structure type or compound family. In this regard Pearson's near-neighbor diagram (NND) analysis is noteworthy [17,18]. An improvement over the original NND approach was achieved by incorporating coordination number effects in a more realistic way [2]. For Laves phases the resulting improved line in the NND perfectly matches the data of alkali metal-alkali metal Laves phases. This perfect match was interpreted as being a manifestation of weak chemical bonding in these compounds. Here, we would like to investigate this point further by examining how atomic volumina and charge transfer change in the K-Cs system.

Results pertaining to atomic volumina are given in Table I, containing experimentally obtained average atomic volumina as well as calculated atomic volumina V_K and V_{Cs}. The corresponding atomic basins, which are called hereafter as QTAIM atoms, are shown in Figure 1 for both compounds.

Table I. Average experimental volumes and calculated K and Cs QTAIM atom volumina in the K-Cs system. Different entries for the latter are due to presence of different Wyckoff positions.

Compound	x (Cs)	V (Å³/atom)	V_K (Å³)	V_{Cs} (Å³)
K (bcc)	0.0000	75.880	75.880	-
CsK₂ (C14)	0.3333	87.503	85.917	
Cs₆K₇	0.4615	90.447	85.388	90.517
			86.634	97.151
				98.675
				101.332
Cs (bcc)	1.0000	115.738	-	115.738

The linear relationship between the average atomic volume and Cs content of a given compound can be derived from the elemental solid volumes as $V(x) = 75.880 + 39.858 *x$, where x refers to Cs concentration. The deviation of the experimental volume from this linear behavior is -1.9% for CsK₂ and -4.1% for K₇Cs₆. These deviations are indeed small, e.g., in comparison to the K-Pb system, where experimental volumina deviate from the linearity by approximately 25%. This sharp contrast between these two systems is, in a way, not surprising. The alkali metal-only compounds are expected to behave similar to elemental alkali metals (no "strong" chemical bonding), whereas the salt-like compounds formed by K and Pb should have strong ionic interactions due to the large electronegativity difference. The calculated QTAIM volumes for K atoms in KPb_2 and K_4Pb_4 (our own calculations) range between 25-29 $Å^3$, considerably smaller than the elemental value of ~76 $Å^3$ (or, roughly about one-third of K volumina in K-Cs compounds). Large K volumina differences between the Cs-K and K-Pb systems reflect differences in charge transfer amounts. K is slightly more electronegative than Cs. The excess charge per K atom is calculated to be 0.14 electrons in C14 CsK_2, 0.12 and 0.13 in K_7Cs_6 (there are two Wyckoff positions for K atoms). Amount of electrons lost by Cs atoms is 0.28 in the Laves phase, and varies between 0.05 and 0.25 in Cs_6K_7. For comparison, in K-Pb compounds K atoms lose 0.70 to 0.76 electrons.

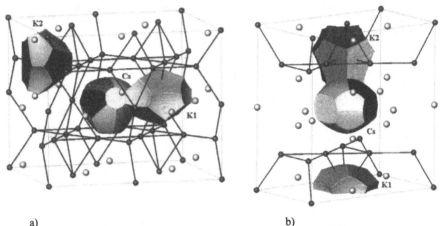

a)   b)

Figure 1. QTAIM atomic basins in CsK_2 (a) and Cs_6K_7 (b). K1 stands for K atoms at (000) positions and K2 stands for K atoms forming the Kagome-like layers. Light circles denote Cs atoms, darker ones K atoms. Because the c lattice parameter of Cs_6K_7 is too long, atomic basins are generated for z up to $c/2$. Therefore, in (b) only half of the full basin of the K atom at a (000) position is shown.

The preceding discussion seems to lend support to the view that chemical bonding does not play a significant role in the compounds formed solely by alkali metal atoms. However, if we recall the basic premise behind Vegard's or Biltz's law type of linear approaches, the elemental atom volumina should be independent of concentration. An examination of Table I shows that this assumption does not hold. Atomic volumes of each element change differently from one compound to another: K remains almost constant from CsK_2 to Cs_6K_7, while Cs shows largely an increase. As a matter of fact, even atoms of the same element located at different Wyckoff positions (implying different chemical environment) have differing QTAIM volumes. These observations suggest that average volume (volume per atom) of a compound being closer to a . linear volume-concentration variation does not necessarily mean that chemical bonding effects can be ignored. It is quite possible that individual atoms are affected fairly differently by chemical bonding and thus end up having different volumes, but the average volume can still turn out to be close to the value expected from a linear behavior. In addition, the results above reveal that the Laves phase CsK_2 does not have a particular position, but fits well the general change tendency of chemical interactions within the binary system.

This case study shows that quantum mechanically computed QTAIM atomic volumes provide useful information and insight regarding general tendencies in change of chemical interactions within a distinct system. More direct access to chemical bonding in a crystal, however, can be obtained by using the analysis with the electron localizability indicator (ELI) [19]. The distribution of ELI in CsK_2 (Figure 2) is a spherical one in the inner shells of potassium and caesium, suggesting that the electrons of these shells do not participate in the valence interactions [20, 21]. The valence region is structured revealing maxima located in the tetrahedral holes formed only by K atoms or by one Cs and three K atoms. These results reveal four-center bonds between (i) only K atoms, (ii) K and Cs atoms. Further work along these lines is necessary for a better and quantum-mechanics-based understanding of the interplay among chemical bonding, atomic volumes, charge transfer and interatomic distances/compound volumes for the Laves phases.

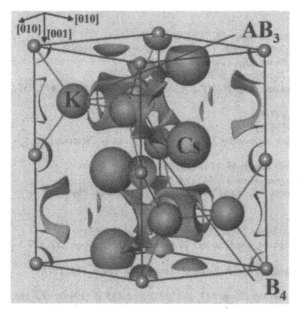

Figure 2. Electron localizability indicator in CsK_2. The isosurface with ELI value of 1.15 reveals four-centre bonding between the atoms. B_4 denotes a four-centre bond between four K atoms. The four-centre bond AB_3 is formed by three K atoms in the Kagome layer capped by a Cs atom.

SUMMARY

The two ordered compounds in the K-Cs system, the C14 Laves phase CsK_2 and the hexagonal Cs_6K_7, are studied by first-principles electronic structure calculations. Atom volumina and charge transfer analyses carried out within the framework of the QTAIM suggest that in alkali-metal-only compounds, where there is only one valence electron per atom, chemical bonding effects, though looking "weak", should be taken into account, and thus the formations of the Laves phases KNa_2, $CsNa_2$, CsK_2 may not be explained on grounds of solely geometrical factors.

ACKNOWLEDGMENTS

This project was realized within the inter-institutional research initiative 'The Nature of Laves Phases' of the Max-Planck Society.

REFERENCES

1. F. Stein, M. Palm, G. Sauthoff, *Intermetallics* **12**, 713 (2004).
2. A. Simon, *Angew. Chem.* **95**, 94 (1983), *Angew. Chem. Int. Ed. Engl.* **22**, 95 (1983).
3. F. Laves and H.J. Wallbaum, *Z. Anorg. Allg. Chem.* **250**, 110 (1942).
4. A. Simon and G. Ebbinghaus, *Z. Naturforsch.* **B 29**, 616 (1974).
5. A. Simon, W. Braemer, B. Hillenkoetter and H.-J. Kullmann, *Z. Anorg. Allg. Chem.* **419**, 253 (1976).
6. J. Hafner, *Phys. Rev.* **B 15**, 617 (1977).
7. J. Hafner, *Phys. Rev.* **B 19**, 5094 (1979).
8. R.F.W. Bader, *Atoms in Molecules: A Quantum Theory*, Oxford University Press, Oxford (1999).
9. K. Koepernik and H. Eschrig, *Phys. Rev.* **B 59**, 1743 (1999).
10. J.P. Perdew and Y. Wang, *Phys. Rev.* **B 45**, 13244 (1992).
11. M. Kohout, program Basin, version 4.2, Max-Planck-Institute for Chemical Physics of Solids, Dresden, Germany (2007).
12. L. Vegard, *Z. Phys.* **5**, 17 (1921).
13. W. Biltz, *Raumchemie der festen Stoffe*, Verlag Leopold Voss, Leipzig (1934).
14. A. Baranov, M. Kohout, F.R. Wagner, Yu. Grin and W. Bronger, *Z. Kristallogr.* **222**, 527 (2007).
15. W. Bronger, A. Baranov, F.R. Wagner and R. Kniep, *Z. Anorg. Allg. Chem.* **633**, 2553 (2007).
16. G.H. Grosch and K.-J. Range, *J. Alloys Comp.* **233**, 30 (1996); *ibid.* **233** 39 (1996); *ibid.* **235**, 250 (1996).
17. W.B. Pearson, *Acta Crystallogr.* **B 24**, 7 (1968).
18. W.B. Pearson, *The Crystal Chemistry and Physics of Metals and Alloys*, Wiley-Interscience, New York, London, Sydney, Toronto (1972) p. 51 ff.
19. M. Kohout, *Int. J. Quantum Chem.* **97**, 651 (2004).
20. M. Kohout, F.R. Wagner, Yu. Grin, *Theor. Chem. Acc.* **108** 150 (2002).
21. F.R. Wagner, V. Bezugly, M. Kohout, Yu. Grin, *Chem. Eur. J.* **13** 5724 (2007).

Mater. Res. Soc. Symp. Proc. Vol. 1128 © 2009 Materials Research Society   1128-U08-03

Microstructural Investigations of the Unusual Deformation Behavior of Nb2Co7

F. Stein[1], M. Palm[1], G. Frommeyer[1], P. Jain[2], K.S. Kumar[2], L. Siggelkow[1,3,4], D. Grüner[3,5], G. Kreiner[3], and A. Leineweber[6]

[1]Max-Planck-Institut für Eisenforschung GmbH, Max-Planck-Str. 1, D-40237 Düsseldorf, Germany
[2]Division of Engineering, Brown University, Providence, RI 02912, U.S.A.
[3]Max-Planck-Institut für Chemische Physik fester Stoffe, Nöthnitzer Str. 40, D-01187 Dresden, Germany
[4]now at: Technische Universität München, Lichtenbergstr. 4, D-85747 Garching, Germany
[5]now at: Division of Physical, Inorganic and Structural Chemistry, Arrhenius Laboratory, Stockholm University, S-106 91 Stockholm, Sweden
[6]Max-Planck-Institut für Metallforschung, Heisenbergstr. 3, D-70569 Stuttgart, Germany

ABSTRACT

Usually, single-phase intermetallics in bulk form can easily be crushed into powder by hammering. It was therefore quite a surprise when we found that a bulk sample of the monoclinic intermetallic compound Nb2Co7 could be extensively deformed at room temperature without shattering or fracturing. In a previous paper, results of microhardness, compression, tensile and bending tests were provided and discussed [1]. In order to understand the observed unusual deformation behavior of this intermetallic phase, its hitherto unknown crystal structure has been studied and the microstructure of undeformed and deformed samples has been analyzed in the present investigation by light-optical, scanning electron and transmission electron microscopy.

Single-phase specimens deformed at very different strain rates (hammering and conventional compression testing) both show the occurrence of microcracks along grain boundaries which, in compression-deformed specimens, are strongly localized in extended shear bands oriented approximately 45° to the compression axis. The grains adjacent to the microcracks are heavily deformed whereas, away from the sheared regions, the samples remain free of any indication of plastic deformation.

INTRODUCTION

In a recent re-assessment of the binary Co-Nb system [2], the formation, thermodynamic stability and homogeneity range of the intermetallic phase Nb_2Co_7 was investigated. Nb_2Co_7 does not melt congruently but forms as a peritectoid reaction product between the Co(Nb) solid solution and the hexagonal C36 Laves phase $Nb_{1-x}Co_{2+x}$ at 1086°C. As the peritectoid formation and dissolution is a very slow solid state reaction (see e.g. [3]), the phase does not occur in as-cast alloys and only forms after heat treatments for sufficiently long times.

Single-phase Nb_2Co_7 specimens showed an unexpected 'tough' behavior during attempts to produce powder for XRD measurements by crushing in a mortar. To study this unusual behavior in more detail, a number of mechanical tests were performed with single-phase Nb_2Co_7 specimens, and the results were reported in a recent paper by Siggelkow et al [1]. Rather intriguing was the observation that in compression (strain rate: 10^{-4} s^{-1}), this phase exhibited "plastic deformation" (non-linear stress-strain curve) of more than five percent strain and a maximum stress of ~900 MPa at room temperature. Beyond the maximum, the stress-strain curve shows some strain softening, a feature that persists even at 600°C. Optical metallography of the specimen deformed at room temperature revealed a "shear band" about 200 μm in width at approximately 45° to the loading axis, and consisting of grain boundary microcracks. Higher magnification images confirmed deformation markings (such as either slip bands or coarse twins) within the grains located in this band. It is conceivable that this localized deformation and micro-cracking is responsible for the observed "work-softening" in the compression stress-strain curves. Perhaps equally remarkable and surprising was that a block of this material could be repeatedly hammered to almost half its height without shattering or fracture. Additionally, Vickers indentation of a polished surface revealed what appeared to be evidence for plastic deformation along the edges of the indent. Flexural tests however did not provide evidence for macroscopic plasticity up to approximately 800°C, and the resulting fracture surfaces then looked predominantly intergranular.

In order to obtain further insight into the unusual deformation behavior of this compound, the microstructure of the undeformed and the deformed material − a hammered specimen as well as a specimen compressed to 5.5 % plastic strain at room temperature − was studied by light-optical microscopy (LOM), scanning electron microscopy (SEM), and transmission electron microscopy (TEM).

EXPERIMENTAL

Several alloys of nominal composition Co-22.2 at.% Nb were prepared from pure Co and Nb (purity for both 99.9 wt.%) by levitation, induction, and arc melting and heat-treated for at least 200 h at 1000°C in an Ar atmosphere. The resulting samples were at least 99 vol. % single-phase Nb_2Co_7.

For metallographic examinations by LOM, the ground specimen surfaces were etched in a solution of 7.5 ml HF, 2.5 ml HCl, 8 ml HNO_3, and 980 ml H_2O.

For the TEM investigations, thin slices, 3 mm in diameter were cut from the undeformed and compression-deformed Co_7Nb_2 specimens using electro-discharge machining techniques. In the case of the deformed specimen, the slices were cut perpendicular to the loading direction. These slices were thinned to less than 100 μm by conventional mechanical grinding, and final thinning and perforation of the foils was carried out by twin-jet polishing in a 10 vol. % nitric acid-methanol solution at a temperature of -25°C and using 15 V. To enhance the electron-transparent area, the specimens were ion-polished for 15 min in a precision ion polishing system (PIPS) at 4 keV and a low angle (~2°). The software, Desktop Microscopist, was used to generate diffraction patterns for the monoclinic compound using the parameters in [2].

RESULTS AND DISCUSSION

Undeformed material

The crystal structure of Nb_2Co_7 was identified on the basis of X-ray powder-diffraction data collected using synchrotron radiation. It was found to be closely related to the monoclinic Zr_2Ni_7 type (first described by Eshelman and Smith [4]) with however a smaller unit cell as

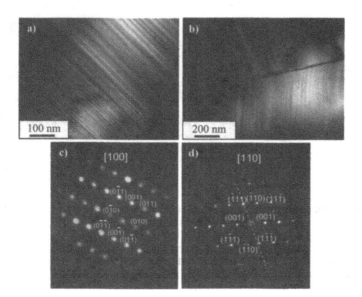

Figure 1. TEM bright-field images (a,b) from the undeformed specimen confirming a single-phase, highly faulted microstructure. Only one variant of the fault appears in each grain. (c,d) Selected area diffraction patterns from two different zone axes indexed using the parameters in [2] that confirm the monoclinic structure of the type Zr_2Ni_7.

proposed by Parthé and Lemaire [5]. The lattice parameters are $a = 0.45874$ nm, $b = 0.81509$ nm, $c = 0.62223$ nm, and $\beta = 107.18°$ [2]. The structure can be described by stacking a certain type of "triple layers", where each triple layer is a stack of a Kagomé layer consisting only of Co atoms sandwiched by mixed, closed-packed Nb/Co layers. A considerable line broadening of certain reflections in the powder diffraction patterns indicates certain types of irregularities in the layer stacking. Details of the crystal structure and the stacking faults will be described elsewhere [6].

A TEM bright-field image of the undeformed material confirms the presence of a highly faulted microstructure (Fig. 1a). The partial dislocations often terminate at the grain boundaries, an example of which is illustrated in Fig. 1b. Selected area diffraction patterns were obtained from several zone axes and two examples are shown in Figs. 1c and d. These could be successfully indexed using the simulated patterns for a monoclinic structure obtained with the software Desktop Microscopist.

Fig. 2a shows a light-optical micrograph of an undeformed specimen. A typical feature of the undeformed material is the occurrence of twinning in many of the grains. In the immediate vicinity of the Vickers microhardness indent, the effect of plastic deformation is visible.

Hammered material

Fig. 2b shows a light-optical micrograph of the surface of a specimen (block of 5x5x5 mm³ original size) which has been reduced to about half of its height by repeated hammering.

Figure 2. Light-optical micrographs of an undeformed (a) and a hammered (b) specimen with Vickers microhardness indents (HV0.3). For details, see text.

Besides a lot of microcracks appearing along the grain boundaries, either slip bands or coarse twins are visible inside the grains. The Vickers microhardness is increased by more than 20% compared to that of the undeformed material (HV0.3(undeformed) = 400 ± 30, HV0.3(deformed) = 510 ± 40, both averages of 5 tests).

Compressed material

In [1], compression tests of single-phase Nb_2Co_7 specimens with strain rates of 10^{-4} s^{-1} up to a maximum strain of 5.5% were described. The room-temperature deformed specimens showed macroscopic shearing at approximately 45° to the loading axis without rupture. In order to address the issue of deformation localization within the shear band reported in [1], TEM specimens were prepared from both a region inside the shear band and another region far away from this shear band. Interestingly, specimens from the latter region are quite similar to those from the undeformed material showing a high density of faults, but being completely free of dislocations. This is very different from the specimens prepared from the shear band region. The bright-field image in Fig. 3a shows the region surrounding a grain boundary microcrack and several dislocations can be seen in the vicinity of the crack. These dislocations may be seen more clearly in Fig. 3b which is taken from another location near a microcrack. It appears that these dislocations may be partials bounding faults supporting the notion that this material has a low stacking fault energy. In addition, twins were frequently observed in this specimen (Fig. 3c). Such features were not observed in the foil from the region away from the shear band, suggesting significantly localized deformation. A second grain boundary microcrack is seen in Fig. 3d and it is evident that the regions adjacent to it are heavily plastically deformed (note that dislocations in the left grain are not in contrast). By combining the observations in the specimens from different regions, we conclude that plastic deformation is extremely localized in this intermetallic phase.

There is still the question if the observed plastic deformation in the grains adjacent to the microcracks is only a consequence of the formation of these microcracks, i.e. if microcracking along grain boundaries occurs first to relieve constraints due to lack of sufficient plastic deformation mechanisms (slip and twinning sytems) and the practically unconstrained single

460

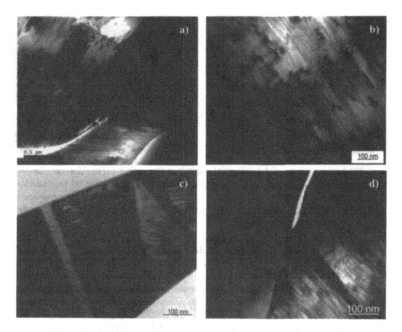

Figure 3. TEM bright-field images from a specimen extracted from a compression specimen deformed to 5.5% plastic strain at room temperature. (a,b,d) dislocation activity in the vicinity of grain boundary microcracks, and c) twinning near a crack tip.

crystals then experience highly localized plastic deformation in the vicinity of microcracks given the complex triaxial stress state in these regions.

SEM micrographs of a compression-deformed specimen reveal plastic deformation in the grains adjacent to microcracks manifested by slip traces in the partially separated grains (Fig. 4a). More importantly, these traces are also present on the fracture surfaces and two specific locations are highlighted by the white and black arrows. These markings on the fracture surface are in concert with those on the specimen surface, suggesting that they likely occurred after the crack formed (as it is not possible to get macroscopic slip steps if the boundary is not a free surface). This lends credibility to the argument that the plastic deformation process occurs after the microcracks have formed, assisted by the removal of constraint. The specimen surface shown in Fig. 4b is reminiscent of twinning and consequent relief on the surface.

CONCLUSIONS

The crystal structure of Nb_2Co_7 is monoclinic and closely related to the Zr_2Ni_7 structure type with however a very high density of stacking faults. The analysis of compression-deformed specimens shows that the deformation is strongly localized taking place in macroscopic shear

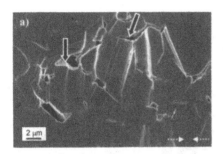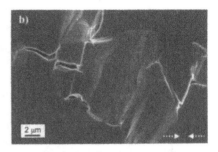

Figure 4. SEM micrographs of the deformed surface in the vicinity of the deformation band in a compression-deformed specimen (the loading direction is indicated by the white, dotted arrows): (a) plastic deformation traces on the specimen surface within an individual grain but also on the microcrack fracture surfaces (white and black arrows), and (b) surface relief that suggests twinning as an active deformation mode.

bands about 200 μm in width. These shear bands consist of a large number of microcracks which form at the grain boundaries. Whereas away from the sheared regions, the samples remain free of any indication of plastic deformation, the grains in the vicinity of the microcracks are strongly deformed. TEM proves the existence of a high number of dislocations in these grains. As deformation markings are not only observed in the grains adjacent to the microcracks but also on the microcrack fracture surfaces, it is deduced that the deformation process starts with the formation of the microcracks along grain boundaries and that highly localized plastic deformation follows in the adjacent grains after the removal of constraints by grain boundary separation.

ACKNOWLEDGMENTS

This work has been supported by the Max Planck Society within the framework of the Inter-Institutional Research Initiative "The Nature of Laves Phases". The SEM and TEM work that was performed at Brown University was supported by the MRSEC Program (Micro- and Nano Mechanics of Materials; Brown University) of the National Science Foundation under award DMR-0520651.

REFERENCES

1. L. Siggelkow, U. Burkhardt, G. Kreiner, M. Palm, F. Stein, *Mater. Sci. Eng. A* **497**, 174 (2008).
2. F. Stein, D. Jiang, M. Palm, G. Sauthoff, D. Grüner, G. Kreiner, *Intermetallics* **16**, 785 (2008).
3. K.S. Kumar, P. Jain, S.W. Kim, F. Stein, M. Palm, in these proceedings.
4. F.R. Eshelman, J.F. Smith, *Acta Crystallogr. B* **28**, 1594 (1972).
5. E. Parthé, R. Lemaire, *Acta Crystallogr. B* **31**, 1879 (1975).
6. A. Leineweber et al. (in preparation)

Mater. Res. Soc. Symp. Proc. Vol. 1128 © 2009 Materials Research Society   1128-U08-04

Site Occupation and Defect Structure of Fe₂Nb Laves Phase in Fe-Nb-M Ternary Systems at Elevated Temperatures

Shigehiro Ishikawa[1], Takashi Matsuo[1,2], Naoki Takata[1,2] and Masao Takeyama[1,2]
[1]Department of Metallurgy and Ceramics Science, Tokyo Institute of Technology, S8-8, 2-12-1, Ookayama, Meguro-ku, Tokyo, 152-8552, Japan.
[2]Consortium of JRCM (The Japan Research and Development Center for Metals)

ABSTRACT

For the Fe_2Nb Laves phase with C14 structure in the Fe-Nb-M (M : Cr, Mn, Co, Ni) systems, the site occupation of M in Fe_2Nb has been examined in terms of XRD Rietveld analysis, particularly paying attention to the two Fe sublattice sites of Fe1 (3^6-net in the triple layer : t) and Fe2 (*kagome*-net of the single layer : s) with the fraction of 0.25 and 0.75, respectively. In any these four ternary systems, the Fe_2Nb Laves phase region largely extends along the equi-Nb concentration direction; for Mn complete solid solubility exists, and the solubility of Cr and Co in Fe_2Nb is more than 50 at.% and that of Ni is 44 at.%. Thus, at least two thirds of all Fe sublattices in Fe_2Nb are occupied by M in all cases. Rietveld analysis revealed that Cr and Mn with which have a larger atomic size than Fe prefer to occupy the Fe1 sublattice site when the amount in solution is less than 0.25 fraction of Fe in Fe_2Nb and the preferred occupation site changes to the Fe2 sublattice site when the amount in solution increases beyond 0.25. In contrast, Co and Ni whose atomic size is smaller than Fe preferentially occupy the Fe2 sublattice site, regardless of the amount. The c/a ratio of stoichiometoric Fe_2Nb increases and becomes closer to the ideal value (1.633) of the cubic C15 structure when the Fe1 sublattice site is occupied by Cr and Mn. However, the degree of symmetries of both tetrahedron and *kagome*-net formed by Fe atoms become better when Fe2 sublattice site is occupied by a certain amount of Ni.

INTRODUCTION

The Fe_2Nb Laves phase with hexagonal C14 structure is a promising strengthener for austenitic heat resisting steels. We have systematically studied the phase equilibria between γ-Fe and Laves phases and the precipitation kinetics of the Laves phase in the γ matrix at elevated temperatures in order to develop a new type of austenitic heat resisting steels strengthened by Laves phases [1-4]. The most important feature of the Fe_2Nb phase is its large homogeneity region, where more than 44 at.% Ni can dissolve in the phase [5-6]. This makes it possible to control the precipitation morphology of the Laves phase through the lattice parameter change by alloying [7]. This feature also suggests a possibility to change its crystal structure to cubic C15, i.e. a higher symmetrical one, through alloying in solution.

The structures of Laves phases consist of the alternating stacking of single layers (s) and triple layers (t). The single layer is a plane of *kagome*-net formed by Fe atoms and the triple layer consists of three planes of a 3^6-net where one plane formed by Fe atoms is sandwiched by two others formed by Nb atoms. The layer stacking of C14 is $stst'$ whereas that of C15 is $st'st'$ [8]. Note that the prime refers to the difference in the stacking direction of the 3^6-nets. The crystal structure of the Laves phase is such that Fe atoms form tetrahedra having mirror symmetry in the

C14 and point symmetry in the C15 with respect to the *kagome*-net, as shown in Figure 1. The tetrahedra in stoichiometoric Fe$_2$Nb is squeezed a little bit along the *c* axis, so that the atomic distance of Fe atoms on the *kagome*-net (Y) is larger than that between *kagome* and 3^6-nets (Z), and that connecting tetrahedra on the *kagome*-net (X) becomes smallest (Y>Z>X). In contrast, the tetrahedra in the C15 are all regular ones, so that the distance of X, Y, Z are all equal (Y=Z=X). Thus, crystallographically there are two distinct Fe sublattice sites in the C14 structure: Fe1 (3^6-net) and Fe2 (*kagome*-net), but no difference between them and the only one sublattice site in the C15 structure exists. Therefore, site occupation behavior of the third elements M in Fe sublattice sites in the Fe$_2$Nb is very important to understand the phase stability as well as phase transformation between C14 and C15. However, almost no information was available for the site occupation behavior of M elements in the Fe$_2$Nb Laves phase.

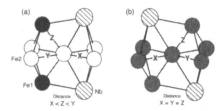

Figure 1. Tetrahedron of Fe atoms and Fe sublattice sites in (a) C14, (b) C15.

In this study, thus, the elements Cr, Mn, Co and Ni, which tend to occupy Fe sublattice sites, and whose atomic sizes are larger and smaller than that of Fe are chosen as a third element M, and the homogeneity region of the C14 Fe$_2$Nb phase with M in solution was first examined and then the site occupation behavior of M and lattice parameters of the C14 Laves phase have been examined by XRD Rietveld analysis.

EXPERIMENTAL PROCEDURES

The ternary alloys with nominal compositions of 66.7(Fe$_{1-x}$ M$_x$)-33.3Nb, where M is Cr, Mn, Co, Ni and x = M/(Fe+M) is in the range of 0 to 1, were prepared by arc-melting of 30 g button ingots under argon atmosphere using 2N9 iron, 4N chromium, 3N manganese, 3N cobalt, 3N7 nickel and 3N niobium metals. Note that the elements except Ni form M$_2$Nb Laves phases. The detailed nominal compositions of the alloys are summarized in Table 1. Each ingot was cut to pieces and they were sealed off in silica capsules back-filled with argon after evacuating to 2.7 x 10^{-3} Pa. These samples were equilibrated at 1373 K~1473 K for 240 h, followed by water quench. Microstructures were examined by scanning electron microscopy (SEM) with backscattered electron imaging (BEI). The phase identification was done by powder X-ray diffraction (XRD) using Cu Kα radiation. The XRD measurements were done from 2θ = 30 degree to 80 degree with fixed time method using crystal monochromator (step width : 0.02 degree, sampling time : 3 sec). The powder with diameter less than 45 μm was selected by sieving after grinding the equilibrated bulk samples, and they were annealed for 0.2 h at the equilibrated temperature in the same way. The preferred occupation site (POS) of M elements, lattice parameters and atom positions of Fe atoms of *kagome*-net in C14 Laves phase were analyzed by Rietveld analysis [9-10]. For determination of the POS of M elements, the S value (goodness-of-fit between experimentally obtained and calculated diffraction patterns) was evaluated by changing the fraction of Fe1 site occupancy of M from 0 to 1 systematically and the POS of the M atom was judged by the minimum S value at a certain fraction of M element. In this case backgrounds, scale factor, profile parameters, lattice parameters and overall isotropic

Table 1. Lattice parameters of Fe$_2$Nb Laves phase in Fe-Nb-M ternary alloys.

Alloys	M	Phase	L.P. (A)			Preferential
	(M + Fe)	present	a	c	c/a	occupation site
Fe-33.3Nb	0	ε (C14)	4.8376	7.8852	1.6300	-
Fe-33.3Nb-16.68Cr	0.25	ε (C14)	4.8639	7.9476	1.6340	Fe1
Fe-33.3Nb-40Cr	0.60	ε (C14)	4.8788	7.9821	1.6361	Fe2
Fe-33.3Nb-16.68Mn	0.25	ε (C14)	4.8468	7.9081	1.6316	Fe1
		μ-Fe$_7$Nb$_6$	*	*	*	
Fe-33.3Nb-33.3Mn	0.50	ε (C14)	4.8807	7.9716	1.6333	Fe1
		μ-Fe$_7$Nb$_6$	*	*	*	
Fe-33.3Nb-50.03Mn	0.75	ε (C14)	4.8736	7.9883	1.6391	Fe2
		μ-Fe$_7$Nb$_6$	*	*	*	
		unknown (white)	*	*	*	
Mn-33.3Nb	1	ε (C14)	4.8897	8.0002	1.6361	-
		unknown (white)	*	*	*	
Fe-33.3Nb-16.68Co	0.25	ε (C14)	4.8084	7.8436	1.6312	Random
Fe-33.3Nb-33.3Co	0.50	ε (C14)	4.8208	7.8429	1.6269	Fe2
Fe-33.3Nb-50.03Co	0.75	ε (C14)	4.8076	7.8403	1.6308	Fe2
Co-33.3Nb	1	C15	6.7694	-	1.6330	-
Fe-33.3Nb-10Ni	0.15	ε (C14)	4.8354	7.8862	1.6309	Fe2
Fe-33.3Nb-20Ni	0.30	ε (C14)	4.8277	7.8599	1.6281	Fe2

* : not determined

thermal parameters were carefully refined, and fractional coordinates x, y, z were fixed. For example, POS of M is the Fe2 sublattice site if the minimum value was obtained at 0 fraction, whereas it is Fe1 when the minimum value was obtained at the fraction of 1. The results are shown later in Figure 3.

RESULTS AND DISCUSSION

Composition homogeneity region of Fe$_2$Nb Laves phase

The detailed analysis of the phase equilibria is given in [11] and the results focusing on the homogeneity region of the C14 Fe$_2$Nb Laves phase in the Fe-Nb-M ternary system are summarized in Table 1 and Figure 2, together with previous results on Fe-Nb-Ni system. In all systems, the single phase region of Fe$_2$Nb extends along the equi-Nb concentration direction from the binary system. In the Fe-Nb-Cr system the fact that an alloy with 40 at.% Cr is C14 single phase, together with the results in [12], suggest that the single phase region expands up to 60 at.%Cr, i.e. approximately 90% of the Fe atoms can be replaced by Cr atoms. In the Fe-Nb-Mn system, a continuous solid solution of C14 Laves phase exists. In the Fe-Nb-Co system the alloy with 50 at.% Co is C14 single phase, indicating that at least 75% of the Fe atoms

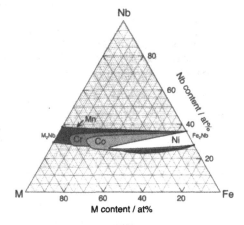

Figure 2. Homogeneity region of the C14 Fe$_2$Nb Laves phase in Fe-Nb-M ternary systems at elevated temperatures.

can be replaced by Co atoms. Note that the single phase regions of Cr_2Nb and Co_2Nb with C15 structure are very limited to the binary edge and only a fraction of Cr and Co atoms can be replaced by Fe in the C15 phase. In the Fe-Nb-Ni system the homogeneity region is the smallest among the elements studied, but even so, about two third of the Fe atoms can be replaced by Ni.

The site occupation and lattice parameters

The POS of M in Fe_2Nb Laves phase determined by Rietveld analysis is summarized in Table 1. As mentioned previously, the POS was determined from the minimum S value by changing the fraction of Fe1 occupancy of M, with respect to the value where all M atoms occupy Fe2 sublattice site (3/4 of the total Fe sublattice site), that is, no occupancy of Fe1 sublattice site. Figure 3 shows the case of Cr where the normalized S values were plotted as a function of the fraction of Fe1 sublattice occupancy. In an alloy with a low Cr content of Fe-33.3Nb-16.7Cr (denoted 1/4Cr in Figure 3), the minimum S value was obtained at the fraction of 0.9. This indicates that most of the Cr atoms occupy the Fe1 site. Thus, the POS is the Fe1 sublattice site. However, in the alloy Fe-33.3Nb-40Cr (3/5Cr) which contains Cr beyond the fraction of Fe1 sublattice sites (0.25) against the total Fe sublattice sites, the minimum S value was obtained at the fraction of 0.3. If the POS remains unchanged regardless of the content, the minimum value should be obtained at the fraction of 1.0 and only excess Cr atoms go to the Fe2 sublattice site. Thus, the POS of Cr is composition dependent and it changes from the Fe1 to the Fe2 sublattice site with increase in the content.

The same analysis was done for all the alloys studied, and the composition dependence of POS for M is summarized in Figure 4. In case of Mn, the minimum S value was obtained at the fraction of 1.0 in the alloy containing Mn up to 33.3 at.% (x=0.5), but it becomes 0 in the alloy with 50 at.%Mn (x=0.75). Thus, the POS of Mn changes from Fe1 to Fe2 with increasing Mn content, similar to the behavior of Cr. On the other hand, in case of Co and Ni, the minimum S values were obtained at the fraction of 0, regardless of the content, except for the alloy with 16.7 at.% Co where the minimum value was on the random distribution line. From these results, it is reasonable to conclude that the elements with an atomic size larger than that of Fe tend to occupy the Fe1 sublattice site (3^6-net) when their amounts are up to the fraction of Fe1 site and they gradually change their POS to Fe2 (*kagome*-net) when the amounts are beyond the half of total Fe sublattice sites, whereas the elements with an atomic size smaller than that of Fe occupy the Fe2 sublattice site (*kagome*-net), regardless of the composition. Thus, the POS of M in the Fe_2Nb

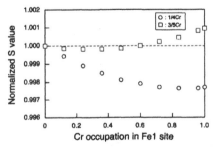

Figure 3. Change in S value with increase in Cr occupancy in the Fe1 sublattice site.

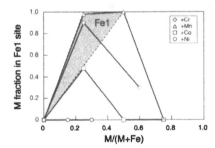

Figure 4. The change in M fraction in the Fe1 site with M/(M+Fe) in nominal compositions.

466

Laves phase is somehow associated with the atomic size of the M element.

The lattice parameters a and c of the C14 Laves phase with M in solution were measured and are summarized in Table 1, and the relationship between a and c axes of the ternary Laves phase is shown in Figure 5. The solid line represents the ideal c/a ratio (c/a=1.633) converted from the C15 structure of Cr_2Nb and Co_2Nb. The c/a ratio of binary Fe_2Nb is 1.630 (solid circle), a little smaller than the ideal value. The c/a ratio becomes closer and almost equivalent to the ideal value when Cr and Mn substitutes the Fe1 sublattice site, in the area encircled by a solid line. However, the ratio exceeds the ideal value when the POS of these elements becomes the Fe2 site. In case of Co and Ni, which occupy the Fe2 sublattice site, the ratio becomes smaller than the value for binary Fe_2Nb.

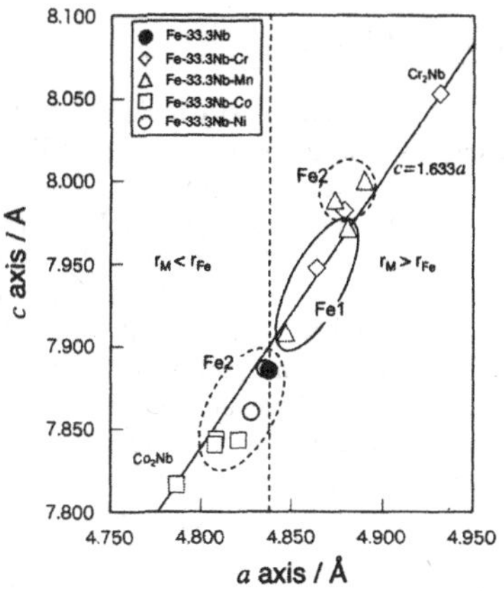

From these results, one can imagine that the occupation of the Fe1 sublattice site by atoms with an atomic size larger than that of Fe is effective in changing the structure of C14 to C15, from the viewpoint of c/a ratio. In addition, the squeezed shape of the tetrahedra along the c axis in binary Fe_2Nb would also be improved to regular tetrahedra by occupation of larger atoms in the Fe1 sublattice site. Thus, in order to understand the relationship between the site occupation and symmetry of the tetrahedra as well as *kagome*-net, the atom positions of the Fe sublattice sites were analyzed.

Figure 6 shows the change in the ratios of interatomic distances X/Y and Z/Y with M content. In binary Fe_2Nb, both ratios are lower than unity and the degree of symmetry of the *kagome*-net (X/Y) is smaller than that of the tetrahedron (Z/Y). Note that X+Y = a. In case of Cr in solution, Z/Y remains unchanged or slightly decreases, whereas X/Y apparently becomes smaller. Thus, unlike the expectation from the change in c/a ratio, substitution of Cr in the Fe1 sublattice site worsens the degree of symmetries of both, tetrahedron and *kagome*-net. On the other hand, in case of Ni in solution, these ratios increase and reach the ideal value 1 at a certain Ni content. This indicates that the symmetries of tetrahedron and *kagome*-net are more controllable by Ni substitution in the Fe2 sublattice site. We do not rule out the importance of the c/a ratio, but control of the occupation site would be a key factor to change the structure from C14 to C15. Further analysis including the Nb sublattice site is now in progress.

Figure 5. The relationship between a and c axis of the Fe_2Nb Laves phase in Fe-Nb-M alloys.

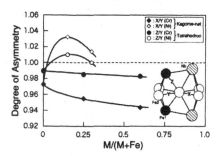

Figure 6. Change in degree of asymmetry of *kagome*-net and tetrahedron in the C14 structure with M/(M+Fe).

467

CONCLUSIONS

The homogeneity region of the Fe_2Nb Laves phase and the preferred occupation site of M elements have been examined in the Fe-Nb-M ternary systems with M = Cr, Mn, Co, Ni and the following conclusions are drawn. The Fe_2Nb phase region extends along the equi-Nb concentration direction by each M element, and at least 60% of Fe atoms are substituted by M atoms in all cases. Cr and Mn preferentially substitute in the Fe1 site at low contents and the preferential site changes to the Fe2 site with the increasing M content. The preferred occupation site of Co and Ni is the Fe 2 site, regardless of the M content. The lattice parameters a and c increase through the preferential substitution of Cr and Mn in the Fe1 site and the c/a ratio increases around the ideal value of 1.633. However, in case of Cr addition, the ratios of interatomic distances X/Y and Z/Y decrease from the ideal value 1 and these ratios increase and reach the ideal value by addition of Ni. The degree of symmetries of both, tetrahedron and *kagome*-net of Fe atoms would be improved by the substitution of Ni in the *kagome*-net.

ACKNOWLEDGEMENT

This research is partly supported by the Grant-in-Aid for Science Research (A) (14205102), JSPS Fellows (20·9769) and "Fundamental Studies on Technologies for Steel Materials with Enhanced Strength and Functions" by Consortium of JRCM (The Japan Research and Development Center of Metals). Financial support from NEDO (New Energy and Industrial Technology Development Organization) is gratefully acknowledged.

REFERENCES

1. M. Takeyama, H. Yokota, M. M. Ghanem, and T. Matsuo, *Journal of Materials Processing Technology*, **117**, 3 (2001).
2. M. Takeyama, *Report of JSPS 123rd Committee on Heat-Resisting Materials and Alloys*, **45**, 51 (2004).
3. M. Takeyama, *Kinzoku*, **76**, 743 (2006).
4. T. Sugiura, S. Ishikawa, T. Matsuo, and M. Takeyama, *Material Science Forum* **561-565**, 435 (2007).
5. N. Gomi, S. Morita, T. Matsuo, and M. Takeyama, *Report of JSPS 123rd Committee on Heat-Resisting Materials and Alloys*, **45**, 157 (2004).
6. M. Takeyama, N. Gomi, S. Morita, and T. Matsuo, *Mater. Res. Soc. Symp.*, **842**, 461 (2006).
7. M. Takeyama, S. Morita, A. Yamauchi, M. Yamanaka, and T. Matsuo, *Superalloys 718, 625, 706 and Various Derivatives*, TMS, 333, (2001).
8. M. F. Chisholm, S. Kumar, and P. Hazzledine, *Science*, **307**, 701 (2005).
9. F. Izumi, *Rigaku Journal*, **31**, 17 (2000).
10. H. Toraya, *Rigaku Journal*, **28**, 3 (1997).
11. S. Ishikawa, M. Yamashita, T. Matsuo, N. Takata, and M. Takeyama, 17th *IFHTSE Congress 2008 submitted*, (2008).
12. N. I. Kaloev, E. M. Sokolovskaya, A. Kh. Abram'yan, L.K. Kulova, and F. A. Agaeva, *Russian Metallurgy, Tranaslated from Izvestiya Akademii Nauk SSSR, Metally*, **4**, 207 (1987).

Mater. Res. Soc. Symp. Proc. Vol. 1128 © 2009 Materials Research Society   1128-U08-05

Composition Dependence of the Hardness of Laves Phases in the Fe-Nb and Co-Nb Systems

S. Voß[1], F. Stein[1], M. Palm[1], D. Grüner[2, 3], G. Kreiner[2], G. Frommeyer[1], D. Raabe[1]

[1] Max-Planck-Institut für Eisenforschung, Max-Planck-Str. 1, D-40237 Düsseldorf, Germany
[2] Max-Planck-Institut für Chemische Physik fester Stoffe, Nöthnitzer Str. 40, D-01187 Dresden, Germany
[3] now at Dept. of Physical, Inorganic and Structural Chemistry, Arrhenius Laboratory, Stockholm University, S-106 91 Stockholm, Sweden

ABSTRACT

Single-phase Fe-Nb and Co-Nb Laves phase alloys were produced by arc melting and levitation melting. By casting the levitation melted alloys in a preheated mould and subsequent slow cooling to room temperature, solid rods of 15 mm in diameter and about 100 mm length of the brittle Laves phases were obtained. Within the extended homogeneity ranges of the $NbFe_2$ and $NbCo_2$ Laves phases, the Vickers hardness was measured in dependence on composition. The results show that the hardness has a maximum at the stoichiometric composition in both systems, indicating defect softening. Nanoindentation measurements on a Co-Nb diffusion couple confirm the dependence of the hardness on composition. In addition, these measurements indicate that the crystal structure of the Laves phase polytype – cubic or hexagonal – seems to have no effect on the hardness. Indentation fracture toughness K_{IC-IF} data for the different polytypes of the Laves phases were evaluated from the Palmquist cracks originating from the edges of the Vickers indentations.

INTRODUCTION

Laves phases are the most abundant intermetallic compounds. With their high strengths and high melting temperatures, transition metal Laves phases are supposed to be interesting materials for high-temperature applications. Until now, only few investigations have been carried out on their mechanical properties.

At low temperatures, Laves phases show a brittle behavior due to their complex crystal structures [1]. At higher temperatures, above the brittle-to-ductile transition temperature (BDTT), Laves phases can be deformed plastically [2, 3]. It is already known, that the mechanical behavior of Laves phases is strongly dependent on the composition [4, 5], although there are contrary reports on this dependence. In [6] defect hardening has been described for several Nb-based Laves phases including $NbFe_2$, whereas otherwise defect softening has been reported for hexagonal C14 $MgZn_2$ [5] and $NbFe_2$ [7]. In both cases anti-site atoms on both, the A- and B-atom rich side, are supposed to be the main type of defect [8, 9].

The present work is focused on the investigation of the composition dependence of the hardness of Laves phases in the Fe-Nb and Co-Nb systems. The $NbFe_2$ Laves phase possesses the hexagonal C14 structure and is stable within a wide homogeneity range of about 8 at.% [11]. In the Co-Nb system all three polytypes of the Laves phase exist in dependence on composition and temperature, i.e. cubic C15, hexagonal C14 and C36 (for a detailed description of the Co-Nb system see [10] in this volume). Therefore, in the latter system, there is the possibility to study the effect of the different crystallographic structures on the hardness.

EXPERIMENTAL PROCEDURE

8 samples of single-phase NbFe₂ of varying compositions were produced by levitation melting. By this method, high-purity samples were obtained with oxygen contents less than 60 wt. ppm. After melting, the material was kept in the liquid state for a few minutes for homogenization and then cast into a cold copper mould. These first attempts sometimes led to extensive cracking due to residual thermal stresses during cooling. Therefore, a preheated alumina crucible was used, where the cast material was held at 1200 °C for several hours and then slowly cooled down with 1 K/min through the BDTT regime. The as-solidified rods were 15 mm in diameter and about 100 mm in length, with a mass of about 300 g (Fig. 2). Another 14 buttons of about 2 g each of single-phase NbFe₂ and NbCo₂ were produced by arc melting under Ar. For homogenization the latter samples were encapsulated in silica tubes filled with Ar and heat-treated at 1100 °C for 720 h. Diffusion couples have been produced by heat-treating rectangular pieces of the pure elements with a polished contact surface and wrapped in a Ta foil under Ar. After the heat treatment the diffusion couples were furnace-cooled to avoid disintegration due to thermal stresses.

Phases were identified by X-ray diffraction (XRD) on a PHILIPS PW3020 diffractometer using CuKα₁-radiation. For light optical and scanning electron microscopy (LOM, SEM) samples were cut by electro discharge machining, embedded in epoxy resin and mechanically polished down to 0.25 μm. The hardness measurements were carried out with a LECO M-400-H1 hardness testing device using a Vickers indenter and 2.94 N (300g) loading for 15 s holding time. For nanoindentation a Hysitron nanoindenter (TriboIndenter) with Berkovich geometry and 1000 μN loading was used. Line scans with a step size of 1 μm were measured along the concentration profile obtained in the diffusion couple.

SEM was performed with a Hitachi S-530 equipped with an energy-dispersive spectrometer (EDS; CDU LEAP system from EDAX). The latter was employed for determining the chemical composition and the homogeneity of the samples. Compositions of certain alloys were also checked by wet chemical analysis. Electron probe microanalysis (EPMA) was carried out on a Jeol JXA 8100 to relate the nanoindentations to certain compositions.

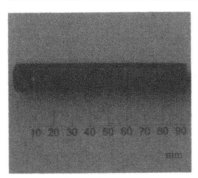

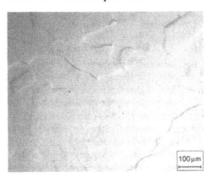

Figure 1. Rod of single-phase C14 Laves phase produced by levitation melting and casting into a preheated mould.

Figure 2. LOM micrograph of single-phase NbFe₂.

470

RESULTS AND DISCUSSION

Fig. 1 shows a rod of about 100 mm length of single-phase Laves phase. By casting into a preheated mould and slow cooling crack free rods were obtained. Fig. 2 shows a light optical micrograph of as-cast stoichiometric $NbFe_2$. By EDS it was confirmed that all samples were chemically homogenous. LOM, SEM and XRD revealed that a few samples contained some small particles on the grain boundaries. By XRD they were identified as having the $NiTi_2$ structure. This structure type is typical for an oxygen-stabilized phase, frequently observed in Fe-Nb alloys [12]. The indentation experiments, which were performed with small loadings (< 3 N), are not affected by these particles, because the indentations itself do not exceed the grain size of the Laves phase grains. The hardness measurements were carried out on grains with diameters above $100\mu m$, which are much bigger than the expected indentations (about $20\mu m$). In most cases the expanding cracks, which form at the edges of the indentations, stayed within the measured grain so that the crack formation was not affected by the grain size.

Figs. 3 and 4 show the results of the hardness tests. Each value is the average of 15 measured data points. It is noted, that the maximum in the hardness is observed at the stoichiometric composition, i.e., both the replacement of Fe by Nb and Nb by Fe results in a decrease in hardness. Figure 3 shows in addition, that differences in the processing of the material do not result in visible differences of the hardness values.

Fig. 4 shows the Vickers hardness in dependence on the Nb content for the $NbCo_2$ Laves phase. Like for the $NbFe_2$ Laves phase, the maximum in the hardness is observed at the stoichiometric composition. Compared to the cubic C15 Laves phase, the hexagonal C36 and C14 polytypes show somewhat lower hardness values. Whether this is caused by the change in the crystallographic structure or by the composition can not be concluded from these measurements as – due to their narrow homogeneity ranges – only a single sample of each of the hexagonal compounds could be produced. In order to clarify this question, a Co-Nb diffusion couple was prepared, which allows measuring the hardness in dependence on composition along a complete concentration profile. After holding for 400 h at 1215 °C the width of the Laves phase zone was about 160 μm, allowing hardness line scans by nanoindentation.

Fig. 5 shows the measured nanohardness in dependence on the Nb content. The homogeneity ranges of the C36, C15 and C14 polytype are marked by the dashed lines. Though there is considerable scatter in the data, they clearly show a maximum approximately at the stoichiometric composition. No marked offset in the hardness is observed when the crystallographic structure of the Laves phase changes from cubic C15 to hexagonal C36. Whether the same is true for the change from C15 to C14 becomes not definitely clear as only few data could be obtained at this transition.

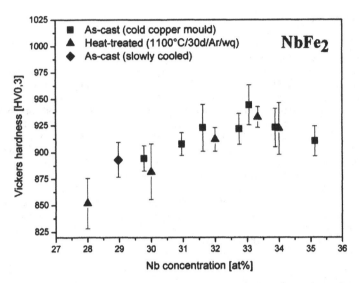

Figure 3. Vickers hardness as a function of the Nb content for binary single-phase NbFe$_2$ Laves phases.

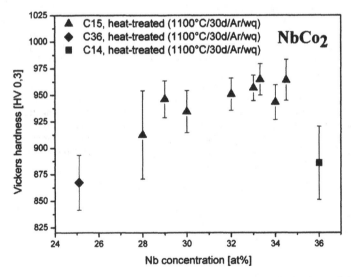

Figure 4. Vickers hardness as a function of the Nb content for binary single-phase NbCo$_2$ Laves phases.

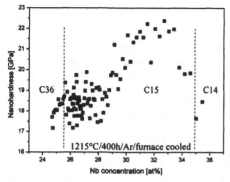

Figure 5. Hardness as a function of the Nb content in the Co-Nb diffusion couple.

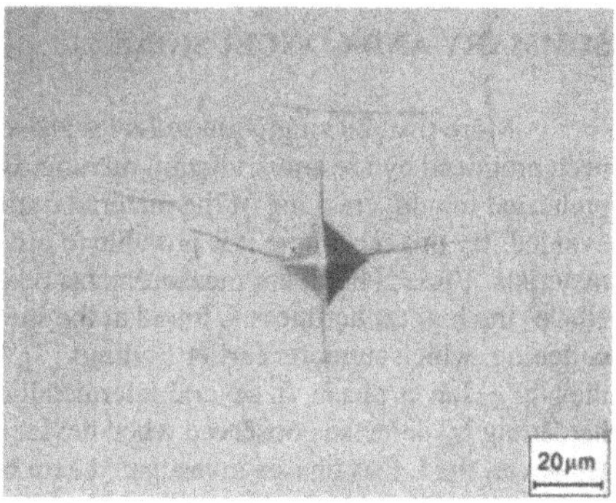

Figure 6. LOM micrograph of a Vickers indentation showing Palmquist cracks.

Laves phases behave very brittle up to high temperatures as can be seen from the micrograph in Fig. 6 where cracks originate near the corners of the Vickers indentation. The formation of such cracks is well known for brittle materials like ceramics. Several mathematical models exist to calculate a value for the critical indentation fracture toughness K_{IC-IF} from the crack length [13, 14]. In order to decide which model has to be applied, the type of crack system has to be determined. In case of the NbFe$_2$ Laves phase it has been found by consecutive polishing of the samples that the cracks are radial, i.e. the so-called Palmquist cracks. It is assumed that the same holds for the NbCo$_2$ Laves phase. From this type of cracks, K_{IC-IF} can be evaluated by the following equation [15] (1) as

$$K_{IC-IF} = 0.0937 \, (H_V * P/4l)^{1/2} \tag{1}$$

where K_{IC-IF} is the indentation fracture toughness in MPam$^{-1/2}$, H_V the Vickers hardness in MNm^{-2}, P the indentation load in MN and l the average crack length in m. The results are shown in Table I. Within the extended composition ranges of C14 NbFe$_2$ and C15 NbCo$_2$, no marked change of K_{IC-IF} in dependence on composition is observed as indicated by the mean variation given in Table I. The authors are not aware of any data for $K_{IC(-IF)}$ reported in the literature. From hardness data and Palmquist crack lengths reported in [16], K_{IC-IF} values were calculated for approximately stoichiometric NbFe$_2$ and NbCo$_2$ Laves phase. While the value for NbFe$_2$ is in good agreement with the present data the value for NbCo$_2$ is higher.

Table I. Indentation fracture toughness K_{IC-IF} for Fe-Nb and Co-Nb Laves phases.

Laves phase	NbFe$_2$ (C14)	NbCo$_2$ (C15)	NbCo$_2$ (C36)	NbCo$_2$ (C14)	NbFe$_2$ (C14) [16]	NbCo$_2$ (C15) [16]
Composition [at.% Nb]	29.8 - 35.1	27.5 - 34	25.1	35	32.7	33.1
K_{IC-IF} [MPam$^{-1/2}$]	0.54 ± 0.05	0.57 ± 0.05	0.69	0.78	0.58	0.77

473

SUMMARY AND CONCLUSIONS

More than 20 single-phase Laves phase alloys in the Fe-Nb and the Co-Nb systems have been produced by arc and levitation melting. By casting the levitation melted alloys into a preheated mould, cracking of the material during solidification due to thermal stresses can be avoided. By this technique it is possible to produce large crack-free samples of these brittle materials. Vickers hardness measurements revealed that for hexagonal C14 $NbFe_2$ and cubic C15 $NbCo_2$ the highest hardness is found at the stoichiometric composition indicating defect softening, which supports earlier findings [7], but contradicts defect hardening claimed in [6] for the $NbFe_2$ Laves phase. In several intermetallic compounds, especially those with a B2 structure, hardening by defects is observed when deviating from the stoichiometric composition [17]. However, the Laves phases investigated here behave differently. The reason for this softening behavior is still not clear. From nanoindentation measurements on a Co-Nb diffusion couple it was found that no offset in the hardness profile appears when the structure of the Laves phase changes from cubic C15 to hexagonal C36. Also first results on mechanical properties have been obtained by evaluating the critical indentation fracture toughness K_{IC-IF} from the crack lengths of the Palmquist cracks originating from the Vickers indentations.

ACKNOWLEDGMENTS

This work was carried out in the Inter-Institutional Research Initiative "The Nature of Laves phases" funded by the Max Planck Society.

REFERENCES

1. P. Paufler, G.E.R. Schulze, *phys. stat. sol.* **24**, 77 (1967).
2. L. Machon, G. Sauthoff, *Intermetallics* **4**, 469 (1996).
3. A.V. Kazantzis, M. Aindow, I.P. Jones, G.K. Triantafyllidis, J.Th.M. De Hosson, *Acta Mater.* **55**, 1873 (2007).
4. P. Paufler, K. Eichler, G.E.R. Schulze, *Monatsber.Dt.Akad.Wissensch. Berlin* **12**, 950 (1970).
5. K. Eichler, S. Siegel, H. Kubsch, P. Paufler, *Wiss. Z. Techn. Univ. Dresden* **20**, 399 (1971).
6. C.T. Liu, J.H. Zhu, M.P. Brady, C.G. McKamey, L.M. Pike, *Intermetallics* **8**, 1119 (2000).
7. G. Leitner, Doctoral thesis, TU Dresden 1971, pp. 129-131.
8. D. Grüner, F. Stein, M. Palm, J. Konrad, A. Ormeci, W. Schnelle, Y. Grin, G. Kreiner, *Z. Kristallogr.* **221**, 319 (2006)
9. G. Kreiner, D. Grüner, Y. Grin, F. Stein, M. Palm, A. Ormeci, in these proceedings
10. C. He, F. Stein, M. Palm, D. Raabe, in these proceedings
11. E. Paul, L.J. Swartzendruber, *Bull. Alloy Phase Diagr.* **7**, 248 (1986).
12. D. Grüner, Doctoral thesis, TU Dresden 2007, pp. 1-276; online available at http://nbn-resolving.de/urn:nbn:de:swb:14-1172078219643-48967.
13. K. Niihara, *J. Mater. Sci. Lett.* **2**, 221 (1983).
14. D.K. Shetty, I.G. Wright, *J. Mater. Sci. Lett.* **5**, 365 (1986).
15. Y.K. Song, R.A. Varin, *Intermetallics* **6**, 379 (1998).
16. A. von Keitz, Doctoral thesis, RWTH Aachen, Shaker Verlag Aachen 1996, pp. 1-139.
17. L.M. Pike, Y.A. Chang, C.T. Liu, *Acta Mater.* **45**, 3709 (1997)

Mater. Res. Soc. Symp. Proc. Vol. 1128 © 2009 Materials Research Society   1128-U08-06

Transmission Electron Microscopy of Fe2Nb Laves Phase With C14 Structure in Fe-Nb-Ni Alloys

Naoki Takata[1,2], Shigehiro Ishikawa[1], Takashi Matsuo[1,2] and Masao Takeyama[1,2]
[1] Department of Metallurgy and Ceram3ics Science, Tokyo Institute of Technology
2-12-1, Ookayama, Meguro-ku, Tokyo 152-8552, JAPAN
[2] "Fundamental Studies on Technologies for Steel Materials with Enhanced Strength and Functions" Consortium of JRCM (The Japan Research and Development Center for Metals)

ABSTRACT

The lattice structure of the C14 Fe2Nb Laves phase with Ni in solution in Fe-Nb-Ni ternary alloys was examined by transmission electron microscopy. Binary stoichiometric Fe2Nb (Fe-33.3 at.% Nb) exhibits a featureless morphology with a low dislocation density. A similar morphology was observed in stoichiometric Fe2Nb containing 20 at.% Ni and in binary Fe-rich Fe2Nb (Fe-27.5 at.% Nb). In contrast, many planar faults parallel to the basal plane of the C14 structure were observed in Fe-rich Fe2Nb with Ni in solution, and the fault density increases with increasing Ni content up to 33.1 at.%. The high resolution transmission electron microscope (HRTEM) analysis revealed that the planar faults are related to the local change in the stacking sequence of the three 3^6-nets (triple layer) of the C14 structure. These results suggest that the presence of both, the point defects (Fe sublattice sites occupied by Ni atoms) and the anti-site defects (Nb sublattice sites occupied by excess Fe atoms), facilitate the formation of the planar faults.

INTRODUCTION

We have systematically investigated the phase equilibria between γ-Fe and Fe2M Laves phases in Fe-M-Ni ternary systems (M: Nb, Ti, Mo) at elevated temperatures in order to seek the possibility for the development of a new type of austenitic heat resistant steel strengthened by intermetallic compounds [1-5]. In those studies, we revealed that the homogeneity region of the Fe2Nb Laves phase with C14 structure extends towards the equi-Nb concentration direction up to 44 at.% Ni [1,2]. This indicates that more than two thirds of the Fe sublattice sites in Fe2Nb can be replaced by Ni atoms. In addition, the hardness of Fe-rich Fe2Nb decreases significantly from 9 GPa to 6 GPa with increasing Ni content [1]. This significant change should be somehow related to the defect structures created by Ni in solution. In the present study, we have examined the lattice structure of Fe2Nb with and without Ni in solution using transmission electron microscopy.

EXPERIMENTAL PROCEDURE

Two series of alloys previously prepared for the phase diagram studies [1,2] were used; one is stoichiometric Fe-33.3Nb alloys containing Ni up to 20 at.% (Fe-33.3Nb-(0-20)Ni) and the other is Fe-15Nb alloys containing Ni up to 40 at.% (Fe-15Nb-(0-40)Ni). These alloys were prepared by arc-melting and equilibrated at 1473 K/240 h. The experimental methods were described in detail elsewhere [1,2]. The stoichiometric alloys exhibited the single phase Fe2Nb, and the Fe-rich alloys showed the two phases γ-Fe and Fe_2Nb [2]. The chemical compositions of the Fe_2Nb Laves phases in these alloys are summarized in Table 1. Note that three types of defects are expected to exist in these Fe_2Nb phases; point defects in stoichiometric Fe_2Nb with Ni, anti-site defects in the Fe-rich Fe_2Nb in Fe-15Nb and the combined defects in Fe-rich Fe_2Nb with Ni. TEM discs were cut from the heat treated samples, and mechanically polished, followed by twin-jet electro-polishing in a solution of ethanol with 12 vol.% perchloric acid at 253 K. Some of the samples were prepared by ion-milling. The microstructures were examined by JEOL JEM-2010HC operated at 200 kV. High resolution TEM (HRTEM) observation was carried out by JEOL JEM-2100F. The HRTEM images of the Fe_2Nb with C14 structure were simulated using a multi-slice method [6].

Table 1 Chemical compositions and lattice parameters of the Fe_2Nb Laves phases studied.

Alloys	Chemical composition (at.%)			Lattice parameter (Å)		Lattice parameter ratio
	Fe	Nb	Ni	a	c	c/a
Fe-33.3Nb	67.5	32.5	-	4.8376	7.8852	1.630
Fe-33.3Nb-20Ni	48.5	31.2	20.2	4.8277	7.8599	1.628
Fe-15Nb	72.5	27.5	-	4.8103	7.8528	1.632
Fe-15Nb-20Ni	57.3	26.7	15.9	4.8006	7.8045	1.625
Fe-15Nb-40Ni	41.2	25.7	33.1	4.7864	7.7798	1.625

RESULTS AND DISCUSSION

Figure 1 shows the TEM images of the Fe_2Nb Laves phases with various compositions. In the stoichiometric alloys, few dislocations were observed in binary Fe_2Nb (Figure 1(a)), whereas some dislocations were observed in Fe_2Nb with 20 at.% Ni in solution (Figure 1(b)). The binary Fe-rich Fe_2Nb also exhibit a featureless morphology with few dislocations (Figure 1(c)). However, in Fe-rich Fe_2Nb with Ni in solution, many planar faults parallel to the basal plane (0001) of the C14 structure were observed (Figure 1(d)). The quantitative analysis of the planar fault density was done by the line intersection method, and the result was plotted as a function of the Ni content, as shown in Figure 2. In stoichiometric Fe_2Nb, no planar faults were observed, regardless of Ni content. In the Fe-rich Fe_2Nb, the binary one does not show any planar faults. However, the density increases tremendously from 0 to 10^5 /m with Ni in solution up to 16 at.%, although the planar faults are not uniformly distributed, as indicated by the scattered data. The density increases further by an order of magnitude at a Ni content of 33 at.%.

Figure 3(a) shows a HRTEM image taken by the incident beam direction $\mathbf{B} = [11\bar{2}0]$ of the faulted Fe_2Nb in the Fe-15Nb-40Ni alloy, together with the selected area electron diffraction pattern. The white dots horizontally aligned parallel to the (0001) plane show some irregularity, as shown by arrows. The irregular regions correspond to the planar faults observed by conventional TEM (Figure 1(d)). A high magnification view in Figure 3(b) clearly shows the difference in layer stacking between the planar fault free area (c) and the faulted area (d). The

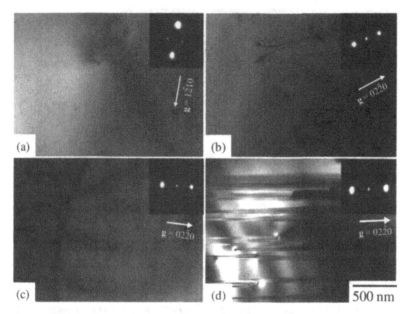

(a)

(b)

(c)

(d)

500 nm

Figure 1. TEM bright field images and selected electron diffraction patterns of C14 Fe$_2$Nb in (a) Fe-33.3Nb (**B** ~ 10$\bar{1}$3), (b) Fe-33.3Nb-20Ni (**B** ~ 11$\bar{2}$1), (c) Fe-15Nb (**B** ~ 11$\bar{2}$0) and (d) Fe-15Nb-40Ni (**B** ~ 11$\bar{2}$0).

observed images squared in Figure 3(b) are shown in Figures 3(c) and (d), respectively, where the atom positions determined by the simulation are superimposed. The lattice parameters measured from the HRTEM images are in good agreement with those measured by X-ray diffraction (Table 1). It is known that the C14 structure consists of single layers (s) and triple layers (t, t') with a stacking sequence of s t s t' [7]. Here, the s layer is a *kagome*-net formed by Fe atoms (Fe2 sublattice), and the t layer consists of three 3^6-nets, where one is formed by Fe atoms (Fe1 sublattice) sandwiched with two others formed by the Nb atoms. t and t' are distinguished by the stacking direction of the Nb planes with respect to the Fe plane. The planar fault free area shown in Figure 3(c) exhibits a complete C14 stacking of s t s t', whereas the faulted area in Figure 3 (d) shows s t's t'. Thus, the observed planar faults give rise to local structure changes from C14 to C15 (s t's t') or C36 (s t s t s t's t') due to the wrong sequence of the triple layer.

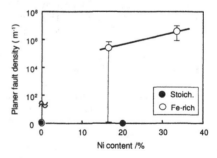

Figure 2. Change in planar fault density with Ni content in stoichiometric and Fe-rich Fe$_2$Nb.

477

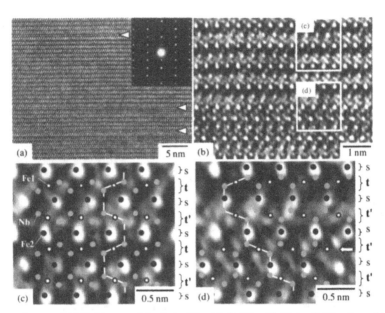

Figure 3. (a) A HRTEM image of the faulted Fe-rich Fe_2Nb with 33.1 at% Ni, taken with $\mathbf{B} = 11\bar{2}$ 0, (b) high magnification view of (a), (c) and (d) simulated atom positions of the planar fault free and faulted area corresponding to the squared regions in (b), respectively.

The occurrence of the planar faults in Fe-rich Fe_2Nb with Ni in solution is associated with the Suzuki segregation effect [8,9]. Fe_2Nb has two types defects; the point defects where Ni atoms substitute Fe sublattice sites and the anti-site defects where the excess Fe atoms occupy Nb sublattice sites. The fact that Fe_2Nb with no planar faults has only the single type defect suggests that the two types of defects would interact with each other. The solute atom segregation would lower the fault energy, leading to the stacking change of the triple layers, just like in the case of fcc/hcp structure. This is likely to occur if the change in the fault energy in the Fe_2Nb is similar to that in bcc Fe, since Fe has a high stacking fault energy (about 600 mJ/m^2) [10]. However, an increase in the Ni content is known to increase the stacking fault energy of austenitic stainless steels [9]. Thus, further study on the effect of the solute element on the fault energy in Fe_2Nb is certainly needed.

Another possibility for the occurrence of the planar faults is the free volume concept proposed by Chu & Pope [11] and Takasugi et al. [12]. The local change in the atomic arrangement due to the defects would create some free volume, since the atomic radius of Ni (124.6 pm) is smaller than those of Fe (127.4 pm) and Nb (146.8 pm) atoms [13]. Then, the fraction of the free volume in the unit cell (V_f) of the Fe_2Nb studied was calculated, and the result is shown in Figure 4. In the calculation, the apparent atomic radii of Fe and Nb atoms were estimated based on the hard sphere model [14] under the following assumptions; (1) Nb atoms touch each other, (2) Fe2 atom is in contact with Fe1 atom but not with Fe2 in Fe the tetrahedron ($Y > X = Z$), as schematically shown in Figure 5. The lattice parameters, a and c used for the calculation are listed in Table1 [2]. The V_f of the binary stoichiometric Fe$_2$Nb is 0.290, and it remains almost unchanged with Ni content. It is interesting to note that, unlike the expectation from the big difference in atomic size between Fe and Nb, the V_f of binary Fe-rich Fe$_2$Nb becomes smaller than that of the binary stoichiometric one, but that it significantly increases with Ni content up to 16 at.%. The trend of V_f is similar to that of the planar fault density shown in Figure 2. The similar trend suggests that the increase in the free volume would enhance the stacking change in the triple layer. In an attempt to evaluate the free volume within the triple layer, we estimated the free distance between the atoms on the 3^6-nets of the triple layer, and the results are plotted as a function of the Ni content in Figures 6(a) and (b), where the vertical axis is the lattice parameter a normalized by the apparent atomic radii of Fe (r_{Fe}) and Nb (r_{Nb}), respectively, obtained from the hard sphere model. In both cases, a/r increases with the Ni content in Fe-rich Fe$_2$Nb, whereas it remains almost unchanged in the stoichiometric one. These trends are similar to the change in the free volume in the unit cell. Thus, the increase in the free distance between the atoms in the triple layer would somehow facilitate the stacking change of the triple layer, presumably through the simultaneous movement of the two 3^6-nets known as synchroshear [7], as shown by the inner schematics in Figure 6.

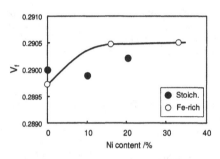

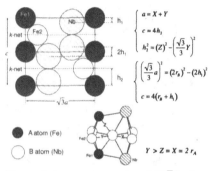

Figure 4. Change in fraction of the free volume in the unit cell (V_f) with Ni content.

Figure 5. Atom arrangements of $(10\bar{1}0)$ of C14 Fe$_2$Nb for the hard sphere model.

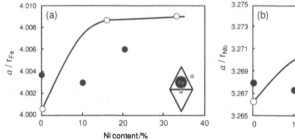

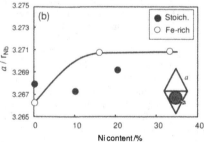

Figure 6. Interatomic distance of two Fe atoms on the 3^6-net, a, normalized by the atomic radius of (a) Fe, r_{Fe}, and (b) Nb, r_{Nb}, as a function of the Ni content in Fe$_2$Nb.

479

CONCLUSIONS

The microstructures of stoichiometric and Fe-rich Fe_2Nb Laves phases with Ni in solution were examined by transmission electron microscopy, and the following results were obtained;

1. Stoichiometric Fe_2Nb exhibits a featureless morphology with a low dislocation density, regardless of the Ni content.
2. Fe-rich Fe_2Nb with Ni in solution exhibits many planar faults parallel to the basal plane of the C14 structure, and the fault density increases with increasing the Ni content.
3. The planar fault corresponds to the local change in the stacking sequence of the three 3^6-nets (triple layer) of the C14 structure.
4. The occurrence of the planar faults is associated with the local change in the atomic arrangement in the triple layer due to the introduction of both, point defects and anti-site defects.

ACKNOWLEDGMENTS

This study was carried out as a part of research activities of "Fundamental Studies on Technologies for Steel Materials with Enhanced Strength and Functions" by Consortium of JRCM (The Japan Research and Development Center of Metals). Financial support from NEDO (New Energy and Industrial Technology Development Organization) is gratefully acknowledged.

REFERENCES

1. N. Gomi, S. Morita, T. Matsuo and M. Takeyama, Report of JSPS 123rd Committee on Heat Resisting Materials and Alloys **42**, 157 (2004).
2. M. Takeyama, N. Gomi, S. Morita and T. Matsuo, Mater. Res. Soc. Symp. Proc. **842**, 461 (2005).
3. S. Ishikawa, T. Matsuo and M. Takeyama, Mater. Res. Soc. Symp. Proc. **980**, 517 (2007).
4. T. Sugiura, S. Ishikawa, T. Mastuo and M. Takeyama, Mater. Sci. Forum **561-565**, 435 (2007).
5. S. Ishikawa, T. Matsuo, N. Takata and M. Takeyama, in this book.
6. K. Ishizuka, Ultramicroscopy **5**, 55 (1980).
7. M. F. Chisholm, S. Kumar and P. Hazzledine, Science **307**, 701 (2005).
8. H. Suzuki, Scientific Report Res. Inst. Tohoku Univ. **A4**, 455 (1952).
9. G. Saada, "Theory of Crystal Defects" ed. B. Gruber (Academic press, 1966) pp. 167-213.
10. H. Suzuki "Ten-iron Nyumon" (Agune, 1967) pp. 355.
11. F. Chu and D. P. Pope, Mater. Sci. Eng. **A170**, 39 (1993).
12. T. Takasugi, M. Yoshida and S. Hanada, Acta Mater. **44**, 177 (1996).
13. E. T. Teatum, K. A. Gschneidner Jr., J. T. Waber, "Compilation of Calculated Data Useful in Predicting Metallurgical Behavior of Elements in Binary Alloy Systems", Report LA-4003, UC-25, Metal, Ceramics and Materials, TID-4500, Los Alamos Scientific Laboratory, (1968).
14. D.J. Thoma and J.H. Perepezko, J. Alloys and Compd. **224**, 330 (1995).

Mater. Res. Soc. Symp. Proc. Vol. 1128 © 2009 Materials Research Society　　　　1128-U08-07

Metastable Hexagonal Modifications of the $NbCr_2$ Laves Phase as Function of Cooling Rate

Jochen Aufrecht, Andreas Leineweber and Eric Jan Mittemeijer
Max Planck Institute for Metals Research, Heisenbergstrasse 3, D-70569 Stuttgart, Germany

ABSTRACT

In as-cast ingots produced by arc-melting, several metastable polytypic modifications of $NbCr_2$ were found additional to the cubic C15 phase stable at room temperature: C14, C36 and 6H-type structures, often highly faulted and/or intergrown. Strikingly, these phases had formed at locations of the specimen which had experienced a relatively low cooling rate, whereas the C15 phase was formed preferentially in regions which had experienced the highest cooling rates.

INTRODUCTION

In the binary Nb-Cr system, the Laves phase $NbCr_2$ exhibits the cubic C15-type structure (prototype: $MgCu_2$) at room temperature [1]. Nevertheless, in as-cast ingots, other Laves-phase polytypes of $NbCr_2$ have been found: Guseva found a "hexagonal structure" [2]. Thoma et. al. reported hexagonal C14 and C36 modifications (prototypes: $MgZn_2$ and $MgNi_2$) in as-cast ingots prepared by arc-melting, together with the stable C15 phase [3, 4]. The amount of the metastable phases was reported to depend on the location within the ingot: more metastable phase at locations adjacent to the water-cooled copper hearth (i.e. the ground plate) of the arc-melting furnace, i.e. at locations where relatively high cooling rates occurred. Splat-quenched $NbCr_2$ samples contained an even larger fraction of C14, but always together with the C15 modification [3]. Kazantzis et al. [5] reported various Laves-phase polytypes in solidified arc-melted alloys together with the C15 modification: Besides C14 and C36 also more complex polytypes (Ramsdell symbols 6H, 8H and 9R) were suggested to be present.

The occurrence of a C14 modification seems to be in agreement with the commonly accepted phase diagram [1], which incooperates a C14-type high-temperature modification of $NbCr_2$ stable between about 1600°C and the congruent melting point at 1770°C. It has thus been supposed that $NbCr_2$ solidifies into this crystal structure, in the first stage of cooling upon solidification, which structure is only partially retained upon continued cooling to room temperature. The high cooling rates occurring after arc-melting or during splat-quenching necessitate very fast kinetics to complete the C14 → C15 transformation, while this type of transformation has been reported to be very sluggish in Laves phases like $TaCr_2$, $HfCr_2$ and $TiCr_2$ [6]. The transformation is assumed to proceed by the "synchro-shear" process, involving mediation by a certain type of complex partial dislocations, changing the stacking sequence of the Laves-phase structures [6, 7]. On this basis the occurrence of stacking sequences of longer period than corresponding with C14 and C15, as 4H (= C36), 6H, 8H and 9R, which could represent intermediate stages, might be understood. Contamination with O and N has also been suggested as possible reason for the formation of these longer-period stacking variants [5].

However, recent work by our group indicates that a reverse C15 → C14 transformation does not occur at high temperatures upon heating [8]; C15 appears to melt directly. Thus, the C14 phase should be of metastable nature at all temperatures, which invalidates the above interpretation as presented in Refs. [2-5]. In the present work, $NbCr_2$ ingots produced by arc-melting were examined using X-ray powder diffraction (XRPD), electron-backscatter diffraction (EBSD), in combination with energy-dispersive X-ray spectroscopy (EDX) and high-resolution transmission electron microscopy (HR-TEM). The occurrence of metastable phases was demonstrated and an alternative explanation for their formation is presented.

EXPERIMENT

$NbCr_2$ ingots within the compositional range from 32.4 at% Nb to 34.5 at% Nb were prepared from high-purity niobium and high-purity chromium (Nb: non-metallic impurities < 10 ppm, Ta ~ 50 ppm, other metallic impurities ~ 80 ppm; Cr: non-metallic impurities 34 ppm, other metallic impurities ~ 20 ppm). Ingots of about 5 g were produced by arc-melting appropriate amounts of chromium and niobium under a titanium-gettered argon atmosphere. Six-fold flipping and remelting of the ingots ensured chemical homogeneity (in the liquid state). In the furnace, the samples were placed in moulds on a water-cooled copper hearth, thus the cooling rate was very high in the arc-melting furnace. However, note that by dedicated rapid solidification techniques (e.g. splat quenching [3]), considerably higher cooling rates are achieved.

For XRPD investigations, the ingots were crushed into several pieces, such that the original locations within the ingots of the obtained single pieces were known. From these pieces, powders were produced by grinding in a mortar; each powder was suspended in isopropanol and sedimented on a (510)-cut Si crystal plate (wafer) using a supporting brass ring during evaporation of the liquid. XRPD patterns were obtained using a "Philips X`Pert MPD" diffractometer equipped with a germanium single crystal monochromator in the incident beam, thus using Cu-Kα_1 radiation.

Metallographic samples for EBSD and EDX analyses were produced by embedding large pieces (about one fourth) of an ingot into a conductive mould and by subsequent grinding and polishing, using a dispersion of colloidal silicon oxide (OPS) for the final polishing step to remove any deformed surface-adjacent material. EBSD analysis was performed using a Zeiss scanning electron microscope (LEO 438VP) equipped with an EBSD system (TSL, EDAX Inc.) and using the software OIM 4.5.

For HR-TEM analysis using a Philips CM 30 (300kV) apparatus, a coarse powder was produced from a piece taken from the top of one of the ingots (initial sample composition of $Nb_{34.5}Cr_{65.5}$). The powder was sedimented on a copper mesh. Due to the brittle nature of the investigated materials, the edges of the powder particles were thin enough to be electron transparent.

RESULTS

The XRD patterns taken from the top part and from the bottom part of ingot obtained by arc-melting of initial composition $Nb_{33.3}Cr_{66.7}$ (see Figure 1) show that besides reflections from the stable C15 room-temperature modification, also reflections pertaining to hexagonal modifications as C14 or C36 are present. Those reflections of these latter structures with the indices $h-k = 3N$ ($N = 0, \pm1, \pm2, ...$) coincide with the C15 reflections. Some of the C36 reflections ($h-k \neq 3N$ and $l = 4M \pm1$, ($M = 0, \pm1, \pm2, ...$)) are severely broadened, so that they

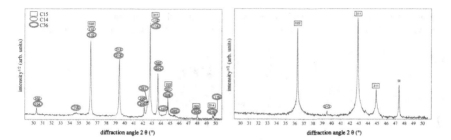

Figure 1. XRPD patterns of material taken from the top (left) and the bottom (right) parts of an NbCr$_2$ ingot upon arc-melting. Whereas the material taken from the bottom part consists predominantly of the C15 modification, the material taken from the top part of the ingot is relatively rich of the hexagonal C14 and C36 modifications. Note that the C36 modification exhibits strongly *hkl*-dependent broadening of the reflections.

are only with difficulties discernible against the background. This selective broadening of reflections can be attributed to specific types of stacking faults occurring in the hexagonal modifications. A quantitative estimation of the relative amount of C14 and C36 is very difficult from the XRPD patterns, since all non-broadened reflections of C36 (reflections with $h - k = 3N$ as well as with $h - k \neq 3N$ and $l = 4M \pm 2$) occur at same locations as C14-type reflections.

Very remarkably, according to the XRD pattern from the bottom part of the ingot, where it was in contact with the water-cooled copper hearth of the arc-melting furnace, a much smaller amount of these hexagonal C14/C36 phases is present at the bottom part, which was found to be the case in this work for all investigated ingots, irrespective of their composition. This is in striking contradiction with results reported in Refs. [3,4], where C14 was found to occur preferably in the bottom parts of the ingots.

Phase distinction by EBSD is difficult: The Kikuchi patterns pertaining to C14 and C36 are very difficult to distinguish and, moreover, the C15 phase often exhibits very fine twins so that patterns originating from two twin variants within one crystallite may overlap, leading to incorrect assignment of C15 regions to C14/C36. Nevertheless, the occurrence of the hexagonal non-C15 modifications preferentially in the top part of the ingot, rather than in the bottom part, could be confirmed by EBSD (Figure 2a). Additionally, the grain size in the top part of the ingot could be shown to be much larger than in the bottom part: The hexagonal crystallites appear to be larger than the cubic C15-type crystallites (Figure 2b). Accompanying EDX measurements did neither reveal a detectable composition gradient from top to bottom of the ingot, nor differences in composition between hexagonal and C15 crystallites.

In a HR-TEM investigation of material taken from the top part of an as-cast ingot (Figure 3), a total of 15 different crystallites were investigated. For only one crystallite, the pure C14 structure could be observed. For two crystallites, the C14 type structure occurred together with the C36 type structure. The latter structure type was the most frequently observed one: In 9 crystallites it occurred exclusively and in two others it occurred together with the C15 modification. The pure C15 modification was found in only one crystallite. Stacking faults occurred in most of the observed structure types (and additionally twins in the C15 structure type). The stacking fault density was highest in the C36 structures type, where as a consequence of the faulting 6H-type stacking sequences (…*cchcch*…, "6H$_1$") were found. Sometimes, intergrowth structures consisting of sequences of C14, C36 and 6H-type stacking units were observed. Stacking faults found in C14 usually involved intergrowths of C36-type sequences.

Figure 2. EBSD orientation (a) and phase map (b). In the phase map, dark areas correspond to the C14 or the C36 structure types; white areas were indexed as C15 structure type. In the orientation map, different grey shades correspond to different crystallite orientations. The crystallite size is much larger in areas the top part of the ingot, where also the hexagonal C14 and C36 modifications occur preferentially.

DISCUSSION

Employing three different techniques it has been shown in this work that additionally to the C15 modification, also C14 and C36 modifications of $NbCr_2$ occur in as-cast ingots. The assumption of C14 being a high-temperature modification stable above around 1600°C has been disproven recently [8]. The appearance of the hexagonal C14/C36 modifications in the top part of the arc-melted ingots is especially surprising, as one would expect that non-equilibrium phases are formed during solidification preferentially at locations where the cooling rate is very high. For samples produced upon arc-melting, the cooling rate in the bottom part of the ingots, where they are in contact with the water-cooled copper hearth, is expected to be significantly higher than in the top part. However, no or only small quantities of C14 or C36 could be detected in the bottom part. Thus, two questions have to be answered: why do metastable hexagonal modifications form from the melt during solidification and why do these structures form preferentially in regions experiencing relatively low cooling rates (top of ingots)?

For the following considerations, two assumptions are made: (i) the Gibbs energies of formation of the competing Laves structures are very similar [10], and (ii) the C36 phase does not crystallize from the melt but forms from an initially solidified C14 phase. Instead of the direct transformation of C14 to equilibrium C15 upon continued cooling, which transformation is supposed to be sluggish as observed in similar systems [6], a (kinetically more favoured) transformation to C36 might occur, In view of Gibbs energy of formation, C36 may be intermediate between C14 and C15. The high density of stacking faults in C36 found in the present investigations can be taken as indication that C36 has been formed from C14 by the

Figure 3. HR-TEM images and corresponding selected-area diffraction patterns (SADPs) obtained from material taken from the top part of an ingot upon arc-melting. In a), a C14 crystallite without stacking faults is shown ([010]-zone. In b), an intergrowth structure consisting of C36 and 6H-type stacking is present ([010]-zone). The faulting gives rise to streaking in the SADPs, i.e. appearance of diffuse intensity along the $h0l$ lattice rows. In c), an image of a C15 crystallite is shown (110-zone).

synchro-shear mechanism rather than by direct crystallization from the melt [9]. Further, the only type of stacking fault found in C14 involves C36-type stacking sequences.

If the C14 phase would be a stable high-temperature modification (i.e. have the lowest Gibbs energy of formation directly below the melting point) of $NbCr_2$, as proposed before [1], the retention of this phase and the formation of C36 from C14 could be understood easily in view of the sluggish C14 → C15 transformation. In the bottom part of the ingot, a higher cooling rate can lead to undercooling of the melt down to temperatures where C15 is the (much more) stable phase, and thus, this phase is present in this part of the ingot. However, such an interpretation is in contradiction with the recent results presented in [8], which invalidate C14 as stable high-temperature phase. (An only very small possible stability range of C14, of some Kelvin below the melting point of $NbCr_2$, could be beyond detectability by the methods as used in [8]).

If C14 is an unstable phase at all temperatures (i.e. has a slightly higher (less negative) Gibbs energy of formation than C15 at all temperatures below the melting point), it could nevertheless be formed during solidification because due to the very similar Gibbs energies of the competing Laves phase structures below the melting point, phase selection upon solidification may be determined by nucleation. A smaller value for the solid/liquid interface energy in case of C14, as compared with C15 could favour nucleation of C14. This will apply especially, when crystallization occurs at temperatures near the melting point, so that nucleation occurs at relatively low undercooling, i.e. in regions experiencing relatively small cooling rates. In the bottom part of the ingot, the melt may be undercooled very fast and thus much more pronouncedly than in the top part of the ingot. At large undercooling the difference between the Gibbs energy of formation of the stable C15 and the metastable C14 will be larger than at small undercooling. Hence, C15 is formed preferentially in the bottom part of the ingot and C14 (and thus C36) is formed preferentially in the top part of the ingot.

As a final note, it is pointed out that no information on solidification at very low cooling rates ("equilibrium solidification") can be given, since the cooling rate also in the top region of the ingot is still high (cooling to room temperature is achieved within some seconds). The present work deals with the significant differences in cooling rate within a solidifying ingot in an arc-melting furnace, for the case of an overall relatively high cooling rate, but smaller than in case of rapid solidification techniques as e.g. splat quenching.

CONCLUSIONS

XRPD, EBSD and HR-TEM characterization of as-cast ingots of the Laves phase $NbCr_2$ produced by arc-melting have shown the occurrence of hexagonal C14 and C36 modifications of $NbCr_2$. These metastable phases were shown to be present, together with the stable cubic C15 equilibrium modification, predominately in the top part of the arc-melted ingots, whereas in the bottom part, which had been in contact with a water-cooled copper hearth, only very small amounts of these metastable phases occur. XRPD and HR-TEM analysis showed that at room temperature (i.e. after completed cooling) the major part of the hexagonal modifications is represented by the heavily faulted C36 phase, and that the C14 phase is a minority phase. The faulting in C36 suggests that this phase had formed from initially crystallized C14 by the synchro-shear mechanism.

The appearance of the metastable C14 phase (and subsequently the C36 phase within the originally C14 phase) is most likely if the undercooling upon solidification is rather small, i.e. in the top part of the ingot where the relatively lowest cooling rates occur. At large undercooling, i.e. in the bottom part of the ingot where the relatively highest cooling rates occur, the difference in Gibbs energy (of formation) of the C15 and the C14 phases can be relatively large, thereby promoting the formation of the C15 phase.

ACKNOWLEDGMENTS

The authors wish to thank Mrs. Viola Duppel, Max Planck Institute for Solid State Research, Stuttgart for HR-TEM analysis and Mr. Ewald Bischoff, Max Planck Institute for Metals Research, Stuttgart for EBSD analysis.

This work has been performed within the framework of the Inter-Institutional Research Initiative "The Nature of Laves Phases" funded by the Max Planck Society.

REFERENCES

1.   M. Venkatraman and J.P. Neumann, *Bull. of Alloy Phase Diagr.* **7**, 462 (1986)
2.   L.N. Guseva, *Inorg. Mater. (USSR)* **1**, 1581 (1965)
3.   D.J. Thoma and J.H. Perepezko, *Mater. Res. Soc. Symp. Proc.* **194**, 105 (1990)
4.   D.J. Thoma and J.H. Perepezko, *Mater. Sci. Eng.*, **A 156**, 97 (1992)
5.   A.L. Kazantzis, T.T. Cheng, M. Aindow and I.P. Jones, *Inst. Phys. Conf. Ser.* **147**, 511 (1995)
6.   K.S. Kumar and P.M. Hazzledine, *Intermetallics* **12**, 763 (2004)
7.   C.W. Allen and K.C. Liao, *Phys. Status Solidi* A **74**, 673 (1982)
8.   J. Aufrecht, A. Leineweber, A. Senyshyn and E.J. Mittemeijer, to be published
9.   J. Aufrecht, W. Baumann, A. Leineweber, V. Duppel and E.J. Mittemeijer, to be published
10. O. Vedmedenko, F. Rösch and C. Elsässer, *Acta Mater.* **56**, 4984 (2008)

Mater. Res. Soc. Symp. Proc. Vol. 1128 © 2009 Materials Research Society   1128-U08-08

Structure and Disorder of the Laves Phases in the Co-Nb System

Guido Kreiner[1], Daniel Grüner[2], Yuri Grin[1], Frank Stein[3], Martin Palm[3] and Alim Ormeci[1]

[1]Max-Planck-Institut für Chemische Physik fester Stoffe, Dresden, Germany
[2]Department of Physical, Inorganic and Structural Chemistry, Arrhenius Laboratory, Stockholm University, Stockholm, Sweden
[3]Max-Planck-Institut für Eisenforschung GmbH, Düsseldorf, Germany

ABSTRACT

A special feature of the Co-Nb system is the occurrence of the three different types of Laves phase with the ideal composition $NbCo_2$. The C36 and the C14 phases are stable only at high temperatures and exhibit small homogeneity ranges, whereas the C15 phase forms with a broad homogeneity range enclosing the ideal composition. In case of C36 and Co-rich C15 the additional Co atoms substitute Nb atoms ($Nb_{1-x}Co_x)Co_2$. In the C36 phase the Co atoms preferentially occupy one of the two crystallographic Nb sites and are locally displaced by approx. 20 pm from the original Nb positions allowing the formation of favorable short Nb-Co bonds. In Nb-rich C14 only one of two crystallographic sites is occupied by Nb. The Kagomé layers of the Co atoms are distorted in the crystal structures of the hexagonal Laves phases. The deviation from the idealized crystal structure is mainly governed by the valence electron concentration. Quantum mechanical calculations show that the distortion is already an inherent feature of the point defect-free structures.

INTRODUCTION

Laves phases [1-4] have been studied intensively to understand the fundamental aspects of phase stability. However, simple factors governing the crystal structure type of geometric (r_A/r_B) and electronic (valence electron concentration, vec, and electronegativity difference, $\Delta\chi$) nature have proven to be helpful in predicting the occurrence and stability of the Laves phases in strictly limited cases [5] only. In general, phase stability and properties of Laves phases are difficult to forecast, especially the origin of the homogeneity ranges and disorder phenomena.

In order to understand the nature of Laves phases, structure and disorder phenomena of Laves phases in the Co-Nb system have been studied in this work. The binary system Co–Nb is particularly suitable to throw light on the stability of the different polytypes due to the coexistence of the C14, C15 and C36 Laves phases. Recently, the phase diagram of the Co-Nb system was reinvestigated [6]. The C36 and the C14 phases form strongly off-stoichiometric at lower and higher Nb concentration, respectively, and only at high temperatures. Both phases exhibit a small homogeneity range while the room temperature phase C15 shows a large asymmetrical homogeneity range enclosing the stoichiometric composition $NbCo_2$. Here, we report on the interplay between chemical disorder, site preferences and the homogeneity ranges and on locally driven and cooperative deviations from the idealized crystal structures.

EXPERIMENTAL DETAILS

Samples of Nb and Co were prepared by arc-melting from the elements (Nb, H. C. Starck, granules, 99.9%; Co, Chempur, foil, 99.995%) on a water-cooled copper hearth in argon atmosphere. To assure homogeneity all samples have been turned over and remelted several times. Finally, the reguli were enclosed in weld-sealed niobium ampoules which were encapsulated in fused silica tubes filled with argon. After annealing at 1100 °C for one month, the samples were quenched in cold water. For the investigation of the sample composition ICP-OES was used. In addition each sample was analyzed for H, N, O, and C impurities by the carrier gas hot extraction or the combustion technique. The microstructures were examined optically and with a SEM. The compositions of the observed phases were analyzed by EDXS and WDXS using elemental Nb and Co as standards. The unit cell parameters at room temperature were determined from X-ray powder diffraction data employing a Huber G670 Guinier-camera (Ge monochromator, Co $K\alpha_1$, Si as internal standard). Single crystals were selected from crushed specimens. The single crystal data collection was carried out with a Rigaku AFC7 diffractometer equipped with a Mercury CCD detector with Mo $K\alpha$ radiation. Single crystal structure analysis was done using the program SHELXL97 [7]. The electronic structures of the compounds have been calculated using the scalar-relativistic LMTO-ASA method within the local density approximation to density functional theory. The program used was TB-LMTO-ASA 4.7 [8]. The basis set consisted of ($5s$, $5p$, $4d$) for Nb and ($4s$, $4p$, $3d$) for Co. The ($4f$) orbitals of Nb have been downfolded. The von Barth-Hedin [9] parameterization of the exchange-correlation functional was used. The radii of atomic spheres have been determined automatically by the program. All Brillouin zone integrations have been carried out by the tetrahedron method. Crystal Orbital Hamilton Population curves COHP(E) and their integrals ICOHP(E_F) have been calculated with a supplemental module of the program package [10]. Electronic structure and phase stability of the C36 and the C14 structure types have been studied by the all-electron, full-potential local orbital (FPLO) [11] minimum basis method within the local density approximation to density functional theory. The basis functions in the FPLO method are nonorthogonal and they are obtained numerically by solving a Schrödinger equation which includes a confining potential for states treated as valence. The confining potential compresses the valence functions in order to make them local. In our calculations the basis orbital set for Nb consisted of $4s$, $4p$, $5s$, $5p$, $4d$, $4f$ states while for Co the states $3s$, $3p$, $4s$, $4p$, $3d$ were used. The semicore states $4s$, $4p$ of Nb and $3s$, $3p$ of Co together with the polarization state $4f$ of Nb were calculated with fixed compression. For the remaining states the compression parameters were optimized during the self-consistency cycles. The Perdew-Wang parameterization of the exchange-correlation functional was used. The Brillouin zone integrations were carried out by the tetrahedron method.

DISCUSSION

C36 $Nb_{1-x}Co_{2+x}$

A Laves phase with C36 structure type forms close to the composition $NbCo_3$ with a narrow homogeneity range from 24.5 to 25.5 at.% Nb [6]. The width of the two-phase field C36 + C15 is smaller than 0.5 at.%. A sample containing a mixture of C15 and C36 phases was investigated by transmission electron microscopy [12]. Selected area electron diffraction (SAED) patterns of the C36 phase reveal diffuse streaks along the c^*-direction. This suggests the existence of stacking faults perpendicular to the c-direction – at least for Nb-rich C36. Stacking faults are clearly visible in the corresponding HREM images. The origin of the off-stoichiometry has been investigated by single crystal X-ray analysis. The off-stoichiometry is caused by a random substitution of about 25% of the Nb atoms by additional Co atoms. Neither vacancies at Nb sites nor Co atoms at interstitial sites have been detected.

The crystal structure exhibits pronounced deviations [13] from an idealized crystal structure with structural parameters taken from the hard sphere model for the Co and the Nb network. A part of the Co network composed of Co_4 tetrahedra is shown in Fig. 1a and a three-layer stack of hexagonal sequence Co3–Co1–Co3 in Fig. 1b. The Kagomé layers of the hexagonal slab display an elongation of the edges B_{11}^c of the basal triangles of the $Co1_3Co3_2$ trigonal bipyramids and a contraction of the edges B_{11}^u of the uncapped triangles. The distortion pattern of the hexagonal slabs is illustrated in Figures 1c and 1d. This type of distortion with $B_{11}^u < B_{13} < B_{11}^c$ is in agreement with the one found for the majority of C14 and C36 Laves phases whose atomic positions are precisely known.

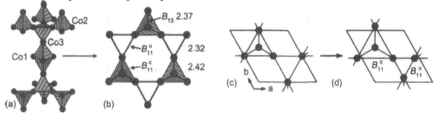

Figure 1. The crystal structure of the C36 phase. (a) Part of the Co network and (b) hexagonal arrangement of three sequential Co layers. (c) and (d) Schematic illustration of the distortion pattern of the Co1 Kagomé layers. Distances are given in Å.

The excess Co atoms occupy preferentially one of two crystallographic Nb sites. Twice as much Co substitutes the Nb2 site as compared to the Nb1 site. The Nb network of C36 is a four-connected net of 4H-SiC topology and shows pronounced deviations from regularity. It may be described as a sequence of double layers equivalent to {Nb2 Nb2}–{Nb1 Nb1}–{Nb2 Nb2}–{Nb1 Nb1} per unit cell, where labels Nb1, Nb2 denote the two different crystallographic Nb sites in the idealized crystal structure. Upon distortion, three different interatomic distances d(M1a–M1a) = 2.935 Å , d(M2a–M2a) = 2.865 Å, and d(M1a–M2a) = 2.902 Å with M1a and M2a labeling Nb1 and Nb2 shifted from the positions of the idealized crystal structure are generated. The distance d(M1a–M1a) is elongated, d(M2a–M2a) shorter and d(M1a–M2a) equal compared to d(Nb–Nb) = 2.902 Å of the model without distortion. A schematic illustration of the distortion of the Nb network is shown in Figure 2a.

The random substitution of approx. 25% of all Nb atoms by the smaller Co atoms triggers a local displacement of the Co atoms on Nb sites. These positions are labelled here M1b and M2b. The split atom positions (M1a, M1b) and (M2a, M2b) correspond to distinct maxima in the difference electron density contour maps as shown in Figures 2c,d. The electron density obtained from X-ray single crystal structure refinements at M1 = (M1a, M1b) is a trigonal pyramid with Nb@M1a at the apical position and a dumbbell at M2 = (M2a, M2b). A network assuming 100% substitution of Nb by Co (Co@Nb) is shown in Figure 2b to display possible distances in the defect structure. Random substitution, preferred occupation of the Nb2 site by Co, and displacements of Co relative to Nb atoms make the real structure exceedingly complicated. The main factor for the displacement of the Co atoms at the Nb network is the formation of Co–Nb pairs with interatomic distances comparable to those between Co and Nb atoms of the respective networks.

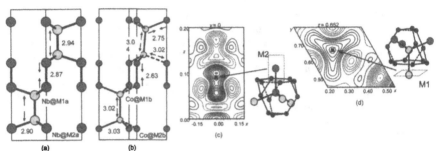

Figure 2. The crystal structure of the C36 phase. (a) Part of the Nb network. Co atoms substitute randomly 16% of the Nb1 and 32% of the Nb2 sites. Artifical Nb network with all Nb atoms substituted by Co atoms. (c) Difference electron density maps around M2 (c) and M1 (d) sites. Distances are given in Å.

C15 NbCo$_2$

The C15 polytype crystallizes at 1100 °C in the range from 26.2 to 34.5 at. % Nb and melts congruently at 1484 °C [6]. Single-phase material has been obtained from 28 to 34 at.% in steps of 1 at.%. The unit cell parameter *a* follows Vegard's rule on the Co-rich side. Here, the origin of the homogeneity range is due to excess Co randomly occupying Nb positions, a fact shown by single crystal structure investigations in combination with chemical analyses. No conclusive information for the Nb-rich side is so far available. Specimens of the C15 phase obtained from the C36 + C15 and from the C15 + C14 two phase fields have been investigated by means of SAED, HREM and convergent beam electron diffraction (CBED) techniques [8]. The findings confirm the space group *Fd-3m* and no superstructure reflections have been observed. Stacking faults for the Co-rich C15 phase were rarely observed while Nb-rich C15 specimens often exhibited stacking faults.

C14 Nb$_{1+x}$Co$_{2-x}$

The homogeneity range of C14 Nb$_{1+x}$Co$_{2-x}$ extends from 35.8 to 37.4 at. % Nb at 1250 °C. The off-stoichiometry is caused by a random substitution of 14% Co by Nb atoms at one of the two crystallographically independent Co sites (Co2 in Figure 3a) according to single crystal structure refinement for C14 Nb$_{1.07}$Co$_{1.93}$. Vacancies at Co sites, Nb atoms at interstitial sites, or an ordered superstructure have not been detected.

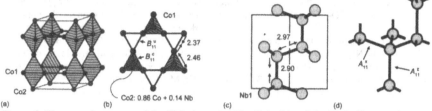

Figure 3. The crystal structure of the C14 phase. (a) The Co(a, b) and the Nb (c,d) network. Distances are given in Å.

The Co and the Nb networks are both distorted with respect to the idealized crystal structure of C14 $NbCo_2$ based on the hard sphere model. The Co network of the C14 phase reveals a similar distortion pattern as that of the C36 phase. The distortion is again characterized by a contraction of the edges of uncapped triangles B_{11}^u and an expansion of the edges of the basal triangles B_{11}^c of the trigonal bipyramids $Co1_3Co2_2$, as shown in Figure 3b. The Nb network of C14 $Nb_{1+x}Co_{2-x}$ is a four-connected net of 2H-ZnS topology and shows pronounced deviations from regularity. It may be described as a sequence of double layers equivalent to {Nb1 Nb1}–{Nb1 Nb1} per unit cell as shown in Fig. 3c. Upon distortion, two different interatomic distances $d(Nb1–Nb1) = 2.972$ Å and $d(Nb1–Nb1) = 2.902$ Å arise, one longer, the other one shorter than the shortest Nb–Nb distance of 2.954 Å of the undistorted model. In order to study the distortion of the Co and the Nb network, the preferential substitution of different crystallographic Nb or Co sites electronic structure calculations have been performed by TB-LMTO-ASA and full potential methods (FPLO). The idealized crystal structures of C14 at stoichiometric composition $NbCo_2$ obtained from the hard sphere model should be unstable with respect to a distortion of the Co and the Nb network according to the full potential total energy calculations. The distortion pattern obtained by the calculations is similar to the observed distortion pattern for off-stoichiometric composition. In case of the C14 phase, the substitution of Co by Nb atoms at Co2 (triangular layers) is energetically preferred compared to a substitution of Co at Co1 by Nb. So far the origin of the differences in the total energy in terms of a local bonding picture is unknown.

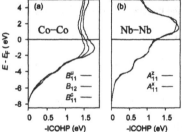

Figure 4. Integrated crystal orbital Hamilton population curves $(–ICOHP(E_F))$ for the different types of bonds within C14 $NbCo_2$ with idealized structural parameters obtained from the hard sphere model. (a) Co-Co from left to right at 0 eV: B_{11}^c, B_{12}, B_{11}^u; (b) Nb-Nb are almost identical.

491

Assuming that the integrated crystal orbital Hamilton population $(-ICOHP(E_F))$ values obtained from LMTO-ASA calculations can be used as a measure of bond strength, the non-uniform bond strength distribution for the Co network in all models may be interpreted as a tendency towards distortion. As an example, a plot of the $-ICOHP(E)$ for Co–Co and Nb–Nb for idealized C14 $NbCo_2$ is shown in Fig. 4. The $-ICOHP(E_F)$ for Co–Co bonds are clearly different although all interatomic distances $d(Co–Co)$ for nearest neighbors are equal in this model. An analysis of various Laves phases $NbTM_2$ with TM = Cr, Mn, Fe, Co shows that the deviation from the idealized crystal structure is mainly governed by the valence electron concentration.

CONCLUSIONS

Substitutional disorder is the origin of the homogeneity ranges of the three different Laves phases in the system Co-Nb. In case of the hexagonal Laves phases preferential site occupation plays an important role. Twice as much Co substitutes one Nb site as compared to the other Nb site in the C36 structure whereas Nb substitutes only one of the two independent Co sites in the C14 structure. In the Co-rich C36 phase the smaller Co atoms at the Nb sites are displaced approx. 20 pm, thus enabling the formation of new Nb-Co bonds with distances similar to that between the two sublattices.

ACKNOWLEDGMENTS

This work has been performed within the Inter-Institutional Research Initiative "The Nature of Laves Phases" funded by the Max Planck Society.

REFERENCES

1. J.B. Friauf, *J. Am. Chem. Soc.* **49,** 3107 (1927); *Phys. Rev.* **29,** 34 (1927).
2. F. Laves and H. Witte, *Metallwirtschaft* **14,** 645 (1935); **15** 840 (1936).
3. G.E.R. Schulze, *Z. Elektrochem.* **45,** 849 (1939).
4. F.C. Frank and J.S. Kasper, *Acta Crystallogr.* **11,** 184 (1958); **12,** 483 (1959).
5. F. Stein, M. Palm and G. Sauthoff, *Intermetallics* **12,** 713 (2004); **13,** 1056 (2005).
6. F. Stein, D. Jiang, M. Palm, G. Sauthoff, D. Grüner and G. Kreiner, *Intermetallics* **16,** 785 (2008); see also paper No. 1128-U05-30 in this volume.
7. G.M. Sheldrick, SHELXL97-2, Program for the Solution and Refinement of Crystal Structures, University of Göttingen, 1997.
8. G. Krier, O. Jepsen, A. Burkhard and O. K. Andersen, Tight Binding LMTO-ASA Program, Version 4.7, Stuttgart, Germany 1998.
9. U. von Barth, L. Hedin, *J. Phys. C* **5,** 1629 (1972).
10. F. Boucher, O. Jepsen, O. K. Andersen, Supplement to the LMTO-ASA-Program Version 4.7, Stuttgart, Germany.
11. K. Koepernik and H. Eschrig, *Phy. Rev.* **B59,** 1743 (1999).
12. T. Yokosawa, K. Söderberg, M. Boström, D. Grüner, G. Kreiner, and O. Terasaki, *Z. Kristallogr.* **221,** 357 (2006).
13. D. Grüner, F. Stein, M. Palm, J. Konrad, A. Ormeci, W. Schnelle, Yu. Grin, and G. Kreiner, *Z. Kristallogr.* **221,** 319 (2006).

Mater. Res. Soc. Symp. Proc. Vol. 1128 © 2009 Materials Research Society   1128-U08-09

An In-Situ Electron Microscopy Study of Microstructural Evolution in a Co- NbCo₂ Binary Alloy

Sharvan Kumar[1], Padam Jain[1], Seong Woong Kim[1], Frank Stein[2] and Martin Palm[2]

1. Division of Engineering, Brown University, 182 Hope Street, Providence 02912, RI, USA.
2. Max-Planck Institut für Eisenforschung GmbH, Max-Planck-Str. 1, D-40237 Düsseldorf, Germany.

ABSTRACT

The microstructure in a Co-rich, Co-15 at.% Nb alloy was characterized in the as-cast condition. A predominantly lamellar eutectic morphology composed of a Co-Nb solid solution and the C15 Laves phase NbCo₂ was confirmed by transmission electron microscopy. The C15 phase was heavily twinned, with only one variant of twins being present in the individual lamella, while the Co solid solution had the face centered cubic structure. In-situ heating to 600°C in the microscope confirmed the decomposition of the metastable Laves phase into a fine equiaxed, ~10-20 nm grain size microstructure, and the product phase is the monoclinic Nb₂Co₇. The individual grains appear faulted. The matrix solid solution retained the fcc structure and no change in structure was observed on cooling to room temperature. Heating to temperatures as high as 1130°C leads to rapid grain growth in the Nb₂Co₇ phase, and the nucleation and growth of a few new grains within the original grains; however, the reverse peritectoid transformation previously reported, was not observed.

INTRODUCTION

Laves phases of refractory metals have been of interest for high temperature applications [1,2] and the Co-Nb binary system has received recent attention [3-5]. The Co-Nb binary phase diagram has been re-assessed and a revised phase diagram was recently published [3] and is reproduced in Figure 1. Noteworthy and relevant to this paper are i) the presence of a C15 Laves phase (NbCo₂) over a range of compositions and congruently melting at 1484°C, (ii) a peritectic reaction between this Laves phase and a more Co-rich liquid to form a highly defective Co-rich C36 Laves phase at 1264°C (see [4] for crystallographic details of this defect C36 structure), (iii) a eutectic decomposition of liquid of composition Co-13.9 at.% Nb into fcc Co solid solution and the C36 phase at 1239°C, and (iv) a peritectoid reaction at 1086°C between the C36 phase and Co solid solution to produce the monoclinic phase Nb₂Co₇. An alloy such as Co-15 at.% Nb (shown by the dashed line in Figure 1) should, under equilibrium freezing conditions, then form a small amount of the C15 phase initially, that at 1264°C should react with the liquid peritectically to form the C36 phase, while the rest of the liquid at 1239°C should decompose eutectically to form the Co solid solution phase and the C36 phase. Subsequently, at room temperature, the microstructure should be composed of the two phases, Co solid solution and Nb₂Co₇. Since peritectic and even more so, peritectoid reactions, are notoriously sluggish, it is unlikely that such a microstructure would result under practical solidification conditions. In such a case, the decomposition of the metastable phase(s) upon subsequent heating may provide an understanding of some of the solid state transformations as for example, decomposition of a metastable C15 or C36 phase to the monoclinic Nb$_2$Co$_7$ phase, and with further heating, its dissolution giving way to the formation of the C36 phase in the alloy Co-15at.% Nb . Such transformations can be followed indirectly by using differential thermal analysis in conjunction with x-ray diffraction and metallography, or alternately, in real time using in-situ heating in a transmission electron microscope (TEM). Additionally, a rather unusual mechanical response of the monoclinic intermetallic phase Nb$_2$Co$_7$ was recently reported [5] in that a rectangular block of this single phase compound in the polycrystalline form could be repeatedly hammered down to almost half its original height at room temperature without shattering it into pieces. In this short note, we restrict ourselves to a description of the microstructure of an as-cast Co-15 at.% Nb alloy and its evolution upon heating a specimen in-situ in the TEM. In a companion paper in these proceedings [6], we discuss the deformation response of the monoclinic phase Nb$_2$Co$_7$.

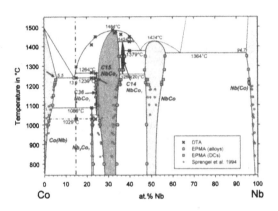

Figure 1: The revised Co-Nb binary phase diagram from [3].

EXPERIMENTAL DETAILS

A Co–Nb alloy with nominal Nb content of 15.0 at.% was prepared from pure Co (99.95 wt.%) and Nb (99.9 wt.%) by crucible-free levitation melting under argon and casting into a cold copper mold. Inductively coupled plasma optical emission spectroscopy analyses showed that the alloy composition was in good agreement with the nominal composition. Thin slices, 3 mm in diameter were cut from the ingot in the as-cast condition using electro-discharge machining techniques. These slices were thinned to less than 100 μm by conventional mechanical grinding and final thinning and perforation of the foils was carried out by twin-jet polishing in a 10 vol. % nitric acid-methanol solution at a temperature of -25°C and using 15 V. To enhance the electron-transparent area, when needed, the specimen was ion-polished for 15 min in a precision ion polishing system (PIPS) at 4 KeV and a low angle (~2°). The microstructure was examined using Philips 420 and JEOL 2010 transmission electron microscopes; the JEOL 2010 has a heating stage capable of reaching 1150°C, facilitating in-situ heating experiments. A video-imaging system was used to document the progression of microstructural events during heating. Periodically, still images of the microstructure were also obtained.

RESULTS

A bright field image of the lamellar eutectic microstructure composed of alternating lamellae of Co solid solution and the Laves phase $NbCo_2$ is shown in Fig 2a. The average lamellae width is of the order of 0.2 μm and fairly uniform. Examination of the Co solid solution at a higher magnification shows the presence of at least two variants of stacking faults (Figure 2b) with the faults terminating at the Co- $NbCo_2$ interfaces. A selected area diffraction (SAD) pattern obtained from the Co solid solution region confirms an fcc structure. The Laves phase $NbCo_2$ is also highly faulted although only a single variant is evident (Figure 2c). SAD suggests that the structure is highly twinned C15. Since the compound supposedly forms from the melt, it is unclear why only a single variant of the twin is favored.

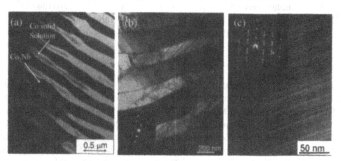

Figure 2: TEM images (a) showing a lamellar eutectic microstructure of $NbCo_2$ and Co solid solution, (b) faulting in both phases and an SAD pattern of the fcc Co solid solution, and (c) a twinned C15 Laves phase structure confirmed by the SAD pattern shown as an inset.

In-situ heating the specimen in the TEM up to 600°C did not transform the matrix from fcc to hcp (for pure Co, the low temperature equilibrium phase is hcp and at 424°C, it transforms to the fcc structure) and images taken at 240°C and 350°C continue to show the two variants of stacking fault found at room temperature in the as-cast condition (Figures 3a,b). Cooling to room temperature after this 600°C excursion confirms the fcc structure once again and multiple variants of stacking faults are again evident (Figure 3c). Thus, the hcp structure was never evidenced in this study. Lastly, some precipitation is noted in the matrix and is likely Nb_2Co_7 as suggested by the phase diagram.

Simultaneously, changes in microstructure were also noted in the C15 Laves phase lamellae during this 600°C heating experiment and these are illustrated in Figure 4a-c. At around 420°C, fine equiaxed grains begin to nucleate in the C15 lamellae, often at the partial dislocations bounding the faults (Figure 4a). With increase in temperature this process evolves, and at 515°C (Figure 4b), the heavily faulted/twinned C15 structure is completely replaced by equiaxed, nanocrystalline grains, many of which show fringe contrast within them (an example is shown by the white arrow in Figure 4b). When the temperature is further increased to 600°C and the specimen is held at temperature for 1.5h, grain growth occurs and the average grain size is of the order of 20 nm; the faults within these grains are more clearly seen and SAD provides a complex ring pattern confirming the polycrystalline nature. Thus a single crystal C15 lamella has transformed into a nanocrystalline region. The diffraction ring pattern could be fitted to the monoclinic Nb_2Co_7 phase, although two additional rings were present that could not be indexed. The specimen that had been cooled to room temperature was reheated in steps to 1130°C in the TEM. Up until 800°C, grain growth was not noticeable. However, at 900°C, grain growth was rapid (Figure 5a). With further increase in temperature, and approaching the equilibrium peritectoid transformation temperature of 1086°C, fault contrast became evident within these coarse grains. At 1100°C, these packets are still evident (Figure 5b), but in addition, as indicated by the arrows in Figure 5b, a few small new grains, 5-7 nm in size appear to have nucleated within the large grains. Specimen drift at these high temperatures precluded any diffraction analysis at temperature. However, after reaching the maximum temperature of 1130°C (maximum safely permitted by the heating stage) and holding for a few minutes, the specimen was quickly cooled (~5 minutes) to room temperature and then analyzed. Large faulted grains were observed (Figure 5c) and SAD patterns obtained from such grains confirmed the monoclinic Nb_2Co_7 to persist, implying that the reverse peritectoid reaction had not occurred.

DISCUSSION

Although the binary phase diagram requires the C36 Laves phase to be in equilibrium with the Co solid solution under equilibrium conditions, it is probable that the first peritectic reaction is suppressed during cooling in a Cu mold; further, the metastable extension of the C15 solvus may intersect the eutectic invariant at a more Co-rich composition than that for the C36. Under these conditions, it will be possible to produce a lamellar structure of Co solid solution and and metastable C15, as observed experimentally. The presence of the fcc structure for the Co solid solution at room temperature and its persistence upon heating and cooling in the microscope is in agreement with the findings of Köster and Mulfinger [7] that Nb strongly stabilizes the fcc structure relative to the hcp structure. The exact mechanism is not well understood but if it is supposed that the fcc-to-hcp transformation proceeds by the motion of 1/6[112] Shockley partials, the presence of Nb in solid solution could discourage (or preclude) the motion of these partials, and substantially delay transformation kinetics (or suppress it).

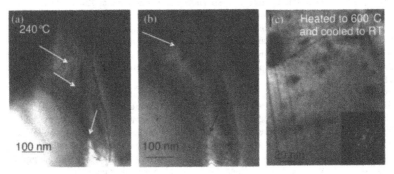

Figure 3: TEM images obtained during in-situ heating showing two variants of stacking faults in the Co solid solution lamella persisting at (a) 240°C, (b) 350°C, and (c) at room temperature following a 1.5h hold at 600°C.

496

The metastable C15 lamellae that are present in the as-cast condition is expected to have a composition that is Co-rich as a consequence of the metastable extension of the C15 solvus (measured compositions in the TEM, using energy dispersive X-ray analysis, range from 22.7 at.% Nb to 26.3 at.% Nb). In those locations where the composition is similar to that for Nb_2Co_7 or close to it, the transformation would be primarily structural without the need for long range diffusion, and therefore on heating, the C15 begins to decompose readily to form the Nb_2Co_7 phase at such low temperatures as 500-600°C. Heating to higher temperatures leads to extremely rapid grain growth at 900°C and above. A reverse peritectoid reaction is expected on heating to temperatures above 1086°C and should involve the formation of the C36 phase and Co solid solution, the latter being in minor amount. Perhaps the fine grains observed in Figure 5b correspond to one of the decomposition phases (Co solid solution phase or C36 $NbCo_2$) but we have not been able to confirm this yet. However, after cooling we confirm the presence of the monoclinic Nb_2Co_7 phase; this implies that even though we were well above the peritectoid temperature of 1086°C, the decomposition process did not proceed to any recognizable extent, quite likely limited by the need for long range diffusion of Co and Nb to produce phase separation. Indeed, in a previous study [3], it has been claimed that although the decomposition of the Nb_2Co_7 phase on heating can be recognized readily using differential thermal analysis, the onset temperature is strongly dependant on heating rate.

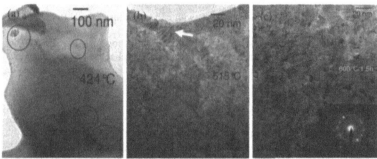

Figure 4: TEM images obtained during in-situ heating showing microstructural changes in the $NbCo_2$ lamella: (a) at 424°C, fine grains nucleate in the heavily faulted C15 phase, (b) at 515°C, the faulted structure is replaced by a nanocrystalline structure and (c) limited grain growth after 1.5h at 600°C and the diffraction ring pattern confirms the polycrystalline structure.

CONCLUSIONS

The microstructure of a binary Co-15%Nb alloy in the as-cast state consists of the fcc Co solid solution and heavily twinned metastasble C15 $NbCo_2$ in a fine lamellar morphology. The C15 phase decomposes on heating at temperatures as low as 500°C to form nanocrystalline Nb_2Co_7. Heating beyond 900°C produces extensive grain growth; even after heating to 1130°C, the Nb_2Co_7 did not decompose into Co solid solution and the C36 Laves phase, likely a consequence of the sluggish kinetics associated with the long range diffusion required to produce the product phases.

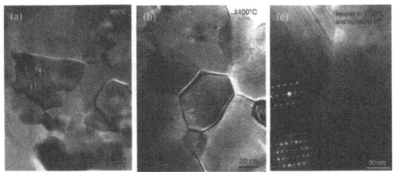

Figure 5: TEM images obtained during in-situ reheating showing microstructural changes in the NbCo$_2$ lamella: (a) at 900°C, significant grain growth has occurred, (b) fault contrast within the grains and fine new grains within the original grains at 1100°C, and (c) large faulted grains at room temperature after a 1130°C excursion.

ACKNOWLEDGMENTS

This collaborative work was performed within the Inter-Institutional Research Initiative "The Nature of Laves Phases" funded by the Max Planck Society. The TEM work was performed at Brown University and supported by the MRSEC Program (Micro- and Nano Mechanics of Materials; Brown University) of the National Science Foundation under award DMR-0520651. We acknowledge A. McCormick for his assistance with the in-situ heating experiments.

REFERENCES

1. J. D. Livingston, Phys. Status Solidi A 131, 415 (1992).
2. B.P. Bewlay, J.A. Sutliff, M.R. Jackson and H.A. Lipsitt, Acta Metall. Mater., 42, 2869 (1994).
3. F. Stein, D. Jiang, M. Palm, G. Sauthoff, D. Grüner and G. Kreiner, Intermetallics, 16, 785 (2008).
4. D. Grüner, F. Stein, M. Palm, J. Konrad, A. Ormeci, W. Schnelle, Y. Grin and G. Kreiner, Z Kristallogr, 221, 319 (2006).
5. L. Siggelkow, U. Burkhardt, G. Kreiner, M. Palm and F. Stein, Mater. Sci. Eng., A497, 174 (2008).
6. F. Stein, M.Palm, G. Frommeyer, P. Jain, K.S. Kumar, L. Siggelkow, D. Grüner, G. Kreiner, and A. Leineweber, in these Proceedings.
7. W. Köster and W. Mulfinger, Zeit. Metallkd., 30, 348 (1938).

Mater. Res. Soc. Symp. Proc. Vol. 1128 © 2009 Materials Research Society

Phase equilibria in the ternary Nb-Cr-Al system and site occupation in the hexagonal C14 Laves phase Nb(Al$_x$Cr$_{1-x}$)$_2$

Oleg Prymak[1], Frank Stein[1], Alexander Kerkau[2], Alim Ormeci[2], Guido Kreiner[2], Georg Frommeyer[1], Dierk Raabe[1]

[1]Max Planck Institut für Eisenforschung, Max-Planck-Str. 1, Duesseldorf, D-40237, Germany.
[2]Max-Planck-Institut für Chemische Physik fester Stoffe, Noethnitzer Str. 40, D-01187, Dresden, Germany.

ABSTRACT

The ternary Nb-Cr-Al phase diagram exhibits extended phase fields of the cubic C15 and the hexagonal C14 Laves phases Nb(Al$_x$Cr$_{1-x}$)$_2$. A number of Nb-Cr-Al alloys were prepared by levitation melting and annealed at temperatures between 1150 and 1450 °C for up to 1500 h. Isothermal sections of the ternary Nb-Cr-Al phase diagram at 1150, 1300 and 1450 °C were obtained from electron probe microanalysis, X-ray powder diffraction and metallographic investigations in order to study the effect of Al on the stability and structure of the Laves phases. The C14 Laves phase in the Nb-Cr-Al system can dissolve up to 45 at.% Al by substituting Cr with Al on the two different crystallographic B-sites 2a and 6h of the C14 AB$_2$ unit cell. The site occupations of the Al and Cr atoms on these two B-sites were determined by Rietveld analysis using the program FullProf. The experimental site occupation factors were compared to site occupation factors computed by a statistical mechanics approach based on first-principles electronic structure calculations. The experimental as well as the calculated site occupation factors indicate a preferred occupation of the 2a site by Al.

INTRODUCTION

Transition-metal Laves phases are promising candidates for novel structural materials used at very high temperatures. A prerequisite for the development of such materials is the knowledge and understanding of the structure and stability of these phases. Laves phases have the ideal composition AB_2 and form the largest group of intermetallic compounds. They crystallize in three crystallographically closely related structure types: cubic C15 (MgCu$_2$-type), hexagonal C14 (MgZn$_2$-type) and hexagonal C36 (MgNi$_2$-type) [1,2]. In Fig. 1 the common features of the C14 and C15 crystal structures can be seen. In all Laves phase structure types, the same fundamental layer, which is built up by four atomic planes, occurs [3]. An important difference between C14 and C15 Laves phases is in the number of available atomic positions. In the C14 Laves phase there are three atom positions - 2a and 6h for the smaller B atoms, and 4f for the bigger A atoms, whereas in the C15 phase there are only two positions - 16d for B atoms and 8a for A atoms (see Fig. 1).

In the binary Nb-Cr system, two types of Laves phase occur: at low temperatures the C15 structure type is observed and at high temperature it transforms to the C14 structure type [4, 5]. Alloys based on NbCr$_2$ have already been studied with respect to their suitability as high-temperature, structural materials because of their high melting temperature (> 1700 °C), good oxidation resistance and relatively low density (7.7 g/cm^3). Additionally, the NbCr$_2$ Laves phase exhibits a wide homogeneity range (up to 7 at.% Nb) and good creep and tensile strength [6, 7]. The major drawback for structural application of the Laves phase is its brittleness at low temperatures [8].

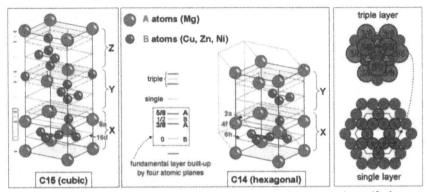

Figure 1. Crystallographically closely related structure types of the Laves phases (for better comparability, the C15 structure is shown in a hexagonal setting with the vertical axis corresponding to the cubic (111) direction)

The present paper focuses on the question of the site occupation in the ternary C14-$Nb(Al_xCr_{1-x})_2$ Laves phases as a function of composition. Lattice parameters and site occupation factors (sof), which give the relative amount of Cr and Al atoms on the two different B sublattice sites of the C14 structure, were determined using Rietveld refinements [9]. Besides the experimental work, the site occupation factors of the $2a$ and $6h$ atom sites of the C14 Laves phase were calculated employing a simple statistical mechanics approach based on first-principles electronic structure calculations. The all-electron, full-potential local orbital (FPLO) method [10] was used in this study.

EXPERIMENTAL DETAILS

Ternary Nb-Cr-Al alloys were prepared by levitation melting from high purity metals (Nb 99.99 wt.%, Cr 99.99 wt.%, Al 99.7 wt.%) and cast into a copper mould of 10 mm in diameter. For heat treatments, cylindrical pieces of 10 mm in length were cut by electro-discharge machining (EDM). Polished samples were encapsulated in quartz ampoules in Ar-atmosphere. The capsules with the samples were heat-treated at 1150 °C for 100 h (some of the samples for 1500 h) and quenched in a 10 % NaCl solution. For heat treatments at 1300 °C (100 h) and 1450 °C (50 h), the samples were wrapped in Nb foil and placed into an alumina crucible with Ti-filings as oxygen getter. The heat treatments were performed in an Ar atmosphere and finally the samples were rapidly cooled in a jet of Ar gas.

The phase equilibria of the ternary system were established by determining the phase compositions in the annealed samples by electron probe microanalysis (EPMA). X-ray diffraction (XRD) from powder produced in a steel mortar was performed using CuK_α radiation (λ = 0.154056 nm) in a 2Θ range from 10° to 110° with a step size of 0.02°. The lattice parameters and site occupation factors were determined by performing Rietveld analysis using the FullProf program in the Windows version WinPLOTR 2006. For these calculations the Pseudo-Voigt profile function as a linear combination of Lorentzian and Gaussian components was used. The final Bragg R-factors were in the range 4 to 8 with the higher numbers resulting for the alloys with the higher amounts of second phases due to some overlapping peaks.

The calculations of the site occupation factors by a statistical mechanics approach were performed under the assumption that the Nb atoms only occupy the $4f$ sites. Furthermore, ideal atomic coordinates and c/a ratio have been used. The N Al atoms per formula unit $Nb_4Al_NCr_{8-N}$ can be distributed in different ways over the $2a$ and $6h$ positions in the unit cell. For each value of N we computed the total energy for all symmetrically distinct configurations. The thermal averages of the occupation factors at $2a$ and $6h$ can then be determined by forming a canonical ensemble.

DISCUSSION

Isothermal sections of the ternary Nb-Cr-Al phase diagram were established for temperatures of 1150, 1300 and 1450 °C from EPMA and XRD data; details are discussed in a forthcoming publication [11]. Figure 2 shows the central part of the isothermal sections containing the Laves phases. It was found, that both the cubic C15 and the hexagonal C14 type exist as stable phases in the ternary system at all investigated temperatures. Both phases have extended homogeneity ranges along a nearly constant Nb content. The width of the homogeneity range of the Laves phase with varying Nb content increases with temperature. With increasing Al content the C14 Laves phase is stabilized and can dissolve up to 45 at.% Al at all investigated temperatures. The solubility of Al in the C15 Laves phase decreases with increasing temperature.

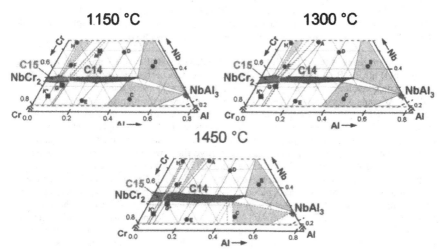

Figure 2. Central parts of the isothermal sections of the Nb-Cr-Al system at 1150, 1300, and 1450 °C (● - analyzed composition, ■* - nominal composition)

Since all investigated samples consist of at least two phases, the investigated ternary Laves phases possess compositions either on the Nb-rich boundary (alloys H, F, A, M, D, B) or on the Nb-poor boundary (alloys K, G, E, C) of the phase field. This means, there are two series of Laves phase samples with varying Cr/Al ratio and approximately constant Nb content. The volumes per atom and the c/a ratios of the Nb(Al$_x$Cr$_{1-x}$)$_2$ unit cells after heat treatment at 1150, 1300 and 1450 °C as derived from the XRD measurements are shown in Figure 3 for these two – Nb-rich and Nb-poor – alloy series. The substitution of Cr by Al results in a continuous increase of the volume per atom of the C15 and C14 Laves phase for a constant Nb content. The same effect can be seen for a constant Al concentration with increasing Nb content (see Fig. 3a). The change of the stable structure type from cubic C15 to hexagonal C14 obviously has no effect on the continuous increase of the volumes per atom.

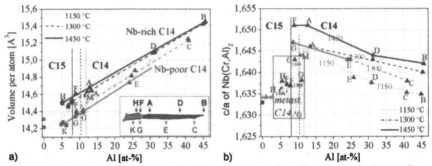

Figure 3. (a) Volumes per atom of the C15/C14 Laves phase (● - values for binary C15 NbCr$_2$ at 1100 °C, taken from [12]) and (b) c/a ratios in the C14 phase determined from XRD data as a function of the Al content and the heat treatment temperature (Δ - values for samples heat-treated at 1150 °C for 1500 h instead of 100 h; ■ - the ideal c/a ratio, see text)

The c/a ratios of the stable C14 Laves phase decrease at all temperatures with increasing Al content (Fig. 3b). Both lattice parameters c and a increase nearly linearly as function of Al content with about the same slope. Some of the alloys with C15 as the stable Laves phase additionally contain some metastable C14 Laves phase. During the preparation procedure these alloys have to cool from the liquid state to room temperature and, therefore, they have to pass through the C14 high-temperature phase field. As a result, the as-cast samples contain certain amounts of metastable C14 Laves phase which dissolves only slowly during the heat treatments. The metastability of the C14 Laves phase in respective alloys was proven by performing heat treatments for different times and comparing the XRD intensities which clearly decreased with increasing time for the C14 peaks compared to the C15 peaks. In Fig. 3b, one can see that the c/a ratios of these metastable C14 Laves phases clearly deviate from the continuous behavior of the equilibrium C14 phase. It seems that the c/a ratios of the metastable C14 approach the ideal value of a close packing of spheres in a C14 structure which is approximately 1.6330 (= √(8/3)); see Fig. 3b. Moreover, this c/a ratio corresponds to the theoretical value of a C15 structure in hexagonal setting (as shown in Fig. 1). Therefore, this behavior can be understood as an indication that the metastable C14 Laves phase is on the way to transform to the C15 equilibrium state.

The central sections of the ternary phase diagram (Fig. 2) show that the C14 Laves phase extends along a nearly constant Nb content with large amounts of Al substituting for Cr. Therefore, the assumption has been made for the performed Rietveld calculations that only Cr and Al on the crystallographic B-sites $2a$ and $6h$ of the C14 unit cell AB_2 substitute each other and Nb atoms solely occupy the $4f$ sites. In addition, we have assumed that the concentration of vacancies is negligible. The wrong Nb content introduces of course a systematic error. However, for a first approximation of the site preferences of Cr and Al, the treatment should be sufficient since the Nb content is constant and the deviation from 33.3 % is small.

Figure 4 shows the site occupation factors for Al as refined from the XRD data by the Rietveld method (Fig. 4a) and as calculated by the statistical mechanics approach (Fig. 4b). Site occupation factors for the metastable C14 Laves phase could not be determined due to the small amount of this phase in comparison to that of the C15 Laves phase and the overlap of the XRD lines of the two Laves phases. Both diagrams clearly show that Al atoms preferably occupy the $2a$ site rather than the $6h$ site in the C14 Laves phase Nb(Al$_x$Cr$_{1-x}$)$_2$ for all investigated temperatures. Both positions are filled up continuously with increasing Al content notwithstanding the preferred $2a$ site occupation. This tendency is observed for both Nb-rich as well as Nb-poor C14 Nb(Al$_x$Cr$_{1-x}$)$_2$ Laves phases. For x ≥ 0.6 the two diagrams do not agree. The reason for that is not yet clear. As the inset in Fig. 4b shows, the preferred occupation of the $2a$ sites means that Al tends to replace Cr on the top (or bottom) of the B atom trigonal bipyramid which form the network of B atoms of the Laves phase structure.

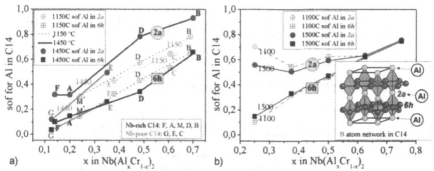

Figure 4. Site occupation factor (sof) for Al on the $2a$ and $6h$ positions of the C14 Laves phase Nb(Al$_x$Cr$_{1-x}$)$_2$ at different temperatures: (a) calculated by means of the Rietveld method (Nb-rich C14 - for the samples F, A, M, D, B and Nb-poor C14 - for the samples G, E, C) and (b) by a statistical mechanics approach

CONCLUSIONS

The ternary Nb-Cr-Al phase diagram was established using EPMA and powder XRD of equilibrated ternary alloys heat-treated at 1150, 1300 and 1450 °C for different times. The results show that Al stabilizes the hexagonal C14 Laves phase, which can dissolve up to 45 at.% Al. The solubility of Al in the C15 Laves phase decreases with increasing temperature. The volume per atom of the C15 and C14 Laves phases increases with increasing Al content at constant Nb content as well as with increasing Nb content at constant Al content. This increase is not affected by the change of the stable structure type from C15 to C14. The site occupation factors of Al and Cr on the $2a$ and $6h$ sites of the C14 unit cell were determined by Rietveld analysis using the program FullProf and compared to site occupation factors computed by a statistical mechanics approach based on first-principles electronic structure calculations. For both methods, preferential site occupation of Al on the $2a$ site in the C14 Laves phase $Nb(Al_xCr_{1-x})_2$ was found at all investigated temperatures.

ACKNOWLEDGMENTS

This work was supported by the Max Planck Society within the framework of the inter-institutional research initiative "The Nature of Laves Phases". The authors are grateful to Mrs. Bente Lebech from Materials Research Department, Technical University of Denmark, for helpful discussions.

REFERENCES

1. J. B. Friauf, *J. Am. Chem. Soc.* **49**, 3107-3114 (1927).
2. F. Laves and H. Witte, *Materialwissenschaft* **14**, 645-649 (1935).
3. K. S. Kumar and P. M. Hazzledine, *Intermetallics* **12**, 763-770 (2004).
4. D. J. Thoma and J. H. Perepezko, *Mater. Sci. Eng.* **A156**, 97-108 (1992).
5. T. B. Massalski (ed.), "Binary Alloy Phase Diagrams" 2nd edition, CD ROM version, *ASM International*, Materials Park, Ohio (1996).
6. H. J. Goldschmidt and J. A. Brandt, *J. Less-Common Met.* **3**, 44-61 (1961).
7. M. Takeyama and C. T. Liu, *Mater. Sci. Eng.* A **132**, 61-66 (1991).
8. M. Yoshida and T. Takasugi, *Intermetallics* **10**(1), 85-93 (2002).
9. L. B. McCusker, R. B. von Dreele, D. E. Cox, D. Louer and P. Scardi, *J. Appl. Cryst.* **32**, 36-50 (1999).
10. K. Koepernik and H. Eschrig, *Phys. Rev.* **B 59**, 1743-1757 (1999)
11. O. Prymak and F. Stein, "Experimental determination of the ternary Nb-Cr-Al phase diagram", *in preparation for Intermetallics*.
12. D. Gruener, "Untersuchungen zur Natur der Laves-Phasen in Systemen der Uebergangsmetalle", *Doctoral thesis*, TU Dresden, Germany (2007), 1–276, http://nbn-resolving.de/urn:nbn:de:swb:14-1172078219643-48967.

Fundamental Aspects of Intermetallics—Phase Stability, Defects, Theory

Mater. Res. Soc. Symp. Proc. Vol. 1128 © 2009 Materials Research Society 1128-U09-01

Dynamics of Atomic Ordering in Bulk and Thin Film Intermetallic Alloys: A Complementary Approach to Atomic Migration

W. Pfeiler[1], W. Püschl[1], Ch. Issro[2], R. Kozubski[3], and V. Pierron-Bohnes[4]

[1] Dynamics of Condensed Systems, Faculty of Physics, University of Vienna, Strudlhofgasse 4, 1090 Vienna, Austria
[2] Department of Physics, Faculty of Science, Burapha University, Chonburi, 20131 Thailand
[2] Interdisciplinary Centre for Materials Modelling, M. Smoluchowski Institute of Physics, Jagellonian University, Reymonta 4, 30-059 Cracow, Poland
[3] IPCMS-GEMME, CNRS-ULP, 23 rue du Loess, BP 43, 67034 Strasbourg Cedex 2, France

ABSTRACT

One of the foremost challenges in today's materials science is the design and development of materials with physical properties customized for technical application. Due to their excellent corrosion resistance and their advantageous mechanical and in many cases also magnetic properties, intermetallic alloys are among the most important materials of the 21st century. Most of their outstanding qualities are linked to long-range order, the fact that unlike atoms are preferred as neighbours, which then segregate to different sublattices. In most intermetallics atomic order persists up to rather high temperatures, if not up to melting. However, connected with the entropy gain, the degree of order depends on temperature and thereby the stability of the designed beneficial materials properties is affected. By monitoring changes in the degree of atomic order an access to atom migration is gained, which is complementary to the usual diffusion experiments, where the degree of order is not changed on average. It is shown in this review on some selected examples how an adequate thermal treatment of the samples in combination with the experimental approach gives detailed information on atom jump mechanisms and structural changes, especially if experiment is combined with up-to-date kinetic Monte Carlo simulations.

INTRODUCTION

In ordered alloys, for example in intermetallics, the alloy atoms, due to a high attractive interaction between unlike atoms, are not distributed at random over the lattice positions; a preferential distribution of each kind of atom on specific sublattices of the crystal lattice is observed, leading to certain superstructures. A strict correlation between the different kinds of alloy atom and specific lattice positions for thermodynamic reasons is however maintained at 0 K only; it decreases with increasing temperature in a characteristic manner. Above the order-disorder transition temperature $T_{O/D}$, which in the cases of directly ordering alloys may be even higher than the melting temperature T_m, this ordering tendency stops and the alloy atoms are then distributed randomly over the possible sites of the crystal lattice. Actually, a slight local ordering tendency (short-range order (SRO)) in many cases remains [1,2] and the completely random atomic arrangement is a high-temperature limit.

It is obvious that this effect of long-range order (LRO) has essential consequences for the physical properties of alloys and thereby for their technical application and performance. Among the advantages of intermetallics are their high temperature strength, which is correlated with the yield stress anomaly, their excellent corrosion resistance and, in some cases, a preferential ferromagnetic orientation, giving hope for a future use in high density magnetic and magneto-optic recording.

Since the degree of LRO depends critically on temperature, the knowledge of details of the kinetic processes, which determine changes of LRO are of enormous importance. To study changes of LRO in bulk alloys we have applied with great success the method of (quasi) residual resistometry (REST) [3-7]. It turned out recently that this method which is very sensitive to small changes of the LRO-parameter is equally well suited for ordered intermetallic thin films and it can be hoped that this will hold also for other nanostructures [8-10].

In the following a review is given on measurement and analysis of bulk intermetallic alloys and thin films with a focus on REST measurement. Finally a short outlook is given to possibilities of studying kinetic processes in nanostructures, experimentally and by advanced kinetic Monte Carlo simulations.

THERMAL TREATMENT AND CORRESPONDING ATOM JUMPS

Crucial for the experimental investigation of kinetic processes is the thermal treatment of the samples. To study a change in the degree of LRO, for example, an equilibrium degree of order has first to be established at a certain temperature by an adequate long-time annealing and then the temperature has to be changed by a certain amount to monitor the process of achieving a new equilibrium state at the new temperature. The new equilibrium, however, can only be attained if the atomic mobility, equivalent to the vacancy mobility, is high enough, so that the process can finish within a reasonable observation time. This leads to essentially 2 procedures of heat treatment: 1) Isochronal annealing and 2) isothermal small step annealing. For isochronal annealing starting from an initially low temperature at which the sample usually is out of thermal equilibrium in a frozen state, the annealing temperature is increased (and then decreased) by equal steps and kept at each temperature for a limited time (isochronal time interval) irrespective of achieving a thermal equilibrium at the new temperature or not. Thermal equilibrium within an isochronal time interval is attained only if the atomic mobility at this temperature is high enough [3,11]. Therefore, an 'isochrone' gives an overview on the kinetics of an alloy system: Where does atomic mobility start, in which temperature range does LRO change reversibly in equilibrium with temperature (equilibrium line), is there a thermal hysteresis when comparing isochronal annealing at rising and falling temperatures, and so on. In contrast, during isothermal annealing starting from a true equilibrium value the sample is held at a new annealing temperature as long as necessary for a corresponding thermal equilibrium to be established. If the temperature step is small enough to neglect changes in vacancy concentration, variation of the measured property (e.g. REST) can be completely attributed to changes of the degree of order and these 'order-order relaxations' may then be analysed with respect to the underlying process kinetics.

Usual measurements of atom jumps in alloys aiming at diffusion properties determine long-range matter transport by the mean-square displacement of a marked atom (the tracer) regardless of the jump type involved, as long as the degree of order is preserved on average. In the $L1_2$ superstructure, for example, the overwhelming contribution to matter transport is from jumps within the majority sublattice, whereas ordering/disordering jumps between the sublattices play a minor role only. When instead measuring order-order relaxation kinetics, we register the net contribution of all atom jumps to a lasting change in the state of order. This is directly achieved by jumps between corresponding sublattices (antisite production/annihilation). The change of spatial distribution of vacancies and antisites by jumps on the majority sublattice is an important ancillary process, however. Figure 1 shows an example of the different jump schemes for the case of an L1$_2$ ordered lattice.

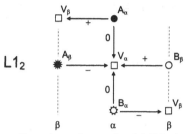

Figure 1. Schematic drawing of atom jumps in an ordered A$_3$B lattice of L1$_2$ superstructure. All possible jumps, those within (intra sublattice jumps, here vertical) and between sublattices (inter sublattice jumps, here horizontal) are shown. Defects: Vacancies V$_\alpha$,V$_\beta$ (□) and antisites B$_\alpha$ (✿) and A$_\beta$ (✳) on the α- and β- sublattices, respectively. (+) antisite generation, (−) antisite annihilation; B-antisites (B$_\alpha$) can easily move via the majority α-sublattice.

EXPERIMENTAL PROCEDURE

To study changes of order in intermetallic alloys extensive REST measurements have been carried out. In this case electrical resistivity is measured by the usual potentiometric method at low temperature so that the phonon contribution can be neglected, resulting in a very high sensitivity for structural changes. There is a comparatively simple relation between changes of LRO and the corresponding changes of electrical resistivity in this case as shown by Rossiter [12], which can be used to transform resistivity changes into changes of LRO-parameter. This is done with very high accuracy for bulk alloys and with some lower but still high resolution for thin film material, and compared to X-ray diffraction (XRD). Measurements of magnetization [13] and Mößbauer spectroscopy [14] (not presented in this paper) have been used as accompanying methods, especially for L1$_0$-ordered ferromagnetic intermetallics.

BULK SAMPLES

In figure 2 as an example for getting information on changes of LRO with temperature in thermal equilibrium the change of the LRO-parameter of Cu$_3$Au is shown versus temperature as obtained from REST and compared with results from XRD and CVM model calculations [15]. Even if REST needs XRD for overall calibration of LRO the superior sensitivity of REST as compared to XRD is impressive: A change in LRO-parameter of a fraction of percent can easily be resolved. Changes of SRO can also be studied using REST, which is almost impossible using any other technique [11,16,17].

In figure 3 we show, taking the CuPt alloy system as an example, how the equilibrium line of LRO (reversible changes of LRO in thermal equilibrium) as shown in figure 2 is determined:

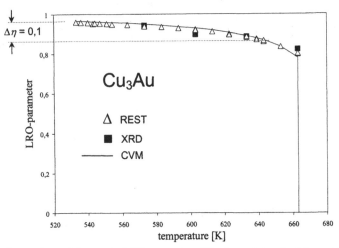

Figure 2. Comparison of changes in LRO-parameter of Cu₃Au as determined from resistivity measurement (REST), X-ray diffraction (XRD) and CVM calculations [15].

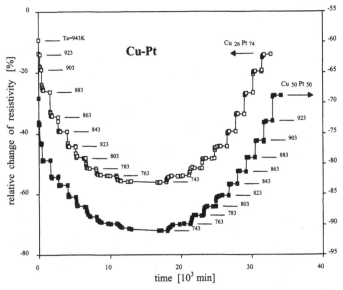

Figure 3. Changes in resistivity of Cu₅₀Pt₅₀ and Cu₂₆Pt₇₄ reflecting changes in the degree of order during a small step annealing treatment ('order-order' relaxations) [18].

A series of small step annealings at decreasing and increasing temperatures until a final plateau value is obtained yield a chain of points which correspond to changes of LRO as a function of temperature [18].

Details of the different order relaxation kinetics can be studied during small step annealing treatments, so called order-order relaxations. An example is shown in figure 4 for Ni_3Al. A careful analysis in this case of a $L1_2$ superstructure showed a two-process kinetics [19,20], which is also found for $L1_0$ structures [21,22].

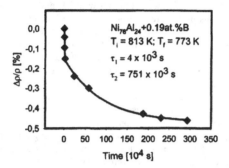

Figure 4. ‚Order-order' relaxation of Ni_3Al at 773 K (initial temperature: 813 K) showing the order relaxation process to be composed of two exponential processes with different relaxation times [19].

The interesting fact of two processes being involved in the $L1_2$ order-order relaxation was extensively studied by Monte Carlo (MC) computer simulation [23,24]. It was proved that the two processes correspond to the fast creation/annihilation of nearest neighbour antisite pairs and the subsequent decoupling/coupling of these pairs by antisite diffusion via the majority sublattice. Figure 5 illustrates these jump schemes for the case of a disordering step (increase of antisites). Due to the fact that the majority of the vacancies are generated on the Ni-sublattice, the

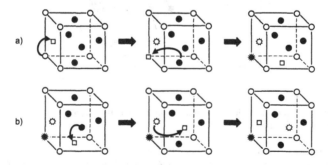

Figure 5. Sequence of correlated atom jumps in $L1_2$ superstructure of Ni_3Al. (a) generation of a nearest neighbour pair of Ni- (✶) and Al- (✿) antisites; (b) uncoupling of such a pair by jumps of the Al-antisite over the Ni-sublattice. O,● regular Al and Ni atoms [24,25].

formation of an antisite pair by correlated jumps of Al- and Ni-atoms is a very effective and fast disordering process: an Al-atom jumps into one of the Ni-vacancies which is immediately followed by a jump of a neighbouring Ni-atom into the just generated Al-vacancy. The other, much slower change of LRO can then occur by diffusion of just produced Al-antisites via the Ni-sublattice thereby decoupling the antisite pairs. On the contrary, fast annihilation of antisite pairs and slow coupling of these pairs by Al-antisite diffusion over the Ni-sublattice are the processes being active during an ordering step. The validity of these jump mechanisms has already been confirmed by MC-simulations for other model binaries with $L1_2$ and $L1_0$ superstructure [26].

A comparison of Arrhenius-plots of relaxation times of various intermetallic systems with different superstructures is shown in figure 6. The activation energies of order-order relaxations show a correlation with specific superlattice structures, which also corresponds to self-diffusion data. So, the high ordering activation energy for $L1_2$ Ni_3Al deviates considerably from the self-diffusion value, but for the $L1_0$ structure it is very similar to that of steady-state diffusion. This corresponds well with the fact that random vacancy diffusion is possible in the $L1_2$ structure just within the majority sublattice, whereas a 3D-diffusion in the $L1_0$ lattice definitely needs jumps between sublattices, which implies ordering/disordering jumps. The situation again is completely different in B2 ordered alloys like NiAl where constitutional triple defects are the dominant defect type. The ordering process in this case shows surprisingly long relaxation times in spite of high vacancy concentration and comparatively low activation energy. A corresponding jump model has already been developed [27] and preliminarily checked by Monte Carlo simulation [6], which explains this fact by a special triple defect ordering/disordering mechanism by which most

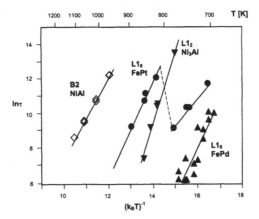

Figure 6. Arrhenius plots of order-order relaxation times as measured for $L1_2$ Ni_3Al [19], $L1_0$ FePd [22] and FePt [9] and B2 NiAl [27].

of the vacancies are trapped and thereby immobilized.

$L1_0$ ordered FePt at a temperature threshold around 800 K shows a discontinuous change of order-order kinetics which is in some correlation with Fe-diffusion in FePt alloys [9,28], but not yet fully understood. This effect is not observed for $L1_0$ FePd alloys [22], probably because temperatures are too low in the range where the new mechanism should get active.

As a sort of bulk limit, a FePd thin foil cold-rolled to 10 μm thickness has been investigated by REST and XRD (figure 7). This is a nice example of getting information on the kinetics of an

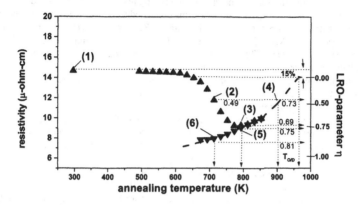

Figure 7. Changes of REST as a function of temperature during isochronal annealing ($\Delta T=20K$, $\Delta t=20min$) of a polycrystalline FePd foil (10μm thickness); initial state: as-rolled. Dashed line: 'equilibrium curve' reflecting changes of LRO-parameter in thermodynamic equilibrium (scaling on the right). Results of XRD measurements (marked by numbers in parentheses) are given by numbers along dotted lines [14].

alloy system from an isochronal annealing treatment. Measurements started with increasing temperatures in the cold-rolled, completely disordered state with frozen atomic mobility (▲). Ordering (resistivity reduction) sets in at about 600 K as soon as atom jumps are enabled; resistivity goes through a minimum when the state of LRO reached corresponds to the current annealing temperature as an equilibrium value. Further temperature increase results in increasing REST values, since the degree of LRO is reduced now. For subsequently decreasing temperatures (▼) a resistivity decrease is observed, reflecting an increase of LRO. There is a certain range of temperatures, where reversible changes can be observed corresponding to a change of LRO in equilibrium with temperature (equilibrium curve, dashed line in figure 7). By considering a resistivity contribution from cold-rolling of about 15% the axis of LRO-parameter η can be scaled correspondingly [14] using $\rho_{LRO} \propto (1-\eta^2)$ [12]. The values of η as determined by XRD (points of measurement marked by numbers in parentheses in figure 7 and given by numbers along dotted lines) coincide very well with the REST curve, confirming once more the good quantitative correspondence between REST and LRO-parameter. A similar correspondence with REST was found for the hyperfine field distribution of Mößbauer spectra (not shown here). Details of the spectra, however, together with a TEM analysis give evidence that the ordered microstructure is complex, containing e.g. twinfree grains and polytwins [29].

THIN FILM MATERIAL

FePd has been used as a model system for investigating details of ordering processes in thin film material. FePd is one of those intermetallics with $L1_0$ superstructure which are ferromagnetic and therefore due to the tetragonal lattice distortion show a high magnetic anisotropy with the tetragonal c-axis being the preferential (easy) axis of magnetization.

Two sorts of thin film sample material were investigated by REST, XRD and magnetization measurement [13]: i) 50 nm sputtered FePd films deposited by dc and rf magnetron co-sputtering on Si(100) substrate at room temperature; ii) Epitaxial FePd thin films (50 nm and 100 nm thickness) deposited on MgO(001) substrate by MBE at a substrate temperature of 773 K.

Electrical resistivity (REST) in the case of thin film material was measured with in-plane van-der-Pauw geometry on samples directly immersed in liquid nitrogen.

The as-prepared sputtered films are polycrystalline and completely disordered fcc, whereasthe MBE-deposited films are nearly single crystalline and show a well developed L1$_0$ LRO with an order parameter of $\eta = 0.83$.

Changes of REST during isochronal annealing of both sorts of film, sputtered and MBE-deposited, (thickness: 50 nm) are compared in figure 8 [8,14]. The resistivity variation is very similar to bulk material: The first decrease (sputtered: 17 $\mu\Omega$cm, MBE: 5 $\mu\Omega$cm) reflects an increase of the degree of LRO, the subsequent increase is due to a reduction of order as soon as the degree of LRO has reached the equilibrium value corresponding to the current annealing temperature. The differences between both curves can easily be explained by the very different initial degree of order in both sorts of film: completely disordered for the sputtered film leading to a big reduction of REST values when ordering is enabled by atomic migration (about 650 K); highly ordered for the MBE-deposited film and therefore only small further reduction of resistivity.

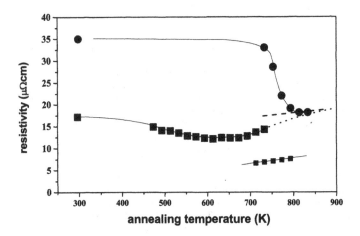

Figure 8. Change of electrical resistivity during isochronal annealing ($\Delta T=20K$, $\Delta t=20$ min) of FePd-film (50 nm) versus annealing temperature. Full circles: sputtered, big squares: MBE-deposited. Points from isothermal small-step annealing of MBE-film from figure 9 (small squares) and presumptive equilibrium line of reversible resistivity change (dashed line) are given for a comparison [30].

For the first time it has been tried to monitor resistivity changes during a small step annealing treatment on thin film material to get information on details of the ordering process under restricted dimensions [30]. The measurements were done on MBE-deposited FePd thin film. Similar to what was observed earlier in bulk FePd [22], for most of these anneals two counteracting processes were observed. The first process is fast and acts in the direction of the corresponding temperature change, the second, slower process acts in the opposite direction. The second process showed a considerable scatter in its contribution to total resistivity change as well as in its relaxation time; to study the fast process carefully, the slow process was eliminated by subtracting it from the as-measured data. The results are shown in figure 9, documenting that plateau values of resistivity are achieved, which obviously correspond to thermodynamic equilibrium values of LRO for the respective temperatures. These findings correspond with a recent theoretical study that only one process (of three) can be attributed to antisite ordering within the ordered antiphase domains [31]. This was also found by a recent MC simulation study in a truly homogeneous $L1_0$ model system [6]. One can therefore argue that the counteracting process may arise from changes in the antiphase volume as already observed earlier for Cu_3Au [32]. The plateau values are additionally plotted in figure 8 (small squares) to indicate the presumptive equilibrium line (dashed line) of reversible resistivity change: In this temperature

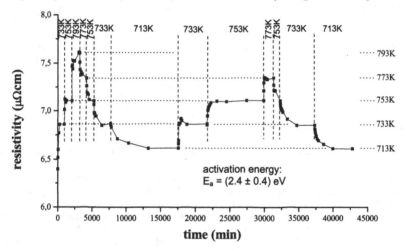

Figure 9. Resistivity change during isothermal small-step annealing treatment of 100 nm FePd film epitaxially deposited by MBE. The first point was measured after annealing 18 hours at 773K. A second, counteracting, process was subtracted. Plateau values of resistivity are definitely obtained corresponding to thermal equilibrium [30].

range it is expected that the state of LRO can be established reversibly by corresponding temperature changes. The plateau values during the small-step annealing treatment are obtained considerably below the minimum of the corresponding isochronal curve (small and big squares, respectively, in figure 8). We therefore conclude that defects grown in during the preparation at the comparatively high temperature of 773 K and the subsequent cooling process have annealed out only after the long-time isothermal annealings before and during the small step annealing procedure.

From an Arrhenius-plot of the relaxation times as determined by single exponential fits to the order-order relaxation curves an activation energy of (2.4 ± 0.4) eV results [30]. Although this value is somewhat smaller than the value of (2.7 ± 0.1) eV which was obtained for the faster process in bulk FePd [22], the relaxation times are in fact *longer* in the thin film. This is surprising because the vacancy migration is easier near the surface, but could be explained by the restricted jump possibilities for vacancies and the numerous vacancy traps present in such films [33,34]. A study of ordering activation energy as a function of film thickness and a corresponding MC simulation study is in progress.

OUTLOOK

Our plans for future research of atomic migration in ordered intermetallics are twofold. Aside from continuing the promising study on thin films, the production of nanostructures (nanowires, atomic clusters etc.) is under way. Nanowires, for example, can be made by filling the pores of anodic oxidized alumina with an intermetallic, e.g. FePd by electro-deposition from $(FeSO_4+PdCl_2)$-solution [35]. After connection with an appropriate metallic layer, changes in resistivity can be measured following an adequate annealing heat treatment. Other techniques to make nanoparticles are sputtering (in progress in Strasbourg) and pulsed laser deposition (PLD), which is currently tried at the University of Pittsburgh [36]. The study of atomic ordering dynamics in nanoparticles will probably need some other, indirect signature of order. In the magnetic systems like FePd and CoPt this can be the magnetic anisotropy [34,37]. Such measurements are planned for the next future in Strasbourg.

A second line of attack concerns our interest to interpret REST measurements in the light of what happens on the atomic scale. This can be satisfied best by MC computer simulations, which describe the jumps of individual atoms on a statistical basis, as in the real crystal many single atom jumps work together to change the structural configuration. As shown above, kinetic MC simulations helped already to explain the two processes observed in the $L1_2$ superstructure and to account for a single time scale in homogeneous $L1_0$ systems. A drawback of usual MC methods, however, is the simplification with respect to the barrier height of the atom jump profile by assuming constant saddle point energy. Calculating *ab-initio* the energy jump profiles for various surroundings of the jumping atom, more realistic atom jump rates may be obtained.

In figure 10 for a comparison the long-range order parameter as determined with variable and constant saddle point energies, respectively, is plotted as a function of MC steps for two annealing temperatures [30]. Both calculations as expected lead to the same equilibrium values, but a striking difference is observed for the kinetic details. More jumps are necessary to achieve a certain change of LRO in the improved model because of a strong correlation effect where atom jumps generating antisites and their back jumps. Further analysis shows that this is a result of slower production of antisite pairs which causes a slowing of the fast process.

A further development concerns the simulation modelling of B2 ordering kinetics in binary intermetallics containing triple defects by studying order-order kinetics with a correlation between vacancy and antisite concentrations [39].

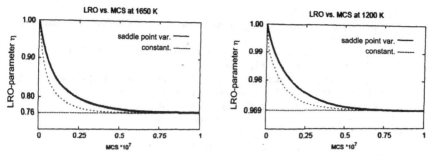

Figure 10. Long range order parameter versus number of MC steps (MCS) for variable saddle point energy (thick line) and constant saddle point energy (dashed line) [30]. Dotted line: final equilibrium value. Annealing temperatures are 1650 K (left) and 1200 K (right). Initial state: completely ordered. Note different scaling.

ACKNOWLEDGEMENTS

Financial support by the Austrian science funds 'Fonds zur Förderung der wissenschaftlichen Forschung' (P1954-N16), COST 535 and COST P19 collaboration supported by the Polish Ministry of Science and Higher Education (COST/202/2006), Polish-French joint program (Polonium 13844QL) and French Agence Nationale pour la Recherche (PNANO-07-NANO-018 ETNAA) is gratefully acknowledged.

REFERENCES

1. E. Kentzinger, V. Parasote, V. Pierron-Bohnes, J.F. Lami, M.C. Cadeville, J.M. Sanchez, R. Caudron, B. Beuneu, *Phys. Rev.* **B61**, 14975 (2000).
2. T. Mehaddene, J.M. Sanchez, R. Caudron, M. Zemirli, and V. Pierron-Bohnes, *Eur. Phys. J.* **B41**, 207 (2004).
3. W. Pfeiler, In: '*Properties of Complex Solids*', Ed. A. Gonis, A. Meike, and P.E.A. Turchi, Plenum Press, London 1997, p. 219.
4. W. Pfeiler, *JOM* **52**, 14 (2000).
5. W. Pfeiler and B. Sprusil, *Mater. Sci. Eng.* **A324**, 34 (2002).
6. R. Kozubski, A. Biborski, M. Kozłowski, V. Pierron-Bohnes, C. Goyhenex, W. Pfeiler, M. Rennhofer, and B. Sepiol, *Adv. Sol. Stat. Phys.* **47**, 277 (2008).
7. W. Pfeiler, W. Püschl, R. Kozubski, and V. Pierron-Bohnes, In: *Solid-to-Solid Phase Transformations in Inorganic Materials 2005* (J.M. Howe, D.E. Laughlin, J.K. Lee, U. Dahmen, and W.A. Soffa, eds.), TMS-Proceedings (2005) volume 1, p. 187.
8. Ch. Issro, W. Püschl, W. Pfeiler, P. Rogl, W. Soffa, G. Schmerber, R. Kozubski, and V. Pierron-Bohnes, *Metall. Mater. Trans.* **37A**, 3415 (2006).
9. R. Kozubski, Ch. Issro, K. Zapała, M. Kozłowski, M. Rennhofer, E. Partyka, V. Pierron-Bohnes, and W. Pfeiler, *Z. Metallkde.* **97**, 273 (2006).
10. R. Kozubski, M. Kozłowski, K. Zapala, V. Pierron-Bohnes, W. Pfeiler, M. Rennhofer, B. Sepiol, and G. Vogl, *J. Phase Equilib. Diffus.* **26**, 482 (2005).
11. Pierron-Bohnes V., Mirebeau I., Balanzat E. and Cadeville M.C., *J. Phys. F: Met. Phys.* **14**,

197 (1984).

12. P.L. Rossiter, *The electrical resistivity of metals and alloys* (Cambridge: University Press, 1987), p. 160-167.

13. Ch. Issro, W. Püschl, P. Rogl, W.A. Soffa, G. Schmerber, R. Kozubski, and V. Pierron-Bohnes, *Scr. Mater.* **53**, 447 (2005).

14. Ch. Issro, W. Püschl, W. Pfeiler, B. Sepiol, P.F. Rogl, W.A. Soffa, Manuel Acosta, and V. Pierron-Bohnes, *Mater. Res. Soc. Symp. Proc.* **842**, S3.10.1-6 (2005).

15. H. Lang, H. Uzawa, T. Mohri and W. Pfeiler, *Intermetallics* **9**, 9 (2001).

16. Balanzat E. and Hillairet J., *J. Phys. F: Met. Phys.* **11**, 1977 (1981).

17. W. Pfeiler, *Acta Metall.* **36**, 2417 (1988).

18. R. Ebner, M. Migschitz, C. Scholz, B. Urban-Erbil, P. Fratzl, and W. Pfeiler, In: *Solid-Solid Phase Transformations*, ed. W.C. Johnson et al., TMS-Proceedings, Warrendale, 1994, p. 401.

19. R. Kozubski and W. Pfeiler, *Acta Mater.* **44**, 1573 (1996).

20. R. Kozubski, *Prog. Mater. Sci.* **41**, 1 (1997).

21. G. Sattonay and O. Dimitrov, *Acta Mater.* **47**, 2077 (1999).

22. A. Kulovits, W. Püschl, W.A. Soffa, and W. Pfeiler, *Mater. Res. Soc. Symp. Proc.* **753**, BB5.37.1-6 (2003).

23. P. Oramus, R. Kozubski, V. Pierron-Bohnes, M.C. Cadeville, and W. Pfeiler, *Phys. Rev.* **B 63**, 174109/1-14 (2001).

24. P.Oramus, R. Kozubski, V. Pierron-Bohnes, C. Massobrio, and W. Pfeiler, *Mater. Sci. Eng.* **A324**, 11 (2002).

25. R. Kozubski, M. Kozłowski, V. Pierron-Bohnes, and W. Pfeiler, *Z. Metallkde.* **95**, 880 (2004).

26. P. Oramus, M. Kozłowski, R. Kozubski, V. Pierron-Bohnes, M.C. Cadeville, and W. Pfeiler, *Mater. Sci. Eng.* **A365**, 166 (2004).

27. R.Kozubski, D. Kmieć, E. Partyka, and M. Danielewski, *Intermetallics* **11**, 897 (2003).

28. M. Rennhofer, B. Sepiol, M. Sladecek, D. Kmieć, S. Stankov, G. Vogl, M. Kozlowski, R. Kozubski, A. Vantomme, J. Meersschaut, R. Rüffer, and A. Gupta, *Phys. Rev.B* **74**, 104301(1-8), (2006).

29. A. Kulovits, J. Wiezorek, W.A,. Soffa, W. Püschl, and W. Pfeiler, *J. Alloys Compd.* **378** 285 (2004).

30. Ch. Issro, V. Pierron-Bohnes, W. Püschl, R. Kozubski, and W. Pfeiler, *Mater. Res. Soc. Symp. Proc.* **980**, 0980-II03-07 (2007).

31. M. Ohno and T. Mohri, *Phil. Mag.* **83**, 315 (2003).

32. H. Lang, H. Uzawa, T. Mohri, and W. Pfeiler, *Diffusion and Defect Data A: Defect and Diffusion Forum* **194-199**, 583 (2001).

33. O. Ersen, C. Goyhenex, and V. Pierron-Bohnes, *Phys. Rev. B* **78**, 035429 (2008).

34. R. V. Montsouka, C. Goyhenex, G. Schmerber, C. Ulhaq-Bouillet, A. Derory, J. Faerber, J. Arabski, and V. Pierron-Bohnes, *Phys. Rev. B* **74**, 144409 (2006).

35. K.J. Bryden and J.Y. Ying. *J. Electrochem. Soc.* **145**, 3339 (1998).

36. A. Kulovits, Ch. Issro, J. Leonard, and J. Wiezorek, private communication.

37. O. Ersen, V. Parasote, V. Pierron-Bohnes, M.C. Cadeville, and C. Ulhaq-Bouillet, *J. Appl. Phys.* **93** 2987 (2003).

38. M. Leitner, D. Vogtenhuber, W. Pfeiler, and W. Püschl, to be published.

39. A. Biborski, L. Zosiak, R. Kozubski, and V. Pierron-Bohnes, in the press for *Intermetallics*.

Mater. Res. Soc. Symp. Proc. Vol. 1128 © 2009 Materials Research Society   1128-U09-02

Lattice-Gas-Decomposition Model for Vacancy Formation Correlated With B2 Atomic Ordering in Intermetallics

A. Biborski[1], L. Zosiak[1], R. Kozubski[1] and V. Pierron-Bohnes[2]
[1]Interdisciplinary Centre for Materials Modelling, M. Smoluchowski Institute of Physics, Jagellonian University, Reymonta 4, 30-059 Cracow, Poland
[2]IPCMS-GEMME, CNRS-ULP, 23 rue du Loess, BP 43, 67034 Strasbourg Cedex 2, France

ABSTRACT

Thermal vacancy formation correlated with atomic ordering was modelled in B2-ordering A-B binary intermetallics. Ising Hamiltonian was implemented with a specific thermodynamic formalism for thermal vacancy formation based on the phase equilibria in a lattice gas composed of atoms and vacancies. Extensive calculations within the Bragg-Williams approximation [1] were followed by Semi-Grand Canonical Monte Carlo (SGCMC) simulations. It has been demonstrated that for the atomic pair-interaction energies favouring vacancy formation on A-atom sublattice, equilibrium concentrations of vacancies and antisite defects result mutually proportional in well defined temperature ranges. The effect observed both in stoichiometric and non-stoichiometric (both A-rich and B-rich) binary alloys was interpreted as a tendency for triple defect formation. In B-rich alloys vacancy concentration did not extrapolate to zero at 0 K, which indicated the formation of constitutional vacancies. Energetic conditions for the occurrence of the effects were analysed in detail. The modelled temperature dependence of vacancy concentration in the B2-ordering A-B binaries with triple defects will be included in the Kinetic Monte Carlo (KMC) simulations of chemical ordering kinetics in these systems with reference to the experimental results obtained for NiAl [2].

INTRODUCTION

The present work opens a project aiming at modelling chemical ordering kinetics in B2-ordered NiAl intermetallic compound by means of atomistic simulations. Despite very high vacancy concentration [3] the NiAl shows surprisingly slow „order-order" kinetics [2]. In particular, the process appears much slower than e.g. the one occurring in Ni3Al containing ca. 10^7 times less vacancies. This effect could be explained in terms of triple defect generation in B2 superstructure [3,4] (Fig.1).

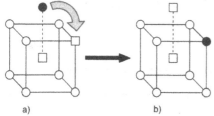

a)   b)

Figure 1. Scheme of triple defect creation in B2 ordering binary alloy: (a) perfectly ordered B2 binary (unfilled circles: A(Ni)-atoms on α sublattice and filled circles: B(Al)-atoms on β sublattice, unfilled squares: two vacancies residing on counterpart sublattices); (b) system with a triple defect: one A(Ni)-antisite and two A(Ni)-vacancies.

In such a case the system disorders by generating predominantly Ni antisite defects, each one associated with two vacancies residing on Ni-sublattice. The vacancies, however, can hardly mediate further disordering as their jumps to nearest-neighbouring (nn) Al-sublattice sites are energetically costly. Consequently, chemical disordering in NiAl is slow and proceeds via correlated generation of antisite defects and vacancies. The simulations must account for this correlation and hence, contrary to the standard works (see e.g. [5]), appropriate algorithms must be implemented with vacancy thermodynamics yielding temperature dependent equilibrium vacancy concentration. In the following, specific thermodynamic model based on a lattice-gas decomposition concept allowing the evaluation of equilibrium concentration of vacancies and antisite defects in B2-ordering A-B binaries is discussed and the conditions for triple defect formation are determined. The Bragg-Williams (BW) solution [1] is now followed by SGCMC simulations yielding results qualitatively similar to the BW ones, but being quantitatively implementable with KMC simulations.

CONFIGURATIONAL THERMODYNAMICS OF B2-ORDERING IN $A_{1-\delta}B_{1+\delta}$ BINARY SYSTEM WITH VACANCIES.

Discussed is an Ising model of a bcc lattice gas composed of vacancies and atoms whose concentrations are in proportions corresponding to the composition of the $A_{1-\delta}B_{1+\delta}$ intermetallic. As a whole, the gas is considered as a closed system with interatomic interactions leading to the finite cohesion and B2-long-range ordering (LRO) at temperatures $T < T_C$ of an $A_{1-\delta}B_{1+\delta}$ intermetallic crystal (see [6] for the original idea); zero interaction is assumed between vacancies. It appears that the lattice gas decomposes into a vacancy-poor and a vacancy-rich phase, where the equilibrium vacancy concentrations C_V and chemical potentials μ_V (generally of non-zero values) are functions of external parameters (T,p). The vacancy poor phase is finally identified with the intermetallic $A_{1-\delta}B_{1+\delta}$ containing the equilibrium amount of vacancies.

The above model consistently treats vacancies as an additional component of the system and their equilibrium concentration is clearly derived from the criterion of phase equilibria. Its Bragg-Williams and SGCMC solutions are presented.

Bragg-Williams approximation

The following "ansatz" free energy of the lattice gas per lattice site is considered [7]:

$$f^{(\delta)} = \frac{F}{N} = \frac{1}{2}\sum_i\sum_k\sum_\mu\sum_\nu Z_{\mu\nu}C_i^\mu C_k^\nu V_{ik} + kT\sum_i\sum_\mu C_i^\mu \ln\left(C_i^\mu\right) + const. \tag{1}$$

where N is the total number of lattice sites, $Z_{\mu\nu}$ is the number of ν-sublattice sites being

nearest neighbours of a μ-sublattice site in B2 superstructure, $C_i^\mu = \dfrac{N_i^\mu}{N^\mu}$, where N_i^μ denotes

the number of i-species ($i=\{A,B,V\}$, V denotes a vacancy) on μ sublattice, N^μ is the number

of μ-sublattice sites ($N^\alpha = N^\beta = \dfrac{1}{2}N$). V_{ik} denote nn pair interaction energies

($i,k = \{A,B,V\}$), where $V_{VV} = 0$.

The "ansatz" (1) is a function of five independent variables: η_{at}, η_{vac}, C_V, T, δ, where $\eta_{at} = C_A^\alpha - C_A$, $\eta_V = C_V^\alpha - C_V^\beta$ are atomic and vacancy LRO parameters and $C_V = C_V^\alpha + C_V^\beta$. Given vacancy concentration C_V, the equilibrium configuration of the lattice gas is determined by the minimum of the "ansatz" (1) with respect to the LRO parameters η_{at} and η_{vac}:

$$\left[\frac{\partial f^{(\delta)}}{\partial \eta_{at/vac}} \right]_{C_V, T, \{V_{ik}\}} = 0. \tag{2}$$

The shape of the resulting curve $f_{eq}^{(\delta)}(C_V, T)$ shows that the lattice gas decomposes into vacancy poor and vacancy rich phases [1]. The vacancy poor phase is identified with an $A_{1-\delta}B_{1+\delta}$ crystal with equilibrium degree of LRO and equilibrium vacancy concentrations C_V^α, C_V^β.

Semi Grand Canonical Monte Carlo simulations

The applied technique [8-10] consists of simulating a configurational and compositional equilibrium in an open AB-V lattice gas (δ=0) with fixed values of chemical potentials μ_i of atoms and vacancies. A sample of B2-ordered AB Ising system containing 20×20×20 unit cells with periodic boundary conditions was simulated by exchanging A-B pairs of atoms with pairs of vacancies. Such exchanges conserved the chemical composition and were executed with a Metropolis probability:

$$\Pi = \begin{cases} \exp\left[-\frac{\Delta E + \Delta\mu}{kT} \right], & (\Delta E + \Delta\mu) > 0 \\ 1, & (\Delta E + \Delta\mu) < 0 \end{cases} \tag{3}$$

where ΔE denotes the exchange-induced change of the Ising energy of the lattice gas and $\Delta\mu = \pm(\mu_A + \mu_B - 2\mu_V)$, k and T are Boltzmann constant and absolute temperature, respectively.

$C_V(\Delta\mu)$ curves determined by relaxing the system for a range of $\Delta\mu$ values and temperatures shoved discontinuities (Inset in Fig. 4a) indicating an equilibrium of two phases differing in vacancy concentration; the vacancy-poor one identified with the AB crystal with equilibrium configuration and vacancy concentration.

RESULTS

Bragg-Williams solution [1]

The calculations were performed for both stoichiometric (δ=0) and off-stoichiometric systems. The scanned space of pair interaction parameters was determined as follows:

$$W = 2 \times V_{AB} - V_{AA} - V_{BB} = -0.08 \text{ eV} \qquad (T_C \text{ (B2}\rightarrow\text{A2)} = 1825 \text{ K)} \tag{4}$$

$$\frac{V_{BB}}{W} = 0.625; \quad V_{VV} = 0 \tag{4a}$$

$$-0.75 < \frac{V_{AV}}{W} < 0; \quad 0 < \frac{V_{AA} - V_{BB}}{W} < 1.5; \quad V_{BV} = -V_{AV} \quad \text{(preference for A-antisite formation)} \tag{5}$$

The calculations yielded temperature dependencies of the four C_i^μ concentrations determining the temperature dependence of a "Triple-Defect-Indicator" (TDI):

$$TDI(T) = \frac{C_A^\beta + C_B^\alpha}{C_V^\alpha + C_V^\beta}(T) \tag{6}$$

$TDI = \frac{1}{2}$ indicating formation of triple defects.

Results for stoichiometric AB system (δ=0):

Calculations have been performed for a wide range of energy parameters generating diverse shapes of $TDI(T)$ curves (Fig. 2a) [1].

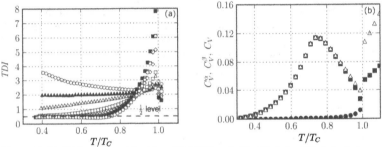

Figure 2. (a) TDI calculated for V_{asym}/W=0.375 and: V_{AV}/W=0 (unfilled circles); V_{AV}/W=-0.0625 (filled triangles); V_{AV}/W=-0.125 (unfilled triangles); V_{AV}/W=-0.25 (unfilled squares); V_{AV}/W=-0.375 (unfilled diamonds); V_{AV}/W=-0.5 (unfilled down-pointing triangles); V_{AV}/W=-0.625 (filled squares); V_{AV}/W=-0.75 (unfilled pentagons); (b) AB system (δ=0), V_{asym}/W=0.875, V_{AV}/W=-0.638, C_V^α (filled squares), C_V^β (filled circles), C_V (unfilled triangles).

Fig. 2a indicates that the plateaux of $TDI = 1/2$ (i.e. the generation of triple defects) occur within wide temperature ranges for $-0.675 \leq V_{AV}/W \leq -0.375$. The effect is accompanied by very high preference for A-vacancy generation and a decrease of C_V at $T \rightarrow T_C$ (Fig.2b).

Results for non-stoichiometric systems

Calculations were performed for $\delta = +0.01$ (Fig.3) and $\delta = -0.01$.

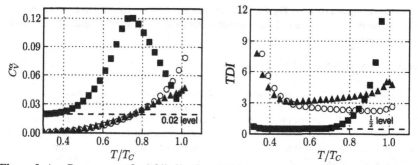

Figure 3. $A_{0.49}B_{0.51}$ system (δ=+0.01), V_{AV}/W=-0.628, V_{asym}/W=0.875 (filled squares); V_{AV}/W=0.0, V_{asym}/W=0.375 (unfilled circles); V_{AV}/W=0.0, V_{asym}/W=0.875 (filled triangles)

Fig.3 shows that in $A_{0.49}B_{0.51}$ (δ= +0.01) modelled with the energies promoting triple defect creation, deviation from stoichiometry is compensated at low temperatures by the existence of constitutional vacancies. No constitutional vacancies were observed in cases without triple defect formation. In $A_{0.51}B_{0.49}$, in turn, no constitutional vacancies were observed at all. Remarkably, triple defect correlation between *thermal* vacancy concentration and *thermal* antisite concentration (*TDI* plateau) was observed both for δ= -0.01 and δ= +0.01.

Semi Grand Canonical Monte Carlo simulations.

The results obtained for an AB binary modelled with pair-interaction energies promoting triple defect generation are shown in Fig. 4. It is clear that the features resulting from the Bragg-Williams solution of the Ising model of the lattice gas (*TDI* plateau and a decrease of C_V^{α} at $T \rightarrow T_C$) are qualitatively reproduced. A new result (unfeasible within the BW solution) is a pair- correlation $AVNNC(m)$ showing fractions of antisites surrounded by m nn vacancies (see the inset in Fig. 4b).

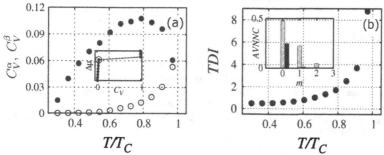

Figure 4. AB system: V_{AV}/W=-0.25, V_{asym}/W=0.875: (a) $C_V^{\alpha}(T/T_c)$ filled squares, $C_V^{\beta}(T/T_c)$ unfilled squares; Inset: typical SGCMC $C_V(\Delta\mu)$ curve of the AB-V lattice gas, the point corresponding to the equilibrium C_V is marked out; (b) *TDI* (T/T_C); Inset: histogram of $AVNNC(m)$ for A-antisites (grey bars) and B-antisites (black bars).

The histogram (Inset in Fig.4b) indicates that the generated triple defects have mostly statistical nature and the system shows no preference for creating nn complexes of 1 A-antisie and 2 A-vacancies.

DISCUSSION AND CONCLUSIONS

Equilibrium vacancy concentration in B2-ordering $A_{1-\delta}B_{1+\delta}$ binary systems was determined within an Ising model basing on a lattice-gas decomposition. The model was solved in a Bragg-Williams approximation and by means of Semi Grand Canonical Monte Carlo simulations. For particular pair-interaction energetics temperature ranges were found, where vacancies were preferentially formed on α-sublattice and their concentration was proportional to the concentration of antisite defects. The effect was interpreted as a signature of the generation of triple defects occurring in selected binary systems ordering in B2 superstructure – e.g. in NiAl. Consequently, decrease of the degree of LRO at temperatures close to T_C caused in these systems such a strong decrease of the dominating A-vacancy concentration C_V^{α} that a decrease of the total vacancy concentration C_V was observed. The model predicted the presence of constitutional vacancies at $T\rightarrow0$ K only in non-stoichiometric B-rich systems. However, triple-defect formation was observed both in B-rich and A-rich systems, which supports the interpretation of the experimental results on ordering kinetics in $Ni_{50.5}Al_{49.5}$ [2] in terms of this process.

SGCMC simulations qualitatively reproduced the Bragg-Williams results. In addition, the analysis of the resulting atom-vacancy nn pair correlations $AVNNC$ showed that the triple defects were *not* generated as nn complexes of A-antisite defects and A-vacancies.

ACKNOWLEDGEMENT

Work pursued within the European Actions COST 535 and COST P19 and supported by the Polish Ministry of Science and Higher Education: Grant no. COST/202/2006, as well as by the governments of France and Poland within the program POLONIUM (grant no. 2008/physique/13844QL). Most of the calculations were performed in the Interdisciplinary Centre for Mathematical and Computational Modelling, Warsaw University, grant no. G31-22.

REFERENCES

1. A. Biborski, L.Zosiak, R. Kozubski, V. Pierron-Bohnes, *Intermetallics* **17**, 46, (2009).
2. R. Kozubski, D. Kmieć, E. Partyka, M. Danielewski, *Intermetallics* **11**, 897 (2003).
3. H.-E. Schaefer, K. Frenner, R. Würschum, *Intermetallics*, **7**, 277 (1999).
4. R. J. Wasilewski, *J. Phys. Chem. Solids* **29**, 39 (1968).
5. P. Oramus, R. Kozubski, V.Pierron-Bohnes, M.C.Cadeville, W.Pfeiler, *Phys.Rev.B* **63**, 174109/1-14, (2001).
6. F. W. Schapink, *Scr. Metall.* **3**, 113 (1969).
7. R. Kozubski, *Acta.Metall.Mater.* **41**, 2565 (1993).
8. K. Binder, J.L. Lebowitz, M.K. Phani, M.H. Kalos, *Acta Metall.* **29**, 1655 (1981).
9. G.L. Aranovich, J.S. Erickson, M. D. Donohue, *J. Chem. Phys.* **120**,11 (2004).
10. D.P. Landau, B. Dünweg, M. Laradji, F. Tavazza, J. Adler, L. Cannavaccioulo, X. Zhu, *Lect. Notes Phys.* **703**, 127 (2006).

Mater. Res. Soc. Symp. Proc. Vol. 1128 © 2009 Materials Research Society　　　　1128-U09-03

First-principles investigations of point defect behavior and elastic properties of TiNi based alloys

Jian-Min Lu, Qing-Miao Hu, Rui Yang
Shenyang National Laboratory for Materials Science, Institute of Metal Research, Chinese Academy of Sciences, Shenyang 110016, China

ABSTRACT

First-principles calculations by the use of a plane-wave pseudopotential method are performed to investigate intrinsic point defect behavior in TiNi. The results show that TiNi is an antisite type intermetallic compound. The calculated interaction energies between the point defects demonstrate that Ti antisites are attractive to each other whereas Ni antisites are mutually repulsive. The attraction between Ti antisites indicates that excess Ti in TiNi may agglomerate so that a Ti-rich phase can easily precipitate. The repulsion between Ni antisites implies that the excess Ni is of certain solubility in TiNi. This result explains well the asymmetric feature of TiNi field in the binary phase diagram. In order to understand the correlation between the composition dependent elastic modulus and martensitic transformation (MT) temperature, the elastic moduli critical to MT, $i.e.$ c' and c_{44}, are calculated as a function of the composition of the off-stoichiometric TiNi and a series of ternary TiNi-X alloys by the use of exact muffin-tin orbital method in combination with coherent potential approximation. It turns out that, generally speaking, the early transition metal (TM) alloying elements in the periodic table increase c' but decrease c_{44}; the middle ones increase both c' and c_{44}, whereas the late ones decrease c' but increase c_{44}. An examination of the theoretical composition dependent elastic modulus and the experimental MT temperature shows that the MT temperature is more sensitive to the variation of c_{44} than to that of c'.

INTRODUCTION

As one of the earliest discovered shape memory alloys, TiNi has already been used successfully in industry. The shape memory effect of this alloy is ascribed to the reversible martensitic transformation (MT) between the high temperature parent cubic B2 phase and the low temperature martensitic orthorhombic B19 or monoclinic B19' phase. However, the fundamental physics underlying the martensitic transformation are still to be clarified. One of the ambiguous issues is the precursor effect appearing just above the MT temperature as demonstrated by the central peak in the elastic neutron phonon spectrum measurement, diffuse scattering (corresponding to tiny domains) and extra diffuse peaks in X-ray diffraction, and remartensitic attenuation in ultrasonic attenuation experiments [1]. With the aid of numerical simulations, Kartha and colleagues suggested that the intrinsic chemical disorder could be one of the driving forces for pretransitional phenomena [2]. The chemical disorder in intermetallic compounds generally manifests itself by the existence of point defects, $i.e.$ vacancies and antisite atoms. Therefore, knowledge of the point defects and their interactions is crucial for our understanding of precursor phenomena in TiNi.

Alloying is a widely used approach to adjust the MT temperature and mechanical properties of TiNi so as to suit its applications in different environments. To enable the efficient design of

TiNi alloys, an understanding of the relation between alloy composition and its MT behavior is demanded. By using the Landau theory, Ren and colleagues [3-5] have shown that the composition dependence of the MT temperature is closely related to the single crystal elastic modulus of the alloys: the softer is the elastic modulus, the higher the MT temperature. Such a model provides the possibility of determining the composition dependent MT temperature by surveying the composition dependent elastic modulus of the alloy. Furthermore, we can explore the physics of the composition dependence of the elastic modulus so as to understand the composition dependence of MT temperature, for which, however, the Landau theory adopted by Ren *et al.*, as a phenomenological model, is not sufficient. Although the model of Ren *et al.* built a connection between the elastic modulus and the MT temperature for TiNi alloys, one cannot know *a priori* which elastic modulus dominates since both elastic moduli, c' and c_{44}, are very small and could potentially affect the MT temperature.

In this paper, we first determine the point defect configuration in TiNi using a statistical mechanics model in conjunction with a first-principles method. The interactions between the point defects and their electronic structure origin are also discussed. With the determined point defect configuration, we calculate the single crystal elastic modulus of B2 TiNi alloys as a function of the Ni content. Furthermore, we calculate the composition dependent elastic modulus of ternary TiNi-X alloys. By examining the composition dependent elastic modulus and experimental MT temperature for these alloys, we attempt to gain insight into the connection between the elastic modulus and the MT temperature.

METHODS

There are four types of intrinsic point defects in TiNi intermetallic compound, vacancies on the α (Ti) sublattice n^V_α, vacancies on the β (Ni) sublattice n^V_β, Ni antisites on the α sublattice n^{Ni}_α, and Ti antisites on the β sublattice n^{Ti}_β. We calculate these four thermal equilibrium concentrations using a statistical mechanics model taking the "raw" formation energies (*i.e.* the energy difference between defected and the perfect TiNi with the same size of supercell) from first-principles calculations as input. We refer the readers to Ref. [6, 7] for details of this method. The interaction energy (ΔE) between point defects A and B is evaluated as:

$$\Delta E = E(N,A+B) + E(N) - E(N,A) - E(N,B) \tag{1}$$

in which $E(N,A+B)$, $E(N)$, $E(N,A)$ and $E(N,B)$ are the total energies of the N-site supercells with one A-B defect pair, perfect TiNi, with one A point defect, and one B point defect, respectively. A positive ΔE indicates that the two point defects are repulsive to each other while a negative one means they are attractive.

The total energies of the defected and perfect TiNi supercells are calculated by the use of the first-principles plane-wave pseudopotential method based on density functional theory, implemented as CASTEP [8, 9]. The plane-wave cutoff energy is set to 400 eV for all calculations, and ultrasoft pseudopotentials with core correction for both atoms are employed. The generalized gradient approximation (GGA) parametrized by Perdew, Burke and Ernzerhof [10] is used to describe the exchange-correlation functionals. To obtain the point defect "raw" formation energies, 16-site supercells with a 4×4×4 Monkhorst-Pack grid for k-point sampling are adopted; and to evaluate interaction energies of point defects, 24-site supercells with a 3×4×4 sampling grid are employed. The lattice parameters and atomic coordinations of supercells are fully optimized.

Supercell calculation is not very convenient for the systems with random compositions. In order to investigate the composition dependent elastic modulus of TiNi based alloys, we employ another first-principles method, exact muffin-tin orbitals method in combination with coherent potential approximation (EMTO-CPA) [11-13]. In this method the effective potential in the one-electron equation is treated with optimized overlapping muffin-tin approximation, and the random distribution of the alloying atoms is described by the CPA. The bulk modulus B is determined by fitting the total energies versus the volumes according to the Murnaghan equation. Then, to obtain the other elastic moduli $c' = (c_{11}-c_{12})/2$ and c_{44}, we first perform volume-conserving orthorhombic and monoclinic deformations, respectively, that is,

$$\varepsilon_o(\delta) = \begin{pmatrix} \delta & 0 & 0 \\ 0 & \delta & 0 \\ 0 & 0 & \delta^2/(1-\delta^2) \end{pmatrix} \quad (2)$$

and

$$\varepsilon_m(\delta) = \begin{pmatrix} 0 & 0 & \delta \\ 0 & \delta^2/(1-\delta^2) & 0 \\ \delta & 0 & 0 \end{pmatrix} \quad (3)$$

in which the strain δ varies from 0 to 0.05 with an interval of 0.01. And then we fit the total energies $E_o(\delta)$ and $E_m(\delta)$ of 5 deformed crystals to get c' and c_{44} by both

$$E_o(\delta) = E(0) + 2V_0 c' \delta^2 \quad (4)$$

and

$$E_m(\delta) = E(0) + 2V_0 c_{44} \delta^2 \quad (5)$$

For all the EMTO-CPA calculations, the one-electron equation is solved within the scalar-relativistic and soft-core approximation. The $3d^2 4s^2$ electrons of Ti atom and $3d^8 4s^2$ electrons of Ni are treated as valence electrons. The $s, p, d,$ and f orbitals are all included in the EMTO basis set. The Green's Function is calculated for 16 complex energy points distributed exponentially on a semi-circular contour. 10 orbitals are used in the one-center expansion of the full charge density while 8 for the calculation of conventional Madelung energy. Here the GGA-PBE is also employed to describe the exchange-correlation potential. An uniform 25×25×25 k-point grid without smearing technique is used for Brillouin zone sampling in the calculations.

Table I. Basic properties of stoichiometric TiNi with B2 structure

Lattice constant (Å)	Heat of formation (eV/atom)	Bulk modulus (GPa)	c' (GPa)	c_{44} (GPa)	Reference
3.024	-0.37	163	14	51	CASTEP (this work)
3.017	-0.38	156	33	56	EMTO-CPA (this work)
2.977	-0.33	156	12	50	Mixed BS (Ref.[14])
2.960	-	168	9	39	PW (Ref. [15])
3.015	-0.35	140	17-19	35-39	Experiment (Ref. [16-18])

RESULTS AND DISCUSSION

Basic properties of stoichiometric TiNi

We first calculate some basic properties, such as lattice constant, heat of formation, etc. of stoichiometric B2 TiNi intermetallic compound by both CASTEP and EMTO-CPA, and compare them with several existing theoretical [14, 15] and experimental [16-18] results (Table I).

Both of our equilibrium lattice constants from GGA calculations agree well with that from experiments and are a bit larger than those from LDA calculations [14, 15], which is typical for density functional theory calculations since GGA-PBE underestimates the binding of the crystal while LDA overestimates it. So is the case with calculated heats of formation. The elastic constants c_{44} from both calculations are overestimated in comparison with the experimental value, but they are in reasonable agreement. For the calculated c', the one from CASTEP is slightly smaller than the experimental data, whereas the other from EMTO-CPA is much larger.

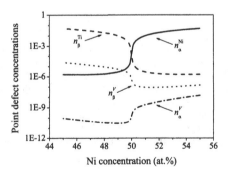

Figure 1. Equilibrium point defect concentrations of TiNi at 923 K, as a function of Ni concentration.

Equilibrium point defect concentrations in TiNi

Figure 1 shows the thermal equilibrium point defects of TiNi at 923 K (from which the TiNi is practically quenched) as a function of the Ni content. On the Ni-rich side, the dominant point defects are the Ni antisite atoms on the Ti (α) sublattice. The concentrations of vacancies on both sublattices are very low. On the Ti-rich side, the major point defects are the Ti antisites on the Ni (β) sublattice. However, the concentration of Ni vacancies is also noticeable. Since excessive Ti in TiNi is easy to precipitate as Ti-rich phases such as Ti_2Ni so that the matrix almost remains stoichiometric, it was believed that the quenched vacancies are responsible for the insignificantly raised MT temperature and broaden the range of temperature from start to finish of the transformation of Ti-rich TiNi relative to the perfect TiNi [19, 20]. In brief, B2 TiNi alloys are antisite type intermetallic compounds.

Interaction between point defects

The interaction energies between two antisites, two vacancies, and between one antisite and

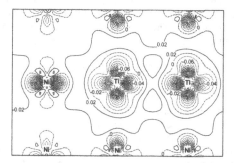

Figure 2. Bonding charge density on the (001) plane of TiNi, with a pair of second-nearest neighboring Ti antisites.

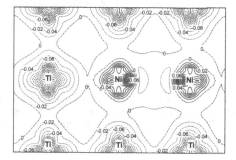

Figure 3. Bonding charge density on the (001) plane of TiNi with a pair of second-nearest neighboring Ni antisites.

one vacancy are calculated according to equation (1). Here we present only those between two antisites since antisites are dominant point defects in TiNi as mentioned above. The calculated interaction energy between one Ni antisite and its nearest-neighboring Ti antisite is -0.47 eV, indicating that they attract each other. For the second-nearest-neighboring interactions, two Ti antisite atoms are attractive (-0.15 eV) while two Ni antisites are strongly repulsive (0.71 eV). Since Ti antisites are attractive to each other, it is expected that they are prone to forming local Ti-rich domains. If the local concentration of Ti is large enough, Ti-rich phases (e.g. Ti_2Ni) will participate so that the matrix remains almost stoichiometric for Ti-rich TiNi. On the contrary, Ni antisites repel each other. Therefore, they prefer to distribute evenly in the matrix and the solubility of excess Ni in TiNi could be significant. These results are in good agreement with experiments [21] and explain reasonably the asymmetric feature of the solid solution limit of TiNi in the Ti-Ni binary phase diagram.

The interaction between the point defects can be understood from our bonding charge density calculations. Figs. 2 and 3 present the bonding charge densities of the supercell with a pair of Ti antisites and a pair of Ni antisites, respectively. As seen from Fig. 2, charge density accumulation is observed between two Ti antisites, which is responsible for the attraction between the Ti antisites. The bond between two Ti antisites is unidirectional, indicating that it is mostly of metallic character. For the Ni antisites, both of them gain electrons, and there exists charge density depletion in between. Therefore, Coulomb static repulsion may dominate the interaction between Ni antisites.

Composition dependence of elastic modulus and its effects on MT temperature

The elastic properties critical to the martensitic transformation in TiNi binary alloys are $\{110\}$ $<1\bar{1}0>$ basal plane shear modulus c' and $\{001\}$ $<1\bar{1}0>$ non-basal shear modulus c_{44}. Experiments have shown that both c' and c_{44} of the high temperature B2 phase soften with decreasing temperature until the MT occurs. With a simple Landau-type model, Ren and Otsuka [5] interpreted the composition dependence of MT temperature from a thermodynamic point of view. They considered the critical elastic modulus at which MT takes place to be independent of the temperature for a martensitic alloy. Therefore, if the elastic modulus is altered by composition, the MT temperature has to be changed to meet the constraint of a constant critical elastic modulus. Using a Landau-type model, they derived a relation between the MT temperature and elastic modulus approximately as follows:

$$\frac{dT_0}{dc} = -\frac{1}{\beta c} \tag{6}$$

with T_0 being the MT temperature and $\beta = dc/(TdT)$ being the temperature coefficient of the elastic modulus c. According to the above relationship, the alloying element that increases the elastic modulus lowers the MT temperature, and vice versa. Such a model provides the possibility of determining the composition dependent MT temperature by surveying the composition dependent elastic modulus of the alloy. However, since both c' and c_{44} of B2 TiNi are quite small and soften with decreasing temperature during MT, one cannot know a priori which elastic modulus dominates the MT if the elastic moduli change oppositely with the composition of the alloy. In order to gain insight into this problem, we calculate the composition

530

dependent elastic modulus of off-stoichiometric TiNi and some ternary TiNi based alloys.

(a) Elastic modulus of off-stoichiometric TiNi

With the point defect configuration reported previously in this paper, we first calculate the c' and c_{44} of the off-stoichiometric TiNi as a function of Ni concentration using EMTO-CPA method, as shown in Fig. 4. On the Ti-rich side, both c' and c_{44} do not change very much with composition. On the Ni-rich side, however, c' decreases significantly with increasing Ni content, whereas c_{44} increases.

The electronic structure origin of the composition dependence of the elastic moduli has been discussed in detail in our paper published elsewhere [22]. As shown previously in the paper, the excess Ni in the Ni-rich TiNi exists as antisite on the Ti sublattice and gains electrons such that Coulomb static repulsion occurs between the Ni antisite and its surrounding Ti atoms. When applying orthorhombic distortion to the Ni-rich TiNi lattice for the calculation of c', the distances between the Ni-antisite and its nearest neighbors increase. Thus, the Coulomb repulsion becomes weaker, which contributes negative energy to the distorted system, slowing down the increase in the total energy of the system with increasing distortion. The higher the concentration of the Ni antisite, the more significant is this effect. Therefore, c' decreases with increasing concentration of Ni antisites. The monoclinic distortion for c_{44} shortens four of the eight distances between the Ni antisite and its nearest neighbors, but dilates the other four. However, a detailed analysis shows that the strengthening of the static Coulomb repulsion due to the shortening of the distances dominates, which contributes positive energy to the distorted system [22]. Consequently, c_{44} increases with increasing concentration of Ni antisites. For the Ti-rich TiNi, the Ti antisites form mainly metallic bonds with their nearest neighbor Ni atoms, so that there is no such Coulomb repulsion as in the case of the Ni-rich alloy. This is why the change of the elastic moduli of the Ti-rich alloy with composition is less significant than that of the Ni-rich alloy.

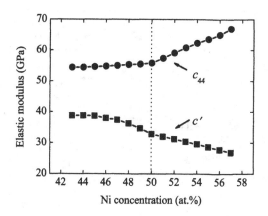

Figure 4. Shear moduli c' and c_{44} as a function of Ni content in TiNi calculated by EMTO-CPA.

(b) Elastic modulus of ternary TiNi-X alloys

The effects of many ternary alloying elements (X) on the MT temperature of TiNi-X alloys have been investigated experimentally. However, the elastic modulus of most of these alloys remains unknown. We calculated the elastic moduli c' and c_{44} of these alloys using the EMTO-CPA method. Figs. 5 and 6 present the change of the elastic modulus of alloys relative to that of the unalloyed TiNi (c_0). It is noted that the hardening/weakening rate, *i.e.* the slope of the

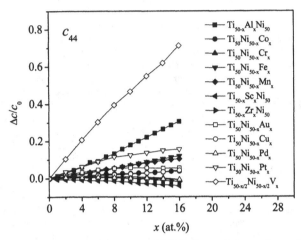

Figure 5. Elastic modulus c_{44} of the ternary TiNi-X alloys relative to the unalloyed TiNi

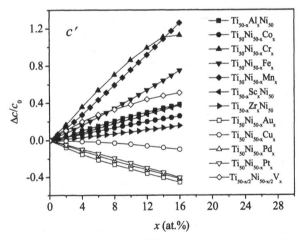

Figure 6. Elastic modulus c' of the ternary TiNi-X alloys relative to the unalloyed TiNi

$\Delta c/c_0(x)$, changes significantly with different alloying elements. Generally speaking, c' is much more sensitive to the composition of the alloy than c_{44}. For most of the elements, the elastic modulus changes monotonously with the concentration of X. However, if we examine the curves more closely, we find that c_{44} of the TiNi-Zr alloy increases slightly with the Zr concentration below 10 at.% but decreases at higher Zr concentrations [23]. This result is consistent with experimental observations (see Ref. [23] for more details).

A notable feature of Figs. 5 and 6 is that the influence of the transition metal (TM) alloying elements on the elastic modulus is more or less related to their position in the periodic table (see also Fig. 7 below). The TM alloying elements located in the middle of the periodic table, such as V, Mn, Fe, Co, *etc.*, generally harden both c' and c_{44}, whereas the late TM alloying elements (Cu, Ni, Pt, Au, *etc.*) harden c_{44} but weaken c'. The early TM alloying elements Sc and Zr increase c' but mainly decrease c_{44}. This fact indicates that the influence of the alloying elements on the elastic modulus is closely related to the electronic configurations of the alloying elements.

(c) Elastic modulus and MT temperature

In order to gain insight into the relation between the composition dependences of the elastic modulus and the MT temperature, we compare the hardening/weakening rates, $\partial(\Delta c/c_0)/\partial x$, of c' and c_{44}, in Fig. 7. Here, the hardening rates are obtained through a linear fit to the $\Delta c/c_0(x)$ curves in Figs. 5 and 6 although some of the curves may deviate from a linear relationship.

For X = Mn, Fe, Al, V, and Co, both c' and c_{44} increase with increasing x. According to the Landau model of Ren, *et al.* [5], these elements should lower the MT temperature. Cr should also decrease the MT temperature since it hardens greatly c' but does not change c_{44}. The alloying of Pd decreases both c' and c_{44}; therefore, it raises the MT temperature. The above results are in good agreement with experimental findings [24-28].

The situation is complicated for X = Cu, Au, Pt, Ni (Ni-rich), Zr, and Sc. For these elements, c' and c_{44} change oppositely with x. For X = Cu, Au, Pt, and Ni, c' decreases whereas c_{44}

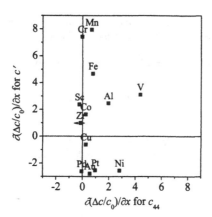

Figure 7. Comparison between the hardening/weakening rates of c' and c_{44} of ternary TiNi-X alloys.

increases. Experiments have demonstrated that Cu [29] and excess Ni [30-34] lower the MT temperature, indicating that c_{44} dominates during the MT although the weakening rate of c' and the hardening rate of c_{44} are almost at the same level. Nevertheless, as shown by the experiments, Au [24, 35] and Pt [36] generally raise the MT temperature. The reason may be that the weakening rate of c' is much lager then the hardening rate of c_{44} for both Au and Pt as seen in Fig. 7, such that c' dominates in this case. For Zr, c' increases linearly with x whereas c_{44} increases slightly at low Zr concentrations and then decreases when the Zr concentration exceeds about 10 at.%. According to the experiments of Feng et al. [37] and Hsieh et al. [38], the MT temperature of a TiNi-Zr alloy first goes down and then goes up with Zr concentrations exceeding 10 at.%, in accordance with the variation of c_{44}, indicating that c_{44} dominates the MT at high Zr concentrations although the hardening rate of c' is somehow larger than the weakening rate of c_{44}. From the above analysis, we may conclude that the MT temperature is much more sensitive to c_{44} than to c'. Only if the hardening/weakening rate of c' is much larger than c_{44} as for the Au and Pt cases, then c' can be the main factor to affect the MT temperature.

SUMMARY

In this study, first-principles calculations by the use of a plane-wave pseudopotential method are performed to investigate intrinsic point defect behavior in TiNi. The elastic moduli critical to MT, i.e. c' and c_{44}, are calculated as a function of the composition of the off-stoichiometric TiNi and a series of ternary TiNi-X alloys, by the use of exact muffin-tin orbital method in combination with coherent potential approximation. The main results are summarized as follows:
1. TiNi is an antisite type of intermetallic compound.
2. Ti antisites in TiNi are attractive to each other whereas Ni antisites are mutually repulsive. The attraction between the Ti antisites is due to the metallic interaction between them, whereas the repulsive force between the Ni antisites is mainly ascribed to the Coulomb static electronic repulsion.
3. Generally speaking, the early transition metal (TM) alloying elements in the periodic table increase c' but decrease c_{44}; the middle ones increase both c' and c_{44}, whereas the late ones decrease c' but increase c_{44}.
4. The MT temperature of TiNi bases alloys is more sensitive to the variation of c_{44} than to that of c'.

ACKNOWLEDGMENTS

The authors acknowledge the financial support from the MoST of China under grant No. 2006CB605104, and from the NSFC under grants No. 50631030 and 50871114.

REFERENCES

1. K. Otsuka and X. Ren, Prog. Mater. Sci. **50**, 511 (2005).
2. S. Kartha, J. A. Krumhansl, J. P. Sethna and L. K. Wickham, Phys. Rev. B **52**, 803 (1995).
3. X. Ren and K. Otsuka, Scripta Metall. **38**, 1669 (1998).

4. X. Ren, K. Taniwaki, K. Otsuka, *et al.* Philos. Mag. A **79**, 31 (1999).
5. X. Ren and K. Otsuka, Mater. Sci. Forum **327-328**, 429 (2000).
6. S. M. Foiles and M. S. Daw, J. Mater. Res. **2**, 5 (1987).
7. J. Mayer, C. Elsasser and M. Fahnle, Phys. Status Solidi B **191**, 283 (1995).
8. M. C. Payne, M. P. Teter, D. C. Allan, T. A. Arias and J. D. Joannopoulus, Rev. Mod. Phys. **64**, 1045 (1992).
9. M. D. Segall, P. L. D. Lindan, M. J. Probert, C. J. Pickard, P. J. Hasnip, S. J. Clark and M. C. Payne, J. Phys.: Condens. Matter **14**, 2717 (2002).
10. J. P. Perdew, K. Burke and M. Ernzerhof, Phys. Rev. Lett. **77**, 3865 (1996).
11. L. Vitos, I. A. Abrikosov and B. Johansson, Phys. Rev. Lett. **87**, 156401 (2001).
12. S. Soven, Phys. Rev. **156**, 809 (1967).
13. B. L. Györffy, Phys. Rev. B **5**, 2382 (1972).
14. Y. Y. Ye, C. T. Chan and K. M. Ho, Phys. Rev. B **56**, 3678 (1997).
15. X. Huang, C. Bungaro, V. Godlevsky and K. M. Rabe, Phys. Rev. B **65**, 014108 (2001).
16. O. Mercier, K. N. Melton, G. Gremaud and J. Hagi, J. Appl. Phys. **51**, 1833 (1980).
17. V. N. Khachin, S. A. Muslov, V. G. Pushin and Y. I. Chumlyakov, Sov. Phys. Dokl. **32**, 606 (1980).
18. T. M. Brill, S. Mittlebach, W. Assmus, M. Mullner and B. Luthi, J. Phys.: Condens. Matter **3**, 9621 (1991).
19. J. E. Hanlon, S. R. Butler and R. J. Wasilewski, Trans. Metall. Soc. AIME **239**, 1323 (1967).
20. R. J. Wasilewski, S. R. Butler, J. E. Hanlon and D. Worden, Metall. Trans. **2**, 229 (1971).
21. J. M. Lu, Q. M. Hu, L. Wang, Y. J. Li, D. S. Xu and R. Yang, Phys. Rev. B **75**, 094108 (2007).
22. J. M. Lu, Q. M. Hu and R. Yang, Acta Mater. **56**, 4913 (2008).
23. Q. M. Hu, R. Yang, J. M. Lu, L. Wang, B. Johansson and L. Vitos, Phys. Rev. B **76**, 224201 (2007).
24. K. H. Echelmeyer, Script Metall. **10**, 667 (1976).
25. K. R. Edmonds and C. M. Hwang, Script Metall. **20**, 733 (1986).
26. H. B. Xu, C. B. Jiang, S. K. Gong and G. Feng, Mater. Sci. Eng. A **281**, 234 (2000).
27. H. Hosoda, S. Hanada, K. Inoue, T. Fukui, Y. Mishima and T. Suzuki, Intermetallics **6**, 291 (1998).
28. H. C. Lin, K. M. Lin, S. K. Chang and C. S. Lin, J. Alloy Comp. **284**, 213 (1999).
29. T. H. Nam, T. Saburi and K. Shimizu, Mater. Trans. JIM **31**, 959 (1990).
30. I. I. Kornilov, Y. V. Kachur and O. K. Belousov, Fizika. Metall. **32**, 420 (1971).
31. S. Mayazaki, K. Otsuka and Y. Suzuki, Scr. Metall. **15**, 287 (1981).
32. K.N. Melton and O. Mercier, Acta Metall. **29**, 393 (1981).
33. S. Mayazaki and K. Otsuka, Metall. Trans. **17**, 53 (1986).
34. M. Nishida, C.M. Wayman and T. Honma, Metall. Trans. A **17**, 1505 (1986).
35. S. K. Wu and C. M. Wayman, Script Metall. **21**, 83 (1987); Metallography **20**, 359 (1987).
36. H. Honsoda, M. Tsuji, M. Mimura, Y. Takahashi, K. Wakashima and Y. Ymabe-Mitarai, Mat. Res. Soc. Symp. Proc. **753**, BB5.51.1 (2003).
37. Z. W. Feng, B. D. Gao, J. B. Wang, D. F. Qian and Y. X. Liu, Mater. Sci. Forum **394-395**, 365 (2002).
38. S. F. Hsieh and S. K. Wu, Mater. Charact. **151**, 41 (1998).

Mater. Res. Soc. Symp. Proc. Vol. 1128 © 2009 Materials Research Society 1128-U09-06

Plastic Deformation Behavior of Ti₅Si₃ Single Crystals

Kyosuke Kishida, Masakazu Fujiwara, Norihiko L. Okamoto, Katsushi Tanaka and Haruyuki Inui
Department of Materials Science and Engineering, Kyoto University
Sakyo-ku, Kyoto 606-8501, JAPAN

ABSTRACT

Deformation behavior of binary stoichiometric Ti_5Si_3 single crystals was examined as a function of the loading axis orientation and temperature. Two different types of deformation modes, namely $\{1\bar{1}00\}[0001]$ prism slip, $\{2\bar{1}\bar{1}2\}1/3<2\bar{1}\bar{1}3>$ pyramidal slip were newly identified to be activated above 1300 °C depending on the loading axis orientation. Critical resolved shear stresses (CRSS) for the $\{1\bar{1}00\}[0001]$ prism slip and $\{2\bar{1}\bar{1}2\}1/3<2\bar{1}\bar{1}3>$ pyramidal slip were estimated to be about 130 MPa and 330 MPa at 1400 °C, respectively. The values of the CRSS for these two slip systems decrease monotonously with increasing the temperature.

INTRODUCTION

The intermetallic compound Ti_5Si_3 has attracted a great deal of interest as a new class of high-temperature structural materials, because of its high melting temperature (2130 °C), low density (4.32 g/cm^3) and good oxidation resistance [1-3]. Many studies have been carried out to utilize this material as a strengthening phase in some two phase composite alloys such as Ti/Ti₅Si₃ [1], Ti₃Al/Ti₅Si₃ [4,5], TiAl/Ti₅Si₃ [6,7], TiN/Ti₅Si₃ [8]. However, very few studies have been reported so far on mechanical properties of Ti_5Si_3 itself investigated using polycrystals [1,3] and single crystals [2]. Ti_5Si_3 has a complex hexagonal-type D8₈ structure as shown in figure 1. Because of this complex crystal structure, Ti_5Si_3 is quite brittle at low temperature. Plastic deformation of polycrystalline Ti_5Si_3 has been confirmed in compression tests above 1000 °C, however, any details on the deformation mechanism has been described [1]. An alternative study using single crystals has reported the operation of deformation twinning of $\{1\bar{1}02\}<\bar{1}101>$-type in compression test with some limited loading axis orientations at high temperature above 1300 °C [2]. Except for the limited activation of the deformation twin, almost nothing has been known about the operative deformation modes in Ti_5Si_3. In the present study, we have studied the plastic deformation behavior of Ti_5Si_3 by compression tests of single crystals with various loading axis orientations in order to understand the mechanical properties and the operative deformation modes of Ti_5Si_3.

EXPERIMENTAL

Single crystals of binary stoichiometric Ti_5Si_3 with a nominal composition of Ti-37.5 at%Si were grown using an optical floating zone (FZ) furnace at a growth rate of 6 mm/h under Ar gas flow. Oriented specimens for compression tests with dimensions of 1.5 x 1.5 x 4 mm^3 were sectioned from the as-grown single crystals by electric discharge machining. All specimens were mechanically polished and then finished with 0.3 μm alumina abrasive powders prior to compression tests. Four loading axis orientations: namely [2 2 0 5], [4 3 1̄ 0], [0 0 0 1] and [2 1 1̄

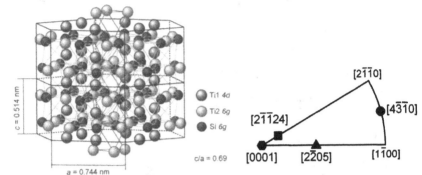

Ti1 4d		$[2\bar{1}\bar{1}0]$	
Ti2 6g	$[2\bar{1}\bar{1}24]$		$[4\bar{3}\bar{1}0]$
Si 6g			
$c/a = 0.69$	$[0001]$	$[2\bar{2}05]$	$[1\bar{1}00]$

$c = 0.514$ nm

$a = 0.744$ nm

Figure 1. Crystal structure of Ti_5Si_3.　　　　**Figure 2**. Loading axis orientations.

Table 1. The highest Schmid factor value for each possible deformation mode in Ti_5Si_3 single crystals with four different loading axis orientations tested in this study.

Deformation mode		Loading axis orientation			
		[2205]	[4310]	[0001]	[21124]
Basal slip	$(0001)1/3\langle2\bar{1}\bar{1}0\rangle$	0.433	0	0	0.175
Prism (I) slip	$\{1\bar{1}00\}[0001]$	0.500	0	0	0.152
Prism (II) slip	$\{2\bar{1}\bar{1}0\}[0001]$	0.433	0	0	0.175
Pyramidal slip	$\{2\bar{1}\bar{1}2\}1/3\langle2\bar{1}\bar{1}3\rangle$	0.233	0.432	0.468	0.500
Twinning	$\{1\bar{1}02\}\langle\bar{1}101\rangle$	0.364	0.324	-	-

24] were selected as shown in figure 2. The highest values of the Schmid factors for the possible deformation modes are summarized in table 1. Compression tests were carried out on an Instron-type testing machine at a strain rate of 1×10^{-4} s^{-1} at temperatures ranging from 1200 to 1500 °C in vacuum. Deformation structures were examined by optical microscopy (OM), scanning electron microscopy (SEM) and transmission electron microscopy (TEM).

RESULTS AND DISCUSSION

Figure 3 shows the temperature dependence of yield stress for four different orientations investigated. The mark × in the figure indicates the occurrence of failure in the elastic region. Plastic deformation is observed above 1300 °C for [2205] and [21124] orientations and above 1400 °C for [4310] and [0001] orientations. The values of the yield stress for these four orientations decrease monotonously with increasing the temperature. Similarly to the case of many transition-metal silicides [9,10] and in covalent semiconductors [11], apparent yield drop behavior is observed in stress-strain curves for all plastically deformed samples (not shown in this paper). This suggests that the density of grown-in dislocations is quite low in as-grown single crystals and also that the initiation of dislocation or deformation twin is more difficult than their propagation in Ti_5Si_3.

Figure 4 shows deformation markings observed on two orthogonal surfaces of Ti₅Si₃ single crystals with four different loading axis orientations. For [2205] orientation, coarse traces are clearly observed on a (5̄504) plane, while traces are unclear on a (1̄120) plane as shown in figure 4a. Trace analysis confirms that the traces are parallel to the (11̄00) prism plane and slip

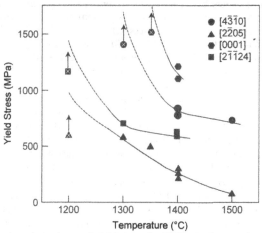

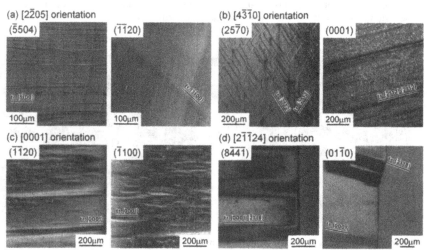

Figure 3. Temperature dependence of yield stress for Ti₅Si₃ single crystals with four different loading axis orientations.

Figure 4. Deformation markings observed on two orthogonal surfaces of Ti₅Si₃ single crystals with four different loading axis orientations deformed at 1400 °C.

direction is parallel to the (1̄120) plane, indicating the operation of (1̄100)[0001] prism slip in the [2205]-oriented sample. The operation of the (1̄100)[0001] prism slip is further confirmed by TEM observation. Figure 5 illustrates bright-field and weak-beam images of dislocations observed in a TEM foil cut parallel to the (1̄100) slip plane. Most dislocations are aligned parallel to [1̄120] or <1̄123>, i.e. they are edge or mixed dislocations. Weak-beam analysis confirms that the dislocations have the Burgers vector of [0001] and are not dissociated into partial dislocations with shorter Burgers vectors (figure 5b). Based on the analysis of the deformation mode, critical resolved shear stress (CRSS) for the (1̄100)[0001] prism slip is estimated to be about 300 MPa and 130 MPa at 1300 °C and 1400 °C, respectively.

For [43̄1̄0] orientation, slip traces corresponding to (2̄112) and (21̄12) are observed as seen in figure 4b, suggesting the simultaneous operation of two types of {21̄12} pyramidal slips. Deformation microstructure in a TEM foil cut parallel to one of the slip plane (21̄12) is examined by TEM as shown in figure 6. As marked in the figure, a dislocation segment marked C is dissociate or decomposed into two dislocations A and B (or D and E). Further contrast analysis reveals that the Burgers vectors of the dislocation segments A, B(D) and C(E) is [0001], 1/3[21̄1̄0] and 1/3[21̄1̄3], respectively. Since the Burgers vector of [0001], 1/3[21̄1̄0] are not on the observed slip plane of (21̄12), they are likely to be generated as a result of climb decomposition of the 1/3[21̄1̄3] dislocation. It is thus concluded that the {21̄12}1/3<21̄1̄3> pyramidal slips are operative in sample compressed along [43̄1̄0] and are climb-decomposed into [0001]- and 1/3[21̄1̄0]-dislocations. CRSS for the {21̄12}1/3<21̄1̄3> pyramidal slip is estimated to be about 330 MPa at 1400 °C, which is about 2.5 times higher than that of the (1̄100)[0001] prism slip (about 130 MPa) and that reported for the deformation twinning of {11̄02}<1̄101>-type [2]. Comparing the value of the CRSS and Schmid factors (table 1), the {11̄02}<1̄101> deformation twinning is considered to be operative in the [43̄1̄0]-oriented samples. It is, however, the {11̄02}<1̄101> deformation twinning is not observed in any samples investigated in the present study. The reason for this discrepancy has not clarified yet.

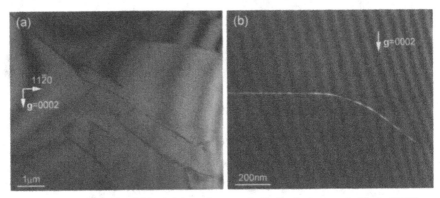

Figure 5. (a) bright field and (b) weak-beam images of a Ti₅Si₃ single crystal with the [2205] orientation deformed at 1400 °C.

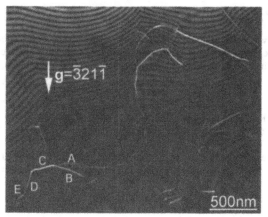

Figure 6. A weak-beam image of dislocations in a Ti₅Si₃ single crystal with the [4310] orientation deformed at 1400 °C.

In contrast to [2205] and [4310] orientations, clear slip traces are not observed for [0001]-oriented samples as shown in figure 4c. Instead, thick and wavy deformation bands, traces of which are inclined less than 10 ° from that of (0001), are observed on the two orthogonal surfaces of [0001]-oriented samples. In order to determine the habit planes of the deformation bands, deformation behavior of additional samples with the loading axis of [2̄1124], which is about 10 °-away from [0001], were investigated as shown in figure 4d. Trace analysis of [2̄1124]-oriented samples reveals that the habit planes are approximately on either (2̄118) or (0001). Preliminarily electron backscatter diffraction (EBSD) analysis reveals that the crystal structure inside both types of the deformation bands is identical with that of the matrix and also orientation difference between the deformation band and matrix is about 20 ° and 3 ° for the former and latter, respectively. These deformation bands are thus considered to be deformation twin or kink band. Details of the deformation bands will be published elsewhere.

CONCLUSIONS

High temperature plastic deformation behavior of Ti₅Si₃ has been investigated by compression tests of single crystals with various loading axis orientations. Two types of deformation modes, namely {1̄100}[0001] prism slip and {21̄12}1/3<21̄13> pyramidal slip are confirmed to be operative at high temperature above 1300 °C. Critical resolved shear stresses (CRSS) for the {1̄100}[0001] prism slip and {21̄12}1/3<21̄13] pyramidal slip were estimated to be about 130 MPa and 330 MPa at 1400 °C, respectively. Deformation bands are developed when the loading axis is close to [0001].

ACKNOWLEDGMENTS

This work was partly supported by the Global COE (Center of Excellence) Program of International Center for Integrated Research and Advanced Education in Materials Science from the Ministry of Education, Culture, Sports, Science and Technology (MEXT), Japan.

REFERENCES

1. G. Frommeyer, R. Rosenkranz and C. Lüdecke, Z. Metallk., 81, 307 (1990).

2. Y. Umakoshi and T. Nakashima, Scripta Metall. Mater, 30, 1431 (1994).

3. L. Zhang and J. Wu, Acta Mater., 46, 3535 (1998).

4. J.S. Wu, P.A. Beaven and R. Wagner, Scripta Metall. Mater, 24, 207 (1990).

5. D. Vojtech, M. Novak, P. Novak, P. Lejcek and J. Kopecek, Mater. Sci. Engng., A489, 1 (2008).

6. F.S. Sun, C.X. Cao, S.E. Kim, Y.T. Lee and M.G. Yan, Metall. Mater. Trans. 32A, 1233 (2001).

7. F.S. Sun and F.H. Froes, Mater. Sci. Engng., A345, 262 (2003).

8. Y. Suehiro and K. Ameyama, J. Mater. Proc. Tech., 111, 118 (2001).

9. H. Inui, M. Moriwaki, K. Ito and M. Yamaguchi, Philos. Mag. A, 77, 375 (1998).

10. H. Inui, M. Moriwaki, N. Okamoto and M. Yamaguchi, Acta Mater., 51, 1409 (2003).

11. H. Alexander and P. Haasen, Solid State Phys., 22, 28 (1968).

Mater. Res. Soc. Symp. Proc. Vol. 1128 © 2009 Materials Research Society   1128-U09-07

Plastic Deformations in Single Crystals of FePd With the L1$_0$ Structure

Katsushi Tanaka, Wang Chen, Kyosuke Kishida, Norihiko L. Okamoto and Haruyuki Inui
Department of Materials Science and Engineering, Kyoto University, Sakyo-ku, Kyoto 606-8501, Japan.

ABSTRACT

Compressive deformations of L1$_0$-ordered single crystals of FePd have been investigated from room temperature to 873 K. The critical resolved shear stress for superlattice dislocations is hard to determine resulting from buckling that occurs after a small amount of conventional plastic deformation. The CRSS for superlattice dislocations determined from yield stress is significantly larger than that of ordinary dislocations. The CRSS for octahedral glide of ordinary and superlattice dislocations are virtually independent of the temperature, and the positive temperature dependence of the yield stress is not observed for both, ordinary and superlattice dislocations, by the present experiments.

INTRODUCTION

Intermetallic compounds and ordered alloys with the L1$_0$ structure are of considerable commercial interests for both, structural and functional applications. Owing to their technological interests, extensive investigations have been performed on the mechanical properties of the compound TiAl [1]. The studies of Ti-56 at.% Al show a positive temperature dependence of the yield stress from room temperature up to a peak temperature at about 900 K and both, ordinary and super dislocations, contribute to the anomaly. The CRSS for the superlattice slip system is slightly larger than that for the ordinary slip system in equiatomic compounds but significantly smaller than that for the ordinary slip system in Al-rich compounds [2]. On the other hand, a study of the deformation behavior of poly-crystalline FePt shows a monotonic decrease in the yield stress over the temperature range from 77 to 873 K [3]. A study of the deformation behavior of FePd poly-crystals shows a positive temperature dependence from room temperature up to a peak temperature at about 650 K [4]. Since the latter two studies had been carried out with poly-crystals, however, the contributions of ordinary and super dislocations to the plastic deformation have not been clarified yet.

In the Co-Pt, Fe-Pd and Fe-Pt systems, the ordered L1$_0$ phase forms in conjunction with a cubic (fcc) to tetragonal (L1$_0$) transformation, resulting in a microstructure with so-called poly-twinned morphology, that is a lamellar structure consisting of {101} thin twin plates. This microstructure makes investigations of mechanical properties of single crystals of these compounds difficult. Recently, one of the authors has found that single crystals of CoPt, FePd and FePt with the L1$_0$ structure form from single crystals with the fcc-disordered phase under an appropriate strength of a compressive stress or of a magnetic field [5]. Using samples obtained by these methods, physical properties of L1$_0$-ordered FePd single crystals have been investigated [6, 7]. Although we have reported our preliminary results of the plastic deformation behavior of single crystals of FePd with the L1$_0$ structure [8], the crystals used in those experiments involve many anti-phase boundaries in them. In order to clarify the deformation properties of single crystals of FePd with the L1$_0$ structure, the experiments with well ordered crystals are desired. In this report, we will show the characteristic features of plastic deformations of FePd single crystals with small amount of APB's in them.

EXPERIMENT

An ingot of Fe-50 at.% Pd was prepared by arc melting from an appropriate mixture of Fe (99.99%) and Pd (99.95%) under an argon atmosphere. A single crystal bar with 25 mm in diameter was grown by a modified Bridgman technique at a growth rate of 6 mm/h. After determination of the crystallographic orientation by the conventional back scatter Laue method, bars with longitudinal direction parallel to the <100> crystallographic orientation were cut by spark machining.

Prior to the heat treatment for ordering, all the specimens were disordered at 1073 K (the order-disorder transition temperature is about 923 K) in an evacuated quartz tube, and quenched into iced brine so as to attain the disordered state. For the ordering treatment, the specimens were heated up to 813 K at a rate of 5 K/min, kept at the temperature for 1 hour and cooled down at a rate of 2 K/min. In order to obtain $L1_0$ single crystals, a magnetic field of 8.0 MAm^{-1} was kept applied along the longitudinal direction of the specimens, that corresponds to the <100> crystallographic directions throughout the heat treatment for ordering. Further ordering treatment to coarsen anti-phase domains (APDs) was applied at 873 K for 1 week in an evacuated quartz tube without a magnetic field. Traces of transformation twins of {101}-type formed at the ordering process were observed in some specimens. We rejected any specimens containing transformation twins.

Specimens for compression tests with orientations of $[\bar{2}53]$ and $[\bar{6}71]$ crystallographic directions, measuring 1.5 mm $\times$1.5 mm in cross section and 4 mm length, were cut from the $L1_0$ single crystals by a spark machining. Such specimens were mechanically polished prior to

Table 1. Schmid factors for possible deformation mode for the three different orientations investigated.

	Slip system	Schmid factor for following orientations	
		$[\bar{2}53]$	$[\bar{6}71]$
$\{111\}\langle110\rangle$	$(111)[1\bar{1}0]$	0.453	0.122
	$(\bar{1}11)[110]$	0.324	0.066
	$(1\bar{1}1)[110]$	0.130	0.056
	$(11\bar{1})[\bar{1}10]$	0	0
$\{111\}\langle101\rangle$	$(111)[\bar{1}01]$	0.315	0.066
	$(\bar{1}11)[101]$	0.084	0.339
	$(1\bar{1}1)[\bar{1}01]$	0.210	0.397
	$(11\bar{1})[101]$	0	0
	$(111)[0\bar{1}1]$	0.148	0.058
	$(\bar{1}11)[0\bar{1}1]$	0.246	0.407
	$(1\bar{1}1)[011]$	0.342	0.454
	$(11\bar{1})[011]$	0	0
$\{111\}\langle112\rangle$	$(111)[\bar{1}\bar{1}2]$	0.097	-0.005
	$(\bar{1}11)[1\bar{1}2]$	-0.094	-0.434
	$(1\bar{1}1)[\bar{1}12]$	-0.320	-0.495
	$(11\bar{1})[\bar{1}\bar{1}2]$	0	0

mechanical testing. The Schmid factors for the possible deformation modes are listed in Table 1 for each orientation. On the Schmid factor basis with the assumption that the CRSSs are identical for all systems, <110> ordinary slip is favored for the orientation of [$\overline{2}$53], while <101> superlattice slip is favored for the orientation of [$\overline{6}$71].

Compression tests were carried out on an Instron-type testing machine at an initial strain rate of 1.0×10^{-4} s^{-1} in the temperature range from room temperature to 823 K. According to the measurement of the lattice constants [7], the change in the long range order parameter of the L1$_0$ ordered arrangement is very small in this temperature range. All the tests were made in vacuum to prevent a surface oxidation. After yield was observed, the test was stopped and then the deformation markings on the specimen surfaces were observed by an optical microscope.

RESULTS AND DISCUSSION

Stress-strain behavior and slip system activated

Figure 1 shows the stress-strain curves obtained at room temperature to 823 K for the specimens oriented to [$\overline{2}$53] and [$\overline{6}$71] together with our previous results in which the specimens involve many APBs in them (size of APD ~ 500 nm). Smooth yielding and subsequent virtually constant deformation stress after yielding is observed for the specimens oriented to [$\overline{2}$53]. The yield stresses of the present results are slightly smaller than our previous results. This decrease in the yield stress is simply explained by decreasing APB hardening in the present specimens. On the other hand, the present specimens oriented to [$\overline{6}$71] were hard to deform in all the investigated temperature range, in contrast to our previous results where the specimens could be deformed at a yield stress of about 400 MPa. This may be due to the fact that the long range order parameter in the present specimens increases by applying the additional heat treatment for ordering. Most of the present specimens buckled after a small plastic deformation of about 0.1% as indicated by the arrows in Fig. 1(b).The shape change of the

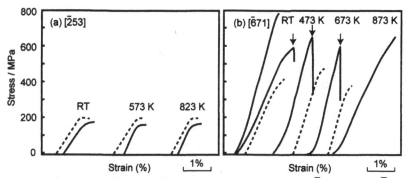

Figure 1. Stress-strain curves of Fe-50.0at.%Pd single crystal with (a) [$\overline{2}$53] and (b) [$\overline{6}$71] orientations at selected temperatures. Dashed curves are our previous results with the specimens involving many APBs [8]. Buckling has been observed at the points indicated by the arrows in (b).

(a)   (b)

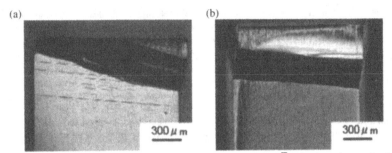

Figure 2. Optical micrograph showing a buckling occurred for [6̄71] oriented crystals after a small plastic deformation. (a) (760) and (b) (6̄7̄85) surfaces.

specimens after buckling is shown in Fig. 2. The buckling has a crystallographic habit plane of about {320}, however further experiments are required to understand the phenomenon.

Deformation markings observed on two orthogonal faces indicate that the slip systems activated in [2̄53]- and [6̄71]-crystals are ordinary slip on (111)[11̄0] and superlattice slip on (11̄1)[011] as expected from the Schmid factors for the temperature range investigated. Since the magnitude of the deformation is small and a buckling occurs, the deformation markings on (11̄1)[011] for the [6̄71]-crystals are not prominent. Since the superlattice dislocation slip system was not activated very well, it is hard to discuss regarding the absolute value of the respective CRSS and its temperature dependence. However, it is clear that CRSS for the superlattice slip system is considerably (about 3.5 times) larger than that for the ordinary slip system at all the investigated temperatures. This agrees with the fact that most of the dislocations observed in a deformed poly-crystalline specimen are ordinary dislocations [4, 9].

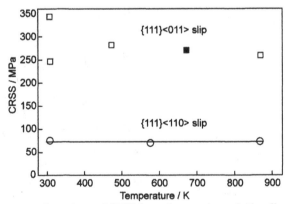

Figure 3. Temperature dependence of CRSS for ordinary and superlattice slip system in Fe-50 at.%Pd single crystals. Filled symbol indicates that the stress is limited by buckling.

546

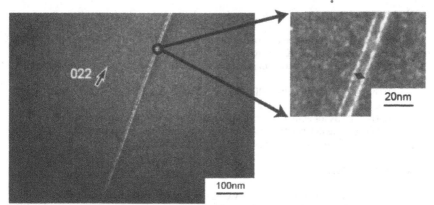

Figure 4. Transmission electron micrographs taken under a weak beam condition ($g = 022$ and $5g$ is excited) showing a dissociation of a superlattice dislocation induced at room temperature.

Dissociation of superlattice dislocations and CRSS for superlattice slip system

The dissociation of superlattice dislocations to superpartial dislocations has been observed in deformed polycrystalline FePd [9]. Since the magnitude of the plastic deformation for superlattice slip is small, it is difficult to find superlattice dislocations in the present deformed specimens. Figure 4 shows an observed superlattice dislocation and its dissociation to a pair of superpartials. The observed superlattice dislocations are almost straight lines aligned to the direction of its Burgers vector, that is, the nature of the dislocations are almost pure screw type. This corresponds to the lowest elastic energy for superlattice dislocations. Only two superpartials are observed with the dissociated width of 7.1 nm. The anti-phase boundary (APB) energy estimated from the observed width of the dissociated superpartials and elastic constants [10] is 91 mJm^{-2} which is slightly smaller than that estimated from the order-disorder transition temperature [11] of 115mJm^{-2}.

From the manner of the dissociation of the superlattice dislocations, it is hard to understand why CRSS for the superlattice slip system is significantly higher than that for ordinary dislocations. One of the possibilities is the fact that all grown-in dislocations have the nature of (1/2)<110> whose length corresponds to that of superpartials resulting from the existence of order-disorder transformation. This indicates that some dislocation reactions to produce a pair of superpartials are required when superlattice slip systems are activated. If grown-in superpartials move with remaining APBs behind, an additional shear stress against the back stress to shrink the fault is required. The estimated additional shear stress to move a superpartial alone is 400 MPa from the APB energy, which is much higher than the observed CRSS for the superlattice slip system. This can be one of the reasons why the number of superlattice dislocations induced is small in Fe-50 at.% Pd. Further investigations are required to make properties of the superlattice slip system clear.

547

CONCLUSIONS

Compressive deformation of $L1_0$-ordered single crystals of FePd have been investigated from room temperature to 873 K. The results are summarized as follows.
(1) The superlattice slip system is hard to activate and buckling occurs after very small plastic deformation.
(2) The CRSS for superlattice slip is about 3.5 times larger than that for ordinary slip in all the investigated temperature range.
(3) The CRSSs for both, ordinary and superlattice dislocation slip, are virtually temperature independent in the temperature range from room temperature to 873 K.
(4) Superlattice dislocations dissociate to two superpartials.

ACKNOWLEDGMENTS

A part of this work was supported by Grant-in-Aid for Scientific Research (A) (19656179) from the Ministry of Education, Culture, Sports, Science and Technology (MEXT), Japan, by the Grobal COE (Center of Excellence) Program of International Center for Integrated Research and Advanced Education in Materials Science from the MEXT, Japan and by Asahi Glass Foundation.

REFERENCES

1. M. Yamaguchi and H. Inui: *Structural Intermetallics*, edited by R. Darolia, J.J. Lewandowski, C.T. Liu, P.L. Martin, D.B. Miracle and M.V. Nathal, TMS (Warrendale, PA), 127 (1993).
2. H. Inui, M. Matsumuro, D.-H. Wu and M. Yamaguchi, *Phil. Mag. A* **75**, 395 (1997).
3. S.H. Whang, Q. Feng and Y.-Q. Gao, *Acta Mater.*, **46**, 6485 (1998).
4. M. Rao and W.A. Soffa, *Scripta Mater.*, **36**, 735 (1997).
5. K. Tanaka, T. Ichitsubo and M. Koiwa, *Mater. Sci. & Eng. A*, **312**, 118 (2001).
6. H. Shima, K. Oikawa, A. Fujita, K. Fukamichi, K. Ishida and A. Sakuma, *Phys. Rev. B*, **70**, 224408 (2004).
7. T. Mehaddene, E. Kentzinger, B. Hennion, K. Tanaka, H. Numakura, A. Marty, V. Parasote, M.C. Cadeville, M. Zemirli and V. Pierron-Bohnes, *Phys. Rev. B*, **69**, 024304 (2004).
8. K. Tanaka, H. Ide, Y. Sumi, K. Kishida, H. Inui, *Mater. Sci. Forum*, **561-565**, 459 (2007).
9. H. Xu, J.M.K. Wiezorek, *Acta Mater.*, **52**, 395 (2004).
10. T. Ichitsubo and K. Tanaka, *J. Appl. Phys.*, **96**, 6220 (2004).
11. T. Ichitsubo, K. Tanaka, H. Numakura and M. Koiwa, *Phys. Rev. B*, **60**, 9198 (1999).

Mater. Res. Soc. Symp. Proc. Vol. 1128 © 2009 Materials Research Society   1128-U09-11

Phase Field Simulation of Coarsening Kinetics in Al-Sc and Al-Sc-Zr Alloys

H. Zapolsky, J. Boisse, R. Patte and N. Lecoq
GPM, UMR 6634 CNRS, University of Rouen, Avenue de l'Université, BP 12, 76801 Saint-Etienne du Rouvray, France

ABSTRACT

The coarsening kinetics of γ' precipitates in binary and ternary $Al_3Sc_{1-x}Zr_x$ alloys is studied by using the two- and three-dimensional phase-field simulations. Our focus is on the influence of diffusion coefficients of Sc and Zr atoms on the transformation path kinetics from disordered f.c.c. matrix to two phases equilibrium state with γ' precipitates and f.c.c. disordered matrix. Our simulation results demonstrate that in the case of binary alloys taking into account the concentration dependence of the mobility of atoms decreases the coarsening rate. In the case of ternary alloys we show that the Al_3Sc particles precipitate first following by appearance of a Zr-rich shell. Our simulations results are in good agreement with experimental observations.

INTRODUCTION

Scandium (Sc) is considered as a promising alloying element in aluminium (Al) alloys, and it has been shown that a small Sc-addition can improve a range of material properties. Al-Sc-based alloys exhibit a unique combination of high strength and plasticity, corrosion resistance and weldability [1,2]. Most of the beneficial effects from the Sc-addition are linked to the formation of the Al_3Sc precipitates. These ordered $L1_2$ (γ') precipitates increase the tensile strength and inhibit recrystallisation. Recently it was demonstrated that these precipitates remain coherent up to sizes of 20 nm in diameter [3]. The low Zr addition lead to the rapid nucleation of $Al_3(Sc,Zr)$ dispesoids (the same $L1_2$ structure) at high density with even higher resistance to the coarsening and recrystallization of Al-Sc alloy [4,5]. Harada and Dunand [6] shown that zirconium addition reduces the lattice parameter of Al_3Sc and concomitantly the interface free and elastic strain energies. Several studies have shown that $Al_3(Sc,Zr)$ dispersoids have a chemical heterogeneous structure [7,8]. The precipitate core is essentially scandium rich and the shell shows a high zirconium concentration. This peculiar chemical structure can be attributed to two complementary effects: the combination of the very different diffusion coefficient of Zr and Sc and the decrease of elastic energy by the presence at the shell Zr atoms.

The aim of the present paper is to study the precipitation kinetics in binary Al-Sc and in ternary Al-Sc-Zr alloys. Phase field approach based on the Cahn-Hilliard and Ginzburg-Landau equation has been used. We examined the morphology and temporal evolution of the microstructure of Al-Sc alloys with three different volume fractions of ordered phase as well as formation of the heterogeneous precipitates in ternary $Al_3(Sc,Zr)$ systems.

MODEL DESCRIPTION

Phase field model has been extensively used for microstructure simulation studies because of its ability to reproduce complicated microstructures without any *a priori* assumptions [9-11]. We consider a model for two-phase system, in which parent phase has a disordered fcc structure and the precipitates have ordered L1$_2$ structure, appearing in four translation variants. In this case the kinetic equations that describe the temporal evolution of the composition variable c(r,t) and the long range order parameters η(r,t) are:

$$\frac{\partial c(\mathbf{r},t)}{\partial t} = \nabla M(\mathbf{r},t)\nabla(\frac{\partial F}{\partial c(\mathbf{r},t)}) + \xi \quad\quad (1a)$$

$$\frac{\partial \eta_i(\mathbf{r},t)}{\partial t} = -L\frac{\partial F}{\partial \eta_i(\mathbf{r},t)} + \varsigma_i \quad\quad i=1,2,3 \quad\quad (1b)$$

where L and M are kinetic coefficients, F is a total non-equilibrium free energy functional, ξ and ζ are the Langevin's noise terms describing the compositional and structural thermal fluctuations, respectively.

The total free energy of the system consists of the chemical free energy F$_{ch}$ and the elastic energy E$_{el}$. Assuming an isotropic interfacial energy, the chemical energy F$_{ch}$ in the diffuse-interface description can be written as:

$$F_{ch} = \int_V (f(c,\eta_j) + \frac{\alpha}{2}(\nabla c)^2 + \frac{\beta}{2}\sum_{j=1}^{3}(\nabla \eta_j)^2)dV \quad\quad (2)$$

where α and β are gradient energy coefficients, f is the local free energy density which can be expressed in Landau polynomial form. In the case of ternary systems it takes the following form:

$$f(c_a,c_b,\eta_1,\eta_2,\eta_3) = \frac{1}{2}A_1(c_a + c_b - c_1)^2 + \frac{1}{2}A_2(c_2 - c_a - c_b)\sum_{p=1}^{3}\eta_p^2 + \frac{1}{3}A_3\eta_1\eta_2\eta_3 + \frac{1}{4}A_4\sum_{p=1}^{3}\eta_p^4$$

$$+ \frac{1}{4}A_5(\eta_1^2\,\eta_2^2 + \eta_1^2\eta_3^2 + \eta_2^2\eta_3^2) \quad\quad (3)$$

where c_a and c_b are the composition variables of scandium and zirconium atoms, c$_1$ and c$_2$ are constants with values close to the equilibrium compositions of the matrix and precipitates, respectively and A$_i$s are constants for the given temperature. The coefficients and constants were chosen such that the free energy curve provides a qualitative description of the thermodynamics of the Al-Sc and Al-Sc-Zr system at 375°C, with the constraint that for the disorder-order phase transition to be first ordered, A$_3$ must be non-zero.

Khachaturyan's model [12] is used for the description of elastic energy arising from the lattice misfit, with a homogeneous modulus approximation. In our calculations we assumed that the lattice parameter a$_γ$ and a$_{γ'}$ of the disordered and ordered phases, respectively, has a linear dependence on solute composition (Vegard's law).

The kinetic coefficient *M* in equation (1a) corresponds to the chemical mobility of elements in alloys. In many cases, for simplicity, the chemical mobility is assumed to be constant and is calculated on the average concentrations in the system. However, *M* is related to the atomic motilities β_{Sc} and β_{Al} of each system's elements and is defined in the lattice reference frame as a composition and temperature-dependent function. Hence, in the case of an Al-Sc binary alloy the chemical mobility can be expressed by the following equation:

$$M = c_{Sc}(1 - c_{Sc})[(1 - c_{Sc})\beta_{Sc} + c_{Sc}\beta_{Al}] \tag{4}$$

where c_{Sc} is the mole fraction of Sc. The atomic mobility β_i of species i can be written as:

$$\beta_i = D_i \exp(-Q_i / RT) \tag{5}$$

where R is the gas constant, D_i and Q_i are diffusion factor and diffusion activation energy of species i in the disordered phase. Typically, these two constants can be obtained from atomic mobility databases.

Unfortunately, these experimental databases for atomic mobilities for ternary Al-Sc-Zr system are not available. Hence, we chose constant mobility for Zr and Sc atoms from binary alloys. However, we took into account that at 375°C Zr diffuse 100 times slower than Sc atoms in Al.

NUMERICAL SIMULATIONS AND RESULTS.

Binary Al-Sc alloys.

Simulations were performed by numerically solving the four nonlinear equations (1a) and (1b), one for each field variable, using the semi-implicit Fourier-Spectral method [13]. In 3D computer simulations, 128^3 discrete grid points are used and periodic boundary conditions are applied. The initial state was a homogeneous solution with small composition fluctuations around the average composition. Required input parameters, expressed in units of 3.35×10^7 J m^{-3}, are: A_1=277.78, A_2=66.67, A_3=-21.21, A_4=22.14 and A_5=0. The simulations were performed at T=1054 K. At this temperature the equilibrium concentrations of the disordered fcc phase and the ordered L1$_2$ are c_1=0.1123 and c_2=0.2211, respectively. The gradient energy coefficient β (assuming α to be zero) was chosen to be 4×10^{-11} J m^{-1}, which gives rise to an interfacial energy γ/γ' of ~ 25 mJ m^{-2}. The Langevin's noises were introduced during the beginning of each simulation to ensure the nucleation of small precipitates. Particles with a radius greater than the critical nucleation radius can then grow. Approximately one thousand particles were formed during the nucleation stage. The precipitate microstructure and coarsening kinetics were then extracted from the simulation results, which were averaged over five simulation runs with different initial randomized Langevin's noises.

The parameter ε_0 used in the elastic energy term may be estimated from the γ/γ' lattice misfit, and $\frac{a_{\gamma'} - a_\gamma}{a_\gamma} \approx 0.0056$ at 375°C [14]. If Vegard's law for the misfit is assumed, at this temperature $c_\gamma = 0.0001$ and $c_{\gamma'} = 0.25$, the estimation for ε_0 gives 0.039. The elastic constants used in our simulation are listed as follows: c_2=0.12, c_{11}=108 , c_{12}=61.3 and c_{44}=28.5 GPa at 1000 K extracted from [15]

To determine the influence of the volume fractions of Al$_3$Sc precipitates on the coarsening rate and on the microstructure, three alloys with different volume fraction of the L1$_2$ phase, 0.35, 0.39 and 0.42, were studied. Therefore, for each alloy, two kinds of simulations were performed: with and without concentration dependent atomic mobility.

Figure 1 shows the morphological evolution of L1$_2$ ordered precipitates. We can conclude that including the concentration dependent mobility does not strongly influence the morphology of microstructure in Al-Sc binary alloys. However, our statistical analysis of temporal evolution for the mean radius of precipitates indicates differences between these two simulations.

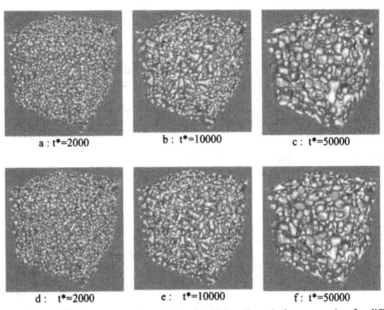

a : t*=2000   b : t*=10000   c : t*=50000

d :   t*=2000   e :   t*=10000   f : t*=50000

Figure 1: .Temporal microstructural evolution for Al-Sc alloys during coarsening for different reduced time t* with : (a-c) constant atomic mobility ; (d-f) concentration dependent atomic mobility. The coherent ordered Al$_3$Sc precipitates are shown in white, the matrix is transparent.

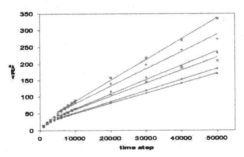

Figure 2: .<R>^{3} versus reduced time step for different concentrations and for constant and concentration dependent atomic mobilities. Solid lines are linear fits of the data from the simulations. From bottom to top, solid lines correspond alternatively to the concentration dependent mobility and constant mobility. The two lower curves are for c = 0.088, the two next ones for c=0.097 and for c = 0.105, respectively.

We find that the cubic growth law provides reasonably good fits in all cases. It is apparent in figure 2 that the slope of the curves, i.e. the value of the coarsening rate constant K, increases with increasing volume fraction which is also in good agreement with experimental data. This behavior was predicted in the modified Lifshitz-Slyosov-Wagner (LSW) theory proposed by Ma and Ardell [16]. But, we observe that coarsening rate in the case of simulations with concentration dependent mobility is smaller than with constant mobility for each volume fraction. This is related to the decrease of atomic mobility with the decrease of Sc atomic concentration in the matrix during growth and coarsening of precipitates.

Ternary $Al_3Sc_{1-x}Zr_x$

To simulate the microstructure evolution in ternary alloy we used the next set of input parameters: $A_1=91.3541$, $A_2=23.99$, $A_3=-32.58$, $A_4=A_5=9.30$, $c_1=0.044$, $c_2=0.12$. The elastic constants were taken the same as in the previous case. The misfit for the Al_3Zr and for the Al_3Sc precipitates was chosen as: $\varepsilon_{0Zr}=0.038$ and $\varepsilon_{0Sc}=0.042$ [1].

Dealing with the Al-Sc-Zr alloys, it was demonstrated that the ordered Al_3Sc particles appear first and exhibit a cuboïdal shape which results from the minimization of strain energy (figure 3). The Zr atoms diffuse slowly and precipitate on the interface between matrix and Al_3Sc particles. The main effect is due to the strain energy and not due to the diffusion velocities of the components.

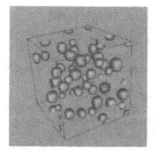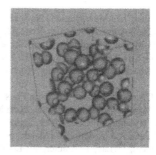

Figure 3. Microstructural evolution from 3D simulation at reduced time t*=10000, showing the heterogeneous structure of γ' precipitates in a γ (transparent) disordered matrix. Left : Iso-concentration of Al_3Sc (yellow); Right : Zr rich-shell (blue) surrounding the previous Al_3Sc precipitates.

The precipitation of $Al_3(Sc,Zr)$ particles has been successfully studied by means of transmission electron microscopy [17] and 3D atom probe analysis [19-21]. It has been shown that ordered precipitates contain a Sc-rich core and a Zr enriched outer shell. Our simulations confirm these observations and show that the Al_3Sc nucleus are formed first, favored by the fast Sc-diffusion rate, followed by later Zr segregation on the γ'/γ interface.

CONCLUSIONS.

Our simulations show that phase field models can succefully reproduce the microstructural evolution in binary and ternary systems. The concentration dependence of atomic mobilty and misfit was integrated in our calculations. It was shown that elastic interaction is responsable for the heterogeneous structure of $L1_2$ precipitates in Al-Sc-Zr alloys.

ACKNOWLEDGEMENTS

The simulations were performed at the CRIHAN center under project 2006007.5

REFERENCES

[1] C.B. Fuller, D.N. Seidman and D.C. Dunand, Acta Mater. 51, 4803 (2003).

[2] V.G. Davydov, T.D. Rostova, V.V. Zakharov, Y.A. Filatov and V.I. Yelagin, Mater. Sci. Eng. 208, 30 (2000).

[3] S. Iwamura and Y. Miura, Acta Mater. 52, 591 (2004).

[4] A. Tolley, V. Radmilovic and U. Dahmen, Scripta Mater., 52, 7, 621 (2005).

[5] Y.W. Riddle and T.H. Sanders, Metal. Mater. Trans. A, 35, 341 (2004).

[6] Y. Harada and D.C. Dunand, Mater. Sci. Eng. A, 329-331, 686 (2002).

[7] B. Forbord, W. Lefebvre, F. Danoix, H. Hallem and K. Marthinsen, Scripta Mater., 51, 333 (2004).

[8] C.B. Fuller, J.L. Murray and D.N. Seidman, Acta Mater., 53, 5401 (2005).

[9] L. Q. Chen, Annu. Rev. Mater. Res., 32, 113 (2002).

[10] V. Vaithyanathan and L.Q. Chen, Acta Mater., 50, 4061 (2002).

[11] J. Boisse, N. Lecoq, R. Patte and H. Zapolsky, Acta. Mater., 55, 6151 (2007).

[12] A.G. Khachaturyan, Theory of structural transformations in solids.(John Wiley, New York) 1983.

[13] L.Q. Chen and J. Shen, J. Comp. Phys. Comm., 108, 147 (1998).

[14] E.A. Marquis and D.N. Seidman, Acta Mater., 49, 1909 (2001).

[15] R. W. Hyland and R.C. Stiffler, Scripta Metall. Mater., 25, 473 (1991).

[16] Y. Ma and A.J. Ardell, Acta Mater., 55, 4419 (2007).

[17] E. Clouet, L. Laé, T. Epicier, W. Lefebvre, M. Nastar and A. Deschamp, Nature Mater. 5, 482 (2006).

[18] E.A. Marquis, D.N. Seidman and D.C. Dunand, Acta Mater., 51, 4751 (2003).

[19] W. Lefebvre, F. Danoix, H. Halem, B. Forbord, A. Bostel and K. Marthinsen, J. Alloys Compd., (2008), in print. http://dx.doi.org/10.1016/j.jallcom.2008.02.043

[20] C.B. Fuller and D.N. Seidman, Acta Mater. 53, 5415 (2005).

AUTHOR INDEX

SUBJECT INDEX

Printed in the United States
By Bookmasters